Lens Design (Second Edition)

Automatic and quasi-autonomous computational methods and techniques

Series Editor

R Barry Johnson a Senior Research Professor at Alabama A&M University, has been involved for over 50 years in lens design, optical systems design, electro-optical systems engineering, and photonics. He has been a faculty member at three academic institutions engaged in optics education and research, employed by a number of companies, and provided consulting services.

Dr Johnson is an IOP Fellow, SPIE Fellow and Life Member, OSA Fellow, and was the 1987 President of SPIE. He serves on the editorial board of Infrared Physics & Technology and Advances in Optical Technologies. Dr Johnson has been awarded many patents, has published numerous papers and several books and book chapters, and was awarded the 2012 OSA/SPIE Joseph W Goodman Book Writing Award for Lens Design Fundamentals, Second Edition. He is a perennial co-chair of the annual SPIE Current Developments in Lens Design and Optical Engineering Conference.

Foreword

Until the 1960s, the field of optics was primarily concentrated in the classical areas of photography, cameras, binoculars, telescopes, spectrometers, colorimeters, radiometers, etc. In the late 1960s, optics began to blossom with the advent of new types of infrared detectors, liquid crystal displays (LCD), light emitting diodes (LED), charge coupled devices (CCD), lasers, holography, fiber optics, new optical materials, advances in optical and mechanical fabrication, new optical design programs, and many more technologies. With the development of the LED, LCD, CCD and other electo-optical devices, the term 'photonics' came into vogue in the 1980s to describe the science of using light in development of new technologies and the performance of a myriad of applications. Today, optics and photonics are truly pervasive throughout society and new technologies are continuing to emerge. The objective of this series is to provide students, researchers, and those who enjoy self-teaching with a wide-ranging collection of books that each focus on a relevant topic in technologies and application of optics and photonics. These books will provide knowledge to prepare the reader to be better able to participate in these exciting areas now and in the future. The title of this series is Emerging Technologies in Optics and Photonics where 'emerging' is taken to mean 'coming into existence,' 'coming into maturity,' and 'coming into prominence.' IOP Publishing and I hope that you find this Series of significant value to you and your career.

Lens Design (Second Edition)

Automatic and quasi-autonomous computational methods and techniques

Donald C Dilworth

Optical Systems Design, Inc., East Boothbay, ME, USA

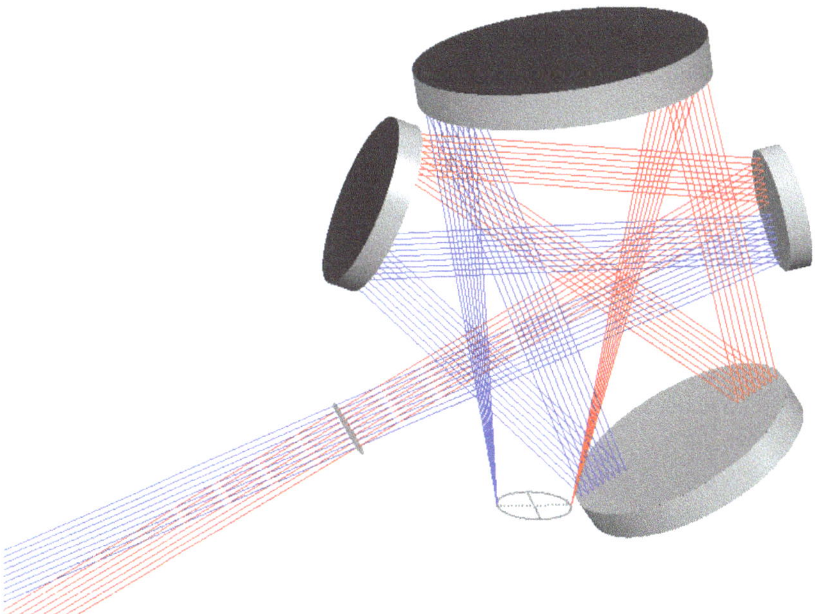

IOP Publishing, Bristol, UK

ISBN 978-0-7503-3695-6 (ebook)
ISBN 978-0-7503-3693-2 (print)
ISBN 978-0-7503-3696-3 (myPrint)
ISBN 978-0-7503-3694-9 (mobi)

DOI 10.1088/978-0-7503-3695-6

Version: 20201201

IOP ebooks

British Library Cataloguing-in-Publication Data: A catalogue record for this book is available from the British Library.

Published by IOP Publishing, wholly owned by The Institute of Physics, London

IOP Publishing, Temple Circus, Temple Way, Bristol, BS1 6HG, UK

US Office: IOP Publishing, Inc., 190 North Independence Mall West, Suite 601, Philadelphia, PA 19106, USA

I cannot help fearing that men may
reach a point where they look on
every new theory as a danger, every
innovation as a toilsome trouble, …
and that they may absolutely refuse
to move at all for fear of being
carried off their feet.

—Alexis de Tocqueville, 1840

Contents

Preface to the second edition

In the rapidly developing field of computer software and computer-aided engineering, any text, however comprehensive and thorough when it was written, becomes dated as soon as the technology improves and new capabilities become available. To meet the needs of lens designers who might benefit from understanding how the tools in SYNOPSYS have evolved in the last two years, we have updated all chapters of this book and added seven new ones

The optimization algorithm and design search tools are frequently updated, and with each iteration, many of the examples in this book return results that differ somewhat from those described in the first edition. While this is to be expected—and does not detract to any serious degree from the usefulness of the text, since the methods of dealing with common problems are still valid—it is less of a burden on the reader to see what the current tools now deliver and how to use them. For that reason, we dedicate this edition to the many very smart readers who have suggested improvements and whose careers have benefited immensely from knowledge of how the latest tools have revolutionized the art of lens design.

Series editor's foreword

Mr Donald C Dilworth began developing in the late 1960s a lens design and analysis software package called SYNthesis of OPtical SYStems (SYNOPSYS™). Over the past 50 years, by himself, he created, developed, and has continued to improve and expand the capabilities of this program to meet the demanding needs of lens design professionals throughout the world. During this period of time, other lens design programs were developed for both internal corporate use and general use by the optics community. Many of these programs left the marketplace while a modest few have stood the test of time. All of the current commercial lens design programs have excellent capabilities that meet the general needs of the optical designer; however, over the past couple of decades, optics has become broadly pervasive, and the demand for people capable of accomplishing optical design rapidly outstripped the supply of trained and talented lens designers. To overcome this lack of supply, lens design software providers worked hard to make it easier for essentially untrained people to use their programs and obtain meaningful designs. One important capability that has been explored by the software providers is to incorporate a means to allow the user to input basic parameters of the optical system and then allow the software program to attempt to create design(s) that meet the designer's requirements. In this book, Mr Dilworth presents how to design and analyze lens systems, and his innovative and unique developments in automated lens design that have had, and will continue to have, significant utility in mitigating the growing lack of professional lens designers as well as improving the efficiency and creativity of the professional lens designers. Achieving the goal of autonomous optical design continues to be an emerging technology for lens design software providers; however, Mr Dilworth has made great strides to accomplish quasi-autonomous lens design.

Arguably, during the 1960s and 1970s Berlyn Brixner (Los Alamos National Laboratory) was the first to suggest and demonstrate starting a design using just flat plates and allow the program to 'explore' alternative paths to generate a design solution. As computer power became available at affordable costs and research in optimization techniques evolved, lens design program developers began to incorporate what is commonly called *global optimization*. The objective is to allow the program to search for a solution that is hopefully superior to that following normal optimization methods. Indeed, the various approaches often did find better designs and often new design forms the human designer had not envisioned. Such a search process is quite time- and resource-consuming, with the result potentially being a design that isn't practical. Nevertheless, the global optimization tool has been of great benefit to the beginning lens designer as well as the professional. Further refinement of potential designs can be obtained by using a simulated annealing feature available in all current programs.

While lens design software developers focused on methods to improve global optimization along similar lines, Mr Dilworth additionally explored and exploited several 'out-of-the-box' approaches to achieving, what might be called, *quasi-autonomous lens design*. There are four such innovations that provide lens designers additional tools that have the potential to create solutions to their design requirements in less time, with more ease, and by exploring new configurations the SYNOPSYS program produces.

The first two innovations to aid the designer were developed initially about 1990, and provided artificial intelligence (AI) capabilities utilizing natural language processing and expert systems. The natural language AI feature provides a very flexible way to interact with SYNOPSYS, and to perform certain tasks not readily accomplished with the normal command syntax or spreadsheet input. Among these tasks are displaying and changing certain lens parameters, defining new commands or character strings, and obtaining plotted parametric curves involving up to three different quantities. Input consists of English sentences comprising subjects, verbs, and conditions. The vocabulary contains many hundreds of words and the flexibility for the user is vast. This is a very powerful and useful tool for designers.

In general, an 'expert systems' program is one that employs a tree-structured logic wherein decisions are derived from the responses of a number of experts in a particular field to a lengthy debriefing. Remarkable performance has been achieved in some areas, rivaling that of a human expert. In contrast, the SYNOPSYS program takes a somewhat different approach in that the expert system feature (XSYS) is presented with a number of finished lens designs that are the products of expert designers and represent the state of the art. Using these lenses as models, the program analyzes the optical properties in great detail, determining for each lens the first and higher-order properties, and the aberrations that are present in the beam before and after each element. In so doing, the program 'learns' how a particular optical problem was solved by its designer. The more examples provided to XSYS, the more it learns. When presented with a new problem or a lens that is not well corrected, XSYS can determine if the current problem resembles one for which it knows a solution and then attempts to use bits and pieces of lenses in its expert-designed-lenses library to create various potential configurations for the user to review, select one to optimize, and then analyze. This feature is very creative and often finds unexpected configurations that work better than having followed a conventional design path. Of course, not having a library of relevant lenses will mitigate the utility of XSYS. But if it is available, XSYS can provide the lens designer with a powerful tool to identify potential lens configurations for further optimization and analysis.

As mentioned previously, a bit over a quarter of a century ago, as computer resources became more capable, the search for the optimum solution of a lens design moved from optimizing a 'point' design to allowing the software program to explore an enormous variety of possible configurations in solution space. Such a search process takes, in general, a huge amount of time in contrast to point design optimization. This is known as global optimization and often includes simulated annealing. A variety of approaches have been developed by researchers to locate the 'true' global optimum solution (minimum merit function) while at the same time minimizing the convergence time. Arguably, progress has been made, but the time needed for finding candidate solutions often remains excessive. Early in this decade, in an effort to dramatically reduce the search time for candidate design solutions,

Mr Dilworth developed a quasi-autonomous method to search for candidate fix-focused lens configurations which he named DSEARCH™ (Design SEARCH). This method is not intended to produce a finished lens design; its purpose is to identify attractive starting points, and each of the designs can then be subjected to further optimization, first with transverse intercepts, then with OPD targets, and lastly with modulation transfer function targets in the merit function. Although the solutions lie in a multidimensional hyperspace that can't be visualized, it is common to visualize the solution space as a rough mountain range where the highest peak corresponds to a flat-plate starting design. The objective is to find the lowest valley, which is certainly a challenge. Since any direction from the mountain peak is downward, the question is which direction to go. Mr Dilworth reckoned that an efficient selection of directions to try would be a 'binary' approach where the flat-plate lenses are assigned either a positive or negative power. For an N element lens, this implies 2^N starting optical configurations or directions to consider. When the valley in a particular direction is found, the question still remains, Is the this the best solution in this direction? To overcome this, simulated annealing is used to explore the surrounding hyperspace to determine if a lower valley exists. Of course, a RANDOM mode can be used to attempt to find the best solution, but it can be far more demanding of time. If the initial flat-plate design is specified to have N elements, and in the event that the best solutions found are not good enough, an automatic element insertion option is available and the prior solution lenses can be reprocessed. Should the solutions be 'too good,' an automatic element deletion option can be used to reprocess the lenses to look for adequate solutions having N-1 elements.

The final innovation to mention is the evolution of DSEARCH to accomplish finding potential solutions for zoom lenses. Design of zoom lenses is significantly more challenging than design of fixed-focused lenses. Mr Dilworth named this new feature ZSEARCH™ (Zoom SEARCH) which was presented at the 2016 SPIE Meeting in San Diego. ZSEARCH works essentially like DSEARCH except that the searching process is much more complex and the results are rather amazing.

Mr Dilworth continues to develop improvements in his quest for *quasi-autonomous lens design* to address the issue of growing dearth of experienced lens designers. Inclusion of additional features in the search and final design selection process that consider manufacturability and cost of the candidate lenses should be most beneficial to lens designers and their employers.

Working alone on his SYNOPSYS program for the past half century, Mr Dilworth has made exceptional contributions in computer-based optical design software, which is used worldwide, and in his development of the aforementioned innovations to assist both novice and expert designers in their work to design high-performance optical systems in minimal time.

Those learning lens design will find the material in this book helpful in mastering how to design lens systems regardless of which lens design software is used. All readers should benefit from the knowledge and wisdom Mr Dilworth has gained over the past half-century. Unlike the traditional books on lens design, the interactive nature of this book provides readers with an unprecedented opportunity to work examples themselves and further explore what happens if they modify the examples while using a provided version of SYNOPSYS. This is an excellent method to learn the subject and sharpen one's skills. Optics has become pervasive throughout most technological areas, and the design and manufacture of optical systems is the fundamental foundation.

R Barry Johnson, D.Sc., FInstP, FOSA, FSPIE

Series Editor, *Emerging Technologies in Optics and Photonics*

Huntsville, Alabama

Foreword

The use of computers in lens design dates back over sixty years. In that time, computers have increased in speed by many orders of magnitude and in memory capacity by even more. This has led to enormous improvements in the capabilities of lens design programs. Analysis of lens performance has not changed significantly for several decades, but optimization has. And this book is, at its root, about optimization.

Optimization algorithms may be divided into three categories: local, regional and global. The first of these categories refers to algorithms that proceed from the given starting point to the nearest local minimum. Regional algorithms attempt to escape from this local minimum and find a nearby one that is better. As the name implies, global algorithms attempt to search the entire design space and provide a solution that is better than any other alternative.

Don Dilworth has made significant contributions to all three categories of optimization algorithms which are included in his program, SYNOPSYS. His advance over DLS is called the pseudo second derivative (PSD) approach. This algorithm uses consecutive derivative matrices to approximate the second derivative matrix and uses this to calculate an improved damping factor for each variable. The result is a tremendous increase in convergence speed from initial designs that are far from optimal. Dilworth's program also has an algorithm that is often able to start with a lens that does not pass all the required rays and tweak it until it does before starting optimization.

In the regional optimization category, SYNOPSYS starts with a standard simulated annealing algorithm, but combines it with PSD to make it far more effective than simulated annealing is in other programs. Masaki Isshiki's global optimization with escape function algorithm is also implemented, but this reviewer does not have sufficient experience with other programs' implementations to offer a comparison. Regional optimization features unique to SYNOPSYS are 'Automatic Element Insertion' and 'Automatic Element Deletion', which either insert or delete a lens element at the optimal location. The former algorithm can be run in a way that is very similar to Florian Bociort's saddle point algorithm.

New global optimization algorithms, DSEARCH and ZSEARCH, which Dilworth has recently added to SYNOPSYS, are most impressive. DSEARCH starts from a very rough description of a lens (the object, wavelengths, F/# and number of elements) along with any other required constraints, and produces several candidate designs that are often close to final designs. ZSEARCH does the same thing for zoom lenses. Both algorithms can provide lens designs even if the designer has no idea of what an initial configuration might look like. Dilworth has published a paper with Dave Shafer, a renowned lens designer, comparing the results of DSEARCH with Dave Shafer's design of a well-designed eleven-element lens. The paper is an entertaining discussion of man versus machine. DSEARCH was able quickly find solutions with eleven, ten, nine, and remarkably only eight elements. Once knowing that there was a potential solution space having fewer elements, Shafer was able to also find designs, but taking much more time to do so. He found one design that DSEARCH missed (with default options), but the algorithm came up with several other designs which surpassed this famous lens designer's.

One could say a lot more about optimization in general, and Dilworth's contributions in particular, but this should be enough to give you a flavor. Reading this book will teach you more design tricks and insights. Many examples are provided that readers can run on their own computer and experiment with by changing parameters and other commands. I'm certain that you will enjoy and profit from this book.

Dr Steve Eckhardt

Eckhardt Optics LLC

White Bear Lake, Minnesota

March 2018

Preface

Why this book? A new paradigm. The lens design landscape. The lens design tree.

When a friend suggested I write a book about lens design, my first reaction was, Why? There are a number of excellent books out there already. Who needs another one?

My second thought was, Why me? I've never taken a course in optics or lens design, and (I hate to admit it) I've never even read most of the aforesaid books. What can I possibly add to the corpus?

My friend pointed out that I have been designing lenses for almost 60 years, and it's likely I've picked up a few pointers along the way. True enough.

That lead to my third thought: This might be a very short book. I picked up one of the best of the current texts, *Lens Design Fundamentals* by Rudolph Kingslake and R Barry Johnson (SPIE PRESS, 2010), and read descriptions of how one does certain tasks *that you don't have to do anymore.* The classic authors and designers had to make do with very primitive tools and were faced with a staggering amount of manual labor. So they devised shortcuts, found insights where they could, and invented ways to get an approximate answer with relatively little work. What I mean by 'little' can be understood in the context of Kingslake's comment:

> When someone applied for a position in our department at Kodak, I would ask him if he could contemplate pressing the buttons of a desk calculator for the next forty years, and if he said 'yes', I would hire him.

In this computer age, most of those tools and shortcuts are no longer needed. In fact, some of them actually get in the way and are better avoided. The classic texts will not tell you that. This book will.

When I first entered this field, there were no books on computer-aided lens design. I first traced rays on a computer filled with vacuum tubes. So I had to figure everything out for myself. This turns out to be an advantage: it puts me in a position to ask, What do you *really* need to know in order to design lenses? Not all that much, it turns out. Certainly, you don't need all the primitive tools used by the old masters, and if I had to use those tools and do all that manual labor, I would have left this rewarding field years ago. Some will be offended that I give little ink to those techniques in this book—but I want to teach what is useful today, not what people had to learn a generation ago.

Today we live in a new world. I estimate that perhaps 90% of what I read in the classic texts is now irrelevant. I could write on two pages most of one needs to know when designing lenses, assuming one has a modern PC and the most powerful of today's lens design programs. (And a basic understanding of math and physics.) A good friend, who teaches lens design at a major university, once boasted that her student could design a five-element lens in only five days. Modern software can design a seven-element lens in less than one second. My friend did not know that. Things are changing.

Many brilliant minds have contributed to this field, and prudence says we should not just discard the knowledge and skills handed down to us by the old masters. So how can I justify saying that very little of that knowledge is relevant today? Here's how: I challenged my friend to design a 90-degree eyepiece, essentially perfect, with the classic tools she had long mastered. She gave up after 100 hours of work. Using software that I have developed over many years, I was able to solve that problem in only 15 minutes—using almost none of those tools. I think experience speaks for itself. We're thinking out of the box here, and the results are wonderful. This is a game-changer.

Before I go too far, however, I have to caution the reader that learning to use lens design software is not for everyone. For one thing, it requires some familiarity with optics and physics. The software is not going to anticipate and solve every problem for you. The aptitude necessary to become a good engineer is not to be found in every human head, which is why engineering schools have entrance exams. You will be obliged to read the instruction manuals and make an effort to understand them. I sometimes get questions from very naïve beginners showing they are starting without this familiarity and are confused by many of the concepts presented in the manuals and in this book. There is a reason people attend college and get a degree in the sciences. With that background, those concepts are simple and clear enough. I urge those without a technical education to take steps to acquire it before diving into this fascinating field.

Some will object that the methods I present in this book rely too much on chance. A throw of the dice. Trial and error. They want the process to be deterministic, guided at every turn by expert knowledge. There is much to be said for that point of view—but consider a conversation I once had with Kingslake (an expert if ever there was one). I said:

> In your book, you show how one could alter a certain radius of curvature in steps and plot an aberration curve, then change a certain thickness and plot a second curve. Where those curves cross is a better design—but how did you know you should change *this* radius and *that* thickness?

He replied: 'That's the only thing that worked. *I tried everything else.*'

It seems that even the experts had to resort to trial and error much of the time, if one makes an objective appraisal of how they worked. What I have done is to make it work much better and much faster. How much better? A colleague once volunteered, 'These search methods work so well it's *scary*!' He knew what he was talking about; he had tried them himself.

But, wait a minute. Two pages? A friend objected: 'we cannot just teach the students to *push a button*!' True enough. There is a whole lot of optics written between the lines on those pages, and the student really should know what happens when you push it. Kingslake observed,

We are losing the ability to design a lens through an effort of the intellect.

And he is right[1]. Grab a qualified person off the street, sit him in front of a modern lens design program, and ask, What does he need to know in order to use it effectively? I hope to put the answers on the pages of this book.

[1] But we have also lost the ability to make flint arrowheads and ride horses, and I don't care.

Acknowledgements

I want to thank the many loyal and very smart customers using my lens design program who have suggested new features and improvements. Together, we are more effective than I could be by myself.

I especially want to thank Dr R Barry Johnson, whose encouragement and ability to see outside the box has been instrumental in the progress of my career and led me to write this book.

Lastly, I wish to thank my wife Sarah, who, in spite of knowing nothing about optics, continues to believe that what I am doing has merit.

Author biography

Donald C Dilworth

Mr Dilworth has been designing lenses since 1961, when he participated in the development of the optical navigation system for the Apollo project, the American program to land men on the moon. He is a graduate of MIT, where he majored in physics, but is self-taught in the fields of optical engineering and lens design. His later work includes designing one of the spy cameras that flew over the Soviet Union during the Cold War, and he has worked in industry for many years, designing infrared FLIR systems (which can see in total darkness), and numerous kinds of lens systems including a submarine periscope. He formed his own company, Optical Systems Design, Inc. in 1976 and has been developing and marketing the lens design program SYNOPSYS™ (**SYN**thesis of **OP**tical **SYS**tems) ever since.

Chapter 1

Preliminaries

I am going to use a lens design program called SYNOPSYS[1] (**SYN**thesis of **OP**tical **SYS**tems) in this book, partly because it will do all the lessons quickly and easily, and also because it happens that I wrote it—but the principles are valid for any code with comparable features. The best way to learn these new techniques and become familiar with the software is to work through many different examples. With this in mind, I have written a set of chapters that exercise many features of the code. To save the labor of typing all of the lens files and optimization MACros in these examples, a copy will be found in the folder DBOOK-ii, which you can download as instructed below. (a 'MACro' is a file containing input commands and data that are recognized by the program.)

To install and run the program you will need two files as well as the example folder, all located at www.osdoptics.com.

SYNOPSYS200_v15.zip,
InstallSYNOPSYS_DLL_IA32.zip, and
DBOOK-ii.

To request a trial license, please use the Trial Request Form at OSDOptics.com, https://osdoptics.com/accessinstallation/, or send a request email to office@osdoptics.com.

The program creates a folder C:\SYNOPSYS, and under that a folder USER. Copy the DBOOK-ii folder under C:\SYNOPSYS as well. Now you should have two directories under C:/SYNOPSYS: USER and DBOOK-ii. The program expects to find this directory structure, so don't move things anywhere else. The previous edition (using folder DBOOK) illustrated the input and output from an earlier release of the software, much of which is altered in the newer versions, and is no longer relevant.

[1] SYNOPSYS™ is a trademark of Optical Systems Design LLC, an Arizona, USA, corporation.

doi:10.1088/978-0-7503-3695-6ch1

The program is frequently updated, and you always want the current revision, so check the website frequently for updates—and sign up for the mailing list and you'll stay informed.

When you start the program, it will tell you if you do not have a hardware key. In that case, just select the option to run in demo mode, with a limit of 12 surfaces.

Then type the characters **CHD** (**CH**ange **D**irectory) into the Command Window (**CW**). A dialog opens. Click the Browse button, shown in figure 1.1.

Select the directory DBOOK-ii (which you should have copied per the above instructions) and click OK. This becomes your new default directory.

Then type **HELP TM** in the CW, and the Tutorial Manual will open. It would be a very good idea to print each chapter and go through the text carefully. I don't want you to get stuck on a simple problem simply because you do not yet know how to use the program. The chapters in this book contain many examples that assume you know the basics—and I will coach you here and there when something is important but perhaps not obvious—but you should be acquainted with the basic features of the program and know how to use them before you go much further.

When you have finished for the day, exit from SYNOPSYS by typing **EXIT** in the Command Window or by clicking the 'X' in the upper-right corner of the frame. The program cleans up many temporary files when it exits, and if you terminate or crash the program, that last step won't take place.

Many of the chapters below refer to lens files and MACros that you will want to open. The former are text files with the extension '.RLE' that contain a description of a lens, and the latter are lists of commands with the extension '.MAC'. When they are appropriate to the lesson, you will be informed by entries such as (**C40L1**), which means the lens is found in file **C40L1.RLE** and can be opened with the command **FETCH C40L1**, and for MACros with entries like (**C40M1**), which refers to a MACro called **C40M1.MAC** and can be opened in an editor with the command **LM C40M1**. (**LM** means 'Load MACro'.)

You can also select from a list of lens files in the current directory with the menu **MWL**, and from a list of MACros with **MWM**. MWL lists the ID line of each lens file and gives you a preview of the lenses. Yet another way to open MACros is by clicking the button  on the editor toolbar. This method shows a preview of the file in the editor window to help you select the one you want.

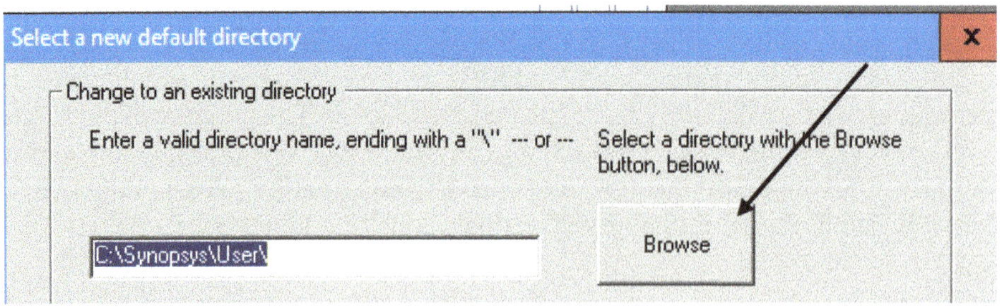

Figure 1.1. The Change Directory dialog.

Some of the instructions have you change the MACro in certain ways after you open them, as you make decisions and work on the lens. It is a good idea, however, to first rename the MACro, so the original is not overwritten when you run it. Click the button -N, and a working copy will be saved with the name DEFAULT.MAC. That way, if you want to review that chapter again, you can open the original file and start out with the contents as it should be rather than as you changed it. (When you click the **Run MACro** button, the file is first saved, and then executed.)

1.1 Why is lens design hard?

Although the art of lens design has matured and now has the benefit of many powerful tools that were not available when I started—and certainly not known to the old masters—it is not, and probably never will be, a simple endeavor. The difficulty arises primarily because of the geography of the lens design landscape. One must not only design a great lens, but also must take into account the dimensions of the housing, the properties, cost, transmission, and availability of the glass to be used—while at the same time avoiding lenses whose tolerances are so tight that nobody can build them. But before you even get to those considerations, you have to find a lens construction that works. Well. That's hard to do.

Why is it hard? Because we are dealing with a design space of many dimensions, where many variables and the image quality are related to each other in a nonlinear manner, and where clumsy boundary conditions apply to most variables. Few engineering disciplines have to reckon with such difficulties.

Traditional methods have long relied on having a good starting point, a design not too far from the desired goals, and then working to improve it. If the starting point was indeed a good one and your skills were sharp enough, you could in that way arrive at a great design. But one rarely has such a starting point, and only a few of us have the required skills. So the job is hard for most, and difficult even for the experts, most of the time. The core problem here is the fact that, except in a few simple cases, there is no *closed-form* solution to the lens design problem. That means there is no formula you can simply plug numbers into and get a great design. You have to think, try things, learn from experience, and iterate. My goal in writing the program has been to put as much of the burden on the computer as possible, freeing you from the most tedious of traditional tasks. I hope you will thank me once you see how much easier the job is now, with these new tools.

1.1.1 The lens design landscape

I often describe the lens design landscape as a mountain range with peaks and valleys all over the place. In that scenario, your job is to find the lowest valley, which corresponds to the lowest *merit function* (**MF**). The merit function is usually defined as the sum of the squares of a set of quantities that represent how far the design is from its ultimate goals; the MF would be zero if all the targets were met exactly—which almost never happens. The lowest valley overall is the best, or 'optimum' design, because it has the lowest MF. How can one find it?

Figure 1.2. Graphical illustration of the lens design tree.

One way is to start at the top of the highest peak, from where you can see all the valleys, select a direction, and head downhill. That is the principle behind **DSEARCH**[2], a tool you will use in many of these chapters. The highest peak, in this view, is a lens with all plane-parallel surfaces; that design can go anywhere. DSEARCH selects a variety of directions according to its own logic, and then heads downhill, evaluating the quality in the lowest valley it comes to in each direction. The algorithm is discussed more fully in Appendix B.

Another way to visualize the task is to imagine climbing a tree, as shown in figure 1.2 Here, one can start at the bottom and climb up—but which branch to take? See the problem? There are usually many solutions to a given task, of roughly equal quality, and when you run a lens optimization program[3], you are climbing up whatever branch you happen to be on. A different starting point will go up a different branch. When you reach the end of the branch you are on, you are at a l*ocal minimum*. Just running the optimization program yet again will not move you from that branch to a better one. You need other tools. How can you get to a different branch? How many branches are there? Those are perplexing questions, and although the techniques described in this book enable one to explore the design tree quickly and easily, they have, even now, not been completely answered.

Here's a helpful way to look at it. Figure 1.3 shows the statistics of the merit function resulting from optimizing a total of 5000 random starting points for a

[2] DSEARCH™ is a trademark of Optical Systems Design LLC, an Arizona, USA corporation.

[3] An *optimization program* is computer code that uses an algorithm to deduce how to change the design variables in a way that drives the value of the merit function down while observing boundary conditions.

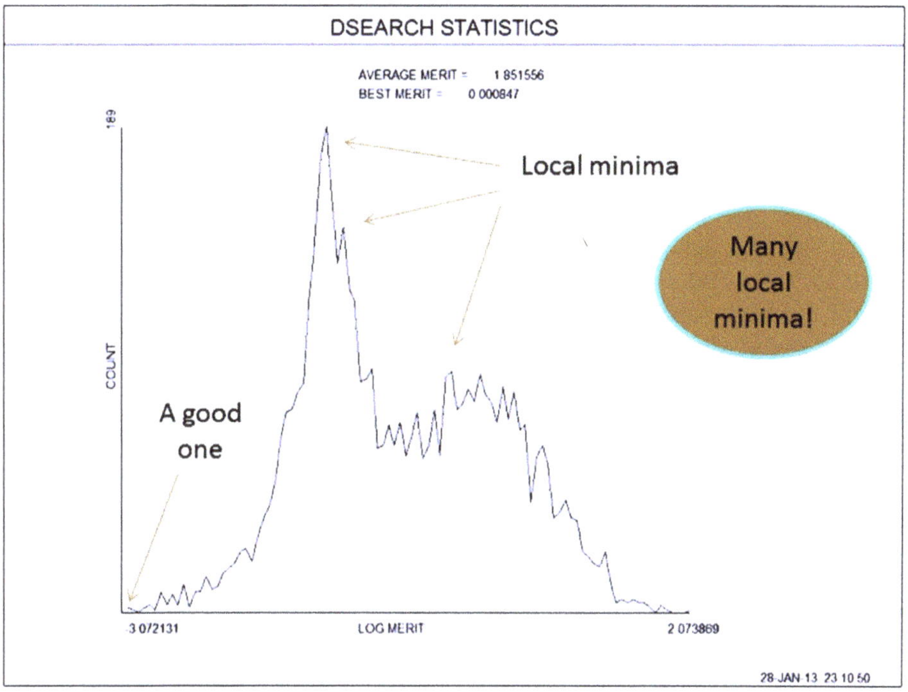

Figure 1.3. Statistics of the results of searching 5000 random branches of the lens design tree.

typical lens design job. The peaks give the number of cases that arrived at a particular value of the MF, and each one corresponds to a different local minimum. Since there is no obvious way to escape from one to another with classical tools, it is evident that the designer has a tough job. The solution you want is way over at the left end of the curve. That is the main reason why lens design is hard: *too many local minima*!

1.1.1.1 Simulated annealing

Most lens design programs today offer what is termed *simulated annealing*, a process that involves making small random changes to each of the design variables and then optimizing, over and over. That technique can jump sideways from one branch to another, although usually not very far. Nonetheless, it is surprisingly effective and is one of the most important tools of the trade today. You will use it in many of the examples below.

Other tools are more powerful yet. SYNOPSYS offers several search routines for exploring the tree. DSEARCH starts at the bottom and can go anywhere, and the tools called **AEI** and **AED**[4] can deterministically select the best place to insert or delete an element. Those tools can jump to an entirely different branch and usually can find a better solution than can a human expert.

[4] AEI™ AED™, and XSYS™ are trademarks of Optical Systems Design LLC an Arizona, USA corporation.

1.1.1.2 Global optimization

Most lens design programs today also offer a form of 'global optimization', which can find a variety of solutions—but most of those programs are not practical because of the very long time required to return their results, often measured in hours or days. DSEARCH, on the other hand, executes very quickly, using the algorithm described in Appendix B, and that is why we use it in these examples.

We should also mention the expert-systems program **XSYS**. Also intended to generate starting designs, this feature utilizes a different principle: instead of exploring a large and complex design tree, it learns about successful designs, which represent the best work of experts, by studying the lens construction, and can then apply that knowledge to construct new lens configurations. Unlike the other search programs, XSYS requires a database of sample designs and must be initialized with those examples before it is able to propose new ones. An example of its use is given in chapter 53.

We now enjoy a new paradigm. It used to be the case than an expert would spend days, weeks, sometimes years making small improvements to a classic design form, always guided by experience, insight, theory, and a large dose of dogged labor. If he succeeded, he was rightly proud of the achievement. Today we do things differently. Instead of inching up a single promising branch of the tree, day after day, hoping the result will justify the effort, we use software that examines hundreds or thousands of branches in a matter of minutes or seconds and returns a set of candidate design forms the user can then evaluate and try to adapt to his current requirements. Those candidate lenses are often so well corrected that they need little or no improvement, and a few are sometimes already at what we call the 'diffraction limit', a condition in which the only significant limit to the resolution is due to the finite wavelength of light. In that case, further improvement to the lens may be unnecessary, and one is concerned mainly with mechanical properties, edge thicknesses, tolerances, the cost of the glass, and so on. This new paradigm is a true revolution, and its importance is only beginning to be recognized. That is perhaps the main reason I agreed to write this book: to spread the word.

A corollary of this deciduous metaphor is the problem of just *how* to write this book. There are many examples in the chapters below that utilize the search tools, and in many cases the reader is instructed to run the simulated annealing program after optimizing the lens. But those methods involve random changes at some stage, and by their very nature those changes are different every time through. So unless special precautions are taken, the user would likely get a result different from that shown here when he works that example. The end point of a run on DSEARCH is very sensitive to the initial conditions and the particular random changes in each annealing step, and would be different every time. This would be disconcerting to the reader, who expects to see on the monitor what is on the paper in front of him.

To resolve this difficulty as far as possible, I have programmed one of the mode-control switches in SYNOPSYS (number 98) to recycle a single random number series whenever randomness is called for. (This is like opening a book of random numbers to the same page every time.) That switch was turned on when these lessons were prepared, and if the user also has the switch on, he will most likely get the same results as those shown here. I say 'most likely', because other effects can influence the path up the design tree as well.

1.1.2 Chaos in lens design

You can get different results if you change some of the other mode-control switches, assign different weights to the requirements, or even use a different release of the software, which is updated regularly. The glass tables change as vendors add or delete glasses from their catalogs. Even a tiny change at any stage of the process can alter the path through the tree and lead to a different end point; the branch of mathematics called *chaos* theory deals with situations like this, and lens design involves more than just a tree: it is a *chaotic* tree. But the saving grace is the fact that, for most problems, many of the branches are equally good. All you have to do is to find a good one. That's good to know. Otherwise most of us would be out of a job. Chapter 27 gives an interesting discussion of the chaos inherent in lens design.

And we caution the reader that, except when following the examples below, you do *not* want switch 98 turned on. The beauty and power of these new tools comes precisely because the results are *different* every time. If you lose your keys, it makes no sense to look for them in the same place over and over. We often run DSEARCH several times when working a design job, selecting the best results from each run, thereby exploring a great many branches of the design tree. Usually the program finds several rather different lens constructions with almost identical performance. Then we can base our selection on considerations of packaging, cost, and so on.

We also observe that, in a previous age, when a designer had to steer a lens to the solution with sometimes heroic effort, when he finally arrived at a good one it is likely he would declare success and look no further. Why go through all that work again? But with modern search tools, the program can suggest many potential solutions very quickly, which adds a whole new dimension to the process. It sometimes finds great designs that an expert would never have thought of, simply because he already found one that seemed good enough and quit looking.

1.2 How to use this book

The chapters that follow present a variety of lens design problems and show how to find solutions using the search tools and other features of the software. In most cases, they advise you to FETCH a given lens, load a given MACro into an editor and run it, and then modify the input to deal with problems that show up. Lens design software has a reputation for solving the problem you asked for, not the one you *should have* asked for. We make our living figuring out how to ask the program to solve the problems we did not anticipate.

Doing so requires modifying the input file, and you are reminded to first *rename* the MACro before you make any changes, by clicking the -N button. It would be unpleasant if you wanted to review an earlier chapter, but the input files were all different and the solution was not the same as before.

Also, you will see instructions like 'optimize and then anneal **(50, 2, 50)**' in many chapters, and it is important that you carry out all of the instructions *in the exact* *order* given. Not because that order or those data are especially important—indeed, you are encouraged to experiment with them—but *not the first time*. Other data combinations will almost always return a different lens, and that is one of the virtues of the search tools described herein. You might find solutions that are better than we give here—but they won't be the *same* solutions, and to properly follow the instructions and learn how to deal with problems that show up, you really want them to be the *same* problems. That will happen only if you follow the steps outlined below exactly.

IOP Publishing

Lens Design (Second Edition)

Automatic and quasi-autonomous computational methods and techniques

Donald C Dilworth

Chapter 2

Fundamentals

Paraxial optics; Snell's law; Lagrange invariant; thin-lens equation; pupils

There is widespread misunderstanding about lenses and images. A customer of mine once objected when I designed an eight-element lens that satisfied his requirements and made a good image; he wanted it done with a singlet (one element)! More enlightened people will understand that, to get a good image, one usually needs more than one lens element. Why is that?

That is the whole domain of lens design: the cause and cure of image defects, which come in a wide variety of types, can be difficult to correct, have to be carefully balanced—but are actually quite easy to understand.

2.1 Paraxial optics

To set the stage properly, let us first discuss *paraxial* optics, also called *first-order* or *Gaussian* optics[1]. That is the domain where the well-known Snell's law is simplified. Consider the lens in figure 2.1.

A ray of light comes in, is refracted at both surfaces, and heads to the image plane. It does not hit the desired image point however, and the image is blurred. The lens surfaces here are portions of spheres, which is by far the most common and usually the most economical situation with lenses.

Now, consider what would happen if we reduce the aperture to a very small value, as in figure 2.2.

The ray now hits much closer to the center of the image. (Galileo, not understanding lens aberrations, found that by 'stopping down' his lens (reducing the aperture diameter) he could sharpen the image). Of course, less light gets in when you do that, but there is a lesson here: if the aperture goes to zero, the angles also go

[1] Strictly speaking, *paraxial* optics refers to analysis with infinitesimal aperture and field. When you scale up the latter, to analyze real systems, it becomes Gaussian optics. The latter can also be analyzed by replacing $n\ i$ with $n \tan(i)$, as mentioned in reference 1 in appendix F.

doi:10.1088/978-0-7503-3695-6ch2

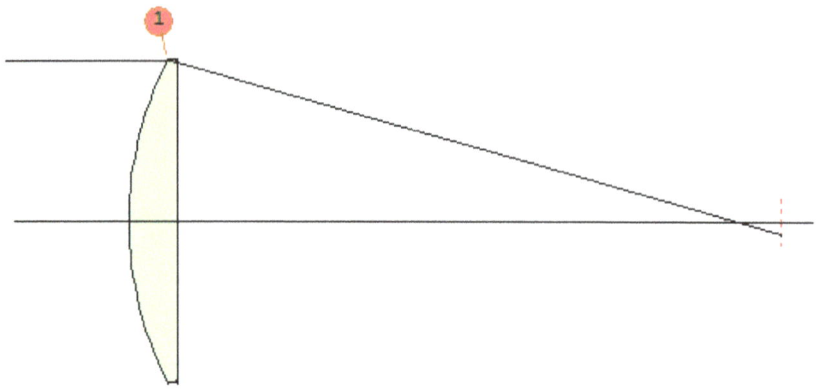

Figure 2.1. A simple lens.

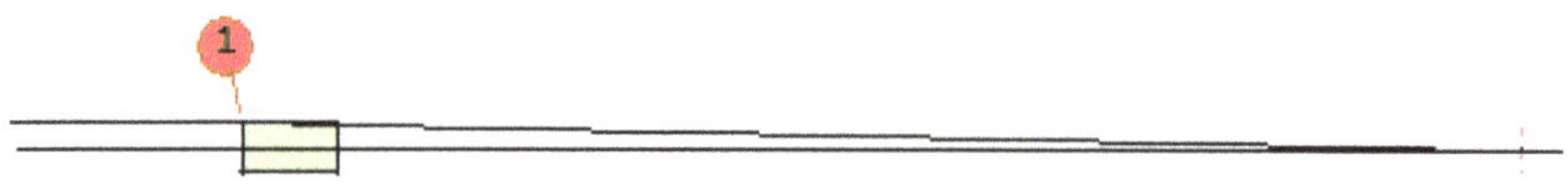

Figure 2.2. The lens with reduced aperture.

to zero, and the sine of a very small angle equals the angle itself. So Snell's law changes from

$$n' \sin i' = n \sin i$$

to

$$n' i' = n i.$$

where n is the *index of refraction* before the surface, n' is the index following it, and i and i' are the angles of the ray relative to the surface normal (a line perpendicular to the surface at the ray point) before and after refraction. All transparent materials slow down the passage of light relative to the speed in air, and the ratio of that speed to the speed inside the material is what we mean by the index of refraction. This geometry is shown in figure 2.3.

So what? Well, it turns out that if the aperture and field both approach zero, the imaging properties of the lens can be described very simply. All the *paraxial* rays, as they are then termed, go to the desired image point, with no aberrations. The value of paraxial optics is that it points the way to the design goals: if real rays wind up just where the paraxial rays do, the lens is close to perfect. When we speak of *aberrations*, we mean the departure of real rays from the image point given by paraxial rays. (The term is often generalized to mean anything that is wrong with a lens, but there is a family of aberrations with characteristic properties and standard names, which we will refer to frequently in this book.) The concept of paraxial optics applies to the common situation in which the lens has no tilted or decentered elements and has a flat image surface; otherwise, the concept may not strictly apply—but it is nonetheless one of the fundamentals you should know about.

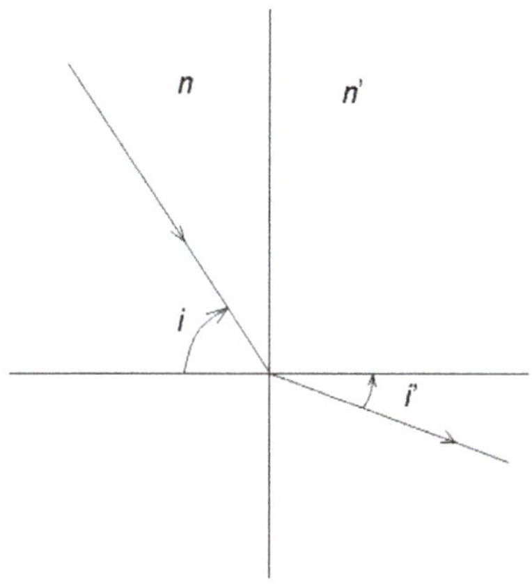

Figure 2.3. Geometry of Snell's law.

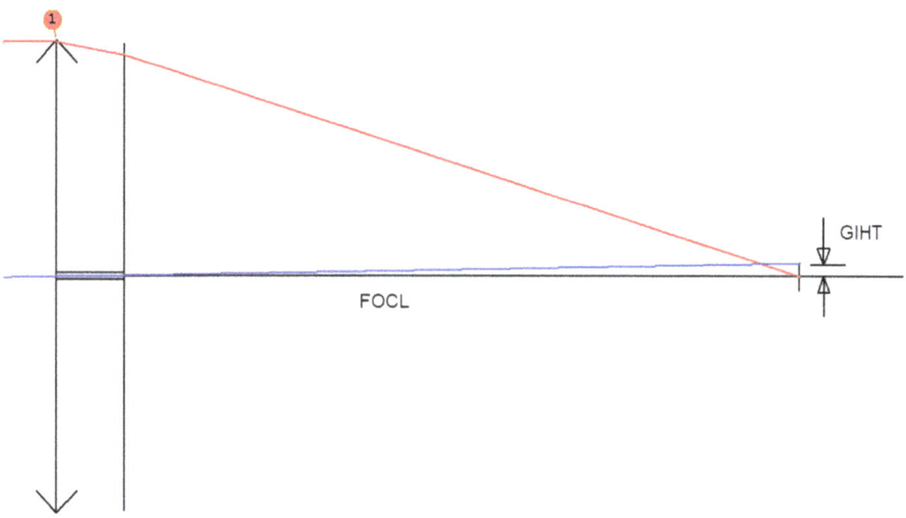

Figure 2.4. Paraxial rays, shown schematically.

And it is very simple: Divide both the aperture and field by a large number, perform a paraxial raytrace—yielding the paths of two rays, using the simpler form of Snell's law. Then multiply the *results* by the same large number. That gives you the paraxial ray paths through the lens, which are sometimes displayed as shown in figure 2.4.

Classic texts usually show the formulas for tracing a paraxial ray through a lens. But all the lens design programs already have those formulas in their code and there is no reason for you to learn them—and certainly no reason ever to trace a paraxial ray by hand. It is enough to know that paraxial rays don't really exist but are a useful concept to help you determine just how good a lens is.

The ray shown in red in figure 2.4 is the *marginal ray*, which comes in from the on-axis object at the edge of the beam, and the ray in blue is the *chief ray*, which starts at the full-field object and comes in at the center of the beam. (We will discuss the fine points of this definition later.)

According to paraxial optics, light from an object at the edge of the field of view forms an image whose distance from the optical axis is called the *Gaussian Image Height*, abbreviated **GIHT**. That distance is just the tangent of the entering chief-ray angle times a constant, called the *focal length* of the lens, or **FOCL**. Another fundamental property is the *F/number*, or **FNUM**, defined as the focal length divided by the aperture diameter. (Those properties are valid for a lens where the object and image are both in air. Your software will correct for cases where this is not the case.)

The above lens is described by a specification listing like this, produced by the command **SPEC**:

```
ID SINGLET
LENS SPECIFICATIONS:

SYSTEM SPECIFICATIONS
------------------------------------------------------------------------------------------
OBJECT DISTANCE      (THO)        INFINITE   FOCAL LENGTH          (FOCL)          8.0886
OBJECT HEIGHT        (YPPO)       INFINITE   PARAXIAL FOCAL POINT                 12.2688
MARG RAY HEIGHT      (YMP1)         2.0000   IMAGE DISTANCE        (BACK)          7.6733
MARG RAY ANGLE       (UMPO)         0.0000   CELL LENGTH           (TOTL)          0.6299
CHIEF RAY HEIGHT     (YPP1)         0.0000   F/NUMBER              (FNUM)          2.0222
CHIEF RAY ANGLE      (UPPO)         1.0000   GAUSSIAN IMAGE HT(GIHT)               0.1412
ENTR PUPIL SEMI-APERTURE            2.0000   EXIT PUPIL SEMI-APERTURE              2.0000
ENTR PUPIL LOCATION                 0.0000   EXIT PUPIL LOCATION                  -0.4153

WAVL (uM) .6562700 .5875600 .4861300
WEIGHTS   1.000000 .0010000 1.000000
COLOR ORDER     2      1      3
UNITS                               INCH
APERTURE STOP SURFACE (APS)           1     SEMI-APERTURE        2.00898
FOCAL MODE                          ON
MAGNIFICATION               -8.08861E-12
GLASS INDEX FROM SCHOTT OR OHARA ADJUSTED FOR SYSTEM TEMPERATURE
SYSTEM TEMPERATURE =     20.00 DEGREES C
POLARIZATION AND COATINGS ARE IGNORED.
SURFACE DATA
------------------------------------------------------------------------------------------
SURF          RADIUS        THICKNESS    MEDIUM              INDEX        V-NUMBER
------------------------------------------------------------------------------------------
  0          INFINITE       INFINITE     AIR
  1           4.18015        0.62992     N-BK7              1.51679       64.17 SCHOTT
  2          INFINITE        7.67331S    AIR
IMG          INFINITE
```

The focal length in this case is given as 8.0886 inches. Now it gets a little more complicated. If you start at the image plane and measure back that distance, you wind up—where? The answer depends on the shape of the lens, and often it is somewhere inside it. If the lens is very weak and has zero thickness (a 'thin lens'), you wind up at the lens itself. Otherwise you go to one of the *principle planes* of the lens. Draw a line along the marginal ray coming into the lens, and another backwards along the refracted ray coming out, as shown in figure 2.5. Where those lines meet is the *second principal plane*, also defined as the plane of unit linear magnification. A related construction yields the first principal plane, and there are in addition two points called *nodal points* that give the points of unit *angular* magnification. We mention these only so the students will know the terms in case they talk

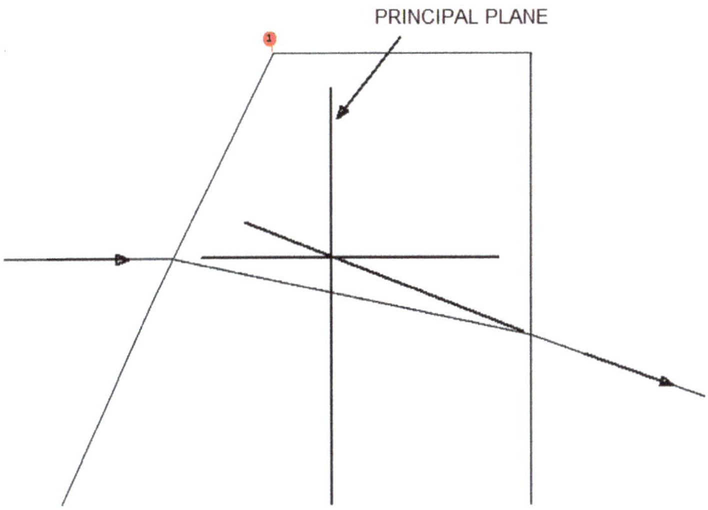

Figure 2.5. Definition of the second principal plane.

to an older generation of lens designers. Personally, I have never had any need for them or reason to look at the values. That is one of the 90% of topics I think one can ignore.

One does sometimes need to look at the *paths* of the paraxial rays, however, and the listing looks like this, produced by the command **PXT P**:

```
ID SINGLET

PARAXIAL RAYTRACE DATA

SURF        Ymarg      U'marg       Imarg        Ychief      U'chief      Ichief
--------------------------------------------------------------------------------
OBJ       0.00000   2.000E-12              -1.74551E+10     0.01746
  1       2.00000    -0.16302     0.47845       0.00000     0.01151     0.01746
  2       1.89731    -0.24726    -0.16302       0.00725     0.01746     0.01151
IMG       0.00000                               0.14119
         GIHT          FOCL        FNUM        BACK         TOTL         DELF
      0.14119       8.08861     2.02215     7.67331      0.62992      0.00000
```

This listing shows the path and angles of the two paraxial rays as well as a first-order analysis of the singlet. (The angles listed are actually the *tangents* of the angles, in keeping with paraxial practice.) Note that the F/number (FNUM) can easily be derived from the final value of the marginal-ray angle U'marg.

$$\mathrm{FNUM} = -0.5/\mathrm{U'marg}.$$

Thus, −0.5/−0.247 26 = 2.0222.

This is useful when a customer specifies the FNUM and you want to know the angle so you can specify a curvature solve, a topic we will introduce in a later chapter.

2.2 Lagrange invariant, thin-lens equation

Another topic we should mention here is called the *Lagrange invariant*. This is a simple concept:

$$\lambda = y_b n\, u_a — y_a\, n\, u_b$$

where y_a is the Y-coordinate of the paraxial ray A, and y_b that of the chief ray, B. Angles u_a and u_b also refer to the paraxial rays, and all quantities are taken at the same surface, somewhere in the lens, where the index of refraction is n. The importance of this equation is profound. Once a beam of light enters a lens, the value of λ is fixed. There is no way to change its value, except by blocking some of the light, and this makes it a great tool for deciding whether a given imaging task is possible. (The quantity λ is sometimes called the 'etendue' of the lens.) If another element changes the value of, say *ua*, you can be sure that the other variables in this equation will automatically adjust themselves in such a way that λ remains constant. We have received more than one request from a hopeful customer to design a lens that violates this rule. (We had to politely decline.) From this simple rule one can deduce, for example, that if a telescope has an eyepiece that magnifies the field by, say 100 X, then the exit pupil, where the observer puts his eye, will be 1/100 of the diameter of the objective lens or mirror. This is a powerful concept. One can derive from this the fact that the paraxial magnification of a system is given by $m = n_0 u_0 / n_k u_k$, where the subscript 0 refers to object space and k is the final surface.

Consider a telescope pointed at a clear blue sky. Ignoring reflection and transmission losses for the moment, what is the brightness of the sky as seen through the telescope? The surprising answer is, as long as the pupil of the eye is filled, the image is *exactly as bright as the sky itself*, regardless of the magnification. Why, you ask? If the telescope magnifies by, say 10X, then the exit pupil is 1/10 of the diameter of the objective lens. So the amount of light passing through the exit pupil from a given patch of sky is 100 times what would pass through the eye of an observer looking directly at that patch. But that patch also looks 10 times larger through the eyepiece, which spreads the light out over an area on the retina that is 100 times larger. So the flux/unit area/unit solid angle is unchanged. A telescope pointed at a star, on the other hand, magnifies the apparent *brightness* of the image by the area ratio, since the size of the image on the retina is not increased by the telescope. (Stars are too far away to be resolved, so the image is always a disk whose size is given by the wavelength and F/number, if the telescope is free of aberrations, and is independent of magnification.) Of course, atmospheric turbulence usually affects the star image as well, so in practice it is somewhat larger.

Among the basic principles you should know about are the thin-lens formulas, which come in two forms. Given the thin lens in figure 2.6, where the focal points are at a distance *f* from the lens and an object at distance *s1* is imaged at distance *s2*, the following relations hold:

$$-1/s1 + 1/s2 = 1/f$$

where *s1* and *s2* are measured from the principal planes (which coincide in a thin lens).

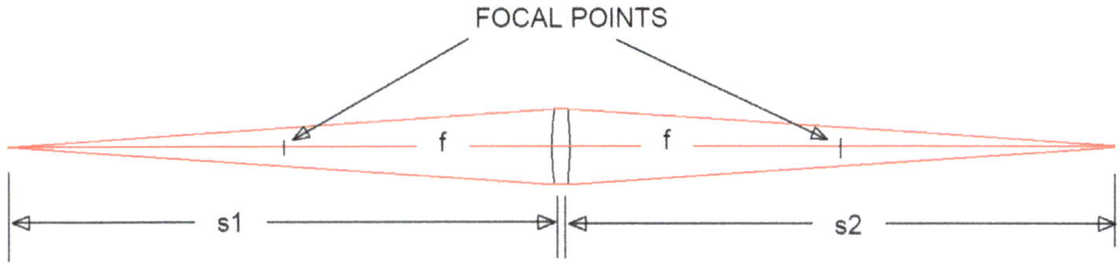

Figure 2.6. Geometry illustrating the thin-lens equation

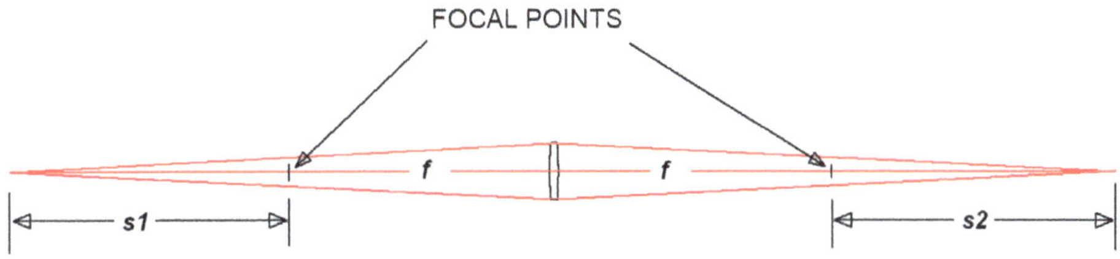

Figure 2.7. Thin-lens equation in Newtonian form.

If one measures *s1* and *s2* from the *focal points* instead of from the principle planes, one gets the Newtonian form, shown in figure 2.7.

$$-s1\ s2 = f^2$$

2.3 Pupils

Now we have to define what we mean by an 'exit pupil'. This idea is also quite simple. There is, in many lens systems, a surface whose purpose is to restrict the diameter of the beam of light that goes through, called the 'aperture stop'. Consider the lens in figure 2.8.

Surface 9 (shown by the arrow) in this case is a dummy surface declared the *stop* of the lens. The elements to the left make an image of that surface that is visible to someone looking in from the left (in 'object space'). That image is called the 'entrance pupil' and is where light has to be *aimed* so that it will actually go through the stop. Similarly, the image of the stop, formed by the elements to the right, as seen from the right (in 'image space') is the *exit pupil*, and light appears to come from that location. In case it is not obvious, a 'dummy surface' is one with no change in index on either side, so light goes straight through. (Those are useful for locating the stop and sometimes for convenience when employing tilted or decentered coordinates.)

In a telescope with an eyepiece, the exit pupil is just the image of the objective, located a short distance beyond the eyepiece, where the observer will put his eye. All the light that is collected by the objective lens therefore passes through that exit pupil and into the eye (ignoring absorption and reflection losses). The stop in this case is at the objective, since that is the most expensive element and one wants to utilize the entire aperture. An example is shown in figure 2.9:

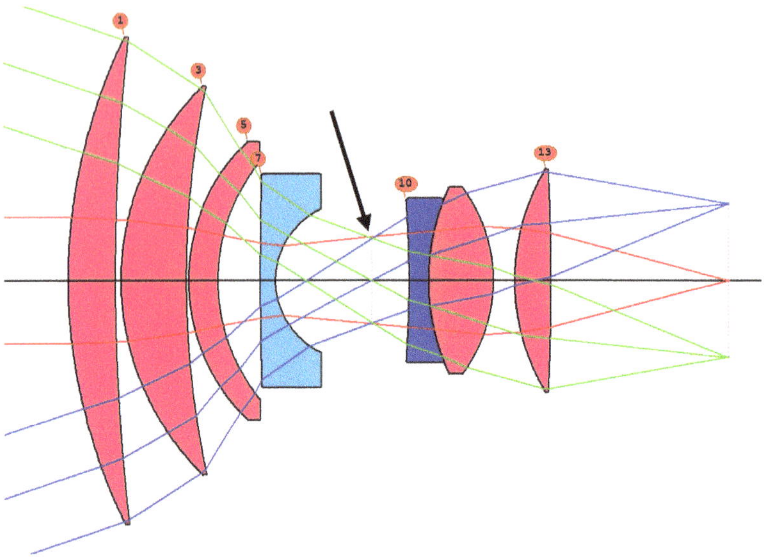

Figure 2.8. Lens with an aperture stop on surface 9.

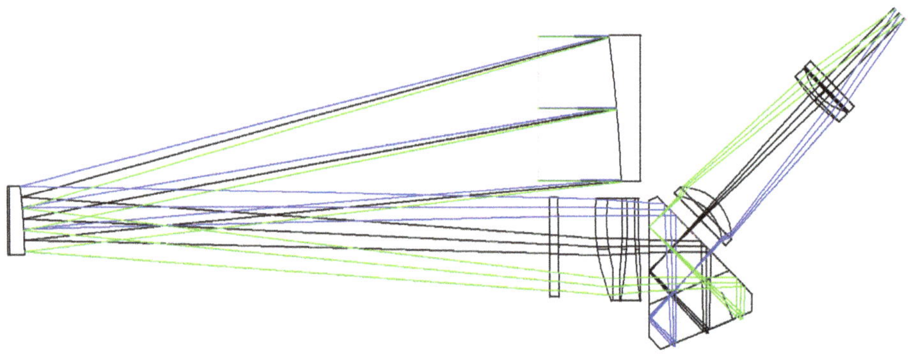

Figure 2.9. A telescope with the stop on surface 1 and an exit pupil formed at the location of the observer's eye, at the upper right.

In this system, the primary mirror is an off-axis paraboloid that collects the light. But the eye point, the exit pupil at the upper right, is just an image of the mirror. So the observer's eye will receive all the light that gets through. (Assuming the *size* of the pupil of the eye is the same or larger than that of the exit pupil, so all the light can get in.) The distance from the last lens element to the observer's eye is called the 'eye relief', and one must make sure it is large enough to allow comfortable viewing, and in some cases even larger, to allow for the use of eye glasses.

The concept of the *pupil*, while simple in principle, is more complicated in practice, and will be dealt with in more detail in chapter 22.

IOP Publishing

Lens Design (Second Edition)

Automatic and quasi-autonomous computational methods and techniques

Donald C Dilworth

Chapter 3

Aberrations

Seidel aberrations; spherical aberration; coma; astigmatism; rayfan curves; correction methods; Abbe sine condition; higher-order aberrations, spot diagrams; wavefront aberrations; OPD; Fermat's principle; MTF; knife-edge trace; encircled energy; chromatic aberration; Strehl ratio

Now we get into the meat of what lens design is all about: minimizing the aberrations that degrade image quality and are almost never zero.

They come in families and are analyzed in one of two ways. The classic approach was to calculate what are called 'third-order' aberrations, an important topic before the middle of the 20th century, before computers were available, when one wanted to know how good a lens was. (By 'lens' we mean a collection of one or more lens elements, each with its own aberration contributions.) In 1856, Ludwig von Seidel published the equations for calculating a very approximate estimate of the quality of an optical image. In those days, even a rough estimate was better than trying to trace a large enough set of rays to get a more accurate answer (using log tables!). But it was not a very good estimate, and the art of lens design split into two schools. In England, designers such as H. Dennis Taylor and Charles Hastings would often work up a third-order solution (which they knew would not be very accurate), and then *grind and polish* that design so they could measure the image quality directly. It was actually faster to make the lens in those days than to trace a representative set of rays. (The physical lens was a great *analog* computer!) In Germany, designers such as Joseph Max Petzval and Ernst Abbe insisted on tracing rays, getting the design right before grinding any glass. Sometimes it took months of ray tracing.

So what, exactly, is a third-order aberration? Well, if paraxial optics uses the first term in the expansion of the sine function (the sine is then equal to the angle), the next term in that expansion is a function of the third power of the angle. Deriving the raytrace equations with this added term yields a different answer than the paraxial result, and that difference is called the third-order aberration. The value can be expressed as a power series involving the position of a ray in the aperture, (ρ,θ, in

doi:10.1088/978-0-7503-3695-6ch3

polar coordinates) and the object point in the field, (*HBAR*), and one should be acquainted with the concept. We can write the image errors in x and y (E_x , E_y) as follows:

$$E_x = SA3\rho^3 \sin(\theta) + CO3\rho^2 HBAR \sin(2\theta) + SI3\, \rho\, HBAR^2 \sin(\theta)$$
$$E_y = SA3\, \rho^3 \cos(\theta) + CO3\, \rho^2\, HBAR(2 + \cos(2\theta))$$
$$+ TI3\, \rho\, HBAR^2 \cos(\theta) + DI3 HBAR^3$$

SA3 is the third-order spherical aberration, *CO3* is the coma, *SI3* and *TI3* are the sagittal and tangential astigmatism, and *DI3* is the distortion.

Here we see that the error due to spherical aberration (*SA*3) is constant over the field (no *HBAR* dependence) and varies as the third power of aperture; coma (*CO*3) varies as the second power of aperture and linearly with field; and astigmatism (*TI*3 and *SI*3) varies as the square of the field height and linearly with aperture. Note that by 'distortion' (*DI*3), which varies as the third power of field, we mean the error in the *location* of the image of a point object; the term does not refer to the *sharpness* of the image, as it sometimes does when used by a layman. All of the third-order aberrations apply only to rotationally symmetric systems.

Although these third-order concepts can be useful when analyzing image errors, they play almost no role when one designs a lens today. (Some older designers still use the methods of Kingslake: changing parameters one at a time and looking at how the third-order aberrations change. Then iterating. Over and over. I do not recommend this procedure.) Since tracing rays with modern computers is so fast, we now deal almost exclusively with the errors of real rays and utilize algorithms that drive the errors as close to zero as possible, largely ignoring third-order aberrations. (There are rare exceptions, which will be noted in the chapters that follow.)

There are also higher orders, fifth, seventh, and so on, which have their own dependencies on pupil and field positions. We mention them only in passing since they are no longer of much interest to the lens designer. (Why are there only *odd* orders you ask? Because the expansion of the sine function has only odd orders.) When one alters a lens, it is the lowest-order aberrations that change the fastest, a fact that makes higher orders more difficult to correct. Fortunately, that task is handled rather well when one defines the merit function with a suitable set of real rays.

3.1 Ray-fan curves

To show the aberrations of a lens, we create so-called *ray-fans* curves, which show the image errors of a fan of rays. Before you can trace them, however, you have to be sure you are tracing the *correct* rays. This brings up the subject of *stops* and *pupils*, which deserves a chapter of its own, chapter 22, and which we will only touch on for now. Assuming you have properly defined the lens and the pupil, the ray fans tell you a great deal about the image-forming properties of the lens.

If you trace a fan of rays through a lens, you get a characteristic pattern in the image. A *sagittal fan*, or **SFAN**, is a set of rays across the aperture in the x-direction,

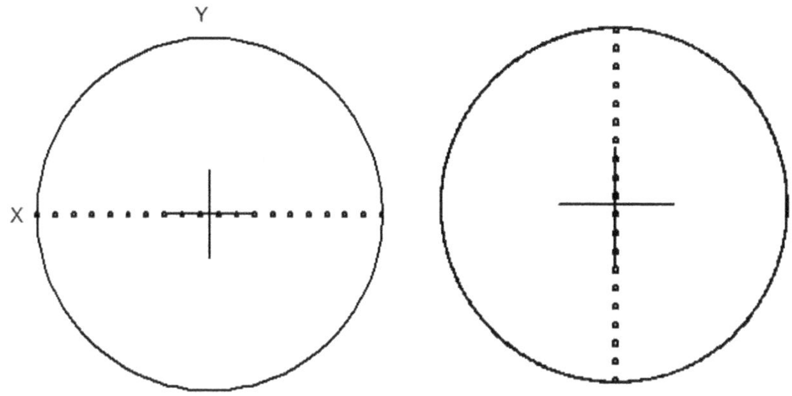

Figure 3.1. Definition of a sagittal fan, on the left, and a tangential fan, on the right.

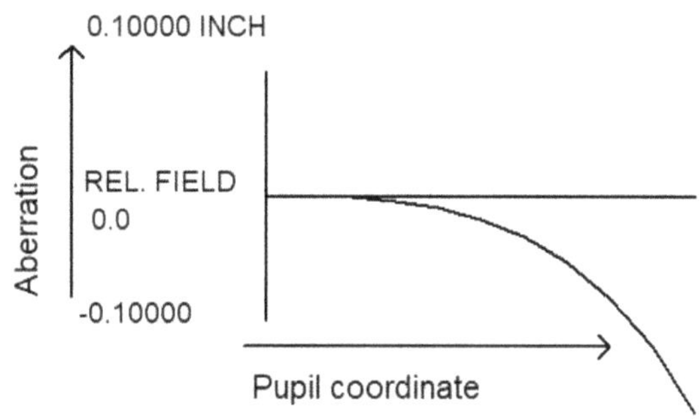

Figure 3.2. Illustration of a sagittal fan showing negative spherical aberration, *SA*3.

shown on the left in figure 3.1, while a *tangential fan*, or **TFAN** goes from bottom to top, shown on the right:

In practice, since a lens is usually symmetrical about the *y–z* plane (the *meridional* plane), one usually traces only one-half of the sagittal fan (because the other half is identical but inverted). If one now plots the *x*-position of a given ray at the image against the *x*-position of that ray in the aperture, one gets, for a lens with undercorrected third-order spherical aberration, an SFAN curve such as shown in figure 3.2:

In this case, the curve is very close to a third-power function of the pupil position, as one might expect. This shape is characteristic of third-order spherical aberration (named because it is the most prominent defect in the image formed by a *spherical* telescope mirror. A *paraboloidal* mirror would make a perfect image at the center of the field because it is free of spherical aberration). Figure 3.3 shows the display on the monitor of a singlet lens along with ray-fans curves at three field points. (This display, provided by the **SketchPAD**[1] feature of the program, is opened with the command **PAD**.) The curves are all very similar, since SA3 is the predominant

[1] PAD™ and SketchPad™ are trademarks of Optical Systems Design LLC, an Arizona, USA corporation.

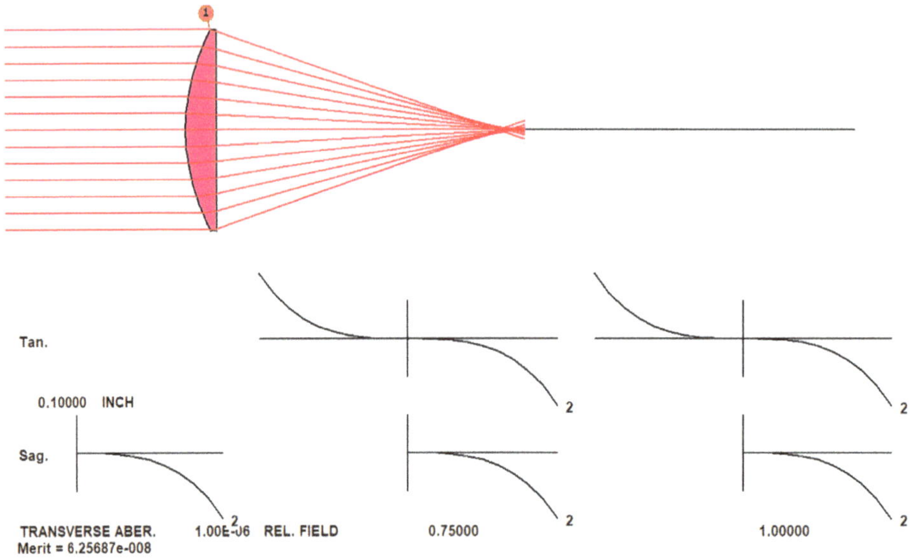

Figure 3.3. SketchPAD display of a lens in which spherical aberration predominates.

aberration in this case and it does not vary with field. The lower three curves are SFANs, while the upper two are TFANS, plotted at the relative field points noted, where full field is at 1.0.

One can ask the software for the values of these aberrations with the command **THIRD**:

```
SYNOPSYS AI>THIRD

ID SINGLET

THIRD-ORDER ABERRATION ANALYSIS

FOCAL LENGTH  ENT PUP SEMI-APER   GAUSS IMAGE HT
       8.089                2.000            0.141

THIRD-ORDER ABERRATION SUMS
                SPH ABERR      COMA  TAN ASTIG  SAG ASTIG    PETZVAL DISTORTION
                    (SA3)     (CO3)      (TI3)      (SI3)     (PETZ)  (DI3(FR))
                 -0.13229  -0.00395   -0.00109   -0.00050   -0.00020 -4.422E-06

PARAXIAL CHROMATIC ABERRATION SUMS
                 AX COLOR  LAT COLOR  SECDRY AX SECDRY LAT
                    (PAC)      (PLC)      (SAC)      (SLC)
                  0.03062  3.849E-05    0.00942  1.184E-05
```

The value of *SA*3 is here −0.132 29, which is just what is shown on the SFAN curves.

You can also evaluate the aberrations from the dialog **MFT**, which lists several other options that are sometimes useful.

A paraboloidal mirror has aberrations of its own, as shown in figure 3.4, except at the center of the field, where it is perfect. This is the common Newtonian configuration.

In this case, there is no spherical aberration, but the tangential fan off axis shows a U-shaped curve, which is characteristic of *coma*, **CO3**. (The fold mirror blocks

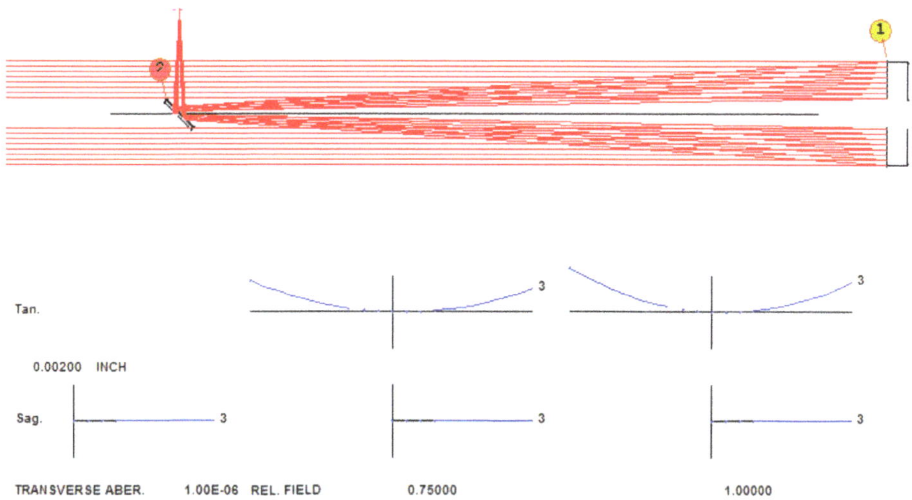

Figure 3.4. A typical Newtonian configuration, showing no spherical aberration at the on-axis point and coma off axis.

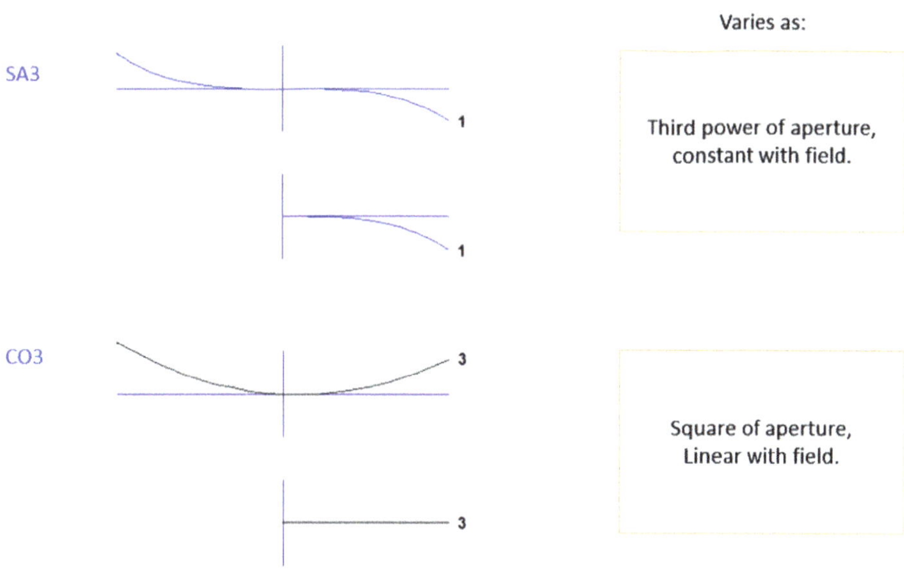

Figure 3.5. Illustration of the shape of ray-fan curves indicating spherical aberration and coma.

some of the light, as indicated by the dots at the center of the fans curves.) Figure 3.5 shows the dependencies of *SA*3 and *CO*3: SA3 varies as the third power of aperture, *CO*3 varies as the second power of aperture and linearly with field, as given by the above equations and illustrated in figure 3.5. Coma can be viewed as a variation of magnification from one zone to another in the lens.

Other third-order aberrations include sagittal astigmatism (**SI3**), tangential astigmatism (**TI3**) and Petzval curvature (**PETZ**). TI3 is shown in figure 3.6; SI3

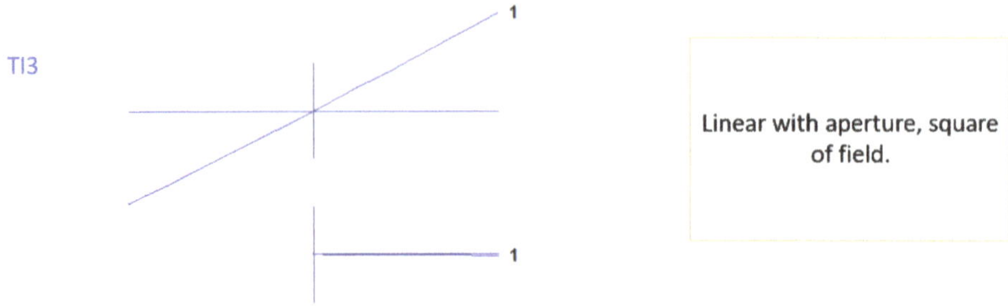

Figure 3.6. Ray-fan curves showing tangential astigmatism.

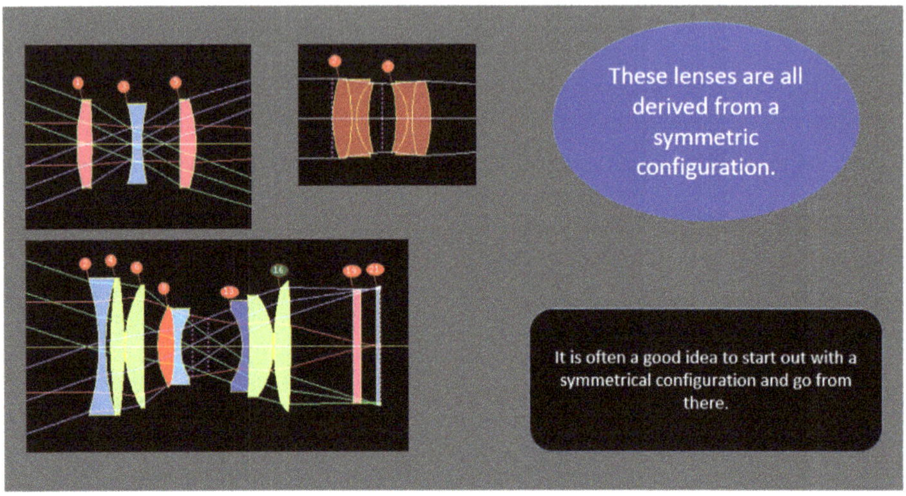

Figure 3.7. Examples of lens configurations derived from symmetrical forms.

is similar but affects the SFAN instead of the TFAN. Petzval curvature affects the flatness of the image surface and is related to both of them. (In case you missed the point, a perfect image would show up as a straight horizontal line on these plots.)

The Petzval curvature is an interesting concept. If all the *other* aberrations of a lens were zero, the sharpest image would be found on a surface with that curvature. If the astigmatism is not zero, the best image surface is found in the space between the S and T surfaces, and is not very close to the Petzval surface, as one might guess. So it does not tell you much in most cases. The formula is very simple:

$$PETZ = -\sum \frac{1}{n_j f_j}$$

where n_j is the index of each element and f_j is the focal length. From this one can learn that two positive elements cannot correct `PETZ` to zero. Adding a negative lens (which has negative f) makes it possible, and this insight lead to the original Cooke triplet, which has a negative element between two positive ones. An example is shown in figure 3.7 on the upper left. A refractive system tends to produce a Petzval surface curved inwards towards the lens, while a reflective mirror of positive power does the opposite, a fact that is sometimes useful. By the way, the *curvature* of a surface is just the inverse of the radius of curvature.

In the past, designers took pains to correct these aberrations, and some still do—but I have learned an important lesson that I want to pass on: When you design a lens, *correct only those things that matter*, not what you learned in an old textbook. **In most cases, you only care about two things: is the image sharp, and is it in the right place?** Third-order aberrations usually get in the way and are not worth calculating, in my opinion. In almost all cases, you need third-order aberrations to balance higher orders, so trying to make them go to zero is a mistake. They don't tell you much about a lens.

With some important exceptions. (I don't want to overstate my case.) It turns out that **any lens that is *exactly symmetric* has no odd-order field aberrations**, so the coma is zero. And a lens construction derived from a symmetrical shape gains some of the same benefit, even if no longer strictly symmetric. Examples are shown in figure 3.7.

Another application of this principle is the case of two paraboloids with a common focal point, as shown in figure 3.8. This combination has no spherical aberration (because the paraboloids have none) and no coma (because the coma from the first is exactly cancelled out by the second.) Does this seem unlikely? Well, Nature is not usually so kind to us lens designers, but it turns out that **the coma from a paraboloid at a given linear image height is a function only of the F/number**, not the focal length. Scale a mirror up by a factor of two, and the coma doubles—at the new full-field image height. But at the old image height, which is one-half of the new one, the coma is one-half of that at the new image, or just the previous value. Perfect balance! So, yes, there is some value in learning about third-order aberrations. But not much beyond these simple cases. And if a customer insists the lens must have, for

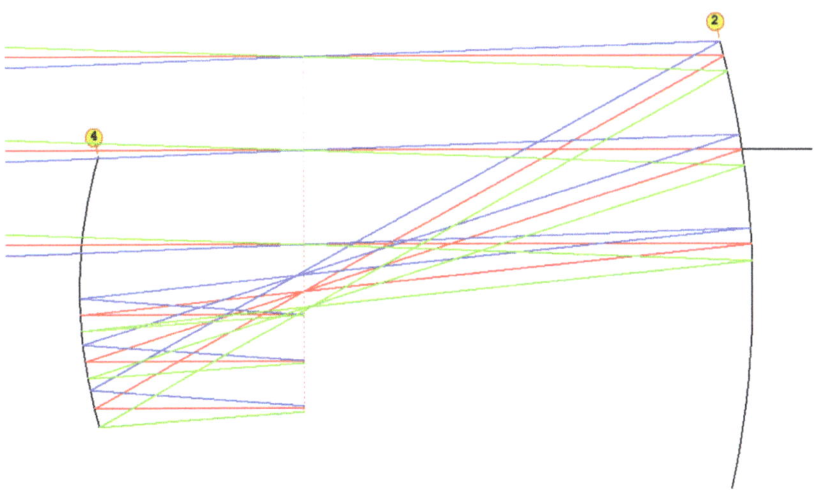

Figure 3.8. Arrangement of confocal paraboloids; this form has no spherical aberration or coma. The output is collimated. One cuts an off-axis portion of the two mirrors in practice.

example, 'no astigmatism', you can safely ignore him. That would rule out many excellent designs in which third-order astigmatism balances out higher orders—which is a concept he does not know about.

The methods used by the old masters for reducing aberrations are still valid, if cumbersome—but today the computer does most of the work for you, so those methods are of little use anymore. That said, here are some basic relationships you should know about.

Be aware that it is impossible to correct spherical aberration in a single element with spherical surfaces if the light comes in collimated (from an object at infinity) unless the power is zero. One can 'bend' the lens, and the aberration changes, but there is no bending that makes it go to zero. Three different bendings of a singlet are shown in figure 3.9, the second at the best position, none of them perfect. (It is,

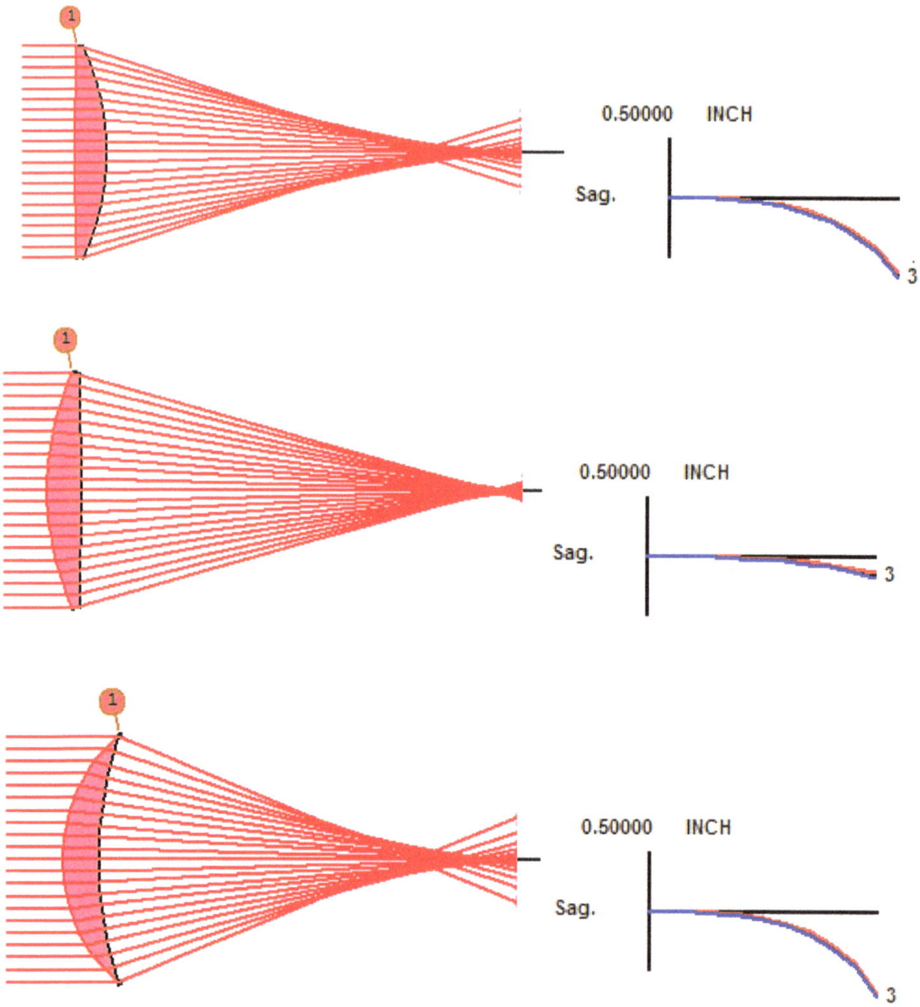

Figure 3.9. Illustration of the variation in spherical aberration with lens bending.

however, sometimes possible to correct spherical aberration by bending a lens if the light that comes in is *not* collimated.)

3.2 Abbe sine condition

Let's also take a deeper look at coma. We want the focal length of all rays in the aperture to be the same, for a given object point. Figure 3.10 shows why this is not the case for the Newtonian telescope. The marginal ray has to go further than the chief ray, so the focal length along that ray is longer and that's why you get coma. Simple.

Bending a lens also bends a nodal surface, and when it makes a sphere about the image point, the coma goes away. This is the basis for the **Abbe sine condition**. In figure 3.11 we corrected $SA3$ by adding an aspheric term to surface 1 and corrected $CO3$ by bending the lens.

When we satisfy the requirement that

$$\mathrm{Sin}(\theta) = A/F,$$

the condition is satisfied and there is no coma. (This equation is valid when the object is at infinity.) A lens that satisfies this condition is termed an *aplanat*. In

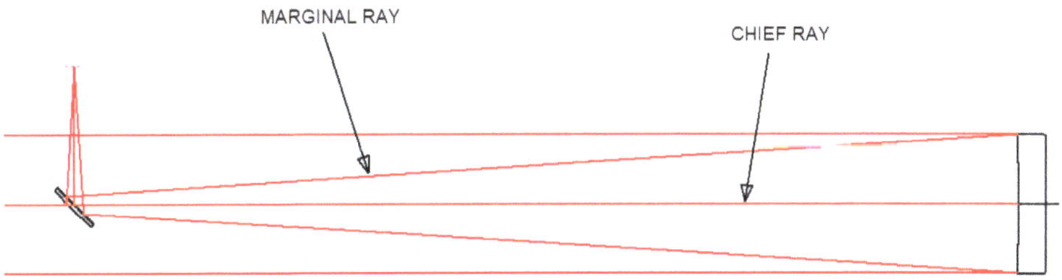

Figure 3.10. Illustration of the origin of coma in a Newtonian telescope.

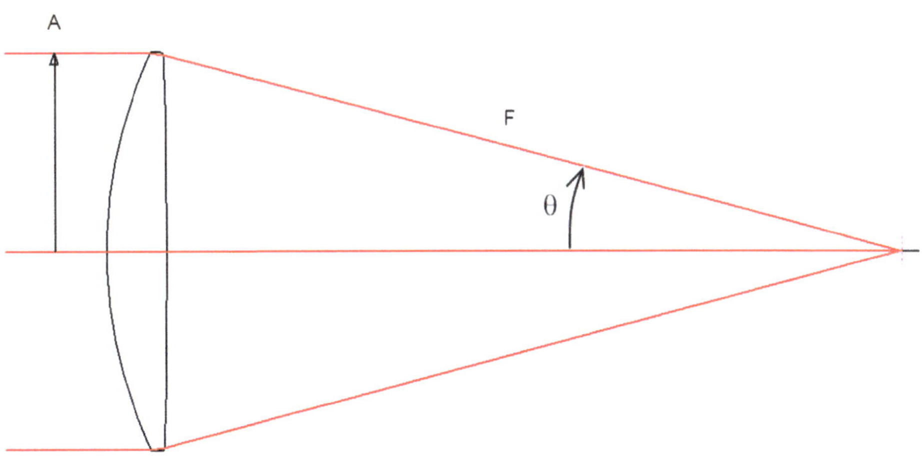

Figure 3.11. An aspheric singlet bent to eliminate coma by satisfying the Abbe sine condition.

practice, one corrects the aberrations of real rays, not third-order aberrations, and the sine condition is satisfied automatically if one succeeds.

A corollary of this rule is the fact that the minimum possible F/number of a lens that is corrected for coma is 0.5. By the rule we gave earlier, $FNUM = -0.5/\mathrm{UA}$, paraxially, and $FNUM = -0.5/\mathrm{n}\,\sin(\theta)$ for real rays. (The quantity n $\sin(\theta)$ is known as the *numerical aperture* of the lens.) Since θ can never exceed 90 degrees, QED. In practice, one rarely approaches an angle as steep as 90 degrees. If someone asks you for a lens at F/0.25, you can be sure it won't have a very wide field (because the coma would be horrible. The focal length would change by a factor of two from the central to the marginal zone!)

3.3 Higher-order aberrations

We've discussed third-order aberrations, but of course there are also higher orders which, in principle, one can calculate by taking successive terms in the expansion of the sine function. In practice, this becomes algebraically intractable very quickly, and the effort to derive and calculate the values is one of the immense tasks our forebears were stuck with but are now seldom warranted.

But the *effect* of higher-order aberrations is still important and is easily seen in the ray fans diagrams. In figure 3.12 one can see that the paraxial defocus is zero (because the curve starts out horizontal at the axis), and the lens has negative third-order spherical aberration (it turns down) and positive fifth-order (it turns up again). This is a common situation when the optimization program balances one aberration against the others.

If you subtract the third-order aberrations from the actual ray fans data, what is left is a good indication of the influence of the higher orders. Generations ago, designers realized that when you change a lens parameter, it is the low-order aberrations that change the fastest. The higher the order, the more slowly it changes —and this is the main reason why high orders are more difficult to correct.

Figure 3.13 shows the curve for a lens with multiple orders. Here you see defocus, third, fifth, seventh, and ninth-order spherical aberration, all balanced so the lens exhibits just over 0.03 waves peak-to-peak wavefront error, as shown in the OPD plot in figure 3.14. (OPD means Optical Path Difference.) This lens operates in the deep UV at 0.226 μm, at a speed of F/0.625, and this is typical of the balance achieved in microlithography lenses of 20 or more elements, which must be functionally perfect. Clearly, balancing higher-order aberrations is a fundamental goal in lens design, and, fortunately, modern software takes care of that task rather well. Lenses such as this represent the pinnacle of the lens designer's art—and constructing them is no less of a challenge, with tolerances so tight that each element must first be centered in its cell and then the cell centered on a rotary table monitored with a laser beam. When completed, the price for such a lens is measured in seven figures or more. The PC on which you run your optimization program would never exist except for these very exceptional lenses, themselves designed on just such a machine.

Whenever you see curves like these, it's a sign that you probably should ask for a grid number larger than the default five on axis and three elsewhere (which you get

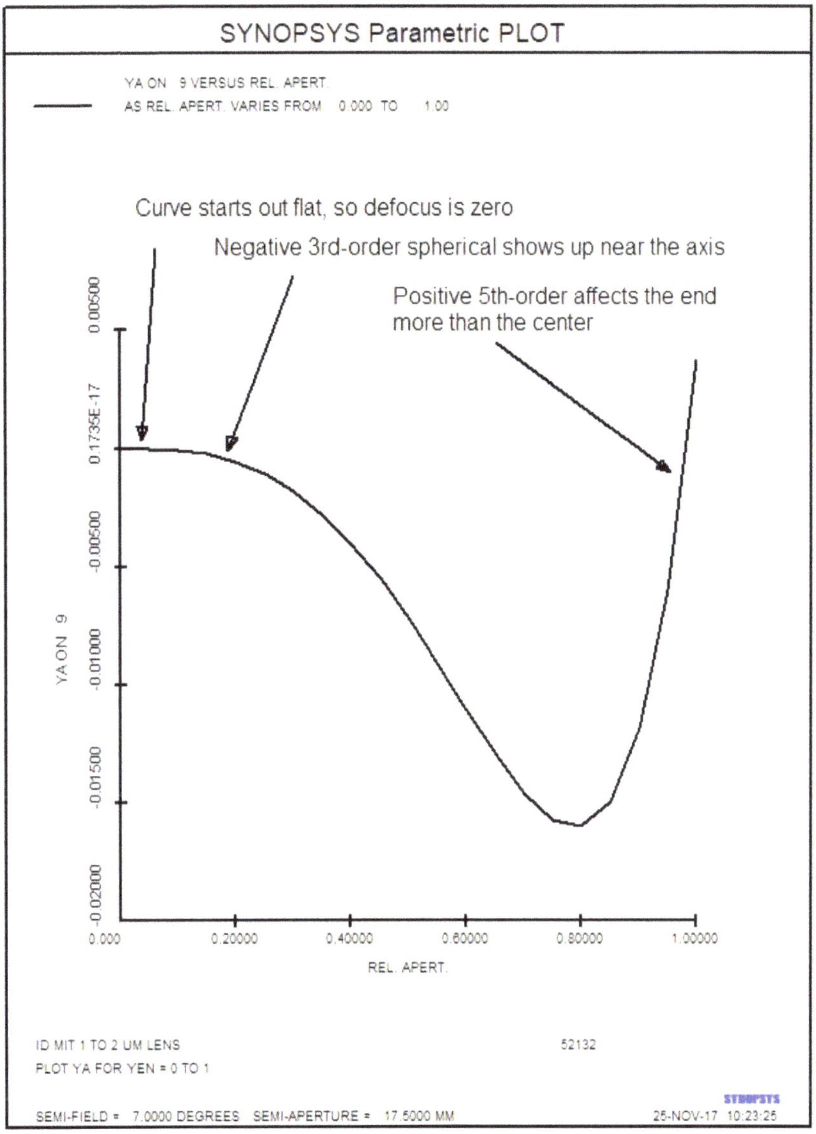

Figure 3.12. Reading aberration contributions from the ray-fans curves.

with ready-made MF number 6, which you will use in some of the chapters below). Otherwise you might happen to get great correction *at* the requested rays but wild swings in between. This can happen with multielement lenses, and especially if aspherics or diffractive optical elements (DOEs) are present.

We should repeat here that the meaning of the aberration balance of a lens depends on exactly how one defines the beam that is traced at each field point. This is the function of the *pupil definition*, and there are subtleties involved. A discussion will be found in chapter 22.

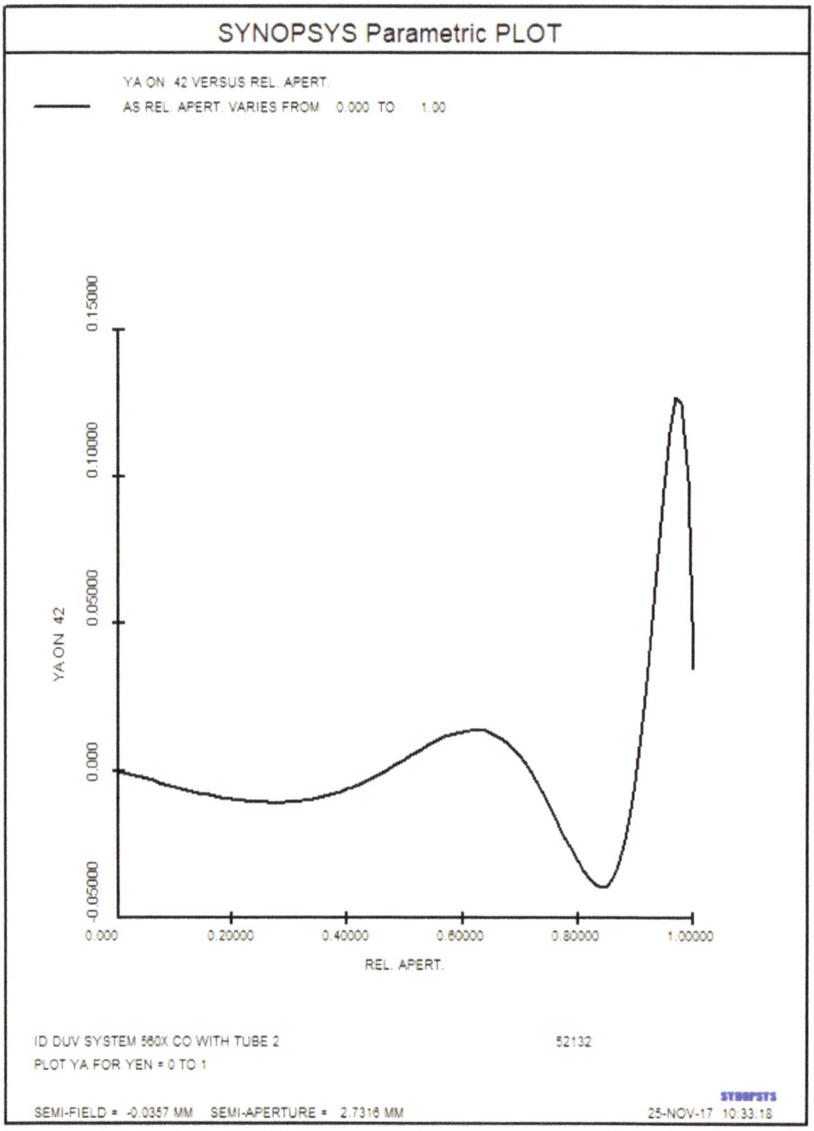

Figure 3.13. Aberration curves for a lens with many orders of spherical aberration.

3.4 Spot diagrams

Today, with fast modern PCs, one can often get a good picture of the quality of an image by tracing a suitable number of real rays and looking at the geometrical image errors. One common method involves creating a *spot diagram*. Consider the lens in figure 3.15. Look at the rays coming out, headed toward the image plane from the full-field object, shown in blue.

Let us magnify the pattern made by a grid of rays when they get to the image plane, shown on the right in figure 3.16.

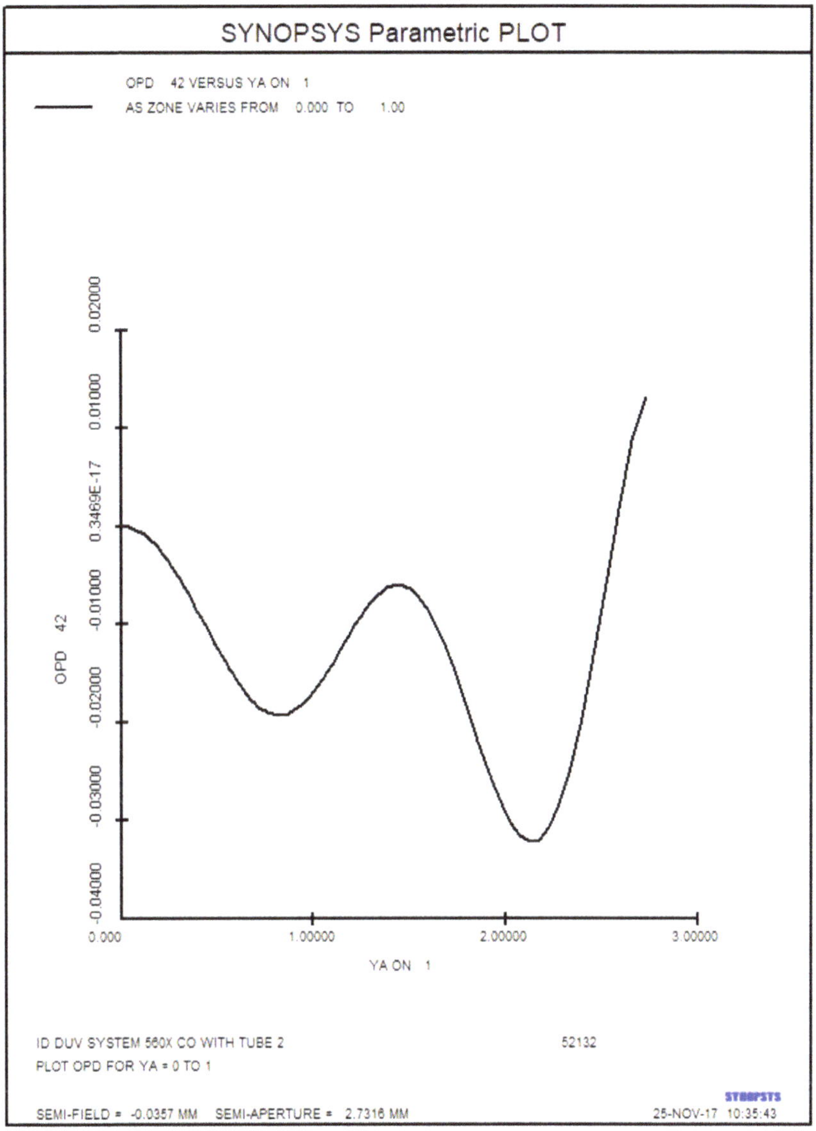

Figure 3.14. OPD aberration for the lens with multiple orders of aberration.

This is certainly not a very sharp image. It was calculated by tracing 1731 rays, in three colors, an effort that would have been impractical only a generation ago (imagine doing that with a Marchand calculator!) This kind of picture gives you a good indication of the image quality, especially if it is not a particularly good image. (The structure evident in the spot diagram is a consequence of the ray pattern in the aperture; here it is a uniform square grid. Other ray patterns alter the appearance, but the overall effect is much the same.)

If it is a *very* good image, on the other hand, one needs to look at the wavefront.

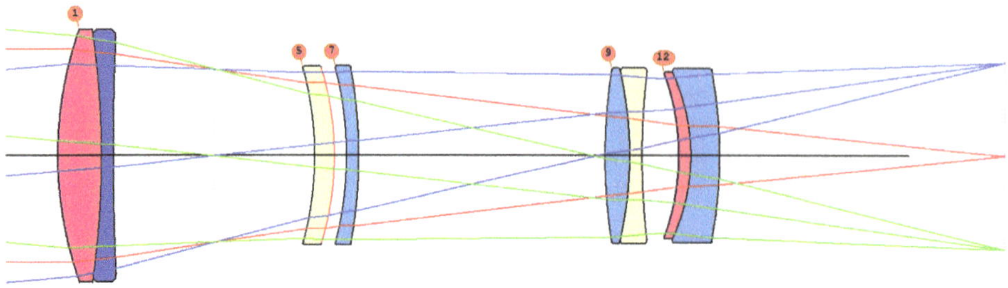

Figure 3.15. Example lens for calculating a spot diagram.

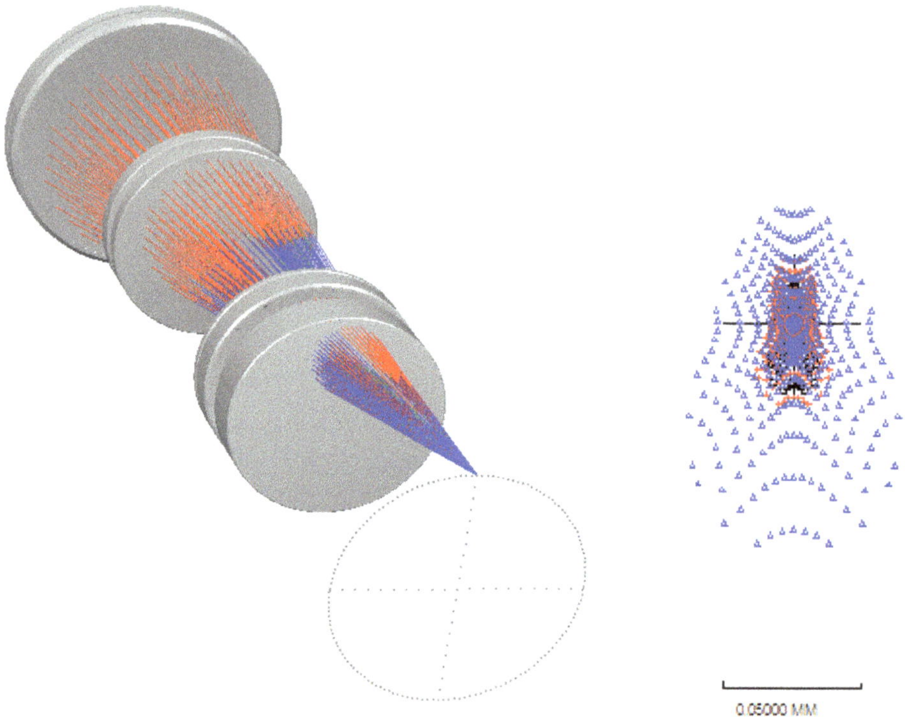

Figure 3.16. Geometry of a spot diagram.

3.5 Wavefronts and aberrations: the OPD

Even if an image were geometrically perfect, it would not focus to a single point due to the finite wavelength of light. Instead of a single point, one gets a *diffraction pattern*. Considering again the Newtonian telescope—and ignoring the obscuration for the moment—the pictures in figure 3.17 show, from left to right, the geometric image on axis, the diffraction pattern on axis, the geometric image at the edge of the field, and the diffraction pattern there. (This telescope has an aperture diameter of 10 inches and a focal length of 80 inches, and thus an F/number of 8. The semi field angle is 0.5 degrees.)

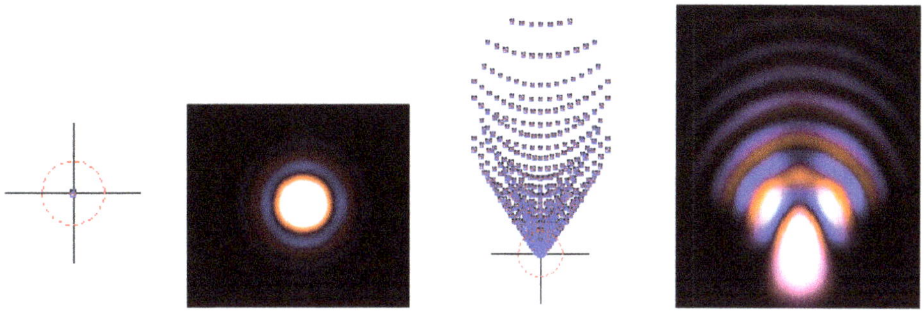

Figure 3.17. Image characteristics of a Newtonian telescope. Diffraction effects enlarge a geometrically perfect image.

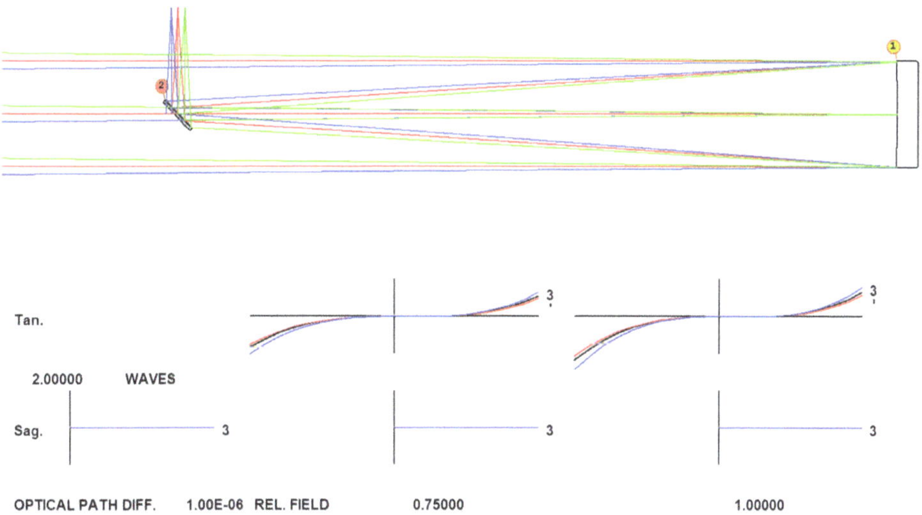

Figure 3.18. Newtonian telescope with OPD fans shown on the bottom of the display.

The red circle on the first picture gives the size of the first dark ring in the diffraction pattern, and the geometric blur is much smaller than that. (It is not zero in the picture due to the size of the symbol representing each ray.) If a lens is corrected so well that the geometric blur is comparable to or smaller than the diffraction pattern, one must usually try to control the Optical Path Difference (**OPD**) of a ray instead of (or in addition to) the geometric blur. You will find examples of this kind of correction in subsequent chapters. And in this case, the image must be evaluated in a way that takes diffraction into account as well. The OPD is just the difference in total path length of a given ray compared to the path of the central ray (the *chief ray*). In a perfect lens, those paths would be identical, a condition known as *Fermat's principle*.

Look at the OPD fans for this telescope, shown in figure 3.18. Here we plot the OPD of each ray instead of the transverse aberration shown in the previous examples.

Again, the on-axis aberrations are zero, but now the TFAN at the edge of the field has a different shape. This is what coma looks like when OPDs are plotted. The relationship is simple: **the OPD plot is the *integral* of the transverse plot** (which we call the **TAP** plot). It can be viewed as giving the *shape* of the wavefront forming that image, relative to a perfect spherical wavefront converging to that image point. You can see here that the wavefront has an S-shaped curve, relative to perfect, which is characteristic of coma.

3.5.1 Image analysis

Any program that can calculate the spot diagram or the diffraction pattern can usually calculate any of a number of other quantities related to image quality. The choice of which kind of analysis to perform depends on the purpose of the lens, what kind of technology will utilize the image (CCD, film, IR sensor) and the preferences of the people involved. The principal techniques are the following:

1. **The MTF**, or modulation transfer function. This is a curve showing the frequency content in the image of a point source, often interpreted as the contrast in a sine-wave image of a given frequency. (The two are equivalent if there are no pupil aberrations.) Looking at a sine-wave target whose spatial frequency at the image is 100 cycles/mm, the on-axis MTF shown in figure 3.19 would have a contrast of about 0.37. When the actual sine-wave image is plotted, shown in red in figure 3.20, one can measure the contrast, as (MAX—

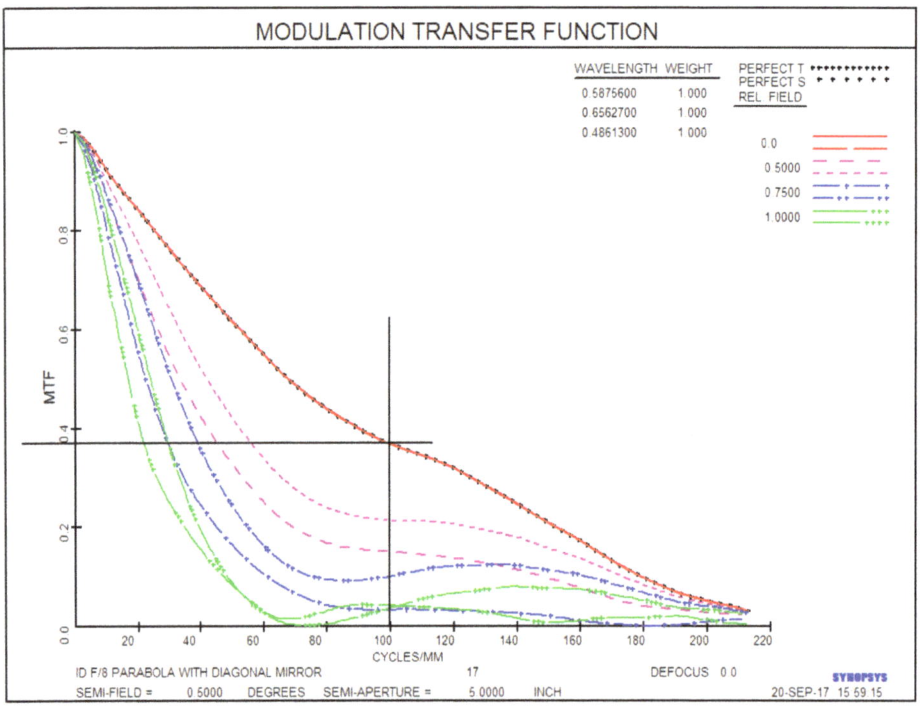

Figure 3.19. MTF curves for several field points with a Newtonian telescope. The On-axis MTF is about 0.37.

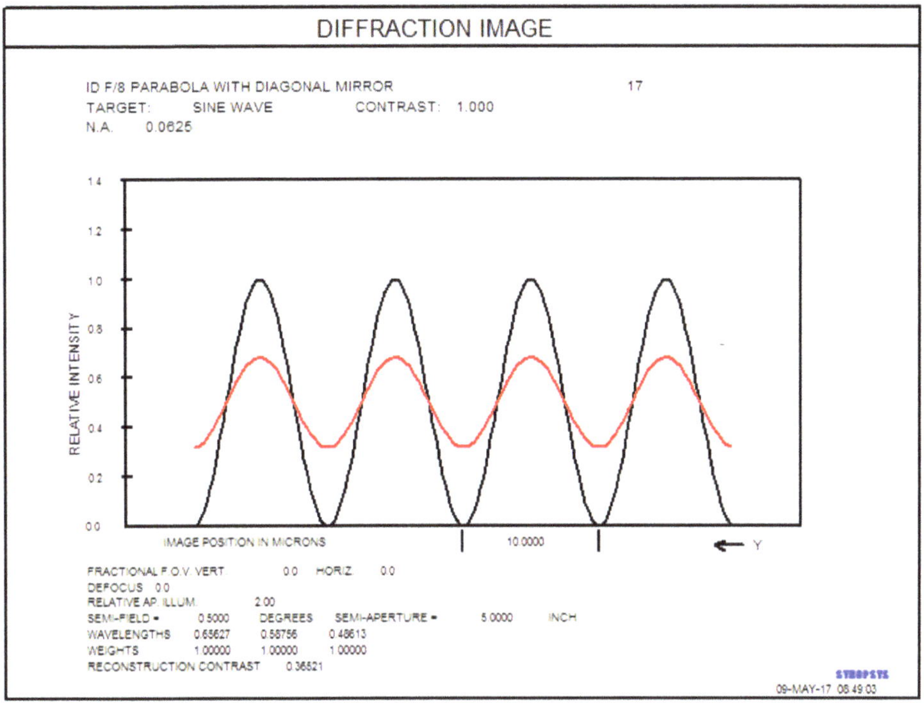

Figure 3.20. Characteristics of the on-axis image of the Newtonian telescope, showing a contrast of just over 0.36 for the image profiled in red.

MIN)/(MAX + MIN). Here, the result is 0.0362, in good agreement with the prediction. This MTF was calculated for the Newtonian telescope above, with the obscuration taken into account. At low spatial frequencies, on the left end of the curve, the MTF is high, and toward the right it goes to zero. The zero point is called the cutoff frequency, and an image requiring a higher frequency cannot be resolved with this system. The MTF is sometimes called the Optical Transfer Function, or OTF, if other components in the system also affect the resolution. In that case, the OTF is composed of both the MTF and the physical transfer function, or PTF, the latter due to all other sources of degradation. The MTF options can be accessed from the dialog **MOP**.

The curves in figure 3.20 were calculated by the partial-coherence program **PARTC**, which can be run from the dialog **MPA**. Enter a filling factor of 2.0, spatial wavelength of 10 μm, and Execute. This filling factor makes the analysis incoherent, rather than coherent, and 10 μm is the spatial wavelength of a sine wave of frequency 100 c/mm.

An obscuration materially affects the image structure and the MTF. Here are the MTF and diffraction point-spread function plots of this F/8 telescope for various sizes of the obscuration, shown in figures 3.21–3.23. These results show why one usually tries to minimize the obscuration as

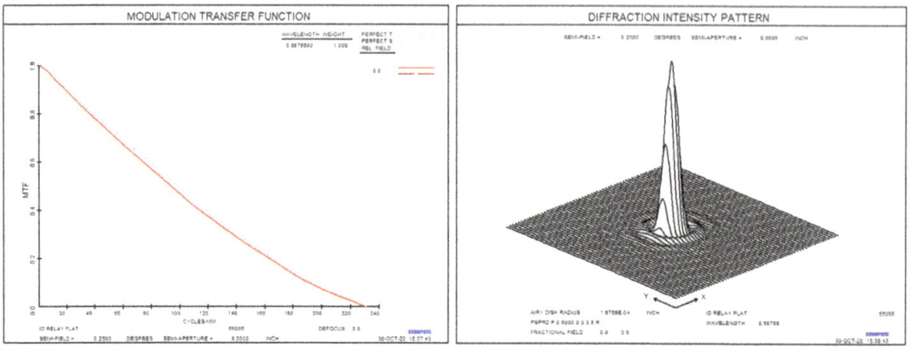

Figure 3.21. MTF and image structure of the telescope ignoring the obscuration.

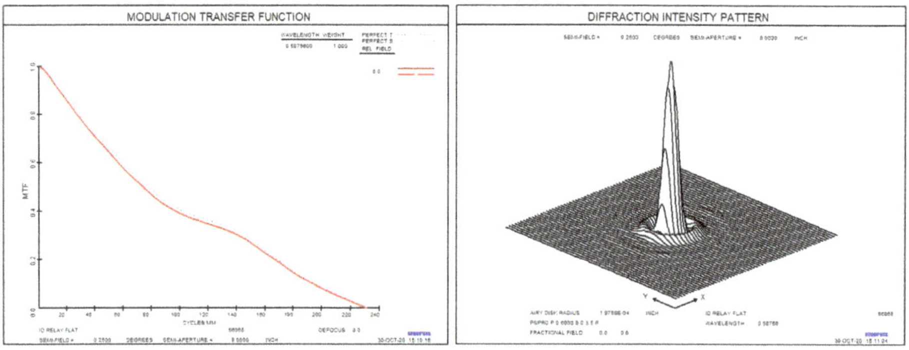

Figure 3.22. MTF and image when the obscuration is 25% of the aperture diameter.

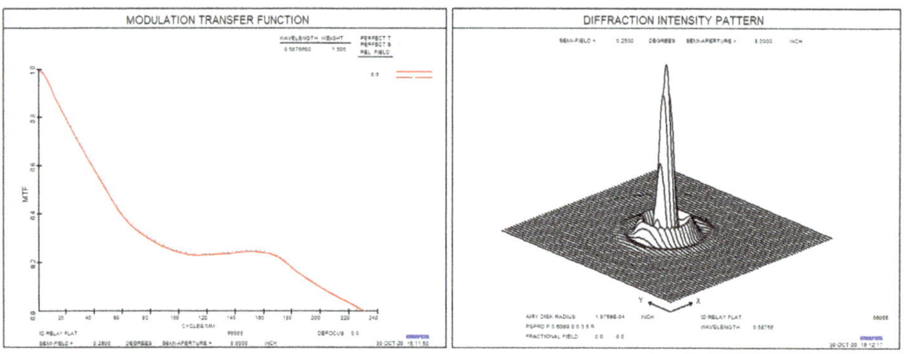

Figure 3.23. MTF and image when the obscuration is 50% of the aperture.

much as possible: image quality declines as the obscuration grows. In each case here, the *geometric* image is perfect, but the diffraction pattern is not if the obscuration is nonzero. The top left MTF is as good as possible for this F/number and wavelength.

Figure 3.24. Diffraction pattern of the Newtonian telescope off axis.

2. **The knife-edge trace**. Figure 3.24 shows the image at the edge of the field of the Newtonian telescope (ignoring the obscuration):

   One can pass a knife edge over this image, from bottom to top, and plot the energy that is uncovered, making the curve shown in figure 3.25. Some lens designers find this kind of analysis useful, and it is popular with lens makers since it is conveniently measured‘.
3. **Encircled energy.** Sometimes one plots the energy through a circle as the radius of the circle increases. The result of course depends on where one centers the circle. Here is an example, in figure 3.26, at the center of the field of the telescope. The diffraction rings are clearly apparent:
4. **Wavefront Variance**. If one traces a large grid of rays equally spaced over the entrance pupil, a simple calculation yields the variance, denoted as σ^2:

   $$\sigma^2 = \langle \mathrm{OPD}^2 \rangle - \langle \mathrm{OPD} \rangle^2 ,$$

   where <> denotes an average over the set of rays.

   This quantity is frequently specified as a design goal. Zero would be perfect.
5. **Strehl ratio.** Another common measure of lens quality is the Strehl Ratio (**SR**), often used when lens performance is close to perfect. It is defined as the ratio between the peak intensity in the diffraction pattern divided by the intensity it would have if all aberrations were zero. If the image is perfect, the ratio is 1.0.

   The SR can be calculated in two ways: approximately with an exponential function of the wavefront variance, or by calculating the diffraction point-spread function (**PSF**) with and without aberrations and taking the ratio of the peaks.

   If the OPD errors are small, the exponential approximation is reasonably accurate. This works because, for small errors, the SR does not depend strongly on which aberrations are present. Finding the ratio via the PSF is somewhat

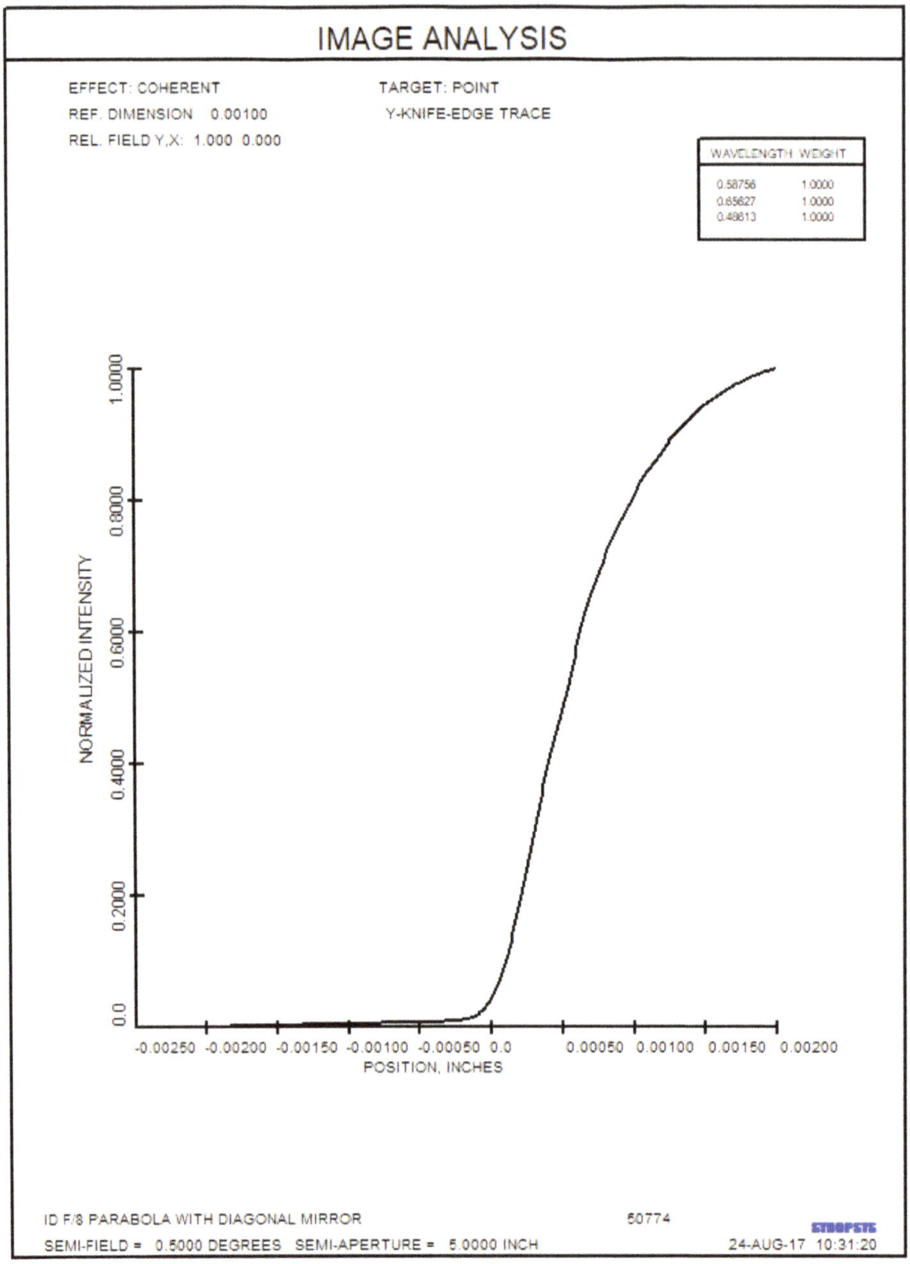

Figure 3.25. Knife-edge trace of the diffraction pattern.

more rigorous, but if the OPD errors are not small, the shape of the pattern can become complex and it may not be obvious just where the peak is. Since it is useful only when errors are small in either case, the exponential approximation is widely used.

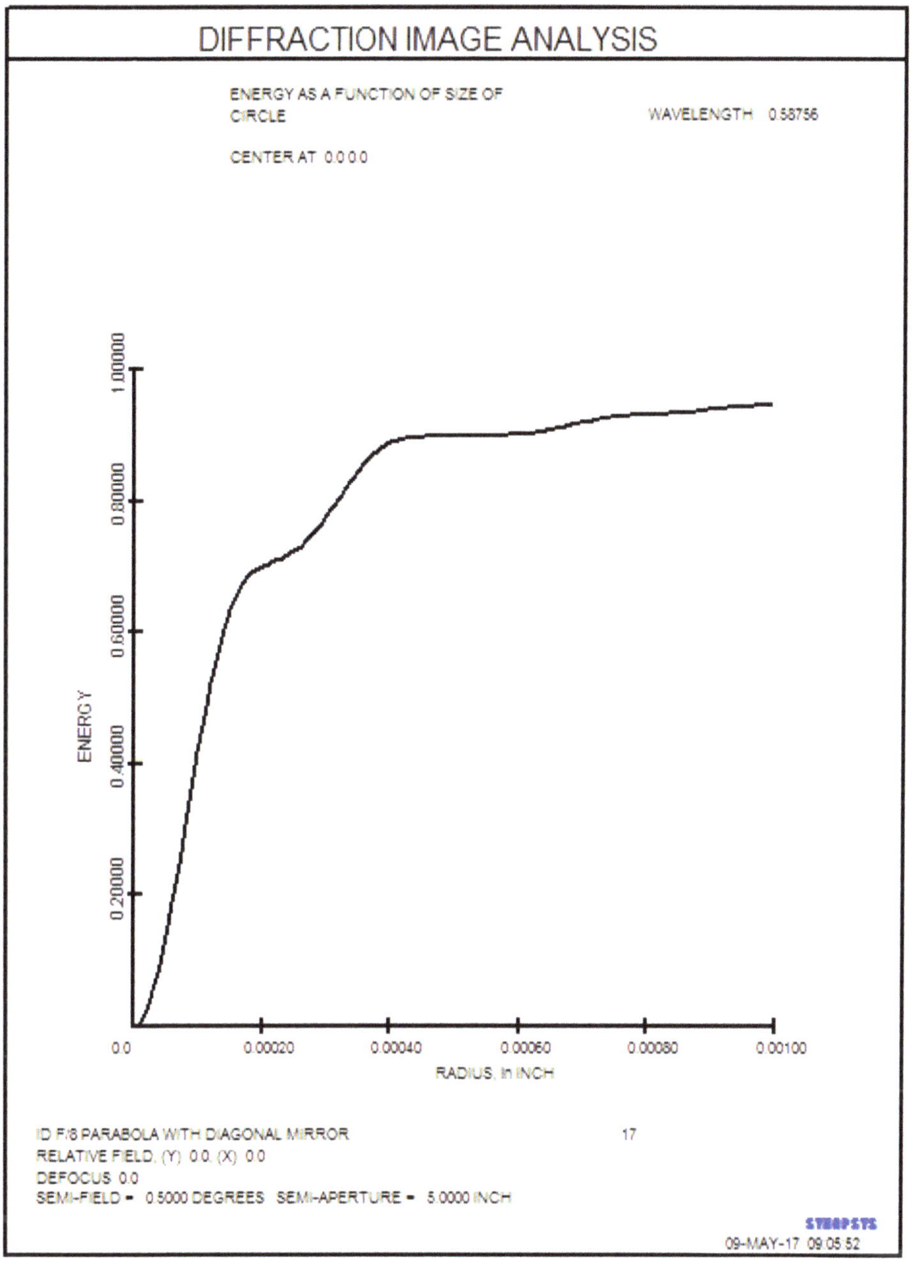

Figure 3.26. Encircled-energy plot for the Newtonian telescope on axis.

The SR can be calculated more quickly than can the diffraction MTF, and since it correlates strongly with the latter, it is a useful measure of image quality. Figure 3.27 shows two curves: the SR and the diffraction MTF at 100 c/mm as a function of defocus for a paraboloid mirror with a central obscuration. For small errors, the two curves track rather well.

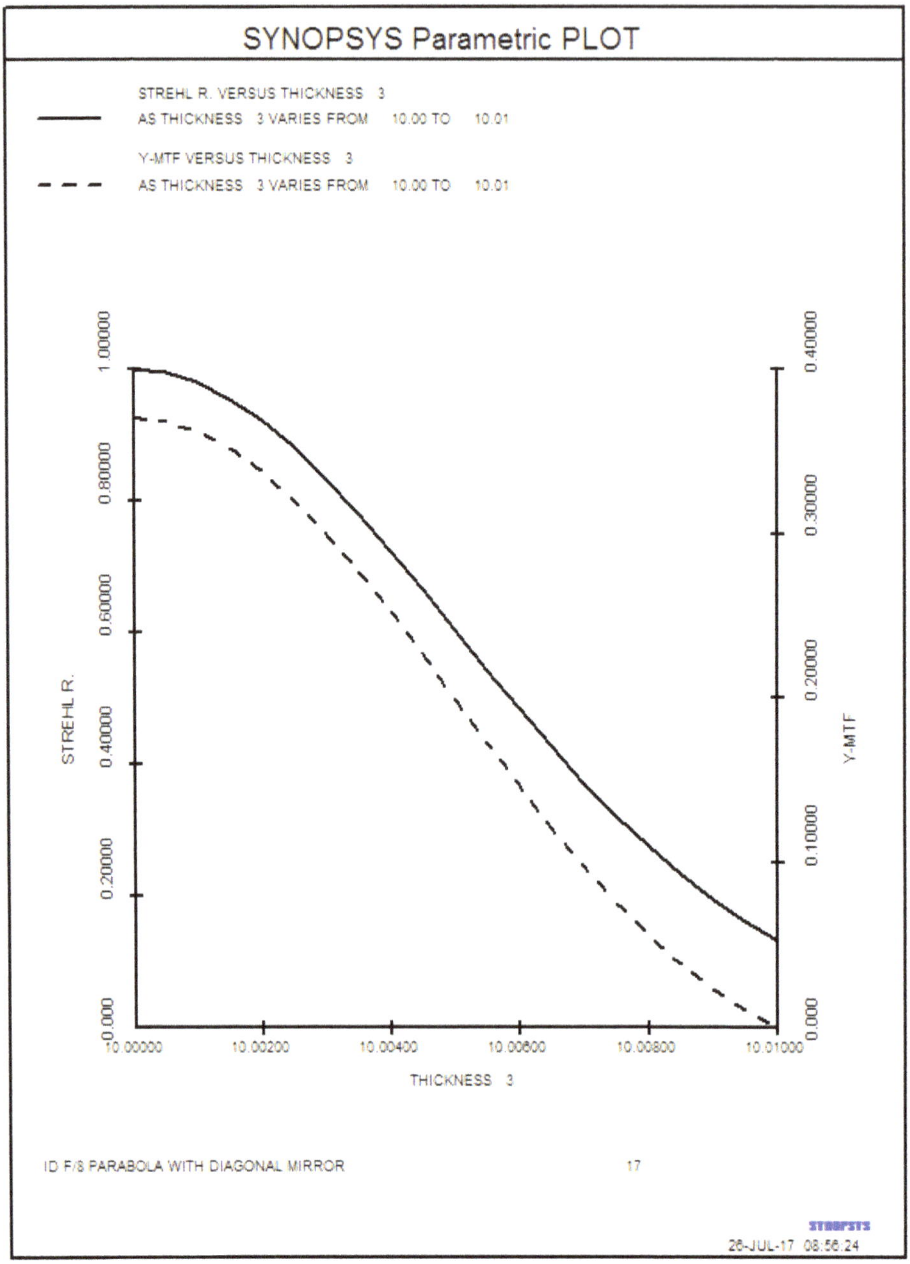

Figure 3.27. The Strehl ratio and MTF at 100 c/mm as a function of defocus. The SR degrades at about the same rate as the MTF, which makes it a useful measure of quality in the design phase.

Here is how to make the above picture:

1. **FETCH NEWTONIAN**
2. Open a MACro editor and enter the lines

```
VARIANCE P 0 600 0 0
ORD = FILE 3
```

3. Open a second editor and enter the lines

```
    MTF P 0 100
   ORD = FILE 1
```

4. Run the first MACro, and then type, in the CW:

```
    MULTI DO MACRO FOR 3 TH = 10 TO 10.01
```

5. Run the second MACro, and then type

```
   ADD DO MACRO FOR 3 TH = 10 TO 10.01
```

6. Type **END**.

3.6 Chromatic aberration

Since the index of refraction of all glasses changes with wavelength, it follows that the size, shape, and location of the image also vary, and an important goal of the lens designer is usually to minimize this effect along with the other aberrations. When choosing the exact wavelengths to correct and the goals, one must take into account the spectral sensitivity of the sensor and the properties of the light source.

When designing optics for the visual spectrum, one must weight the aberrations according to the sensitivity of the eye. Figure 3.28 shows this sensitivity, which is a function of the light level; in bright light one sees in color according to the curve on the left, while in dim light one sees a monochromatic image following the curve on the right, the peak of which is shifted toward the blue end of the spectrum. The three triangles at the bottom are located at the at the C, d, and F Fraunhofer lines, with wavelengths of 0.6563, 0.5876, and 0.4861 micrometers, which are often chosen when designing optics for visual use. The reasons are historical, having to do with the availability of monochromatic gas lamps that emit at those wavelengths for use in laboratory testing.

The ray fans in figure 3.29 are calculated for these wavelengths, as is the spot diagram to the right. In this image, for this lens, you see that the error at the image in color 2 is fairly small, while color 3 shows a larger error.

The actual size of the image can be deduced fairly well from the shape and size of the curves on the left. This kind of analysis can be done with many fewer rays than can a spot diagram and is useful for evaluating the progress of a lens design in the initial stages. Indeed, it is fun to watch the curves change during optimization, as the

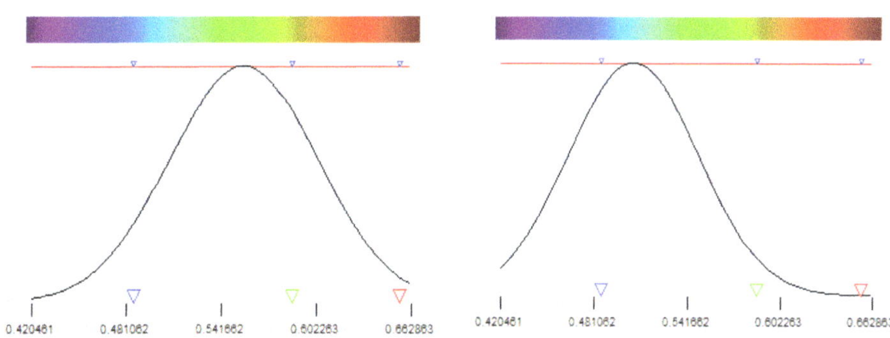

Figure 3.28. Eye sensitivity, bright light on the left, and dim light on the right.

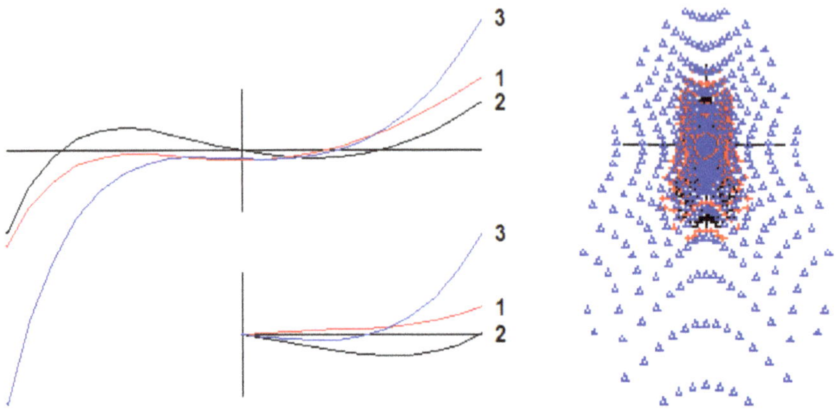

Figure 3.29. Effects of chromatic aberration on image structure. This image is fairly well corrected for axial color, but spherical aberration in blue light (color 3) is greater than in red. This effect is called *sphero-chromatism.*

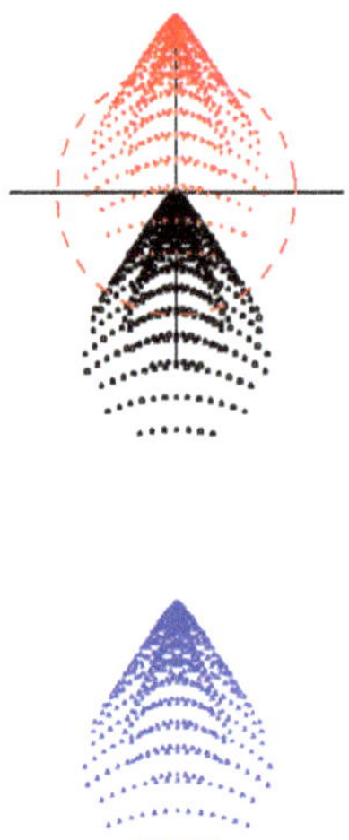

Figure 3.30. Example of off-axis image showing both coma and lateral color. The dotted red circle shows the size of first dark ring in the diffraction pattern for this system, for comparison.

software makes tradeoffs and balances one aberration against another. In these curves, we see both *primary axial color*, or **PAC**, and *secondary axial color*, or **SAC**. The software calculates PAC by taking the difference between the transverse image intercepts at the marginal ray in the long and short wavelengths, while SAC is calculated by taking the differences at the long and central wavelengths, all in the third-order approximation.

Lenses also can suffer from *primary lateral* color, or **PLC**, which is the change in image height between the long and short wavelengths, along with *secondary lateral color*, or **SLC**, which is the change between the long and central wavelengths. An example of an image with both coma, PCL, and SLC, is shown in figure 3.30.

Chromatic aberration is usually corrected by combining two or more elements made of glasses whose *dispersion*, the change of index with wavelength, differs substantially. Examples will be given in chapter 12 of lenses for which chromatic aberration correction is of particular importance and even *secondary* color is a major issue. That is the case when longer and shorter wavelengths focus in the same place, but intermediate wavelengths do not.

3.6.1 Cemented doublets

A classical approach to color correction was to *cement* the elements of a doublet together. A fringe benefit of that solution was the reduction of reflection losses, since two air-glass interfaces are replaced with one glass-glass interface. In the days before antireflection coatings were available, this was an important advantage. Today, we have very good AR coatings and that benefit is not as important—but there is another that should not be overlooked: one can avoid total internal reflection with a cemented interface in those cases where rays arrive with a steep angle. Figure 3.31 shows a lens with a cemented interface at surface 11. If we change that to a thin air gap between the elements, rays near the edge of the beam at full field will not trace, due to Total Internal Reflection (**TIR**) at surface 11, as shown in figure 3.32. The search programs DSEARCH and ZSEARCH, which are used in many of the chapters below, do not investigate systems with cemented surfaces, but you can request the Automatic Element Insertion (AEI) program to include them in its search algorithm. This can sometimes open new solution regions one would likely not discover otherwise. If TIR problems are not a problem, on the other hand, you

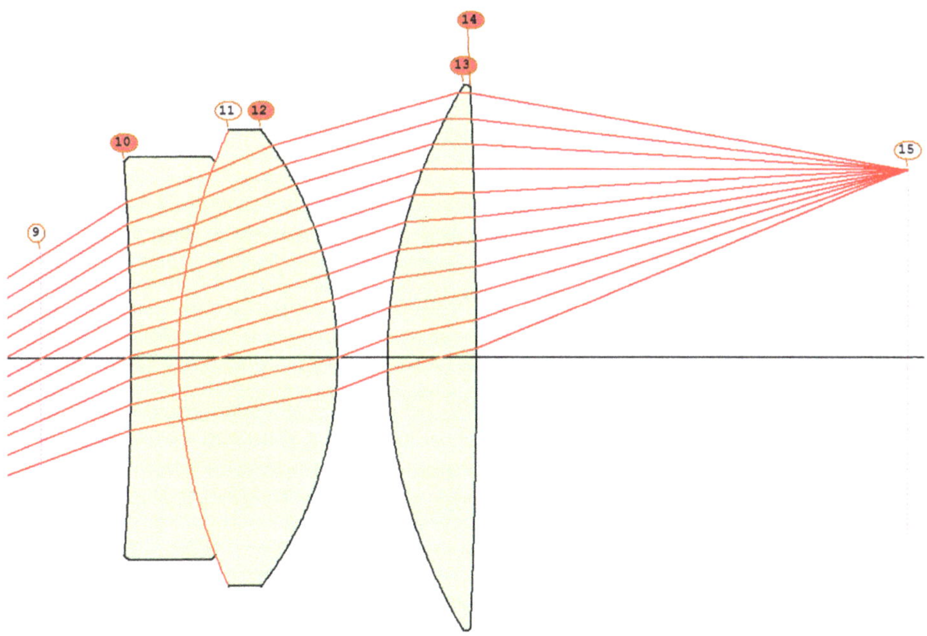

Figure 3.31. Example cemented interface at surface 11. All rays in the tangential fan trace properly.

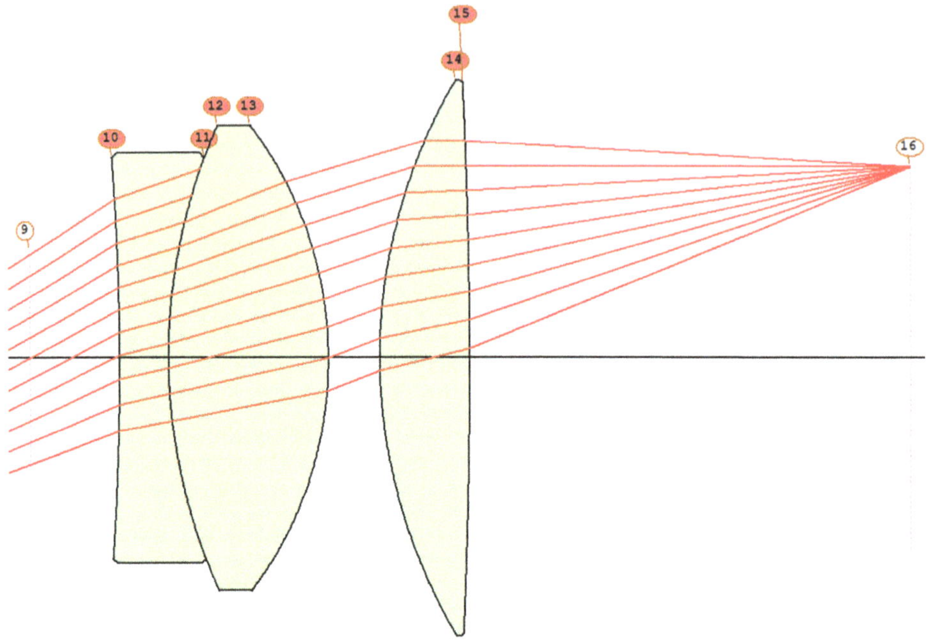

Figure 3.32. Changing the cemented interface to a glass-air interface causes total internal reflection (TIR) for some rays, which therefore cannot get through.

usually want airspaced rather than cemented elements. Each case then provides two additional degrees of freedom for the optimization program, and the MF is usually better as a result.

Where does TIR come from? From Snell's law. If a ray is refracted such that it comes out of the lens at exactly 90 degrees to the surface normal, just grazing the surface itself, we say it is at the *critical angle*. Any steeper, and it gets reflected back into the lens.

There is sometimes another reason for cementing elements as well. If an element has strong optical power, the centering tolerance can be very tight. But if two strong elements of opposite power can be cemented together, the combination will have less power and a looser tolerance. The tight tolerance of each separate element is held during the cementing process, where it is easier to do than when mounting the elements separately into a lens cell, and of course there is then no airspace requiring its own tolerance as well. But be careful when asking for large elements to be cemented; if the temperature changes, and if the coefficient of thermal expansion (TCE) is different between the two different glass types, the cemented pair undergoes mechanical stresses that can warp the two or in some cases break the cement bond or the glass itself.

3.6.2 Secondary color

Secondary color has a history of being especially difficult to correct, and some older designers may insist that one needs at least one material with abnormal dispersion characteristics—as was the case decades ago. Calcium Fluoride was a popular choice, especially for microscope objectives, where the lenses are physically very small and natural crystals were available.

But modern glass companies now have a wider assortment of glass types, and today one can correct secondary color with ordinary glasses if they are properly

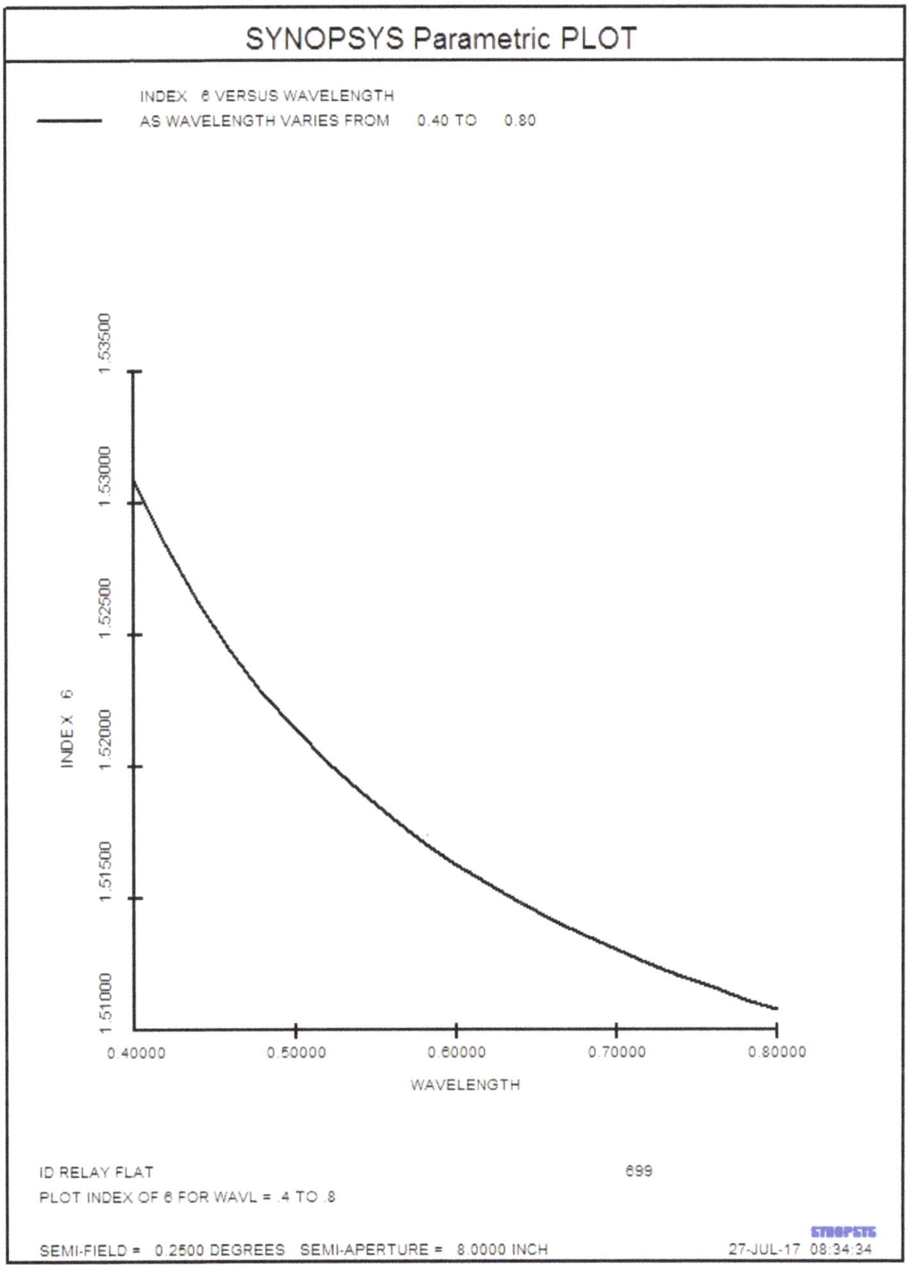

Figure 3.33. Index of glass BK7 as a function of wavelength.

selected. Figure 3.33 shows how the index of glass type BK7 varies over the wavelength range 0.4–0.8 micrometers. The curve is nonlinear, and that is the cause of much of the problem.

By combining positive and negative elements with different dispersions (different *slopes* of the index curves), one can make the change in focal point stationary with wavelength at a chosen wavelength. But because the glass curves also tend to exhibit different *curvatures*, the correction is imperfect at other wavelengths. Chapters 12 and 34 explain how one can minimize this effect by clever choice of three different glass types, and chapter 47 shows a design in which the program has found an excellent combination all by itself.

IOP Publishing

Lens Design (Second Edition)

Automatic and quasi-autonomous computational methods and techniques

Donald C Dilworth

Chapter 4

Using a modern lens design code

Sign conventions; stop and pupil definitions; wide-angle pupil; ray aiming; paraxial solves; the WorkSheet; the process of lens design

One cannot teach lens design today without also teaching the use of the software, since the latter actually does most of the work. But before you can use it, you have to learn how. If you have previously used a different program, you must first *unlearn* the procedures that apply there, before learning the capabilities of SYNOPSYS. The effort will be amply rewarded, however, as the power of the new tools becomes apparent.

Different programs have different sign conventions, and in SYNOPSYS they are as shown in figure 4.1.

This is a left-handed coordinate system, which is the default.

We spoke of paraxial optics in chapter 2, and a correct analysis requires a rigorous definition of the first-order properties of the entering beam. This is defined by four quantities, which we call YPP0, TH0, YMP1, and YPP1, defined in figure 4.2. All these dimensions are in the y–z plane. (A similar set of definitions can be assigned in the x–z plane, for cases when the lens or field is not rotationally symmetric. We will ignore those in this book.) All of these first-order pupil parameters can be entered explicitly by the user or implied by something else.

You can tell the program to calculate a value for YMP1 so that the marginal ray hits the stop surface exactly at the edge of the clear aperture, or calculate YPP1 so the real chief ray hits the center of the stop, and even adjust the object height YPP0 to yield a desired GIHT; these options are explained below. There is also an option for describing a wide-angle object, for which the input angle can exceed 90 degrees; this is object type **OBD**, discussed in chapters 41 and 45.

In figure 4.3, the lens has a stop where indicated, and the paraxial chief ray, which follows the path appropriate to the *paraxial* value of YPP1, goes to the center of the stop surface as it should.

But in this lens, to get a *real* ray, drawn with a dotted line in figure 4.3 (and which obeys Snell's law exactly) to go there, it has to be decentered coming in or else it will

doi:10.1088/978-0-7503-3695-6ch4

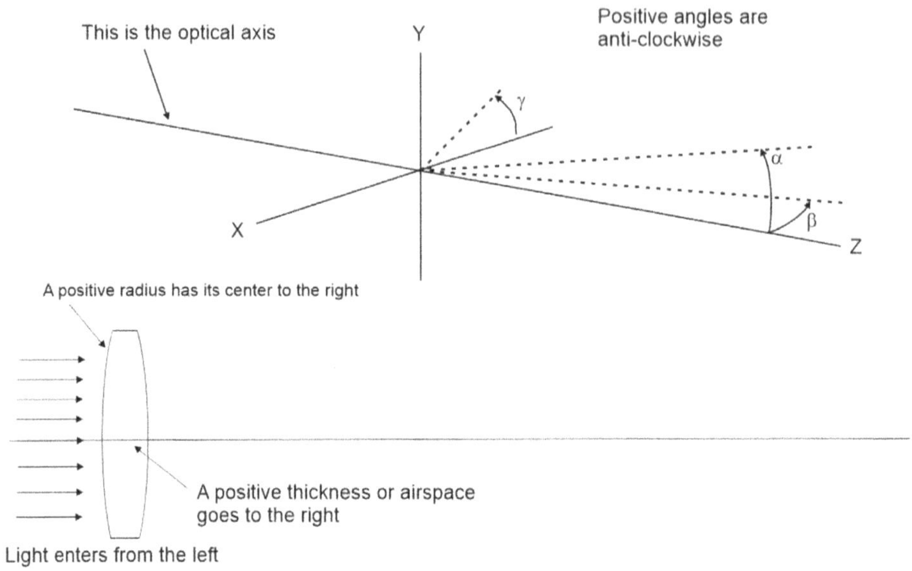

Figure 4.1. Sign conventions of lens geometry.

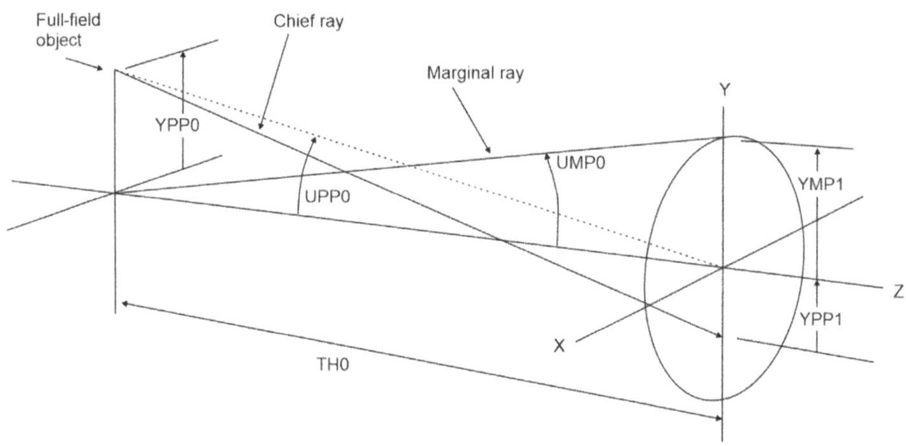

Figure 4.2. Definition of object specifications.

miss the center of the stop. So a real YPP1 will often differ from the paraxial. If you declare a surface to be a real stop, the program then finds the real chief ray by iteration whenever it is required. (To declare it, enter a *negative* surface number on the **APS** input line in the RLE file.)

4.1 Wide-angle pupil options

We briefly mention here the three wide-angle pupil options, called WAP 1, 2, and 3; a more detailed discussion is given in chapter 22. The default is WAP 0, which does no adjustments to the entering paraxial beam.

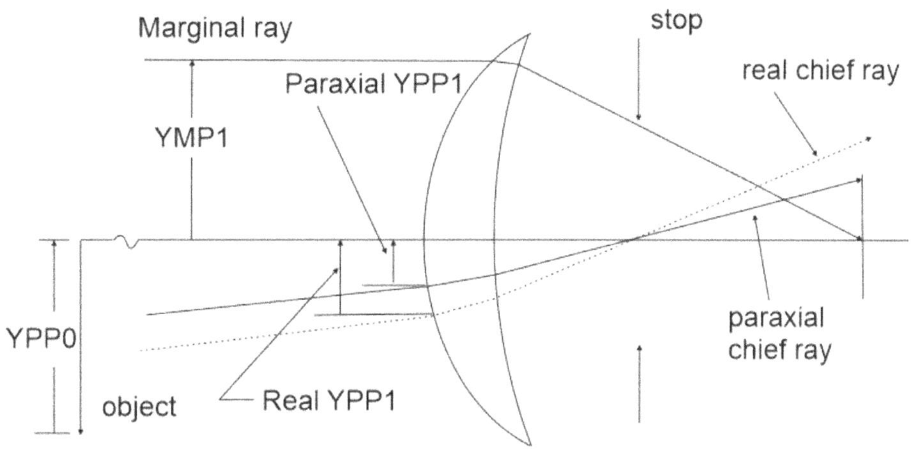

Figure 4.3. The case where a real chief ray does not coincide with the paraxial.

WAP 1 keeps the diameter of the entering beam constant for all field angles, measured normal to the chief ray. This is described in chapter 41.

If WAP 2 is turned on, the program will calculate a value of YMP1 so the marginal ray hits the stop surface at the edge; an example is given in chapter 22.

Sometimes the stop is not the only surface blocking the beam, however, and one must consider *vignetting* when rays that would otherwise get through the stop are blocked elsewhere. One then has the choice of simply letting the image evaluation programs delete those vignetted rays, or else defining the entrance pupil so it is reduced in size to encompass only the rays that actually do get through. That last option uses Wide Angle Pupil WAP 3, the geometry of which is summarized in figure 4.4.

If WAP 3 is turned on, the pupil size is reduced as shown on the lower right. That option should be saved for analyzing a finished design, since it is usually not practical when a design is still being optimized, and even then, only when vignetting is an issue at surfaces other than the stop.

4.2 Ray aiming

The above discussion relates to what is broadly termed *ray aiming*. When analyzing an image, one must ensure that an appropriate bundle of rays is traced. Some programs implement this by creating a grid at the stop and then finding, by iteration, a ray that hits every point in that grid. While this works, it is very slow. SYNOPSYS instead can define the pupil by iterating only five rays to find the outline and then filling the pupil so defined with a uniform grid. This greatly speeds things up since there is no iterating once the pupil has been found.

So which pupil option should you use? The simplest one that gets the job done. The paraxial pupil is the fastest, a real pupil is slower, and the WAP options are slower still. It is unwise to always specify a sophisticated option, such as WAP 3,

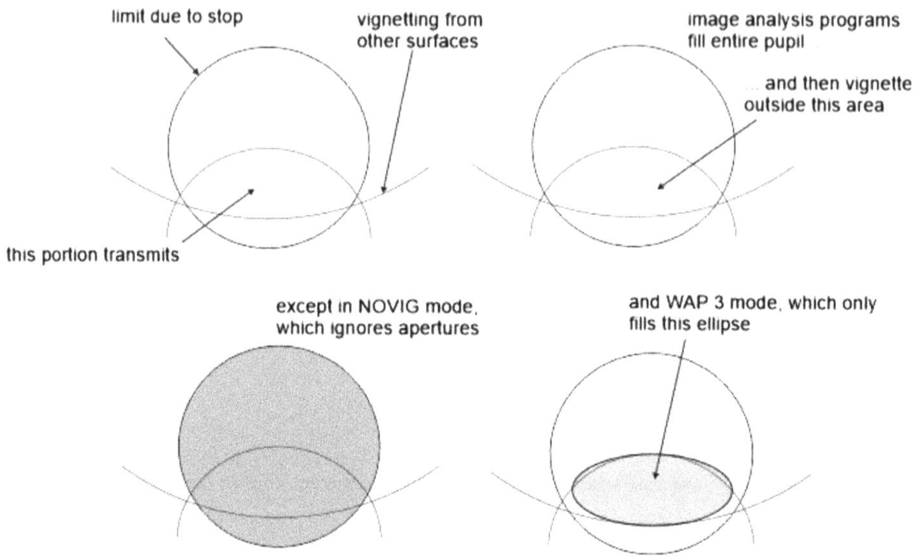

Figure 4.4. Geometry of wide-angle pupil options.

when it is wholly unnecessary. So start with paraxial and look at the path of the chief ray (the central ray in the bundle) to see if it passes near the center of the stop. If not, activate the real pupil option. Then check whether the stop is properly filled. If not, consider the WAP 2 option. We will revisit this topic in chapter 22.

4.3 Paraxial solves

One other item we should mention here is that of paraxial *solves*, an important concept that will be used frequently. When a solve is defined, the program will calculate the actual curvature or thickness so as to satisfy a paraxial requirement, and you do not then give it a value yourself. Fetch the lens **SINGLET.RLE**.

Suppose we want the singlet, shown in figure 4.5, to operate at an F/number of 4.0; at the moment, it is about F/2. Enter a change file as follows:

```
CHG
2 UMC .125
END
```

And the lens is changed, as shown in figure 4.6.

The paraxial angle is just 0.5/FNUM, with a negative sign since the marginal ray is heading downwards.

From the table below, you see that a UMC solve will control the paraxial angle of the marginal ray. This lens also has a YMT solve on surface 2, so the image plane also moves when the F/number is changed. The several kinds of solves are listed in table 4.1; a complete discussion is given in the help file of the program.

```
CHG
2 UMC -.125
END
```

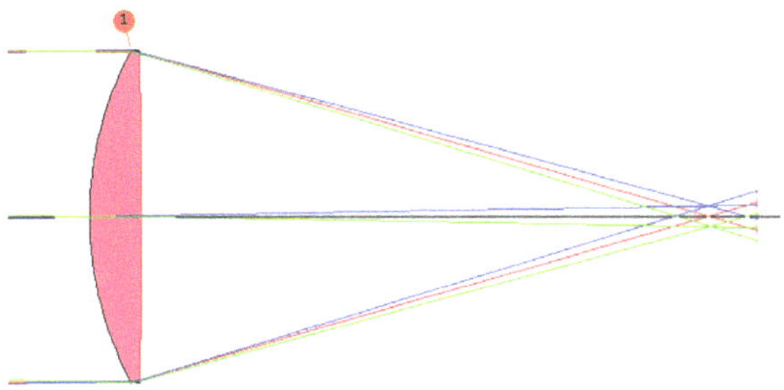

Figure 4.5. An F/2 singlet.

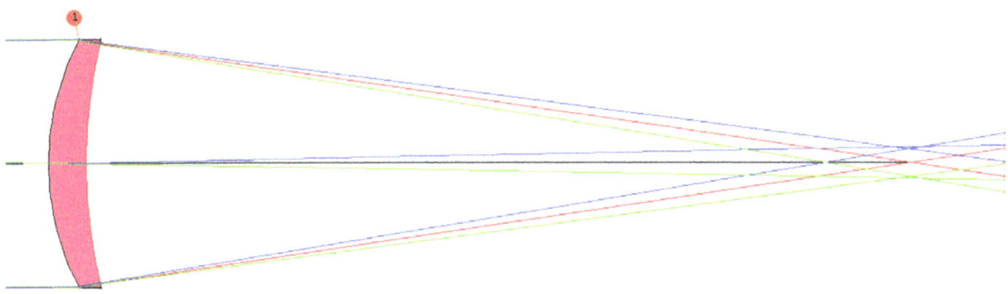

Figure 4.6. Singlet with F/number controlled by a UMC solve, yielding F/4.

Table 4.1. Curvature and thickness solves.

UMC NB	**Curvature solves:**
UPC NB	U is a paraxial angle
YMC NB	Y is a paraxial height
YPC NB	M is the marginal ran
AMY	P is the principal ray (the chief ray)
CCY	C designates a curvature solve
	T is a thickness solve
YMT NB	Thickness solves
YPT NB	

4.4 The WorkSheet

You made the change to the singlet above by typing a CHG file in the Command Window (**CW**), which is the window where you can type commands and where your printed output shows up. But you will frequently make use of the WorkSheet[1] for editing a lens.

[1] WorkSheet™ is a trademark of Optical Systems Design LLC, an Arizona, USA corporation.

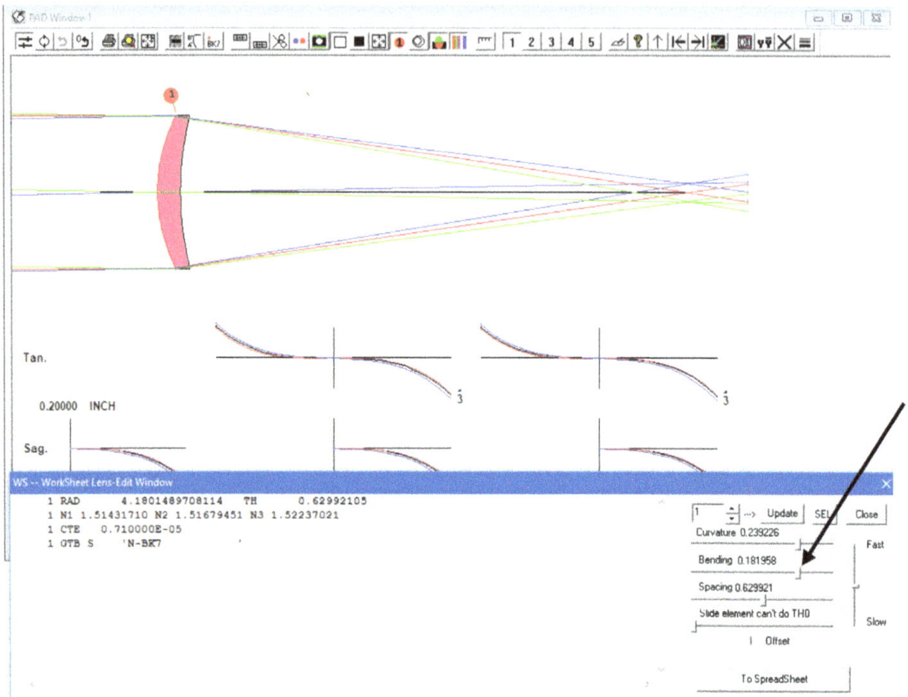

Figure 4.7. Using the WorkSheet slider to change lens bending.

In the CW, type **`WS`** to open the WorkSheet, and then click on surface 1 in the lens drawing. The edit pane shows the lens data for surface 1, as in figure 4.7.

Move the Bending slider to the left and right, watching the ray fans, and stop when the curves are as flat as possible. You will do most of your lens editing in the WorkSheet, because it has useful features like this one that are not available in a spreadsheet. The image is improved by this bending, as shown in figure 4.8.

WS has sliders for the curvature, thickness, and bending; you can slide an element between its neighbors, and even assign a slider to any numeric quantity in the edit pane. It also links to the **Lens Layout Tool**, an example of which is shown in figure 4.9. There you can click and drag the black squares (the 'handles') and the lens construction changes accordingly, even changing the bending and powers of the elements. To see how this works, type **`HELP LLT`** in the CW. You can even change the size of the edit handles.

4.4.1 The process of lens design

Much has changed since the classic texts on lens design were written. Even the more recent authors, when they mention the power and importance of computer optimization, dwell at length on the details of how to do a first-order layout, how to calculate the magnifying effect of the second group in a telephoto lens, stress the importance of coming up with a good third-order design before going to the

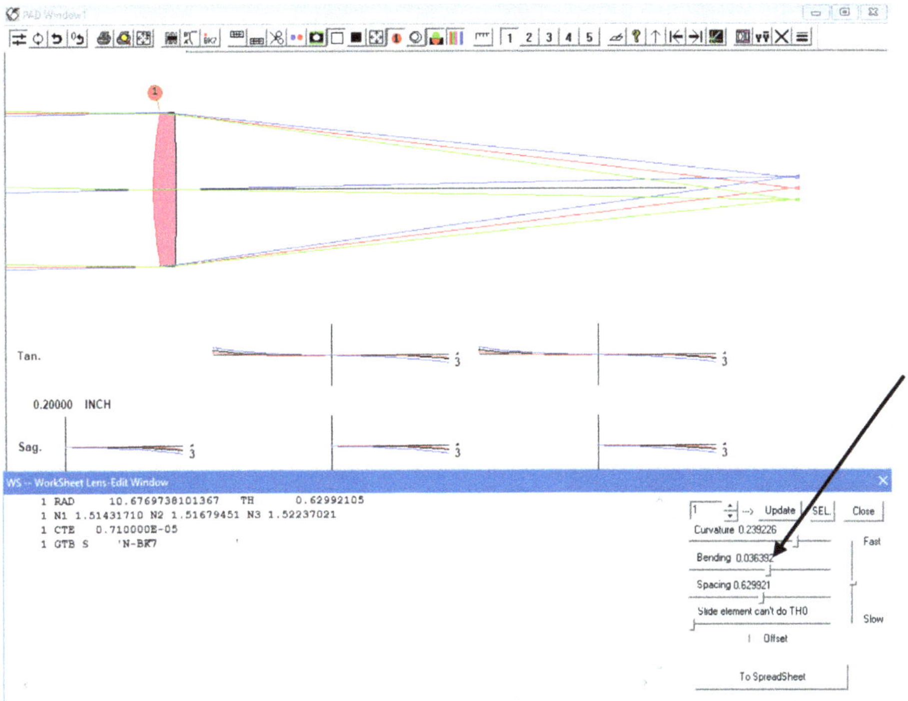

Figure 4.8. Lens with bending adjusted to minimize spherical aberration.

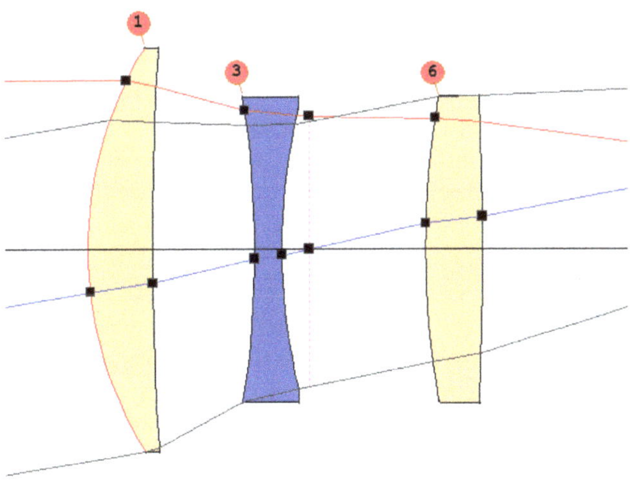

Figure 4.9. The Lens Layout Tool (LLT) showing edit handles on the surfaces of a triplet lens.

computer, and so on. But, except for the first-order requirements, almost all of that is now irrelevant.

Not to overstate my case, it is still necessary for you to understand the purpose of the design, the mechanical envelope, the project budget, transmission requirements, the effect of the environment on the polished lens surfaces, and so on—and to have an understanding of the problems faced by the fabricators because of tradeoffs you make early on. No computer program can make those decisions for you. You will make many design choices based on your understanding of the whole project, not just the image quality requirements—but the latter are arguably the most important single consideration, since if not satisfactory, none of the other matters matter.

Lens designers in the past always tried to fit a new project into the context of their experience, often using a previous design, either from their files or from a publication somewhere, as a starting point for a new one. There are many classic forms, such as the Dagor, Double-Gauss, Petzval lens, and so on, and designers who could simply modify one of those well-known forms, with an eye to some new requirements, often succeeded in a reasonable amount of time. That was the old-school method. If there was no classic design that would do, the problem was vastly more difficult.

Today, you can still do things that way if you want to—but today a new approach is called for, outlined below:

1. Understand the problem. If the customer wants to violate a law of optics, you might persuade him to adopt a more reasonable set of goals. If he cannot, you must decline the job. If the goals are possible but his approach is not feasible, lengthy consultation may yield a consensus on the right way to proceed. It is unethical to agree to work on a problem and collect a fee when you know it is unlikely you will succeed. Your reputation depends on total honesty. If you cannot do the job, say so. Often the customer will express his goals with incorrect terminology, such as asking for a certain focal length when he really means back focus distance. Be sure you and he agree on just what the goals really are.
2. Work out a first-order design. Yes, it is still necessary for you to determine the basic requirements, such as focal length, F/number, image location, and whatever else is important to the customer. No computer can do that part of the job for you, and those results will be your input to the computer.
3. Once the goals are laid out and the boundary conditions expressed, unless the job is very simple try to enter those goals into one of the search programs DSEARCH, ZSEARCH, or FFBUILD[2]. Create and run that input, and then evaluate all of the returns that look practical. Optimize each of the better ones, adjusting the merit function definitions as you discover issues that are not well addressed in the lenses returned by the program.
4. Most of your effort will consist of modifying the merit function. It's been said we don't design lenses anymore; we design merit functions. Go cautiously, increasing weights here and there if a particular image error is not well balanced with the others. If a design does not converge as well as you would like, try another of the designs returned by the search programs. They only consider what they have been given as input, and sometimes there are other goals not easily described there. The SPECIAL AANT

[2] FFBUILD is a trademark of Optical Systems Design LLC, an Arizona, USA corporation.

section of the search input can often include those extra goals, if you know of them in advance.

5. Unless your design is absolutely wonderful, run the search program again, perhaps several times (with switch 98 turned off!). Then the program will investigate a different set of branches of the design tree every time when it reaches the annealing stage. Or you can change the input data slightly, which will also search in different directions. Then you can compare the results from each run to see if there is a better configuration in the new set. In this way, I once came up with 23 designs that were as good as or better than one found by an expert using older tools[3]. (He was astonished.) Chaos can be an asset as well as a problem. Figure 4.10 shows an example of the variety of lenses that were found by running DSEARCH many times, varying some of the input parameters slightly. It is apparent that there is more than just one excellent solution.
6. When the design is nearing completion and if resolution near the diffraction limit is desired, add some OPD targets to the MF. (The search programs DSEARCH and ZSEARCH can both consider OPDs right from the start, with user input OPD or TOPD, if you know in advance they will be required.) When you add OPD targets, take into account the very different weighting factors that are then appropriate. A one-wave error is usually much better than a one-mm (or one-inch!) transverse image error, so the weights should be accordingly lower. Also, if the MTF of the finished lens is a critical requirement, try optimizing with GSHEAR operands. Those minimize the difference in the OPD for pairs of rays distributed over a grid and often improve the MTF in ways the other methods cannot.
7. If a design looks satisfactory, assign real glass types with the automatic real glass program **ARGLASS**, which can be run from the menu **MRG**. Make a checkpoint, and carefully select the glass properties, cost, chemical stability, and so on. If your choices are too narrow, the lens might not be as good as before. In that case, restore your checkpoint and try easing up on the requirements. That is how tradeoff studies are done, and now you can tell the customer he can get a certain resolution with inexpensive glasses but can do better if he is willing to pay more. Just as the designer must consider the project goals and constraints early on, the reverse is also true: the project has to be flexible enough to accommodate new information that becomes apparent during the design phase. If `ARGLASS` does not return a good enough glass selection, try the more powerful GSEARCH[4]. Examples of its use are given in chapters 38 and 47.
8. Always try to match the design to the tooling of the selected vendor. This saves the cost of new tools, and, crucially, has an effect on the tolerance budget. Surfaces that are matched to existing testplates are easier to hold to tight tolerances because they can be measured more quickly. If a surface

[3] Dilworth D C and Shafer D 2013 Man versus machine; a lens design challenge *Proc. SPIE* **8841** 88410G-1

[4] GSEARCH is a trademark of Optical Systems Design LLC, an Arizona, USA corporation.

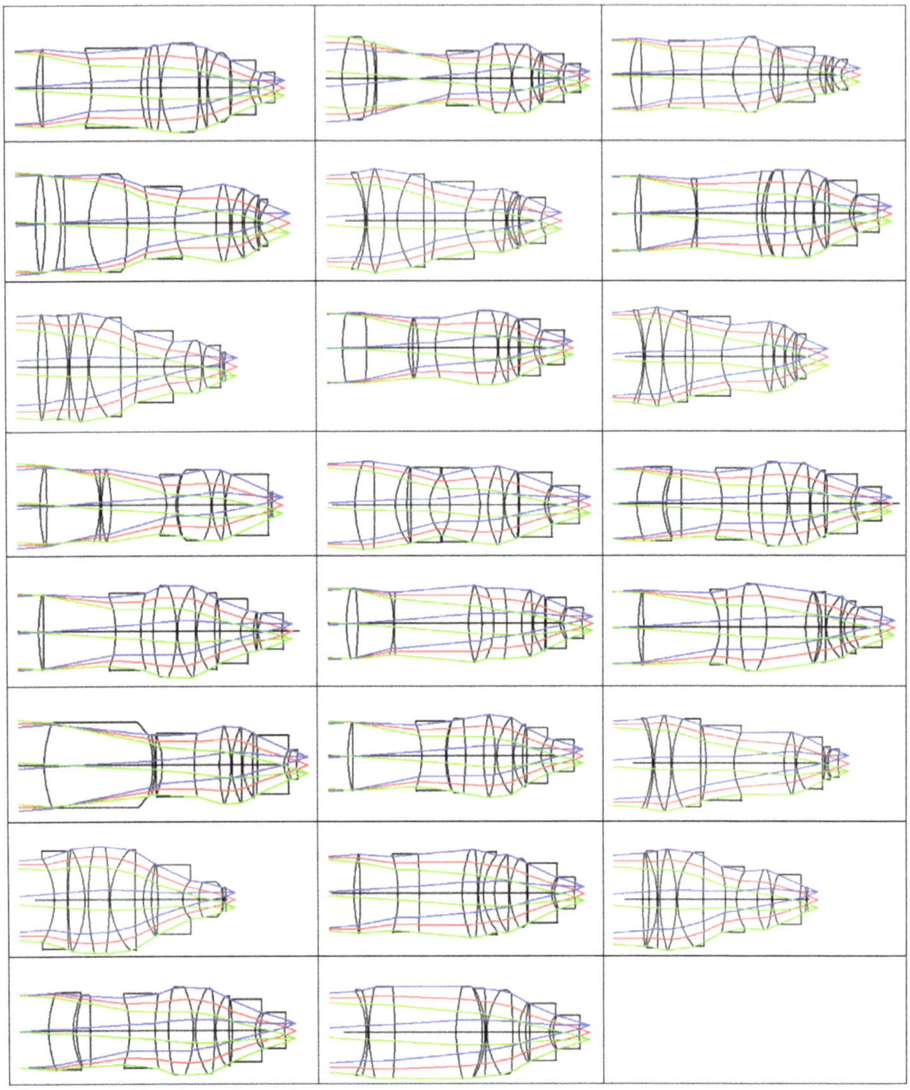

Figure 4.10. A variety of lenses found by DSEARCH for an imaging problem. Each of these lenses is as good as or better than one found by an expert using classic tools.

cannot be matched—because no tooling is close enough to the design value —*do not* simply give a radius tolerance and assume the vendor will make a new tool that is within that tolerance. Although this is commonly done, it has a serious consequence: the statistics of a batch of lenses will come out wrong. One wants the *mean value* of the lens parameter to equal the design value, and the standard deviation should be a function of the tolerance—as the budget assumes. But if the vendor makes new tooling, the mean value will be that of the new tool, itself different from the design value, and the lenses will in turn get that value instead of the correct one, and the statistics of the entire budget will be thrown off. The design can be matched to a testplate list automatically with **TPMATCH**, which can be launched from the dialog **MMT**. Section 50.1 gives an example.

9. When the above steps are completed, it is time to do the tolerance analysis. The main tolerance program in SYNOPSYS is called BTOL, and by following the instructions in the User's Manual you can derive a budget for the entire lens that will ensure that a desired performance level is reached. The resulting tolerances can then be added automatically to the element drawing program ELD and on assembly drawings made with DWG. The statistics can be verified with the Monte-Carlo program MC. Both BTOL and MC can account for the effects of adjustments that you plan to perform on the finished lenses. In the event that tolerances come out impossibly tight after all this work, it is time to look into tolerance desensitization. Read Section 10.1 of the User's Manual, add the appropriate requirements to the MF, and iterate. An example of this extra step is given in chapter 10.
10. If the lens is to be used in the thermal infrared, be sure to examine the *narcissus* effect. Chapter 30 explains how to control it. It is also a good idea to evaluate the ghost-image properties of your lens. That too can be controlled in the merit function, but first you have to know about it. Chapter 36 explains how.
11. In the event that your project has important requirements that cannot be input to the search programs, you have to work somewhat harder. Remember, those programs work by evaluating a number of potential configurations, usually starting with elements of very weak positive or negative power. If necessary, you can emulate that approach yourself. Just as there are many good designs returned by the search programs, it is likely there will be several that you can find in this manner. So the process may take less time than you expect.

IOP Publishing

Lens Design (Second Edition)

Automatic and quasi-autonomous computational methods and techniques

Donald C Dilworth

Chapter 5

The singlet lens

First steps with the software; aberrations of a singlet

The best way to learn lens design is by actually doing it, and you can start by reading the chapters below, which present a variety of problems and their solutions, working the problems as instructed. Most of these lessons involve lens input files (with the extension 'RLE'), and MACro files (with the extension 'MAC'). To save typing, you can open those files as instructed in section 1.1.

Whenever you encounter a mnemonic whose meaning you do not know, be sure to look it up in the help file. You will become a power user in no time.

5.1 Entering data for the singlet

Here's how to enter a lens with the editor:

When the program opens, type, in the Command Window (**CW**),

```
EE
```

to open the Excellent Editor, and type the text as shown in figure 5.1:

Click the run button, , and then click the PAD button in the top toolbar, or type **PAD**. The lens picture shows up along with TAP ray fans curves, as shown in figure 5.2.

The meaning of the entries in the RLE file should be self-evident, except perhaps for the OBB line, which declares the object coordinates: the three arguments give the entering marginal ray angle UMP0 (which is zero for an infinite object), 5 degrees semi-field angle UPP0, and 12.7 mm for the semi aperture YMP1. This is all explained in the menus **MPW** (Menu, Pupil Wizard) and **MOW** (Menu, Object Wizard), which you should examine before going much further. Like most input files, the RLE file must end with the **END** line.

This lens is in the DBOOK-ii directory with the name **C5L1.RLE**. Type **FETCH C5L1** to open it.

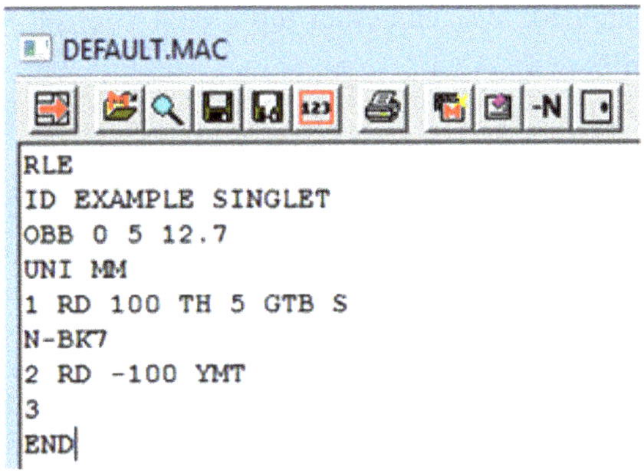

Figure 5.1. Entering a lens file in the EE editor.

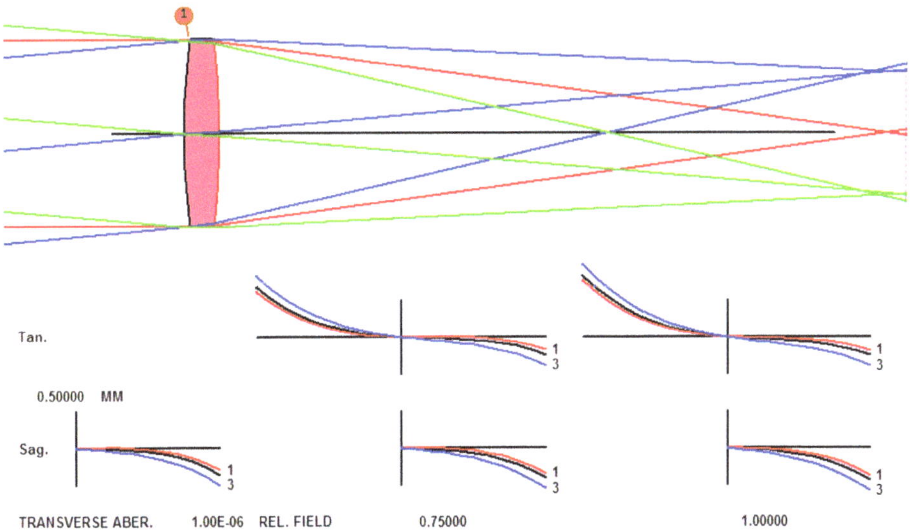

Figure 5.2. The SketchPAD display of the singlet lens.

Some people prefer to enter lens data with the SpreadSheet, which you can open with the command **SPS** or with the button in the top toolbar—but we find that entering the data with the EE editor as shown above is many times faster and easier. To enter this simple singlet with the SpreadSheet, one must execute no less than 17 different mouse clicks and type into several boxes. Users of other programs sometimes think the spreadsheet should be open all the time—as it is with those codes—but once you learn how to use the WorkSheet, you'll not want to go back. Use SPS when it makes sense, and close it afterwards.

We did not enter any wavelength data in this lens file, so the program applies the default, which calls for three wavelengths at 0.656 27, 0.587 56, and 0.486 13 microns, equally weighted. You can enter those data yourself in the RLE file, for example, with the line

```
WAVL CDF
```

Since those wavelengths are the commonly used C, d, and F Fraunhofer lines, the program accepts this shortcut. The dialog **MMC** also gives you an easy way to edit the wavelengths.

The screenshot in figure 5.2 shows the **SketchPad** display, which interacts with WS, and if your display does not show those formats, click the button on the PAD toolbar to restore the default display settings. PAD is the main graphical interface of the program, and it can display any combination of lens profile drawings, perspective drawings, paraxial drawings, ray fans, OPD fans, astigmatism curves, or spot diagrams. Explore the buttons on the PAD toolbar to learn how to manage the display. PAD can update itself during optimization too, making it easy to monitor the progress of your design.

Let's examine the performance of this lens, which of course is not very good. What does the image look like? Type **MGI** to open the Geometric Image menu, select the Multicolor option on the **SPT** section, and then click that button. Your spot diagram shows up, shown on the left in figure 5.3. Now press the Enter key, and the **MGI** dialog opens again. Click the switches button, turn off switch 27, and click Apply. Now click the **SPT** button again. The picture on the right appears. There are many mode switches that affect how the program works, and you can experiment with them to learn how to customize the program for your needs.

Now type, in the CW,

```
PLOT BACK FOR WAVL = .4 TO .8.
```

Figure 5.4 shows how the back focus changes with wavelength, evaluated with the artificial-intelligence program. (By the way, you always have to press the **Enter** key

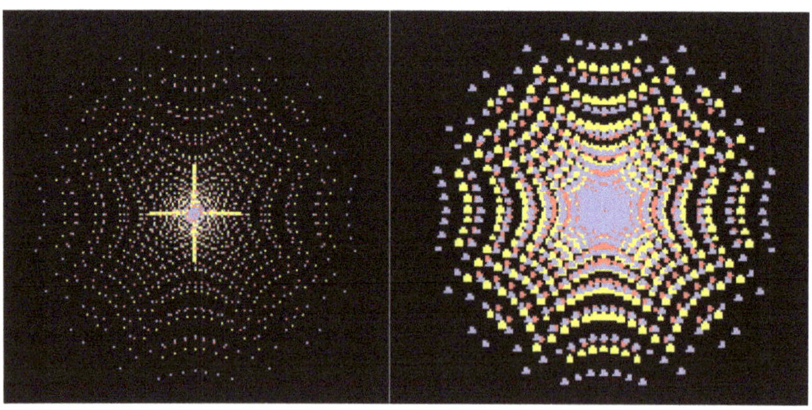

Figure 5.3. Spot diagrams. Switch 27 is on for the left picture, and off for the right.

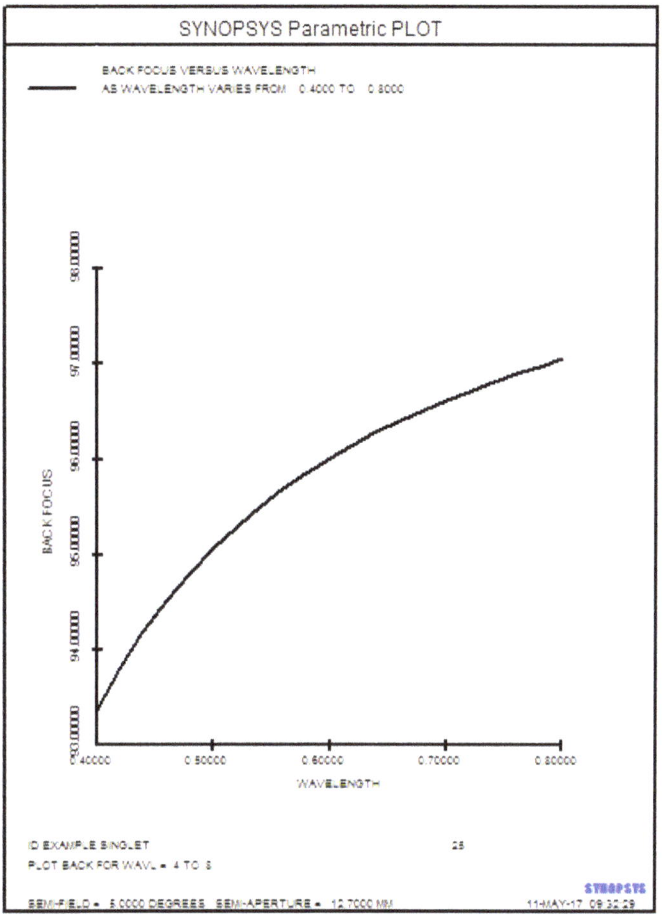

Figure 5.4. Example of an AI plot, showing how the back focus varies with wavelength.

after typing a command or AI sentence.) You have just typed an English-language sentence that is understood by **AI**. We will use AI again in other lessons.

IOP Publishing

Lens Design (Second Edition)

Automatic and quasi-autonomous computational methods and techniques

Donald C Dilworth

Chapter 6

Achromatizing the lens

Color correction with differing dispersions

Now we will add a second element and try to control the chromatic aberration of the singlet from chapter 5 (**C5L1**). First, click the Checkpoint button on the PAD toolbar. It is often a good idea to save a checkpoint before changing a lens. That way, if things don't work out, you can instantly go back, with the button. (You can also press the **F3** key to cycle back through previous versions, but we prefer the button since then we have total control.) Now click the WorkSheet button, click in the drawing on surface 1, and then 2, and see how the RLE data for the selected surface show up in the edit pane. This is where you will do most of your lens editing.

Now add a second element. Click the **Insert Element** button button on the toolbar near the top of the window. Then click in the lens drawing on the PAD display just behind the singlet. A second element shows up, shown in figure 6.1.

The program has removed the YMT solve from surface 2 and added an index pickup on surface 3 (**`3 PIN 1`**). Click on surface 3, and type **`3 GLM 1.6 44`** in the edit pane, as shown in figure 6.2. Then click the Update button.

We are going to vary the glass types in order to correct color, and we don't want surface 3 to always pick up the index values from surface 1. So we assigned it a glass model of its own, somewhere in the flint part of the glass map, to override the pickup. Now type, in the edit pane.

```
4 UMC .125 YMT
```

and click Update again. The lens gets a curvature solve on surface 4, yielding an F/number of 4.0, and surface 5 will be at the paraxial focus because of the YMT solve. Simple enough. Make a new checkpoint and close WS.

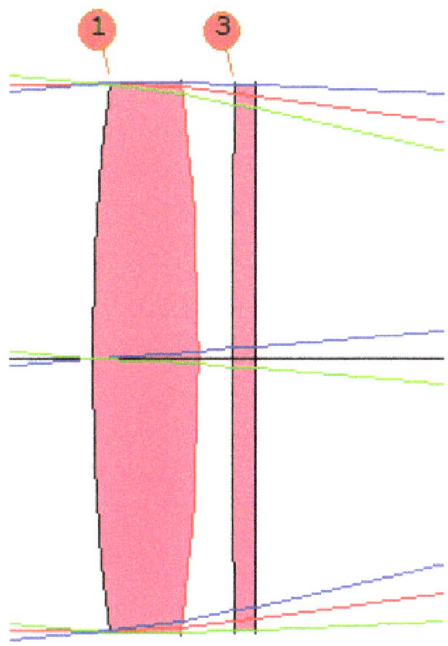

Figure 6.1. A second element has been added with the WorkSheet.

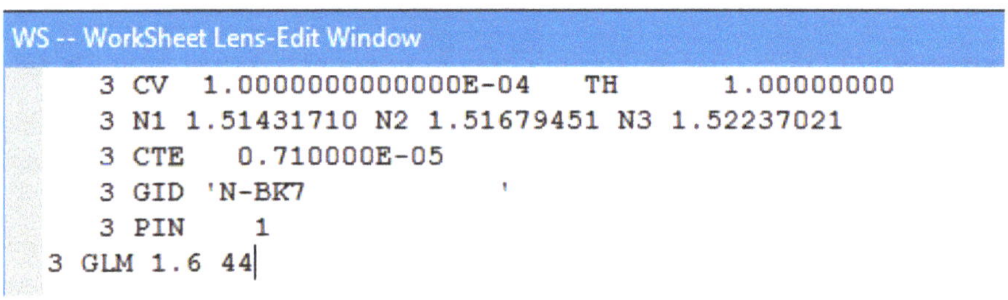

WS -- WorkSheet Lens-Edit Window

```
   3 CV  1.0000000000000E-04    TH       1.00000000
   3 N1 1.51431710 N2 1.51679451 N3 1.52237021
   3 CTE   0.710000E-05
   3 GID 'N-BK7                 '
   3 PIN    1
3 GLM 1.6 44
```

Figure 6.2. The WorkSheet edit pane for surface 3, with GLM data added

Now type **EE** to go back to the editor and click the Clear MACro button button to erase what you typed earlier. You are going to make an optimization MACro. Type in the editor the following lines:

```
PANT
VLIST RAD ALL
VLIST TH ALL
VLIST GLM 1 3
END

AANT
←

END
SNAP
SYNO 20
```

Optimization in SYNOPSYS requires a parameter file (**PANT**), a merit function file (**AANT**), and then some optimization commands. Put the cursor *in the blank line after the AANT command* as shown by the arrow above, and click the **Ready-Made Raysets** button button on the editor toolbar. This opens a dialog where you can select one of nine prepared merit functions. Number 6 is the default and is the one we want, so just click the **Back to MACro editor** button. Now your MACro looks like this:

```
PANT
VLIST RAD ALL
VLIST TH ALL
VLIST GLM 1 3
END

AANT
AEC
ACC
GSR .5 10 5 M 0
GNR .5 2 3 M .7
GNR .5 1 3 M 1

END
SNAP
SYNO 20
```

The editor has created a simple set of optimization requests. You can read the meaning of these lines in the help file. For example, if you type in the CW

HELP AEC

you will learn about the Auto Edge Control monitor (**AEC**). Now just *select* the characters **GSR** in the editor window and look down at the tray, near the bottom of the screen, shown in figure 6.3. If you select a word in the editor that is a command or a common entry in the merit function, the program displays the format of that entry in the tray.

The TrayPrompt gives you an instant summary of the syntax of the GSR ray-grid request. Press the **F2** key when you see something in the tray if you want more

Figure 6.3. The TrayPrompt[1] when the characters GSR are selected in the editor.

[1] TrayPrompt™ is a trademark of Optical Systems Design LLC, an Arizona, USA corporation.

information, and you go right to the help file on that topic. That would probably be a good thing to do right now.

So much for the basics. Now click the Run button button on the editor and watch the display change, as in figure 6.4.

Ooops! What happened? The image looks much better, but the lenses are way too thick! There's a lesson here: the program will do *absolutely anything* it can to bring down the value of the merit function, which is mostly just the sum of the squares of the items in the AANT file. Sometimes it does things that you don't want (and did not tell it about)—but it's just doing its job. The AANT file includes an **ACC** monitor, which is intended to prevent lenses from getting too thick, which they often want to do, but the default limit is one inch, or 25.4 mm. That's what you got. Now edit the AANT file, changing that monitor to **`ACC 4 1 1`**, and run it again, giving a target of 4 mm. Much better, as shown in in figure 6.5!

Type **`HELP MONITORS`** to read about the 17 different kinds of monitors you can use to keep your systems reasonable.

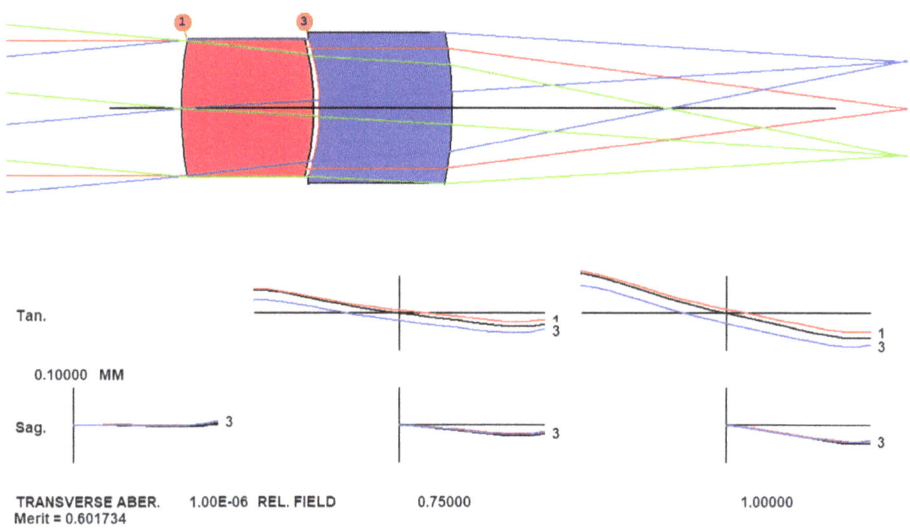

Figure 6.4. Results of the first optimization of the doublet.

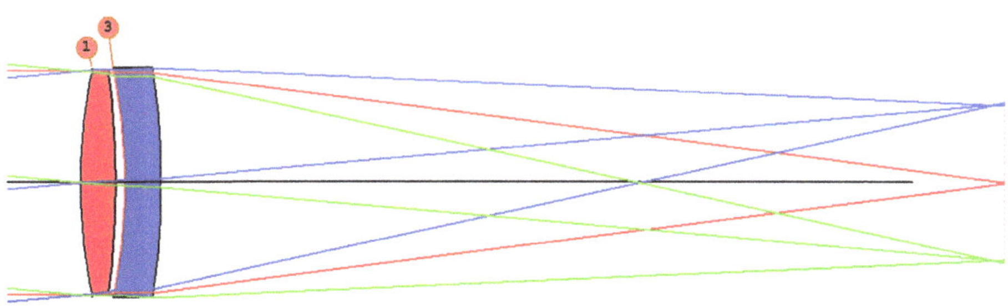

Figure 6.5. Result when a smaller target is assigned to the ACC monitor.

Now ask AI to show the color correction again, but this time define that sentence as a *symbol.* Type the line below in the CW.

```
QQ: PLOT BACK FOR WAVL = .4 TO .8
```

Here, we have defined the characters 'QQ' to equal the rest of that input line. (One to three characters followed by a colon and a space define a symbol.) Now just type **QQ,** and the program processes that sentence. The color correction is much better, as shown in figure 6.6.

The program has balanced the color correction against the change in spherical aberration with wavelengths, so color itself is not corrected completely. Lens design is all about compromise. When you define a symbol this way, you only have to type it once and it remains defined until you exit from the program.

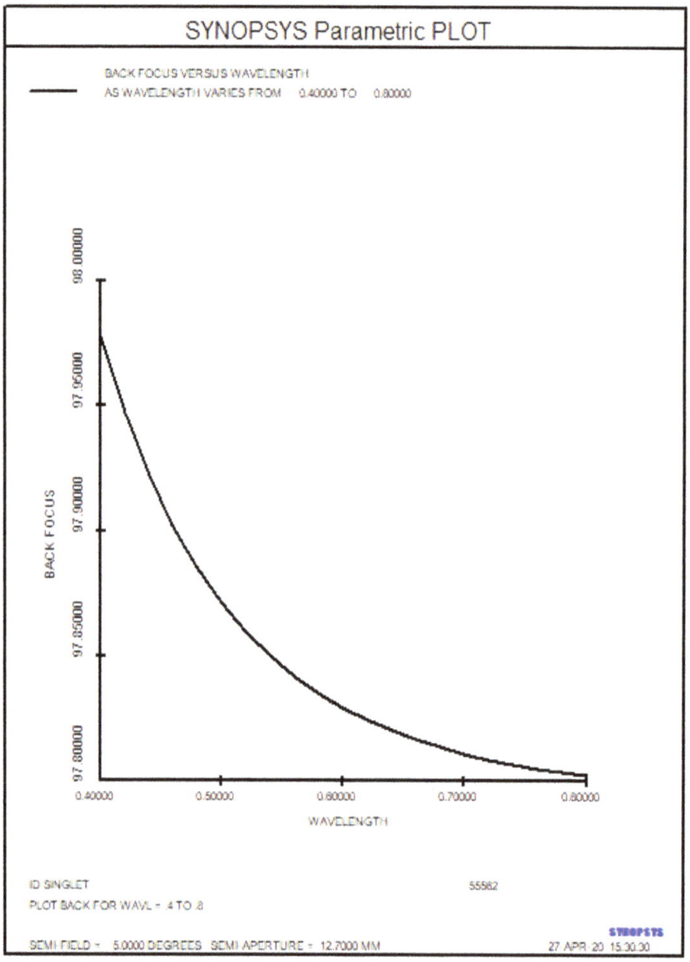

Figure 6.6. Color correction curve after optimization. It is not perfect, but other aberrations are much larger and the program came to this solution.

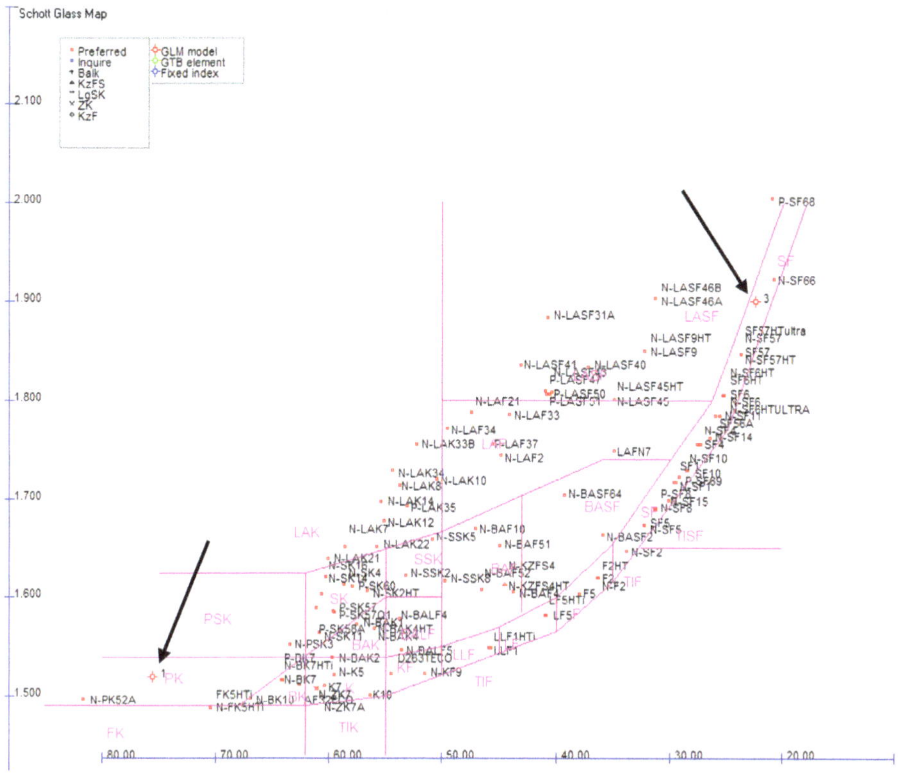

Figure 6.7. Model glass selections.

What kind of glass did the program come up with? Type **MGT** (Menu, Glass Table) in the CW, select the Schott catalog, and click OK. Surface 1 has been assigned a glass model in the PK section, and surface 3 is in the SF section, as you can see in the glass-map picture in figure 6.7. Chapter 12 will develop a three-element apochromat design with even better color correction. But first, let's save this lens.

Type the command

SAVE MYDOUBLET.

That will save an RLE file of the lens data on disk with the name MYDOUBLET. RLE, so you can open it again later if you wish.

You can also store your lens in the lens library. Put this lens in location 4 with the command **STORE 4**. The command **PLB** will list the contents of the library, and **MLB** will open a dialog that gives you another way to access it. The library is a good place to store up to 10 lenses that you are currently working on, and you can save any number of lenses with the **SAVE** command.

(When you run this exercise yourself, you may get somewhat different results, since the exact place where you added the second element will likely differ. That underscores the fact that any change in initial conditions, however small, sends the program in a different direction.)

IOP Publishing

Lens Design (Second Edition)
Automatic and quasi-autonomous computational methods and techniques
Donald C Dilworth

Chapter 7

PSD optimization

Optimization starting with plane-parallel plates; using the PSD III algorithm

Modern software can optimize a lens many times faster than can a human expert using classical tools. This chapter gives a notable example.

There has long been contention in the lens design industry between theorists and ‘number crunchers’. The former endeavor to understand their lens and steer the design in a cogent way following their deep knowledge of aberration theory. The latter employ enough optics knowledge to establish the goals in a cogent manner, but then turn the job of meeting those goals over to the computer. Today, for many problems, the number crunchers can far outpace the theorists. And it is not even close. That is why much of what theorists struggled to understand is no longer as important as it once was.

Figure 7.1 shows a design problem that starts with a very bad lens, where all surfaces are flat, all thicknesses and airspaces are equal, and all glasses are in the middle of the glass chart. Let us see how a good optimization algorithm can quickly turn that bad design into a rather good one.

Here is the optimization MACro (**C7M1**).

```
AWT: .1
OFF 67
RLE
ID START FROM FLAT
UNI MM
OBB 0 20 12.7
1 TH 5 GLM 1.6 50
2 TH 5 AIR
3 TH 5 GLM 1.6 50
4 TH 5 AIR
5 TH 5 GLM 1.6 50
```

doi:10.1088/978-0-7503-3695-6ch7

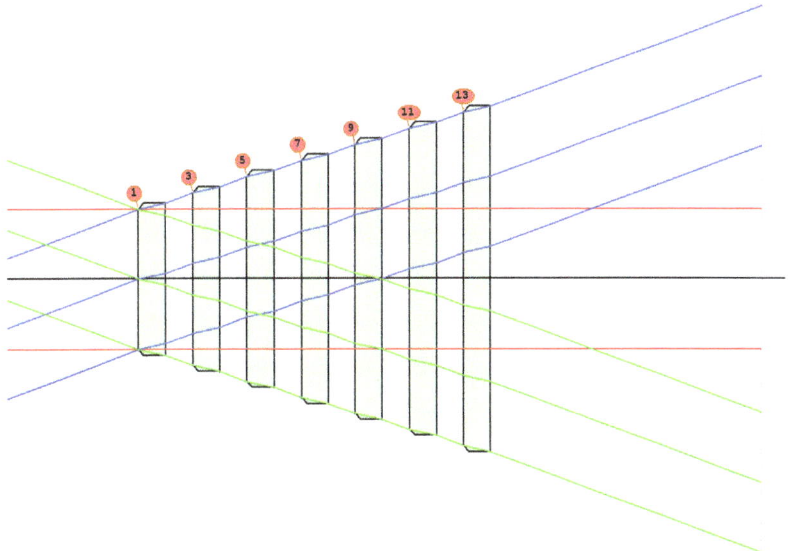

Figure 7.1. Starting design, with all flat surfaces.

```
6 TH 5 AIR
7 TH 5 GLM 1.6 50
8 TH 5 AIR
9 TH 5 GLM 1.6 50
10 TH 5 AIR
11 TH 5 GLM 1.6 50
12 TH 5 AIR
13 TH 5 GLM 1.6 50
14 TH 50 AIR
15
APS 1
END

STO 9
TIME
QUIET
PANT
VY 1 YP1
VLIST RAD 1 2 3 4 5 6 7 8 9 10 11 12 13 14
VLIST TH ALL EXCEPT 14
VLIST GLM ALL
END

AANT
```

```
AEC
ADT 7 .01 1
ACC
M 33 2 A GIHT
GSR AWT 10 5 M 0
GNR AWT 2 3 M .7
GNR AWT 2 3 M 1

END

DAMP 1000
SNAP 20
SYNO 10
SYNO 90
LOUD
TIME

RMS M 0 600
Z1 = FILE 1
RMS M .5 600
Z2 = FILE 1
RMS M 1 600
Z3 = FILE 1
= (Z1 + Z2 + Z3)/3.0
```

Run this MACro, and you get the lens (**C7L1**) in figure 7.2 after about one second.

Whenever you see an entry in these MACros you don't understand, be sure to look it up in the UM immediately. For example, the above file contained the line **`ADT 7 .01 1`**. This is a useful item to consider including; sometimes the results are better with it and sometimes better without. It says that the ratio of the diameter to thickness of lens elements should be at least 7, in this case with a low weight. Including this specification, or varying the parameters, will often cause the search programs to explore different branches. You will want to experiment with it often.

In the CW you see the lines

```
...
--- = (Z1 + Z2 + Z3)/3.0

 The composite value is        0.00646766
```

from the end of the MACro. Those use the AI program to calculate the average RMS spot size at three field points. It comes out to just over 6 micrometers.

Now let's change things a bit. The second parameter on the ray-generation directives currently is 0.1 (the characters assigned to the symbol **AWT**). This applies an aperture-dependent weighting on each ray that says to weight the center rays a little more heavily than those around the edge of the beam, and if we decrease the value to, say 0.0, then rays near the center of the pupil will be weighted the same as

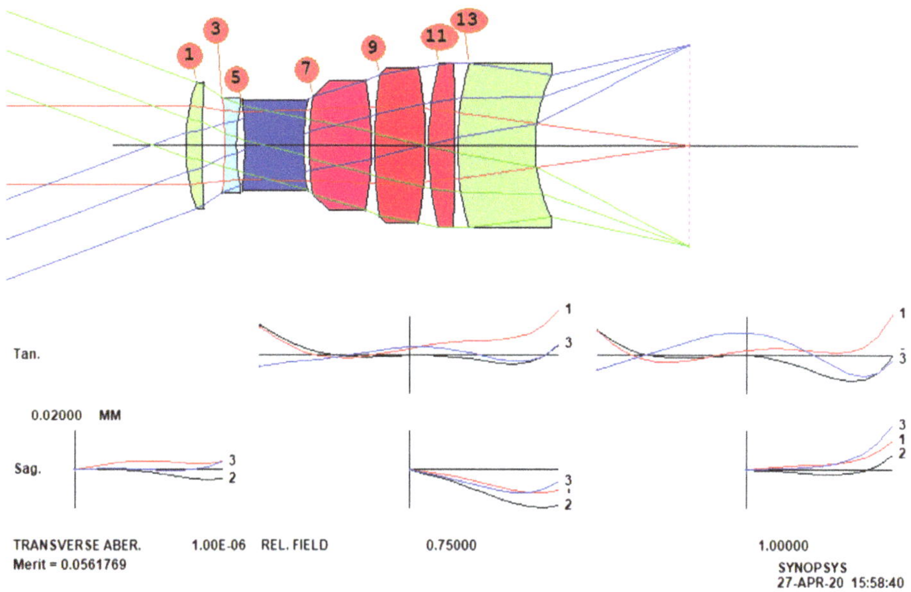

Figure 7.2. Result of PSD III optimization of the very bad starting lens.

rays at the edge. (Remember to click the button **-N** so you won't overwrite the original file.) Edit the MACro, changing the value of the symbol AWT:

```
AWT: 0.0
```

and then run it again. You get a different lens. The symbol **AWT** gets replaced with the characters 0.0 in this case. And the score is slightly better, at 0.006 07 mm. It's another very good lens, and this exercise shows an important insight: When you start with flat surfaces, the PSD algorithm can go *anywhere*—and a very slight change in the starting point or requirements can send it down a different path. This parameter is another one that you will want to experiment with often and is an example of the chaos we mentioned in chapter 1. Try higher and lower values and select the one that works best.

At this point we would normally run the simulated annealing program, by clicking on the button in the top toolbar, temperature 55, cooling 2, passes 50 (**55, 2, 50**). This brings the score down to 5.7 μm. A good lens indeed, shown in figure 7.3. (**C7L2**)

This lesson illustrates the chaos inherent in lens design, which is the principle behind the design search feature, **DSEARCH**. That program creates a set of candidate designs by assigning initial powers according to the bit in a binary number, as explained in Appendix B. Type **HELP DSEARCH** in the Command Window to learn about it. Each case ascends the design tree in a different direction, and one expects that at least a few of the branches it explores will turn out to be just what you are after.

It is sometimes helpful to run the optimization program more than once when starting out with a lens that is far from a good configuration. The reason has to do

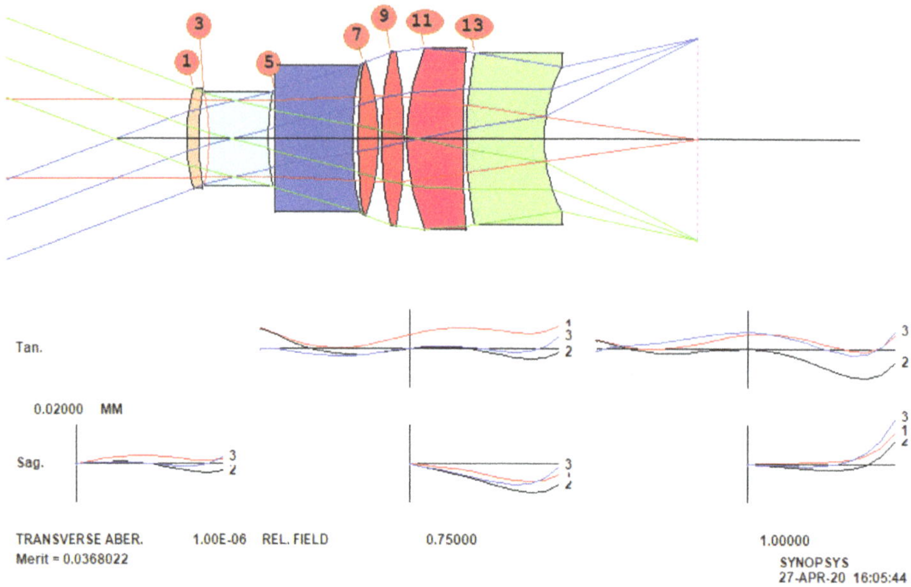

Figure 7.3. Lens optimized and annealed with lower aperture weighting.

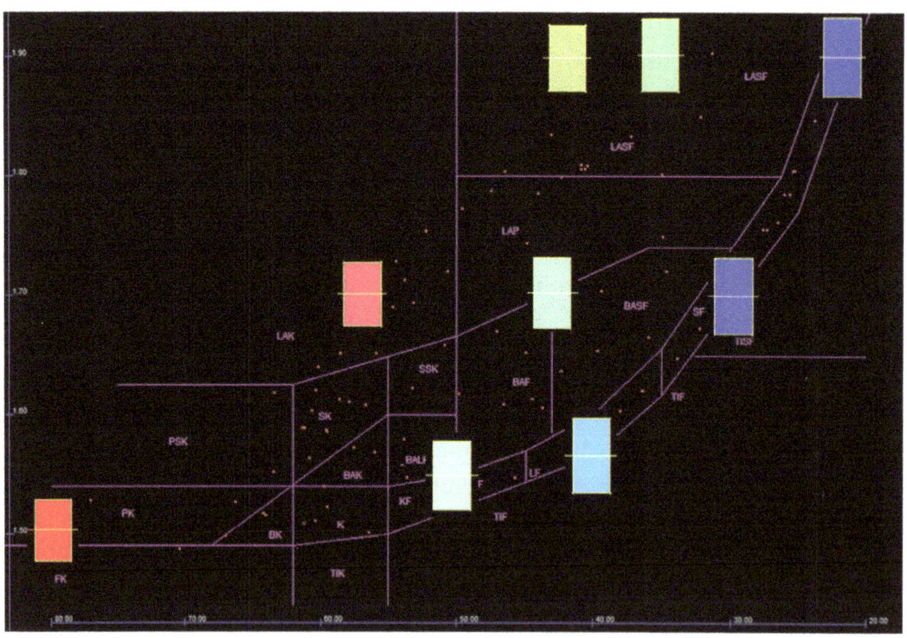

Figure 7.4. Location on glass map as indicated by colors of elements.

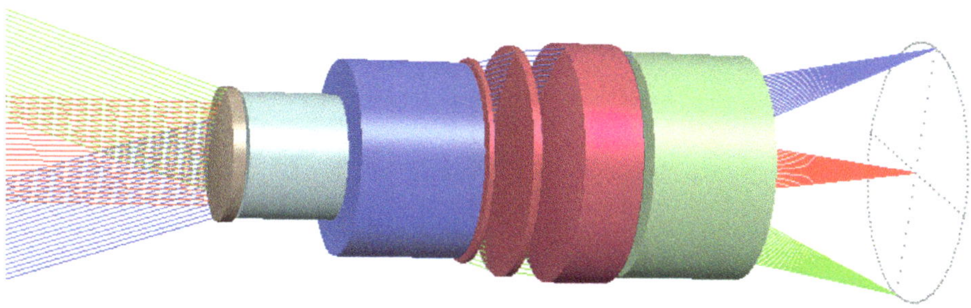

Figure 7.5. RSOLID drawing of lens, colored as in the PAD display.

with the way glass model variables are implemented. Since the lenses in this chapter start out with flat surfaces, it is not obvious right away what the final powers will turn out to be, and along the way the program will likely move the Abbe numbers over to the crown or flint boundary of the glass map, where they become pinned. But after more passes, the powers may well have changed, and a glass might then work better if it left that boundary. Starting the optimization again frees all GLM variables that have become pinned to a boundary so they can leave it if necessary. Running the simulated annealing program does the same thing, freeing GLM variables before it alters the lens. This topic is discussed more fully in chapter 26.

We should mention the colors displayed on the lens drawings in PAD. Those are informative as well as decorative, and give you an idea of where the glass models are located on the glass map. Bright red means the glass is over on the crown side on the left, while blue is displayed for flint glasses on the right. Other colors are intermediate, as shown in figure 7.4.

Similar colors are displayed in the rotating solid model **RSOLID**, found in the dialog **MPE**, if the Autocolor option is selected, as shown in figure 7.5.

Chapter 8

The amateur telescope

A variety of small telescope designs

This chapter is intended for those interested in designing or analyzing the kind of small telescopes popular with amateur astronomers.

8.1 The Newtonian telescope

Probably the most popular kind of amateur telescope is the Newtonian, about the simplest system there is except for a shaving mirror. Here is the input RLE file for a typical telescope (**C8L1**):

```
RLE
ID F/8 PARABOLA WITH DIAGONAL MIRROR
WAVL .6562700 .5875600 .4861300
 APS              1
 GLOBAL
 UNITS INCH
 OBB  0.000000  0.50000   5.00000   0.00000   0.00000   0.00000   5.00000
 MARGIN        0.050000
 BEVEL       0.010000
   0 AIR
   1 RAD   -160.0000000000000    TH     -70.00000000 AIR
   1 CC     -1.00000000
   1 AIR
   1 EFILE EX1      5.050680      5.050680      5.060680      0.000000
   1 EFILE EX2      4.900000      4.900000      0.000000
   1 EFILE MIRROR    2.000000
   1 REFLECTOR
   2 EAO       1.34300000      1.90000000      0.00000000     -0.10000000
   2 CV      0.0000000000000    TH       0.00000000 AIR
   2 AIR
   2 DECEN       0.00000000       0.00000000       0.00000000    100
   2 AT      45.00000004      0.00000000    100
   2 EFILE EX1      1.950000      1.950000      1.960000      0.000000
   2 EFILE EX2      1.950000      1.950000      0.000000
   2 EFILE MIRROR   -0.300000
   2 REFLECTOR
   3 CV      0.0000000000000    TH      10.00000001 AIR
   3 AIR
   3 DECEN       0.00000000       0.00000000       0.00000000    100
   3 AT      45.00000004      0.00000000    100
   3 TH      10.00000001
   3 YMT      0.00000000
   4 CV      0.0000000000000    TH       0.00000000 AIR
   4 AIR
 END
```

doi:10.1088/978-0-7503-3695-6ch8

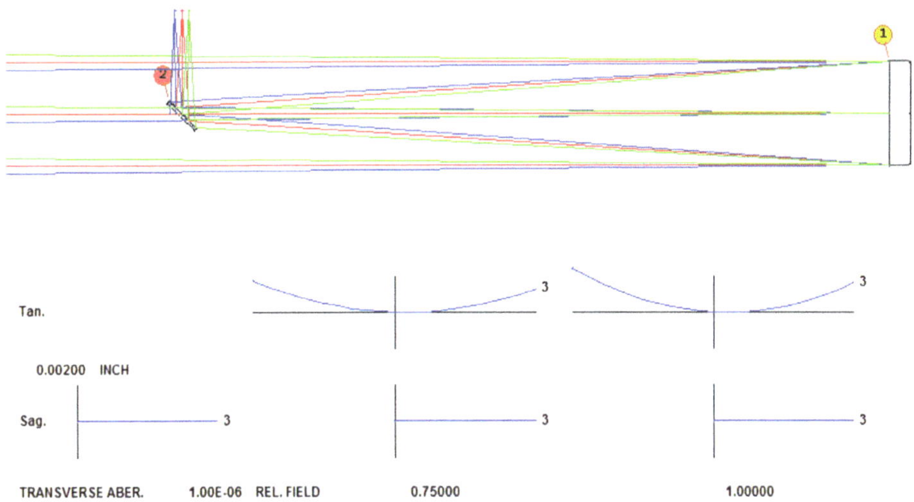

Figure 8.1. The Newtonian telescope.

Figure 8.2. The TrayPrompt display showing the arguments of the OBB object format.

And this is the system, as shown by the PAD display, in figure 8.1. At the moment, the telescope has a fold mirror but has not been assigned an obscuration.

The field of view was set at ½ degree from the axis by the OBB line:

```
OBB  0.000000  0.50000  5.00000  0.00000  0.00000  0.00000  5.00000
```

where the arguments are as shown in figure 8.2. The second entry is the semi-field angle.

To show this information in the TrayPrompt, simply open the WorkSheet and *select* the characters 'OBB' in the editor. The program then looks up the format for you. In this input,

- **ump0** is the incident marginal ray angle, zero for an object at infinity. (The OBB format is used mostly for that situation.)
- **upp0** is the incident chief-ray angle, here 0.5 degrees
- **ymp1** is the incident marginal ray height, here five inches, making a 10-inch diameter entering beam.

The other arguments are all irrelevant here; **yp1** is the chief-ray height on surface 1, zero since that surface is the stop, and the rest refer to the X-Z plane, which we ignore here since the input is axially symmetric. If you're confused already, just open the Object Wizard[1] (**MOW**) to see everything spread out and explained for you there.

[1] Object Wizard™ is a trademark of Optical Systems Design LLC, an Arizona, USA corporation.

The RLE input is designed to be easy to read without any explanation. Surfaces 1 and 2 are declared reflectors, and the conic constant on the primary mirror is –1.0, making it a paraboloid. The **EFILE** data are somewhat cryptic, however; those data are used to define the edge geometry of lenses, and for mirrors the substrate thickness. They make no difference in the paths of rays, but a proper edge makes a nice picture and is important if you want to make drawings of the mirror for the shop. This subject is discussed in more detail in chapter 40.

The above file is produced in response to the commands **LEO** (**LE**ns **O**ut), or **LE** (**L**ens **E**dit). It contains a complete description of the system, including all of the defaults—and you could leave those out if you want to enter the data yourself. In that case, the file would be shorter. **LEO** prints the file in the CW, while **LE** loads it into the editor.

The image is, of course, perfect on axis, but there is lots of coma, a well-known defect of this simple system. How serious is that coma? In PAD, select view 2, (click that number in the PAD toolbar 1 2 3 4 5) and then click the PAD Bottom button. In the dialog that opens, select the OPD Fan Plots option and then OK, as shown in figure 8.3. The OPD aberrations are shown in figure 8.4.

Yes, there appears to be about two waves of coma at the edge of the field.

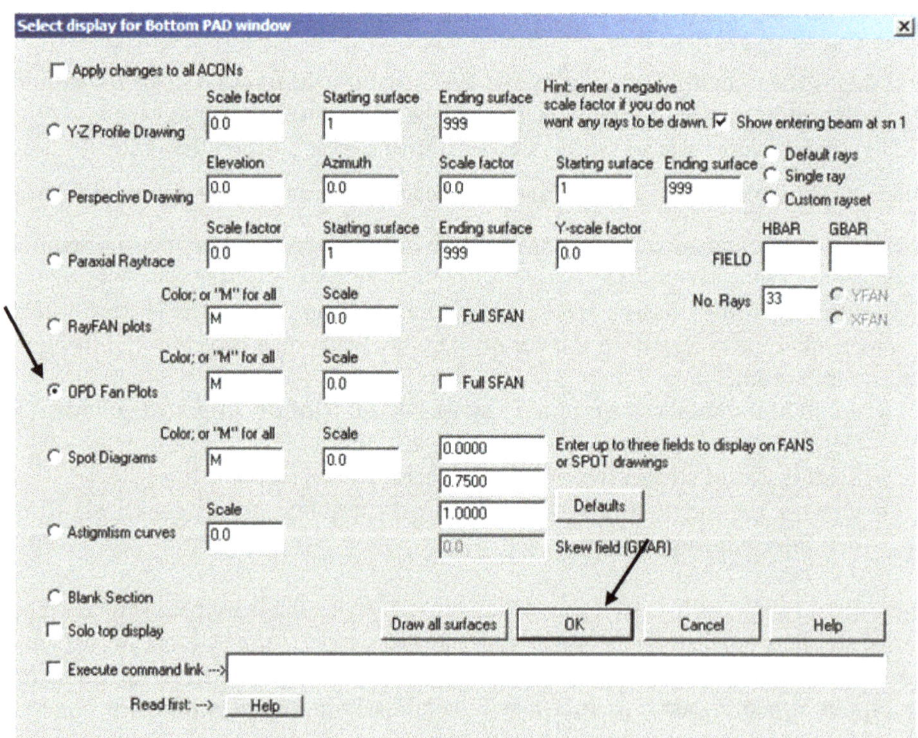

Figure 8.3. Option to select OPD plots in the PAD display.

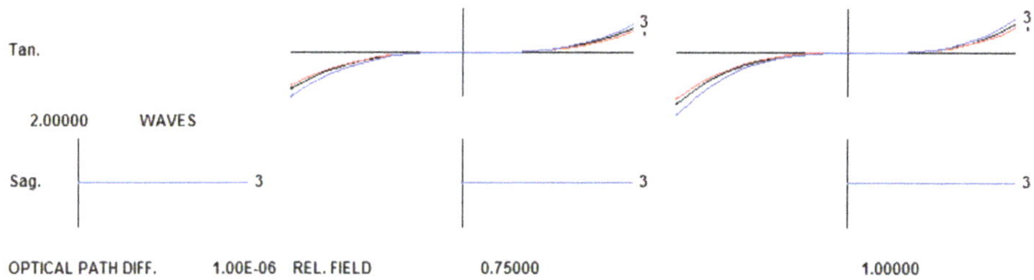

Figure 8.4. Wavefront aberrations of the telescope.

Here is how to get a listing:

```
SYNOPSYS AI>OPD                    ! The next command will be in OPD mode

SYNOPSYS AI>TFA 5 P 1 ! tangential fan, five rays, primary color, full field

ID F/8 PARABOLA WITH DIAGONAL MIRROR
TANGENTIAL RAY FAN ANALYSIS

FRACT. OBJECT HEIGHT                   HBAR      1.000000    GBAR       0.000000
COLOR NUMBER                              2
 REL ENT PUPIL   WAVEFRONT ABERR
      YEN             OPD (WAVES)
________________________________
    -1.000            -2.355059
    -0.800            -1.271960
    -0.600            -0.583027
    -0.400            -0.200234
    -0.200            -0.035356
     0.200            -0.005883
     0.400             0.035526
     0.600             0.212506
     0.800             0.613233
     1.000             1.325667
```

As usual, you can get the same thing without typing the command yourself if you want to. Go to the dialog **MRR** (Menu, Real Rays) or navigate the menu tree, and make your selection there. In this case, typing the command is faster, so I usually do it that way.

Let's look at this image in some detail. There are several tools for the purpose, but I like the Image Tools (MIT) dialog. Type **MIT**, and then make the selections shown in figure 8.5.

This is a rather beautiful example of third-order coma.

Just for fun, try this with the Geometric, and then Diffraction selection in the Effects section. The Coherent analysis is somewhat smoother. It uses a 2-D FFT algorithm, while the Diffraction method evaluates the diffraction integral, clipping at about six times the radius of the Airy disk. This image is slightly larger than that —and since a point source is always coherent with itself, the coherent choice is generally best for that case and definitely better here.

How does the image quality vary with the value of the conic constant? Close MIT and look at the PAD display. Click the Checkpoint button, [icon], and then open the WorkSheet. Click on surface 1 (or enter that number in the number box and click Update). Now, with the mouse, *select* the *entire* number giving the conic constant, as shown in figure 8.6. and then click the **SEL** button. The top slider now controls that

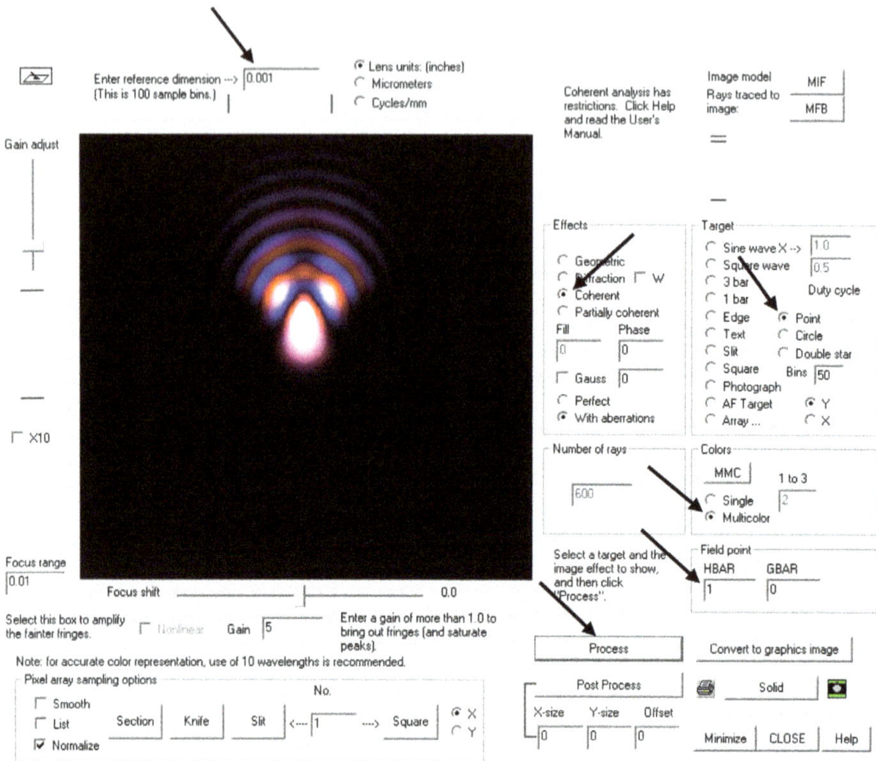

Figure 8.5. MIT display showing off-axis coma of the telescope.

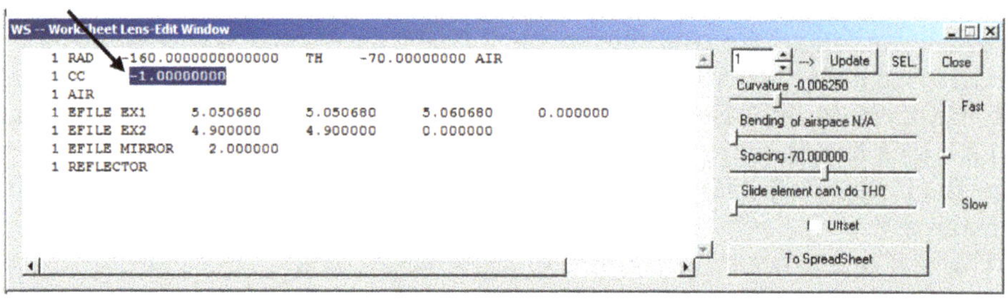

Figure 8.6. Selecting text in the WS edit pane. This assigns the top slider to that quantity with the SEL button.

string of numbers. Slowly drag it to the left and right, watching the PAD display. Those sliders give you a convenient way to evaluate the effects of changing nearly anything in the lens. Restore your checkpoint.

Now evaluate the image quality on axis, using a different tool—but first you need to tell the program about the obscuration, which was not given in the input file. With **WS** still open, type into the edit pane

```
1 CAI 1.4
```

and then click the Update button. (**CAI** means Clear Aperture, Inside.) Now, a hole shows up in the primary mirror. Click the Checkpoint button again. Type **CAP** in the CW, and you see the CAI data listed:

```
SYNOPSYS AI>CAP

ID F/8 PARABOLA WITH DIAGONAL MIRROR

CLEAR APERTURE DATA
(Y-coordinate only)

SURF   X OR R-APER.    Y-APER.    REMARK     X-OFFSET    Y-OFFSET  EFILE?
--------------------------------------------------------------------------
  1        5.0007              Soft CAO                              *
  1        1.4000             *User CAI                              *
  2        1.3430     1.9000  *User EAO       0.0000     -0.1000     *
  3        1.2378              Soft CAO
  4        0.7006              Soft CAO
```

This system has mostly default apertures, although there is now a user-entered inside aperture (a CAI) on surface 1 along with a decentered outside elliptical aperture (EAO) on surface 2, the diagonal mirror. Let's create a footprint plot on the primary mirror. Navigate with the menu tree to MFP (or type **MFP** in the CW). Make the selections in figure 8.7 and click Execute.

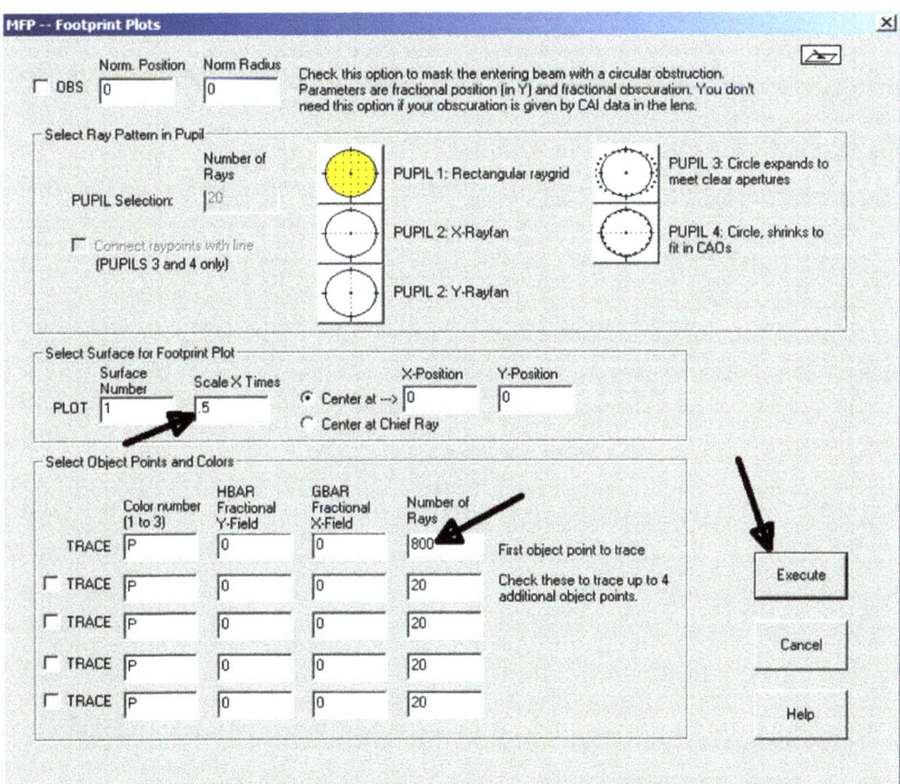

Figure 8.7. Selecting options for a footprint plot.

Now you see the inside aperture, which is free of rays, shown in figure 8.8. Here's a handy trick: suppose you don't know where the rays are being vignetted (that happens sometimes with complex lenses). Here's how to find out: First press the <Enter> key to go back to **MFP**. (When you have used a dialog to run anything, you go right back there when you press that key. That's a time-saver). Now, click the Switches button, and click the radio button to turn on switch 21. SYNOPSYS has about 100 mode-control switches, and that one causes several features to display the surface number where a ray stops. Click Apply, and then run the footprint command again. It makes a new picture,

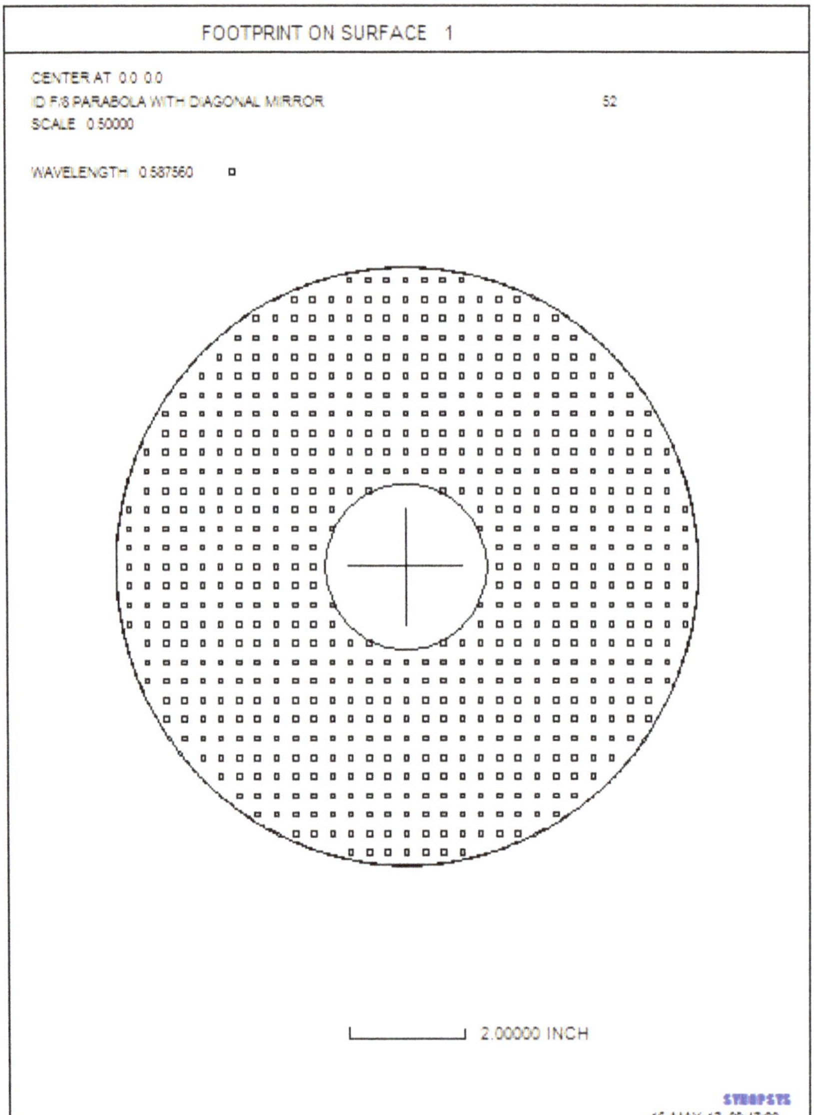

Figure 8.8. Footprint plot showing the obscuration on the primary mirror.

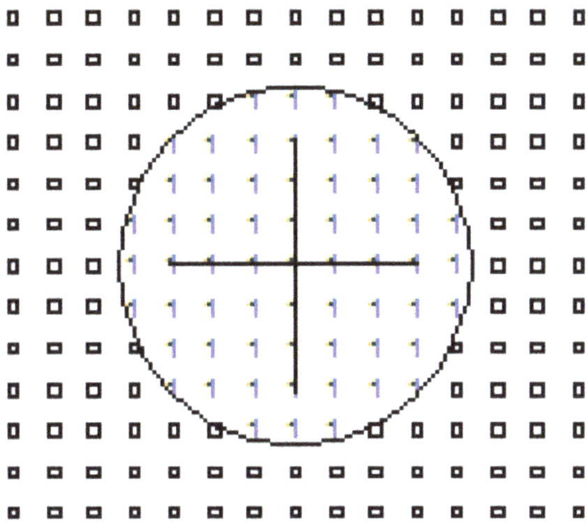

Figure 8.9. Portion of the footprint plot when switch 21 is turned on, showing where vignetted rays were stopped.

and in the center you see numerals where each ray would have gone, showing that it stopped at surface 1, as shown in figure 8.9.

Okay, it's time for some image analysis. Go to the MOP dialog, either with the menu tree or with the command **MOP** (**M**TF **OP**tions). Select the Multicolor option for the MTF, and click the MTF button. The result is shown in figure 8.10. (The grid lines are shown because we turned on switch 87, and the data points are indicated because we turned off switch 27.)

The obscuration has indeed produced a dip in the MTF at mid frequencies, as everyone knows should happen with a Newtonian telescope.

Let's discuss that elliptical aperture on surface 2. In the WS, select surface 2 and then click the button, , to open the **Aperture Dialog**. The option **User-entered elliptical aperture** is selected; click that button to display another dialog where you can change the numbers if you wish. Diagonal mirrors are usually edged with an elliptical shape, and that is where you can enter the data. Or you can just edit the numbers in the WS edit pane, once you identify them.

8.2 The Schmidt–Cassegrain telescope

The classical Schmidt telescope is a great illustration of the power you get when you understand basic optics. That form consists of a spherical mirror and a spherical focal surface, with the stop at the common center of curvature. Since the system is symmetric about that center, there is no unique optical axis anywhere and every field point has the same aberrations. Correct the spherical aberration, and the system has no off-axis aberrations either. This is a brilliant insight—but you have to correct the spherical aberration first, which is done with a thin aspheric plate located at the common center. You don't get perfect correction when you do that, since off-axis beams see the corrector foreshortened, but it is pretty good nonetheless. A serious drawback is the fact that you wind up with a curved image surface, which is tricky to

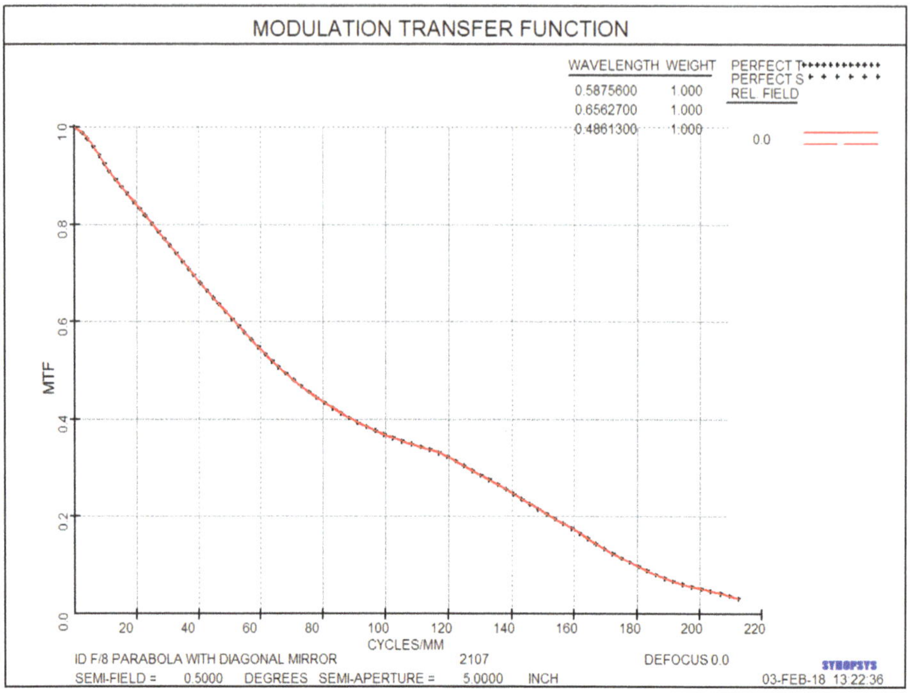

Figure 8.10. MTF curve when the obscuration is accounted for.

manage with glass photographic plates. Still, this is a classic design form that is widely used in astronomy. It has the drawback that the primary mirror is much larger than the entrance pupil, which is at the corrector, if the field of view is wide, as it usually is.

Adding a secondary mirror opens up more possibilities, and the system is then called a Schmidt–Cassegrain telescope. This is a highly-corrected form used with a small field. The lens file below gives the input for this example (**C8L2).**

```
RLE
ID CC SCHMIDT CASS ZERNIKE
 FNAME 'SCT.RLE                                              '
WAVL .6562700 .5875600 .4861300
 APS               1
 GLOBAL
 UNITS INCH
 OBB 0.000000  0.40800  5.00000  0.00000  0.00000  0.00000  5.00000
 MARGIN       0.050000
 BEVEL       0.010000
   0 AIR
   1 CV      0.0000000000000   TH       0.25000000
   1 N1 1.51981155 N2 1.52248493 N3 1.52859442
   1 GTB S    'K5         '
   1 EFILE EX1     5.050000     5.050000     5.060000     0.000000
   1 EFILE EX2     5.050000     5.050000     0.000000
```

```
  2 CV       0.0000000000000    TH      20.17115161 AIR
  2 AIR
  2 ZERNIKE       5.00000000       0.00000000       0.00000000
   ZERNIKE    3      -0.00022795
   ZERNIKE    8       0.00022117
   ZERNIKE   15 -2.00317788E-07
   ZERNIKE   24 -3.81789104E-08
   ZERNIKE   35 -3.47468956E-07
   ZERNIKE   36  3.76974435E-07
  2 EFILE EX1      5.050000      5.050000      5.060000
  3 CAI      1.68000000       0.00000000       0.00000000
  3 RAD    -56.8531404724216    TH     -19.92114987 AIR
  3 AIR
  3 EFILE EX1      5.204230      5.204230      5.214230      0.000000
  3 EFILE EX2      5.204230      5.204230      0.000000
  3 EFILE MIRROR     1.250000
  3 REFLECTOR
  4 RAD    -23.7669696838233    TH      29.18770982 AIR
  4 CC     -1.54408563
  4 AIR
  4 EFILE EX1      1.555450      1.555450      1.555450      0.000000
  4 EFILE EX2      1.545450      1.545450      0.000000
  4 EFILE MIRROR    -0.243545
  4 REFLECTOR
  4 TH       29.18770982
  4 YMT       0.00000000
   BTH       0.01000000
  5 CV       0.0000000000000    TH       0.00000000 AIR
  5 AIR
END
```

Note how the vignetted rays are identified on the fans plots in PAD in figure 8.11. Switch 21 is honored there as well.

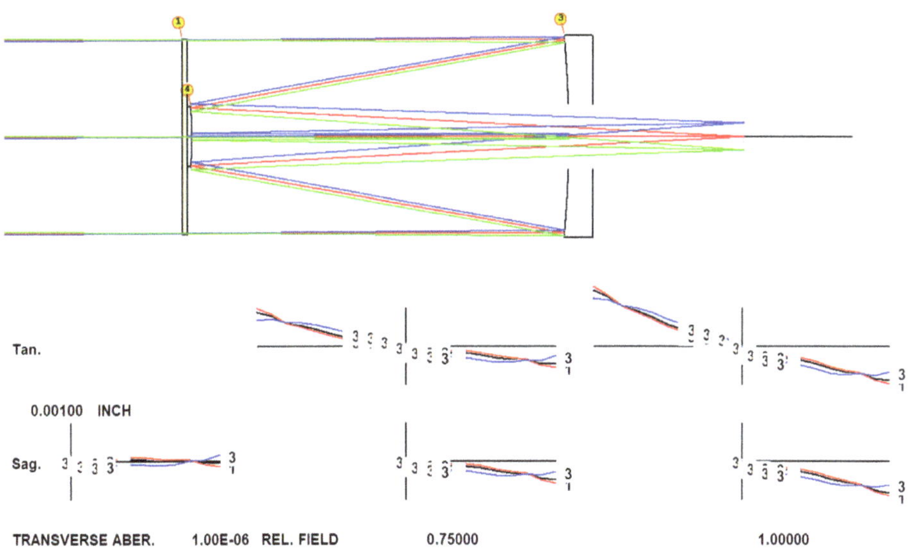

Figure 8.11. PAD display showing where vignetted rays were stopped because switch 21 is turned on.

On the **SPEC** listing you see that surfaces 2 and 4 are aspheric, denoted by the 'O' after the radius column. (You can also access this and other lens listing options from the dialog **MLL**.)

```
SYNOPSYS AI>SPEC

ID CC SCHMIDT CASS ZERNIKE                    272           07-JAN-13    12:40:34
LENS SPECIFICATIONS:

SYSTEM SPECIFICATIONS
-------------------------------------------------------------------------------------
OBJECT DISTANCE     (THO)         INFINITE  FOCAL LENGTH       (FOCL)          98.1614
OBJECT HEIGHT      (YPP0)         INFINITE  PARAXIAL FOCAL POINT               29.1777
MARG RAY HEIGHT    (YMP1)           5.0000  IMAGE DISTANCE     (BACK)          29.1877
MARG RAY ANGLE     (UMP0)           0.0000  CELL LENGTH        (TOTL)           0.5000
CHIEF RAY HEIGHT   (YPP1)           0.0000  F/NUMBER           (FNUM)           9.8161
CHIEF RAY ANGLE    (UPP0)           0.4080  GAUSSIAN IMAGE HT(GIHT)             0.6992
ENTR PUPIL SEMI-APERTURE            5.0000  EXIT PUPIL SEMI-APERTURE            2.0218
ENTR PUPIL LOCATION                 0.0000  EXIT PUPIL LOCATION               -10.5157

WAVL (uM) .6562700 .5875600 .4861300
WEIGHTS    1.000000 1.000000 1.000000
COLOR ORDER     2    1    3
UNITS                              INCH
APERTURE STOP SURFACE (APS)          1    SEMI-APERTURE      5.00000
FOCAL MODE                          ON
MAGNIFICATION              -9.81862E-11
GLOBAL OPTION                       ON

BTH OPTION ON, VALUE =        0.01000
GLASS INDEX FROM SCHOTT OR OHARA ADJUSTED FOR SYSTEM TEMPERATURE
SYSTEM TEMPERATURE =   20.00 DEGREES C
POLARIZATION AND COATINGS ARE IGNORED.
SURFACE DATA
---------------------------------------------------------------------------------------
SURF          RADIUS       THICKNESS    MEDIUM             INDEX        V-NUMBER
---------------------------------------------------------------------------------------
  0         INFINITE       INFINITE    AIR
  1         INFINITE        0.25000    K5                 1.52248      59.49 SCHOTT
  2         INFINITE O     20.17115    AIR
  3        -56.85314      -19.92115    AIR                <-
  4        -23.76697 O     29.18771S   AIR
IMG         INFINITE

KEY TO SYMBOLS
--------------
A  SURFACE HAS TILTS AND DECENTERS       B  TAG ON SURFACE
G  SURFACE IS IN GLOBAL COORDINATES      L  SURFACE IS IN LOCAL COORDINATES
O  SPECIAL SURFACE TYPE                  P  ITEM IS SUBJECT TO PICKUP
S  ITEM IS SUBJECT TO SOLVE              M  SURFACE HAS MELT INDEX DATA
T  ITEM IS TARGET OF A PICKUP

SPECIAL SURFACE DATA
-------------------------------------------------------------------------------------
 SURFACE NO.    2 -- ZERNIKE POLYNOMIAL
 APER. SIZE OVER WHICH ZERNIKE COEFF. ARE ORTHOGONAL (AP)       5.000000
 TERM       COEFFICIENT      ZERNIKE POLYNOMIAL
    3         -0.000228      2*R**2-1
    8          0.000221      6*R**4-6*R**2+1
   15      -2.003178E-07     20*R**6-30*R**4+12*R**2-1
   24      -3.817891E-08     70*R**8-140*R**6+90*R**4-20*R**2+1
   35      -3.474690E-07     252*R10-630*R8+560*R6-210*R4+30*R2-1
   36       3.769744E-07     924*R12-2772*R10+3150*R8-1680*R6+420*R4-42*R2+1

 --------------
 SURFACE NO.    4 -- CONIC SURFACE
 CONIC CONSTANT (CC)      -1.544086
 SEMI-MAJOR AXIS (b)      43.682407   SEMI-MINOR AXIS (a)     -32.221087

THIS LENS HAS NO TILTS OR DECENTERS
SYNOPSYS AI>
```

Surface 2 is defined as a Zernike polynomial aspheric. Let's see what that surface looks like. Type

ADEF 2 PLOT

and you get the plot in figure 8.12.

The black curve shows how the surface departs from the closest-fitting sphere (**CFS**), which in this case is very close to flat. That tells you how to figure the corrector, once you have generated a surface with the radius of the CFS.

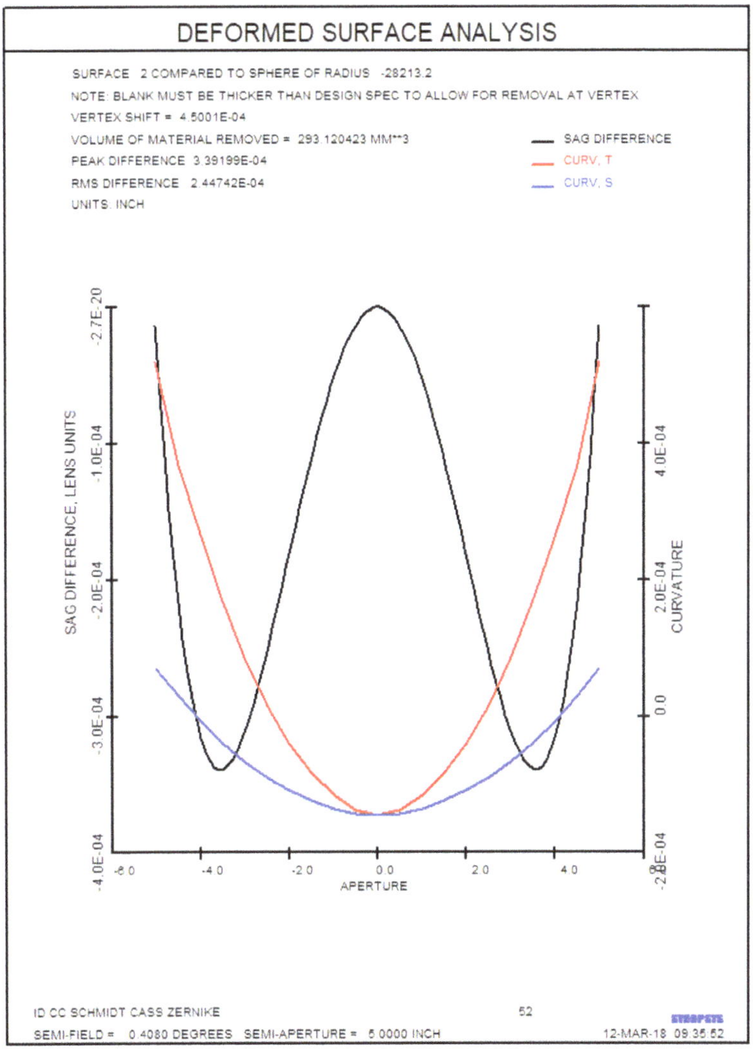

Figure 8.12. Analysis of Zernike surface showing difference in sag with respect to a reference sphere and curvatures in x and y.

The ray fans curves in PAD show the system free from coma and spherical aberration, although there is a tiny bit of spherochromatism. The striking thing is the strong field curvature, indicated by the nearly parallel S- and T-fan curves shown in figure 8.11. A young viewer with a good eyepiece would see a sharp image all over the field, since young eyes can accommodate for the focus shift. Without refocusing, on the other hand, an older viewer would be faced with about 1.5 waves of defocus. Let's see what that would look like over the field.

This time, start at the menu tree (**Diffraction** in the top toolbar) and go down to MDI (Diffraction Image Analysis), where you get many choices. Select MPF, near the bottom (or just type `MPF` in the CW). Select **Show visual appearance** and click **Execute**. You get the results shown in figure 8.13.

The image at the lower left is the on-axis image, and it is essentially perfect, while the upper right shows the image at the edge of the field. That's not too sharp. Let's examine it in a different format. Back in MPF, select the option to **Show as surface**, and change the Height from the default 1 to 0.

Indeed, the image at the edge of the field is pretty smeared out, as shown in figure 8.14. Of course, a young observer would see a much sharper image than shown here, since he can adjust the focus of his eyes.

You can edit the Zernike terms most easily just by changing the value in the WS, but there is also a dialog that lists them by polynomial, which you can reach from the WS by clicking the **Curvature Dialog** button and following the trail as you did

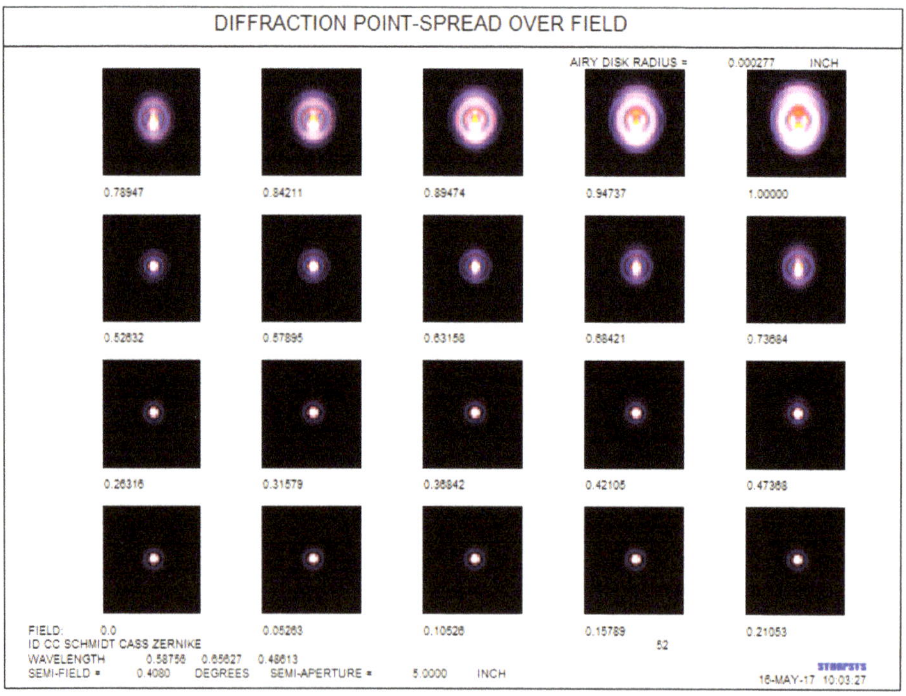

Figure 8.13. Analysis of diffraction point-spread over the field of a Schmidt–Cassegrain telescope.

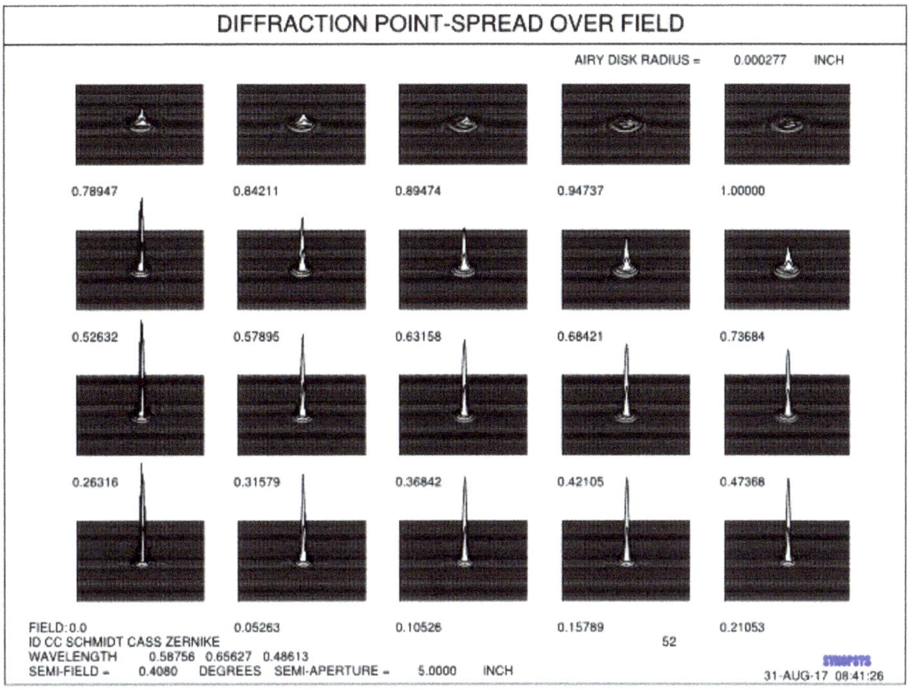

Figure 8.14. Oblique-perspective plots of the PSF over the field.

the aperture dialog above. You arrive at the dialog shown in figure 8.15, where you can change things if you want.

To design this kind of system, vary the Zernike terms with the general-purpose 'G' variables. For example, the PANT entry **VY 2 G 8** would vary term number 8 on surface 2. The definition of the G terms depends on the current shape definition on the surface.

8.3 The relay telescope

This example is a relay telescope that the author built in his basement many years ago. An early version was described in *Sky & Telescope* in 1977, but this one has an extra relay lens and is better corrected. It comes bundled with the program as file 4. RLE, and you can open it with the command

```
FETCH 4
```

You can also open **MWL** (Menu, Window, Lens) to see all of the lens files in the current user directory, with a preview pane for any file you click on, and select the file there.

The telescope, shown in figure 8.16 has a 16-inch diameter mirror, and all surfaces are spherical, making it easy to build compared to an aspheric design.

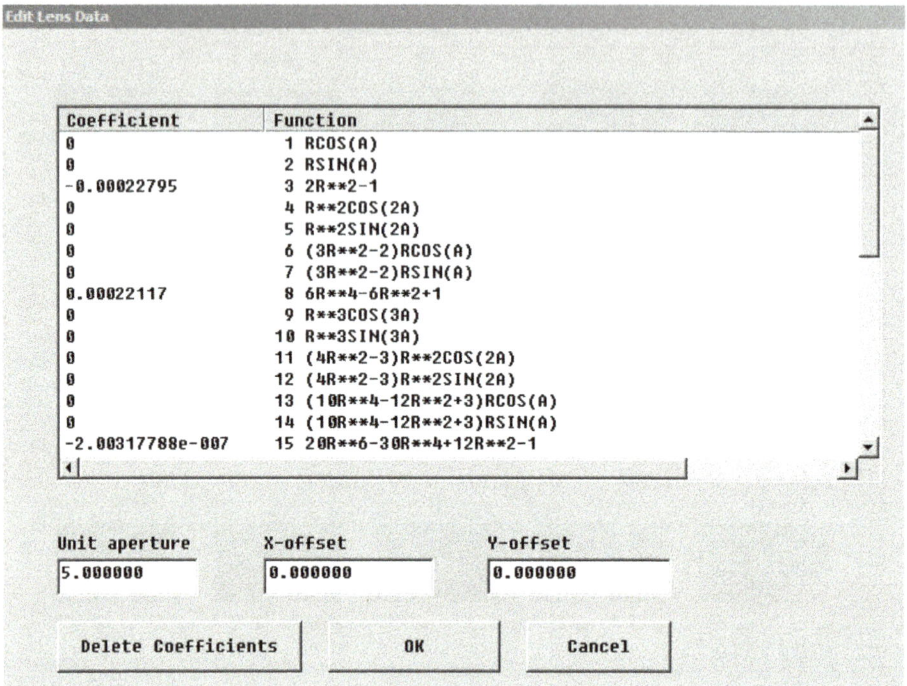

Figure 8.15. Zernike coefficients shown with the Curvature Dialog, where the values can be edited.

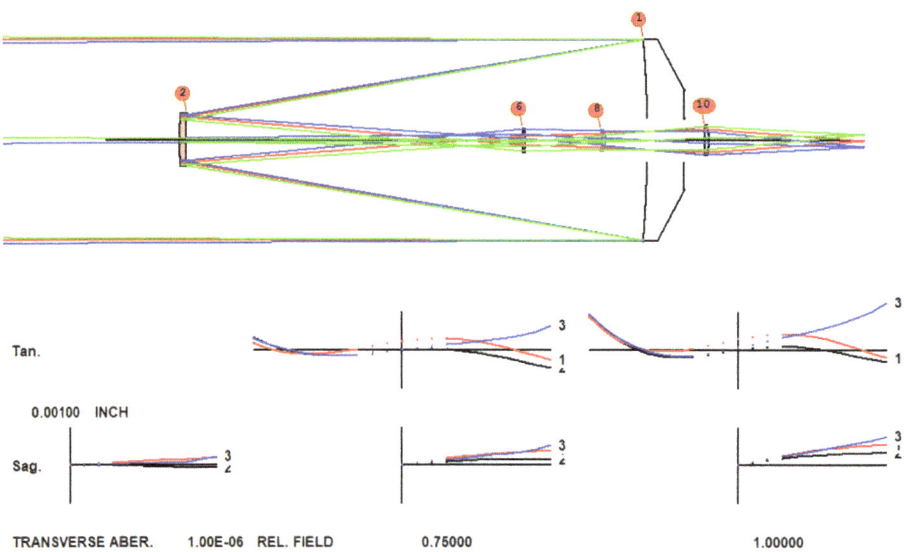

Figure 8.16. The relay telescope.

By now you should be able to enter, modify, and evaluate designs like this without much coaching. The interesting point of this design is the use of a Mangin mirror, from surfaces 2–4. Surface 3 is a reflector, and surface 4 is coincident with surface 2, so the light goes through the element twice, reflecting from the back side at surface 3. With this kind of element, both spherical aberration and secondary color can be well corrected. This telescope has given me and my friends many years of excellent viewing of the heavens.

When the file is opened, type **LEO** in the CW to examine the input file. Note how the Mangin is set up:

```
2 N1 1.58014165 N2 1.58312558 N3 1.58994952
2 CTE 0.640000E-05
2 GTB S 'SK12 '
2 EFILE EX1 2.110000 2.110000 2.110000 0.000000
2 EFILE EX2 2.110000 2.110000 0.000000
3 RAD 536.5921599999994 TH 0.50000000
3 N1 1.58014165 N2 1.58312558 N3 1.58994952
3 CTE 0.640000E-05
3 GID 'SK12 '
3 EFILE EX1 2.110000 2.110000 2.110000 0.000000
3 EFILE EX2 2.110000 2.110000 0.000000
3 REFLECTOR
3 PTH -2 1.00000000 0.00000000
3 PIN 2
4 TH 23.05965000
4 AIR
4 EFILE EX1 2.110000 2.110000 2.110000
4 PCV 2 1.00000000 0.00000000
...
```

In case it is not obvious, surface 4 is the same as surface 2, and both must be present since the light by default goes through all surfaces in sequence. (There is also a nonsequential mode, which is more complicated and is never used for designs as simple as this.) Surface 3 has a pickup of thickness 2, with a sign change, and surface 4 picks up the curvature of 2. When you design systems like this, be sure to assign the appropriate pickups, so the geometry will stay correct as the design variables change.

The shape of the primary mirror, which was ground to a cone shape on the back, was entered with **EFILE** data, which are used to describe edges of elements. In PAD, click the button,  to open the **Edge Wizard** (or type **MEW**, Menu, Edge Wizard), and select surface 1, if it was not selected in WS, as shown in figure 8.17.

This dialog is where you will define up to five points on lenses and mirrors, identified by the small diagram. In the case of a reflective surface, two of the edit boxes are assigned to the mirror thickness (here three inches) and cone angle on the

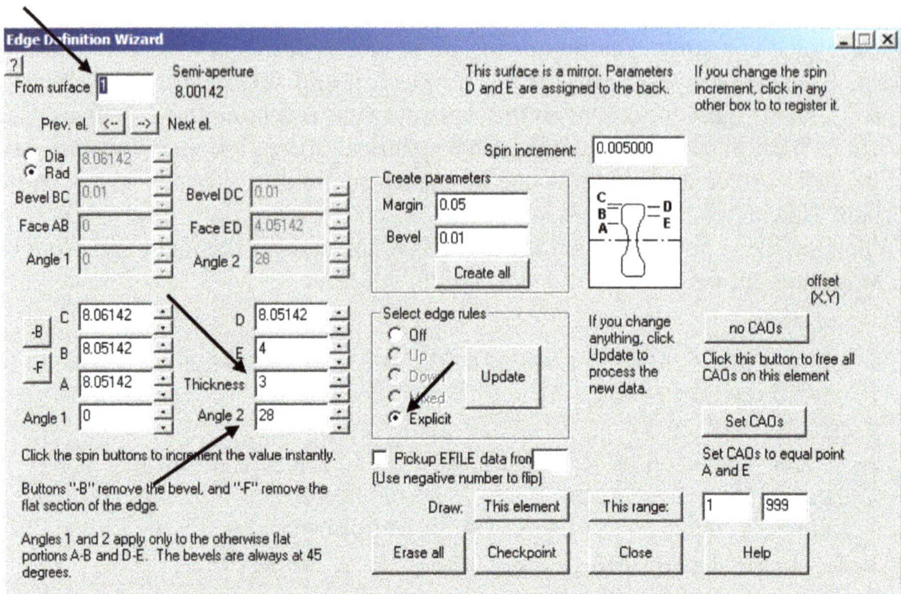

Figure 8.17. The Edge Definition Wizard (MEW) showing data for surface 1.

back (here 28 degrees). Point E in this case marks the start of the cone, four inches from the axis. Click the **Next el.** button, and the program jumps to the first side of the next lens. Click in the PAD display to select other surfaces and see how the parameters A through E define the shape of the lens edges. Then click the button, ?, to read about all the things you can do with the edge definition, or EFILE data. We will discuss the Edge Wizard again in chapter 40.

8.4 How good is good enough?

When designing telescopes such as these, one naturally wants to know when they are good enough, and there are some simple rules of thumb that you should know about: If the peak-to-peak wavefront error is less than ¼ wave, the image will appear nearly perfect—but there are complications. To get this level of performance, a reflective surface has to be twice as accurate, or 1/8 wave, since the OPD error is doubled on reflection. And if there are lenses involved, their errors will add to the mix, so the mirror must be even better than that. And some very fussy astronomers will insist that the wavefront be corrected to 1/10 wave instead of ¼. As noted in chapter 13, tolerances will usually be very tight and many iterations of fabrication adjustments will be required in such systems.

IOP Publishing

Lens Design (Second Edition)
Automatic and quasi-autonomous computational methods and techniques
Donald C Dilworth

Chapter 9

Improving a lens designed using a different lens design program

Adding and removing lens elements to improve the design

In this chapter, we start with a lens designed on another program and then apply some of the newest tools to see if we can improve its performance.

Here is the starting lens (**C9L1**), shown in figure 9.1, along with modulation transfer function (MTF) curves at three field points, in figure 9.2.

(Type **MMF**, select the Multicolor option, then click Execute to make the MTF plot.)

This lens operates in the near IR at a speed of F/3.5, and must be telecentric, have low distortion, and be diffraction limited. At first look, this design is not bad, with less than ¼ wave of aberration. (*Telecentric* means the central ray at all field points should be parallel to the axis, so the exit pupil is at infinity.)

Maximum distortion over the field is just over ½ micrometers, and the maximum departure from telecentricity is about 0.01 radians. Not bad at all—but if we can improve the baseline performance, that will give us more leeway for tolerances, so it is worth a try.

The lens at the moment uses the WAP 3 pupil, which is not a good idea at this stage, so we first make some system changes and then optimize. We will also let the glass types vary, since we don't know if the previous designer used those variables properly—except for the windows at front and back, which we assume the customer wants to keep as is. Here is our MACro (**C9M1**):

doi:10.1088/978-0-7503-3695-6ch9

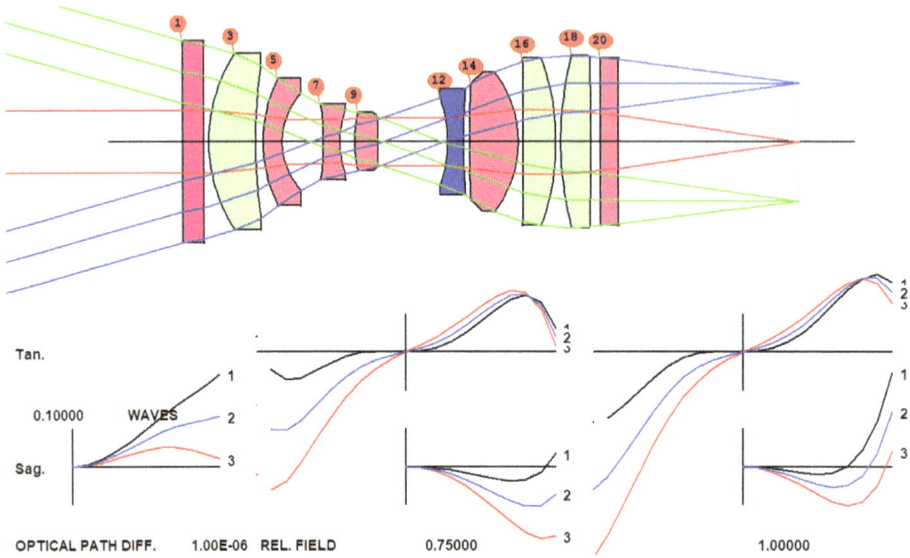

Figure 9.1. Lens to be improved.

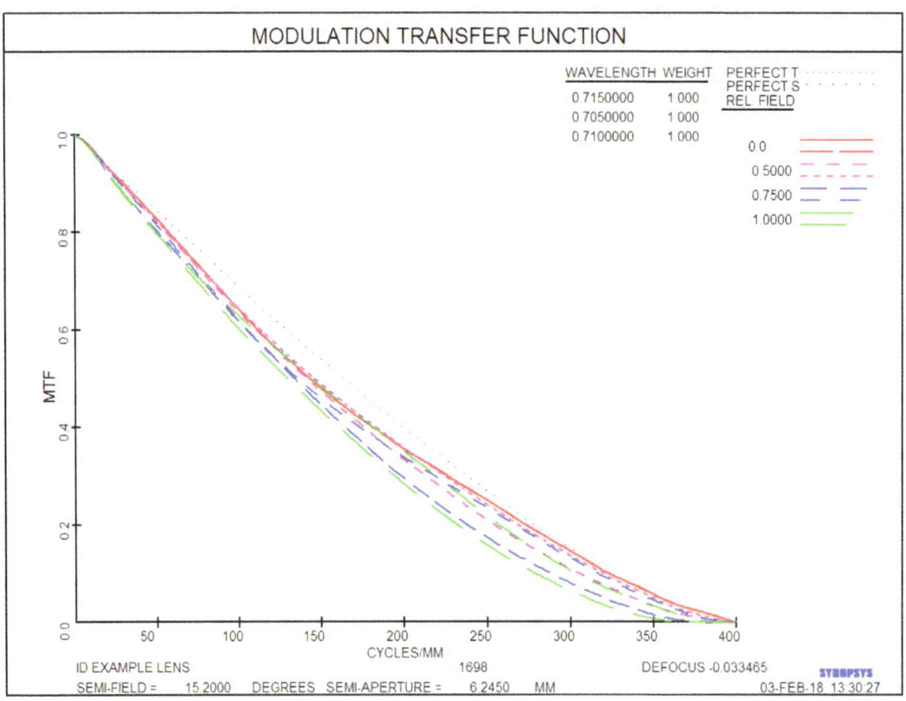

Figure 9.2. MTF of the starting lens.

```
CHG
WAP 1                  ! keep entering beam diameter constant over field
19 UMC -0.14286        ! maintain F/number
CFREE                  ! remove the clear aperture at the stop
END

PANT
VLIST RAD ALL ! all radii will change except 19 and the flat windows
VLIST TH ALL EXCEPT 1 LB2  ! and all thicknesses except 1 and 20
VLIST GLM 3 5 7 9 12 14 16 18
END

AANT
AEC                    ! monitor feathered edges
ACC                    ! and keep thicknesses less than 25.4 mm
M 89.6 1 A TOTL        ! keep total lens length constant
M 0 50 A GIHT          ! control distortion at full field
S P YA 1

M 0 50 A GIHT          ! and at 0.8 FIELD
MUL CONST 0.8
S P YA .8

M 0 50 A GIHT          ! and at half field
DIV CONST 2
S P YA .5

M 0 20 A P HH .7       ! control telecentricity at 0.5 field
M 0 20 A P HH 1        ! and full field

GSO 0 0.1 5 M 0        ! correct OPDs of ray grids at three fields
GNO 0 0.05 4 M .7
GNO 0 0.05 4 M 1
END

SNAP                   ! get snapshot every iteration
SYNO 30                ! optimize for 30 cycles.
```

The simplest way to create this set of ray-grid aberrations is with the **Ready Made Raysets** button in the MACro editor. In this case, select set number 8, which creates both transverse and OPD targets, and then delete the transverse targets and increase the weighting of the OPD targets at full field. The Bare-bones **Rayset** dialog can also do it, with more options then available. ( )

Optimize with this file, and then anneal (**55, 2, 50**). The lens is improved, shown in figure 9.3.

This display shows the lens materials with a cross-hatch pattern instead of the solid colors shown before, which is an option in PAD activated by the button in the PAD toolbar. You will find it a handy way to keep track of which elements have real glass and which are assigned a glass model. Section 13.3 of the User's Manual explains how the colors relate to the location of glass types on the glass map: The blue of the second element, for example, tells you it is in the flint section, over to the right side, while the red at surface 14 is a crown glass near the left side. Figure 7.4 illustrates this color distribution.

Now let's apply some powerful tools. First, run the **Automatic Element Deletion** feature. This finds the element that can be removed with least degradation to the

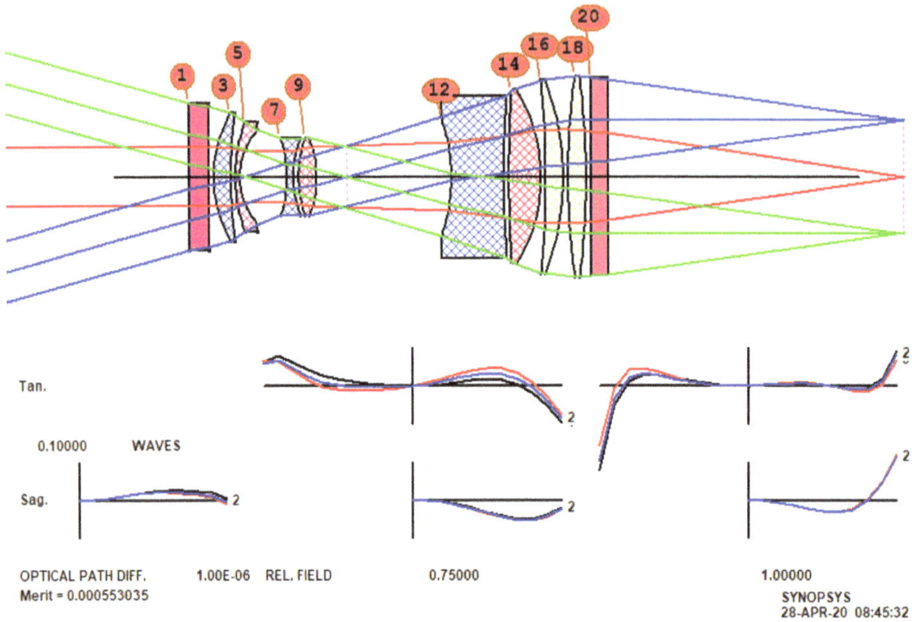

Figure 9.3. Lens as reoptimized.

merit function. To run it, first rename the MACro, and then delete the CHG file at the start of the MACro. Then add the line

```
AED 3 Q 3 18   ! find which element to delete between surfaces 3 and 19.
```

to the MACro before the **PANT** command and reoptimize. The program reports that the lens at surface 16 can be removed. Allow it to remove that element, and then comment out the AED line so you don't remove yet another element. Change the GLM variable declaration to VLIST GLM ALL. The surfaces are renumbered now, and this entry will vary all elements that are already glass models so you don't have to keep track of the numbers yourself.

Optimize and anneal again, and you get the result shown in figure 9.4.

The lens is not quite as good as before, as expected, but still not bad (and definitely better than the starting lens). Now we will use the **Automatic Element Insertion** feature to see if going back to the previous number of elements gives better results than the original lens.

To do this, change the AED line to

```
AEI 3 3 17 0 0 0 20 1   ! insert one element between 3 and 17.
```

and run the MACro once more. (If you have a multicore PC, you should also add the line

```
CORE nb
```

at the top of your MACro, so AEI runs faster, where **nb** is the number of cores you have.)

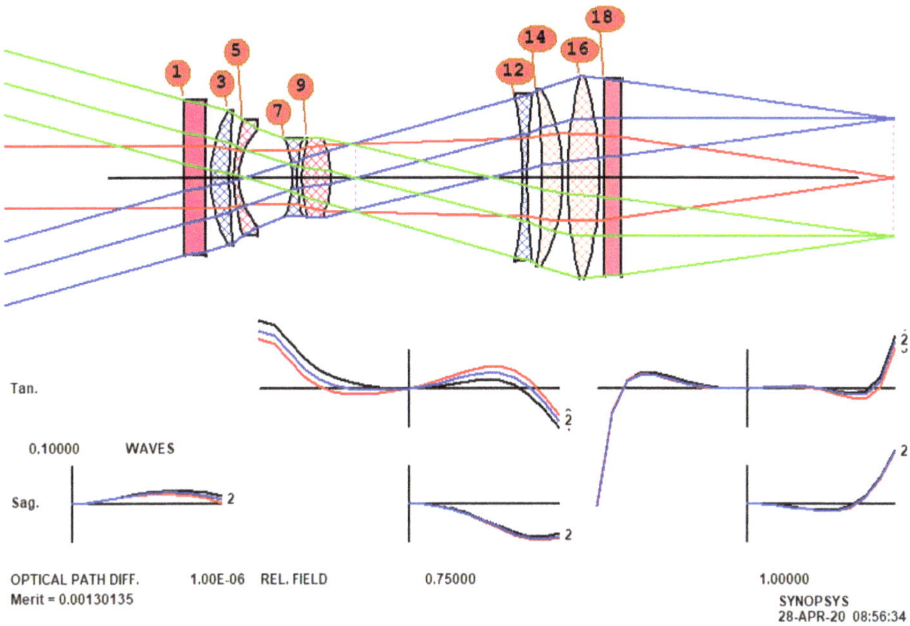

Figure 9.4. Lens with an element removed by AED, optimized and annealed.

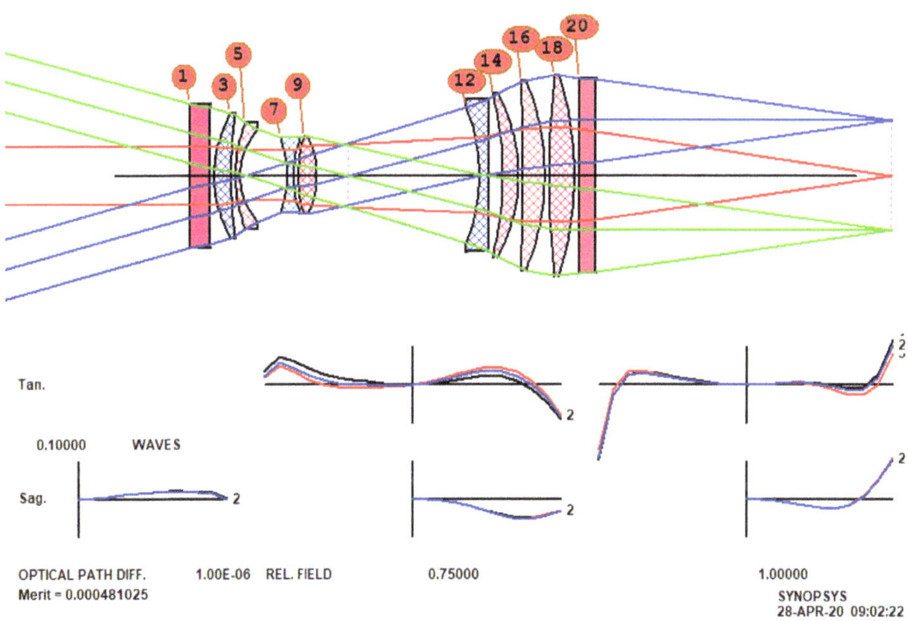

Figure 9.5. Lens after AEI has inserted a new element, then optimized and annealed.

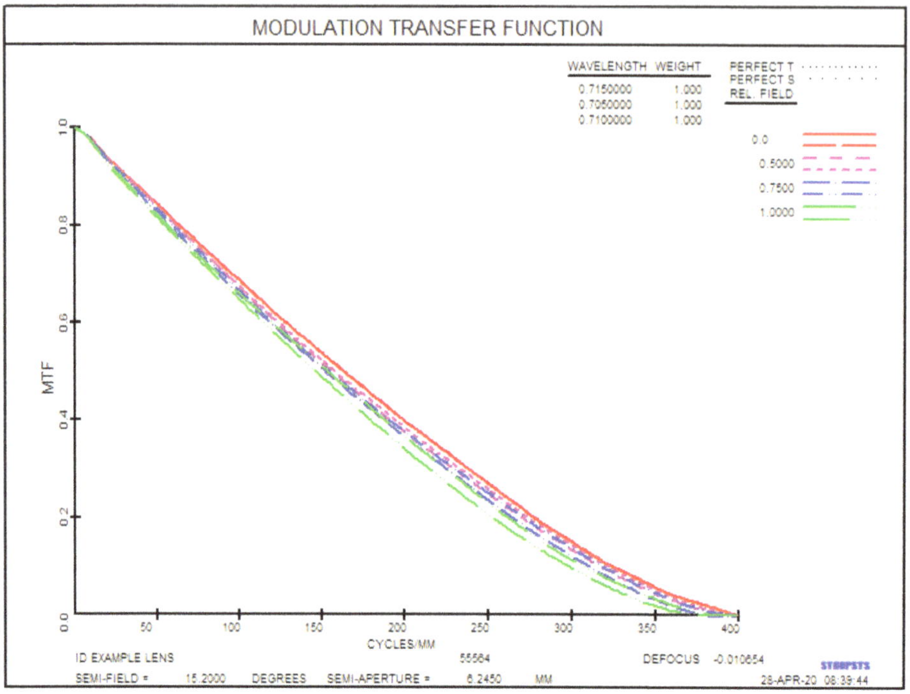

Figure 9.6. MTF of the redesigned lens.

The program inserts an element at surface 16. Now comment out the AEI line, reoptimize and anneal. The result is shown in figure 9.5.

A quick run on **MRG** replaces the glass models with a real glass from Ohara and gives the MTF in figure 9.6. (**C9L2**) (MRG is described in more detail in chapter 44; we will use it often.)

Yes, this exercise was definitely worth doing. The program removed the original lens element at 16 and then replaced it, having found there is no better place to add one. This is often a good exercise when you want to check whether the lens construction is as good as possible. Maximum distortion is now about 0.03 micrometers, compared with 0.6 before, and the maximum departure from telecentricity (given by the tangent of the chief-ray angle) is now about 0.0005, compared with 0.01 before. The performance is much improved over the original, and tolerances will most likely be looser as a result.

We haven't addressed the question of element thicknesses in this chapter, which is a task we often leave for later. That would be the next step in this design, because some elements are a little too thin to be practical. Some of the chapters below show how carry out that step.

Chapter 10

Third-order aberrations

Use and misuse of 3rd-order aberrations; tolerance desensitization

Many students of lens design, and many managers who hire lens designers, are adamant that aberrations have to be very well controlled. They are partly right—but those requirements invariably refer to third-order aberrations, which should all be zero in the opinion of the manager. This is unwise. The point of this chapter is that third-order aberrations are in fact *not* very important—although they still have some uses.

The reason they are not important is because most lenses also have higher-order aberrations, and all orders must be properly balanced. People without deep knowledge of optics do not know that.

Let me illustrate why they are not important. Fetch the lens **C10L1.RLE**. This is a five-element lens with rather good correction, shown in figure 10.1.

(If the PAD display still shows the layout from the previous chapter, with OPDs on the bottom, just click the Defaults button on the PAD toolbar.)

Make an optimization MACro (**C10M1**) that will strongly control the third-order aberrations.

```
PANT
VLIST RAD ALL
VLIST TH ALL
VLIST GLM 1 3 6 8 9
END

AANT
M 1 1 A FNUM
M 7.8 1 A BACK
M 0 1 A DELF
M 0 1 A SA3
```

doi:10.1088/978-0-7503-3695-6ch10

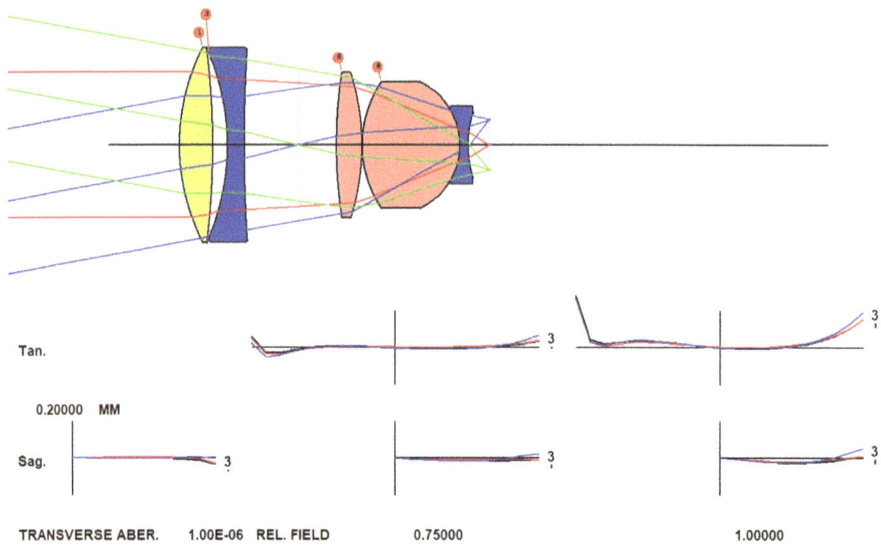

Figure 10.1. Lens with well-balanced aberrations.

```
M 0 1 A CO3
M 0 1 A TI3
M 0 1 A SI3
M 0 1 A PETZ
M 0 1 A DI3
M 0 1 A PAC
M 0 1 A SAC
M 0 1 A PLC
M 0 1 A SLC
END

SNAP
SYNO 30
```

This MACro will vary all the design variables and control the F/number, defocus, and back focus distance, all the while correcting the third-order aberrations to a target of zero. The input **VLIST RAD ALL** will vary all radii, and **VLIST TH ALL** will vary all thicknesses and airspaces, but we could not use the form **VLIST GLM ALL** in this case since that form will only vary those materials that *already* have a glass model, and in this example lens none of them have. So we have to declare the surfaces individually here.

Run this MACro and … the lens is horrible, as shown in figure 10.2. The scale of the fans plots is 25 times larger.

What happened? Did the optimization fail? Ask for the third-order aberrations with the command **THIRD**.

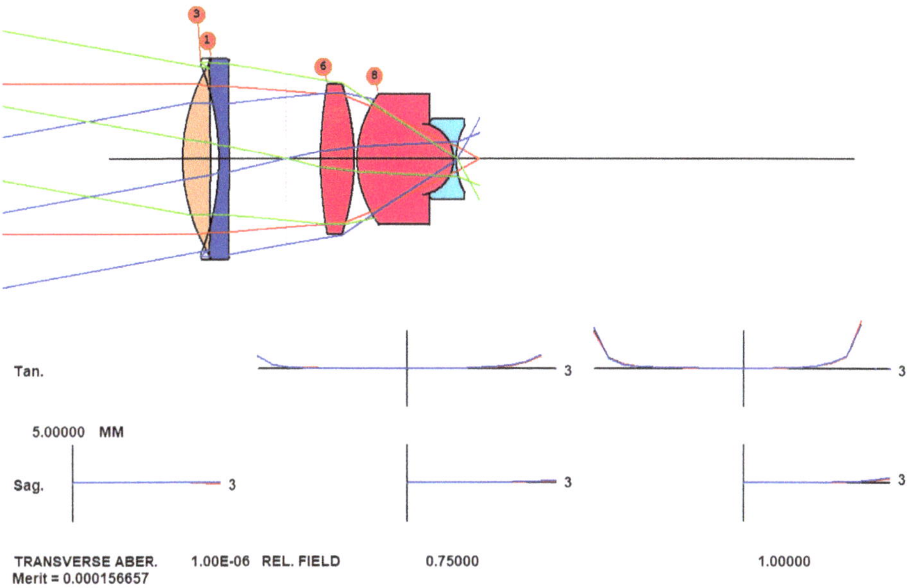

Figure 10.2. Result of correcting third-order aberrations to near zero.

```
SYNOPSYS AI>THIRD

 ID FIVE-ELEMENT LENS                         2146            07-FEB-18    09:05:39

 THIRD-ORDER ABERRATION ANALYSIS

 FOCAL LENGTH   ENT PUP SEMI-APER   GAUSS IMAGE HT
       50.804              25.400            8.958

 THIRD-ORDER ABERRATION SUMS
              SPH ABERR         COMA  TAN ASTIG  SAG ASTIG    PETZVAL DISTORTION
                  (SA3)        (CO3)      (TI3)      (SI3)     (PETZ)  (DI3(FR))
              3.015E-05 -1.732E-05 -1.115E-05 -2.172E-05 -2.701E-05   -0.00040

 PARAXIAL CHROMATIC ABERRATION SUMS
               AX COLOR  LAT COLOR  SECDRY AX SECDRY LAT
                  (PAC)      (PLC)      (SAC)      (SLC)
               -0.00320 -2.982E-05    0.01209    0.00041
 SYNOPSYS AI>
```

Indeed. Those aberrations are very small. How about the starting lens?

```
ID FIVE-ELEMENT LENS

THIRD-ORDER ABERRATION ANALYSIS

FOCAL LENGTH   ENT PUP SEMI-APER   GAUSS IMAGE HT
      50.799              25.400            8.957

THIRD-ORDER ABERRATION SUMS
             SPH ABERR         COMA  TAN ASTIG  SAG ASTIG    PETZVAL DISTORTION
                 (SA3)        (CO3)      (TI3)      (SI3)     (PETZ)  (DI3(FR))
              -0.01812     -0.03732   -0.04232   -0.08743   -0.10998   -0.01754

PARAXIAL CHROMATIC ABERRATION SUMS
              AX COLOR  LAT COLOR  SECDRY AX SECDRY LAT
                 (PAC)      (PLC)      (SAC)      (SLC)
               0.02653    0.01293    0.01665    0.00407
```

Wow! Those aberrations are much larger, and the larger third-order aberrations gave a better lens! Lesson learned. Do not try to out-guess the program when it comes to aberration balancing.

Let me repeat: When you design a lens, you are usually only concerned with two things: **Is the image sharp, and is it in the right place?** If people start talking about aberrations, be polite and smile.

10.1 Tolerance desensitization

We mentioned earlier, however, that those aberrations still have a use. The most important of these deals with *tolerance desensitization*. This is because, when lenses are improperly manufactured, it is the third-order aberrations that change the fastest. To help you keep those aberrations small, the program has a set of eight quantities that can be put into the AANT file. If these are small, tolerances tend to be looser. These quantities are

SAT COT ACD ACT ECD ECT ESA ECO

SAT The sum of the squares of the surface contributions to spherical aberration, SA3.
COT The sum of the squares of the surface contributions to coma, CO3.
ACD The sum of the squares of the amount by which CO3 varies as each surface is decentered.
ACT The sum of the squares of the amount by which CO3 varies as each surface is tilted.
ECD The sum of the squares of the amount by which CO3 varies as each element is decentered.
ECT The sum of the squares of the amount by which CO3 varies as each element is tilted.
ESA The sum of the squares of the element contributions to spherical aberration, SA3.
ECO The sum of the squares of the element contributions to coma, CO3.

Here is an example of how you might use these aberrations to loosen tolerances. The lens shown in figure 10.3 (**C10L2**) was optimized, and then a tolerance budget was created by BTOL with a target wavefront quality of 0.05. (BTOL is discussed in more depth in chapter 13.)

Some of the tolerances came back very tight, as shown in the table below, in which the nominal data are for this lens. (YDC means Y-decenter.)

	3 TH	6 Wedge	7 Tilt	5 YDC	7 YDC	9 YDC	12 YDC
Nominal	0.034	0.23 min	0.24 min.	0.0042	0.0034	0.0053	0.0086
Case A	0.091	0.67	0.42	0.011	0.009	0.011	0.011
Case B	0.112	0.87	0.89	0.015	0.018	0.025	0.014

It would be expensive indeed to hold lens positions to these tight values. (Look at the centration tolerance on surface 7: 3.4 micrometers!) Proceed as follows:

1. Run the command **`THIRD SENS`**, to see the current values of these parameters.

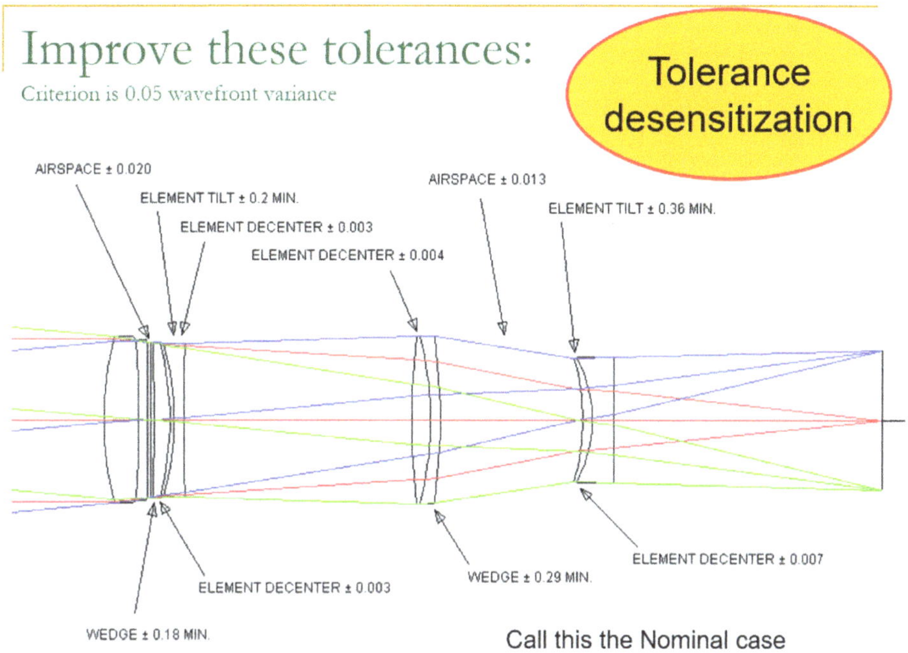

Figure 10.3. Lens with tolerance sensitivity problems.

```
THIRD SENS
ID 8-ELEMENT TELEPHOTO

NORMALIZED 3RD-ORDER ANALYSIS OF TOLERANCE SENSITIVITY

SS OF SA3 BY SURFACE (SAT) =            85.107903
SS OF CO3 BY SURFACE (COT) =            21.404938
SS OF CO3/YDC BY SURFACE (ACD) =         0.007657
SS OF CO3/TILT BY SURFACE (ACT) =       73.889722
SS OF CO3/YDC BY ELEMENT (ECD)  =        0.003941
SS OF CO3/TILT BY ELEMENT (ECT) =       31.259708
SS OF SA3 BY ELEMENT (ESA) =             1.944190
SS OF CO3 BY ELEMENT (ECO) =             0.492351
```

2. Since we are mainly concerned about centration errors, we might try to reduce the value of ECD, the change in CO3 when an element is decentered. To do so, add to the AANT file (in **C10M2**) the line

```
M .001 100 A ECD
```

Since ECD is already a small number (compared to the others in the list) we give it a high weight so it makes a difference to the merit function. Keep in mind that you cannot simply target all of these values to zero, since lens elements in general cannot

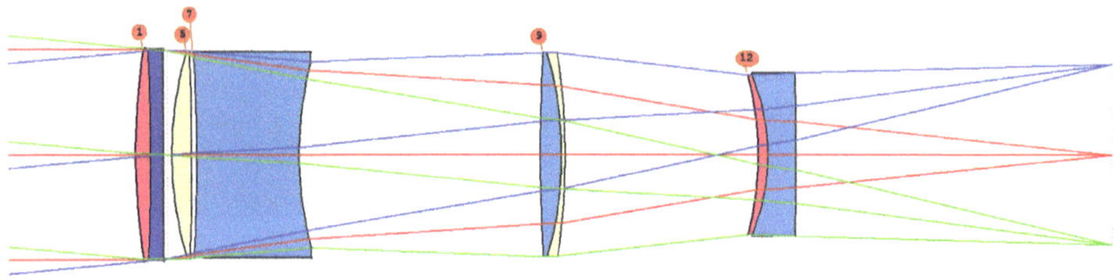

Figure 10.4. Lens with ECD reduced.

be designed without any aberrations and still have any optical power. Also, these quantities are coupled in obscure ways. If, for example, you reduce the value of SAT, you will probably find that COT also got smaller. You could not give an independent value to both of them and expect that the program could find such a combination. So it is wise to proceed with one at a time until you find the parameter that works best with your lens. In this example, by controlling the value of ECD, then optimizing and annealing, we get the lens shown in figure 10.4.

```
THIRD SENS

 ID 8-ELEMENT TELEPHOTO                     1166            24-SEP-17
14:38:04

 NORMALIZED 3RD-ORDER ANALYSIS OF TOLERANCE SENSITIVITY

 SS OF SA3 BY SURFACE (SAT) =              7.039158
 SS OF CO3 BY SURFACE (COT) =              4.886167
 SS OF CO3/YDC BY SURFACE (ACD) =          0.001647
 SS OF CO3/TILT BY SURFACE (ACT) =        19.595239
 SS OF CO3/YDC BY ELEMENT (ECD)  =         0.001064
 SS OF CO3/TILT BY ELEMENT (ECT) =         8.608645
 SS OF SA3 BY ELEMENT (ESA) =              0.185937
 SS OF CO3 BY ELEMENT (ECO) =              0.127815
```

Notice how all of the values returned by **THIRD SENS** have changed, even though we only targeted ECD. Tolerances of this lens are listed as Case A in the above table. Clearly, the tolerances are much looser now, although still a challenge for the shop. Let us experiment some more. This time we will target the value of ACT to the value 7.0, which is 1/10 of the nominal value, and remove the target for ECD.

```
M 7 1 A ACT
```

Optimize and anneal, and the lens now looks like this, in figure 10.5:

The tolerances are listed as Case B, above. This looks like a better budget. (We ignore for this exercise the issue of manufacturability: some elements are much too thin and should be controlled with the ACM or ADT monitors.)

The quantity that you elect to control depends on which tolerances you want to affect. Airspace tolerances, for example, may respond to control over the quantity ESA. Lens thickness tolerances, on the other hand, may respond better to SAT. You

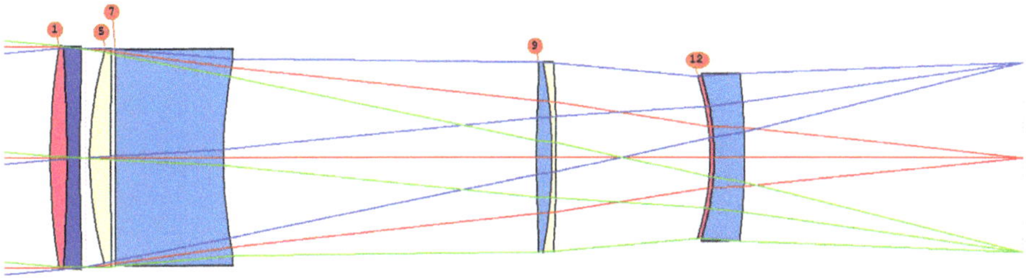

Figure 10.5. Lens with ACT controlled.

will have to understand your lens, and experiment with these tools, to find the best targets and the best BTOL budget.

Sometimes the effect of these quantities is to *increase* the value of the merit function. Normally that would not be a good idea, since if the image gets worse, tolerances generally get tighter. But the relaxing effect of the tools in this lesson can sometimes outweigh that effect, giving tolerances that are looser anyway. This only works up to a point, of course, and if the merit function gets too large, you should require a less-demanding value of that parameter in your AANT file.

There is no guarantee that any of these aberration targets will work in any given case, but experience has shown that they are certainly worth a try. Your tolerances may be relaxed by a factor of from 2 to 10.

There is yet another way to control the sensitivity of an individual element: with the SECTION aberrations. Whereas the quantities discussed above apply to all surfaces or elements and are therefore very easy to use, the SECTION aberrations apply only to the surface range you specify. If an element gets assigned a very tight centration tolerance, even after you try the targets given in this section—which can happen if some tolerances get much looser but the problem element gets tighter—you might control only the coma or spherical aberration of *that* element. That gives you precise control over the aberrations where you need it, and is sometimes worth the extra step. For example, if the element at surfaces 13 and 14 were very sensitive, you might control the spherical aberration of that element with

```
M 0 .1 A SECTION SA3 13 14
```

and experiment with the target and weights until you get the best results. This request will minimize SA3 for the section from surfaces 13–14, and its alignment may then be somewhat less sensitive.

Chapter 11

The in and out of vignetting

Pupil definition; reduction due to apertures; adjusting ray targets for reduced pupil

By 'vignetting' (pronounced *vin-YETTing*) we mean any property of a lens that blocks some of the rays that would otherwise pass through the stop; it is a topic that different programs handle in different ways. Of course, one would often prefer that the beam size remain constant everywhere in the field, since the effective transmission then does not fall off with field on that account. But sometimes the best tradeoff is to accept some vignetting in order to avoid the cost and weight of a more complex lens. In such cases one must know how to manage the varying beam size during optimization and how to set up the lens apertures to model that amount of vignetting for image analysis when the design is finished. That is the subject of this chapter.

Here is a sample lens, in figure 11.1, a triplet with substantial vignetting (**C11L1**).

Notice how the size of the beam at the upper and lower field points (in blue and green) is much smaller than the on-axis beam (in red). A look at the RLE file for this lens shows a real stop on surface 3 (to activate ray aiming for the chief ray) and wide-angle pupil option 3 in effect (WAP 3). (Type **LE** to see this file.)

```
RLE
ID COOKE TRIPLET F/4.5                         747
 WAVL .6562700 .5875600 .4861300
 APS                 -3
 WAP                   3
 UNITS MM

OBB  0.000000     20.00000    5.55500
  0 AIR
  1 CAO      4.69068139       0.00000000       0.00000000
  1 RAD     21.4939500000000   TH      2.00000000
  1 N1 1.61726800 N2 1.62040602 N3 1.62755182
  1 CTE   0.630000E-05
  1 GTB S    'SK16              '
```


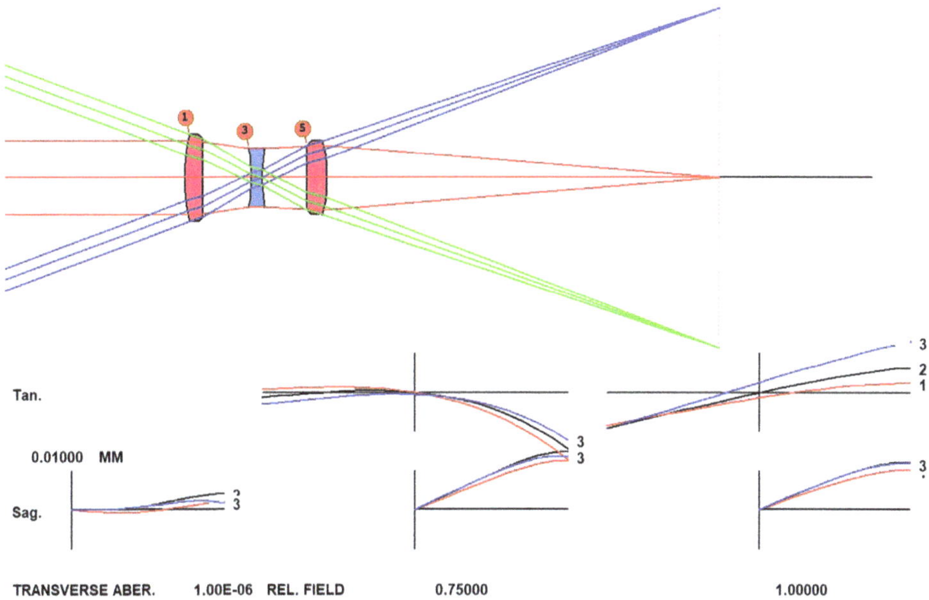

Figure 11.1. Triplet illustrating vignetting.

```
   2 CAO       4.25560632       0.00000000        0.00000000
   2 RAD    -124.0387000000000    TH       5.25509000 AIR
   3 CAO       3.19251725       0.00000000        0.00000000
   3 RAD     -19.1051800000000    TH       1.25000000
   3 N1 1.61163844 N2 1.61658424 N3 1.62846980
   3 CTE    0.830000E-05
   3 GTB S     'F4                 '
   4 CAO       3.15978037       0.00000000        0.00000000
   4 RAD      21.9794700000000    TH       4.93473000 AIR
   5 CAO       3.48158127       0.00000000        0.00000000
   5 RAD     328.3317499999989    TH       2.25000000
   5 N1 1.61726800 N2 1.62040602 N3 1.62755182
   5 CTE    0.630000E-05
   5 GID 'SK16                 '
   5 PIN     1
   6 CAO       4.00000022       0.00000000        0.00000000
   6 RAD     -16.7537700000000    TH      43.24303731 AIR
   6 TH       43.24303731
   6 YMT       0.00000000
   7 CV       0.0000000000000    TH       0.00000000 AIR
 END
```

The WAP 3 option adjusts the entrance pupil dimensions so that the marginal rays at each field point just clear *all* defined lens apertures and feathered edges, if any. Every surface except the image (at surface 7) has been assigned a hard[1] clear aperture (with

[1] A 'hard' aperture is one entered by the user, while a soft aperture is calculated by the program. Only the former can vignette rays.

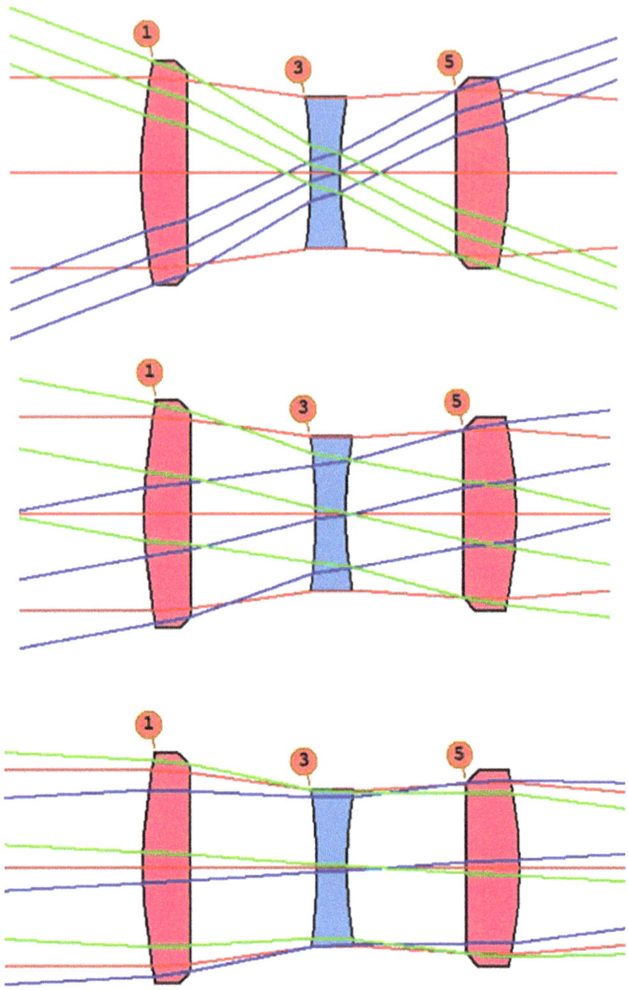

Figure 11.2. The size of the entering beam varies with field, as shown by the PAD Scan button.

CAO data). That is one way to implement a desired amount of vignetting. Note how the size of the entering beam changes with field angle. This is most easily shown with the PAD Scan button  as seen in figure 11.2.

But the WAP 3 option is not the only way, and often is not the best way, to deal with vignetting. During optimization, the size of the beam can change at each surface as the lens changes, and it makes no sense to assign a hard CAO to a surface when you don't even know what that size will be when you are finished. So never use the WAP 3 option during optimization, and use it afterwards only when necessary.

Instead, approach vignetting in stages. To illustrate, let us first remove all of the CAOs and the WAP declaration as well.

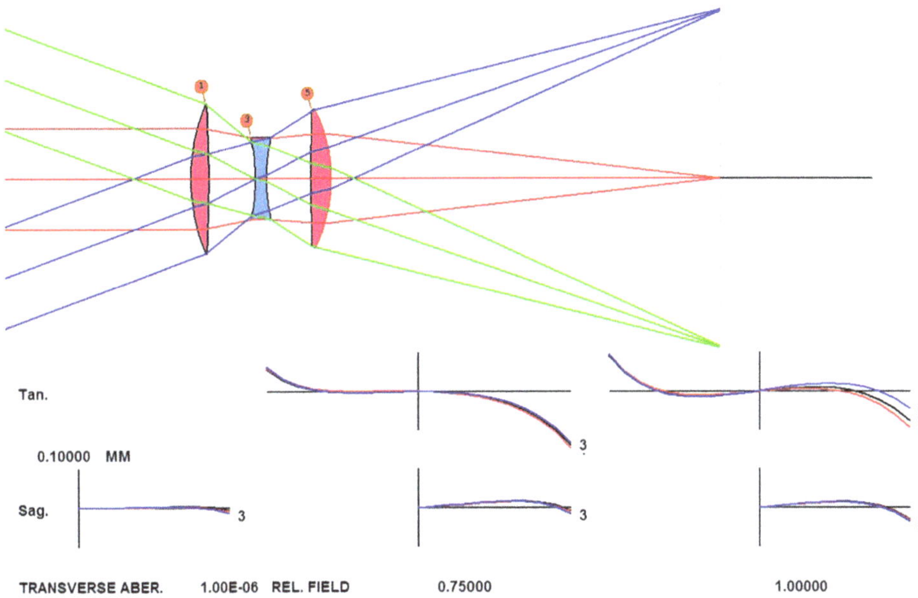

Figure 11.3. Triplet with default apertures and no vignetting, making a poorer image.

```
CHG
CFREE
WAP 0
END
```

Ouch! The image is awful, shown in figure 11.3. Maybe that's why the starting design used those options. Perhaps it made sense under the original conditions. What can we do with this lens?

A look at the current options (with the command **POP**) shows a YMT solve on 6 but no curvature solve. Let us add one. The lens works at F/4.5 paraxially, and the value of a UMC solve is therefore 0.5/4.5 or −.1111, the minus sign because the marginal ray is going down at the image. Change the lens and store a copy in the lens library for future reference.

```
CHG
6 UMC -.1111
END
STORE 3
```

Now open a new editor with the command **AEE** and create an optimization MACro.

```
LOG
PANT
VLIST RAD ALL
VLIST TH ALL
END

AANT
AEC
ACC
GSR .5 10 5 M 0
GNR .5 2 3 M .7
GNR .5 1 3 M 1
END

SNAP
SYNO 30
```

Here we have used merit function 6 from the **Ready Made Raysets** button, .

Make a checkpoint and run this MACro. The result is shown in figure 11.4.

Not too good. The aberrations are out of control, especially at full field. We have to do better. Tell the optimization program to vignette the beam size to 40% of the on-axis value at full field. This is done by adding to the AANT file a **VSET** directive:

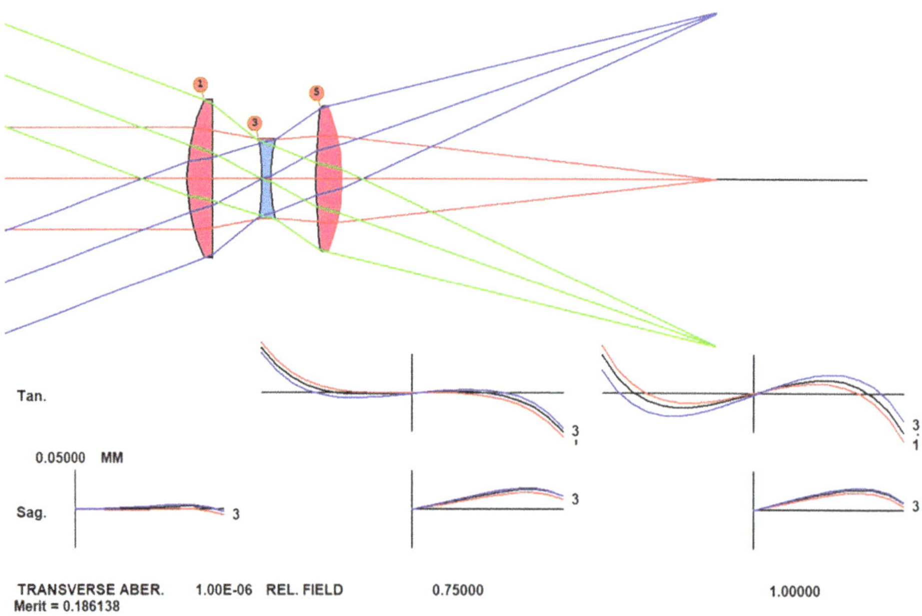

Figure 11.4. Triplet reoptimized to eliminate feathered edges.

```
AANT
AEC
ACC
VSET .4
GSR .5 10 5 M 0
GNR .5 2 3 M .7
GNR .5 1 3 M 1
END
```

Now run it again. The result is in figure 11.5.

The *edges* of the TFAN got worse, which is not surprising, since they are not corrected anymore. But if we were to vignette this beam at 40% of the size shown in the PAD display, the image would look much better. Let's assume this is the solution we are after. Now we have to model the elements so that amount of vignetting actually occurs.

And that's easy to do. Open the WorkSheet (**WS**), enter into the edit pane the directive **CFIX**, and click Update. All surfaces now get a hard CAO assigned, at the same aperture as the default CAO that was currently in effect. Now, click on surface 6 in the lens drawing. The data for that surface are shown in the edit pane. *Select the CAO radius* with the mouse and then click the **SEL** button. That assigns the top slider to that aperture radius. Move the slider thumb to the left, reducing the aperture. Watch the TFAN at full field, and stop when the unvignetted portion on the left side appears to be about at the 40% position, as shown in figure 11.6.

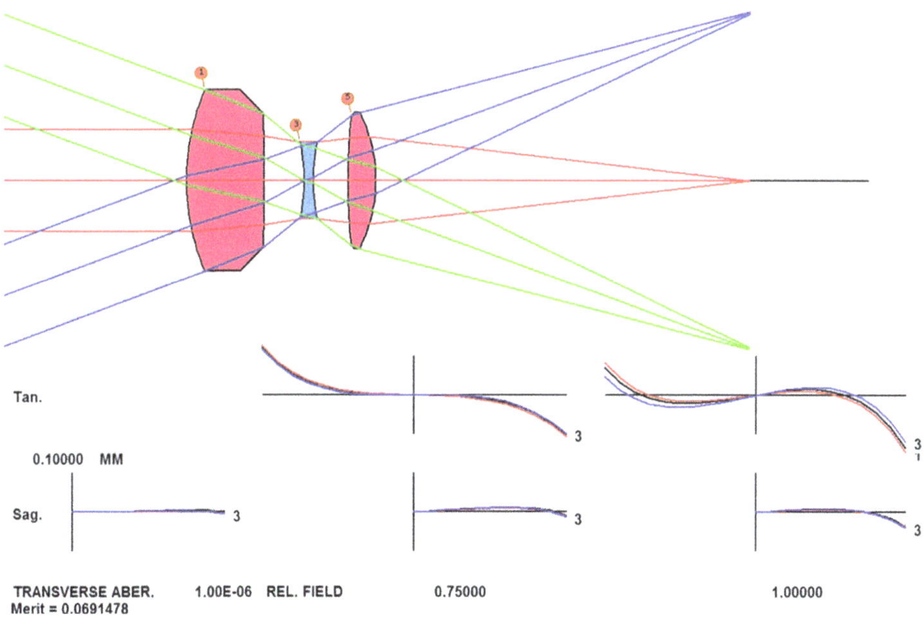

Figure 11.5. Triplet reoptimized with anticipated vignetting to 40% of the aperture.

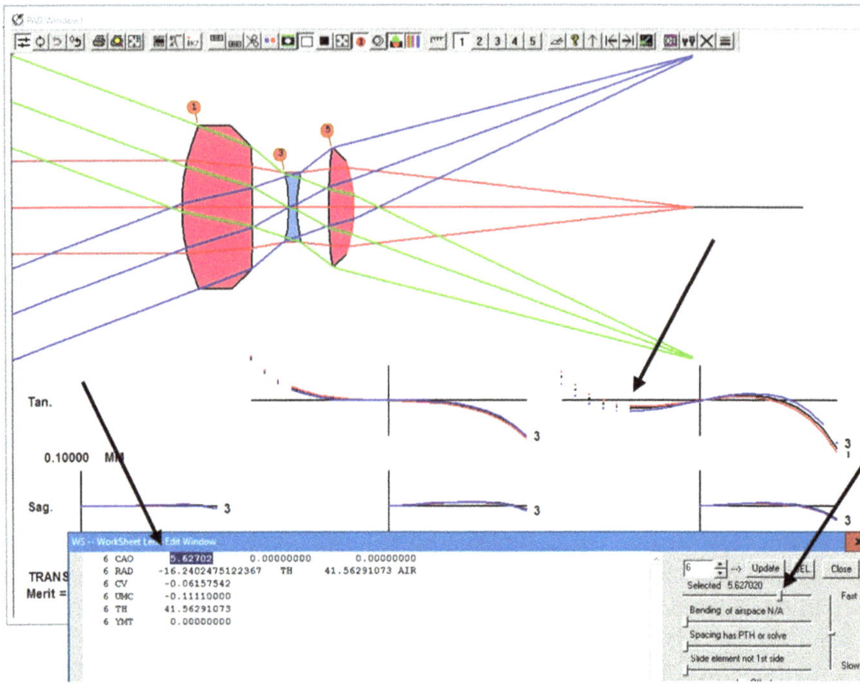

Figure 11.6. Lens with aperture on 6 adjusted to yield the desired vignetting on the left of the TFAN.

This is about where the beam should be vignetted on that side. Do the same at surface 1. Now the beam is vignetted on both sides, shown in figure 11.7.

Why does the drawing on top of the PAD display still show the original, unvignetted beam? Well, there is an option to change that too, which you can activate by turning off mode switch 65. But since that makes the picture look like what would happen with the WAP 3 option turned on—and it is *not* turned on at the moment—we prefer to leave this switch turned on so we don't get confused later.

Okay, we are almost done. We can activate WAP 3 by adding that directive in the edit pane. That's one way to proceed.

But here's another: Declare a set of **VFIELD** parameters. Close WS and type into the CW:

```
FVF 0 .5 .8 .9 1
```

The program calculates the vignetting factors that just clear the apertures we have entered, at five points in the field. Now the display shows the vignetted beam as it should, shown in figure 11.8.

Except the other apertures are still where we left them earlier, because they are all hard apertures. In the WS edit pane, type **CFREE** and click Update. Now the lens has default apertures again, this time calculated according to the VFIELD pupil, as shown in figure 11.9.

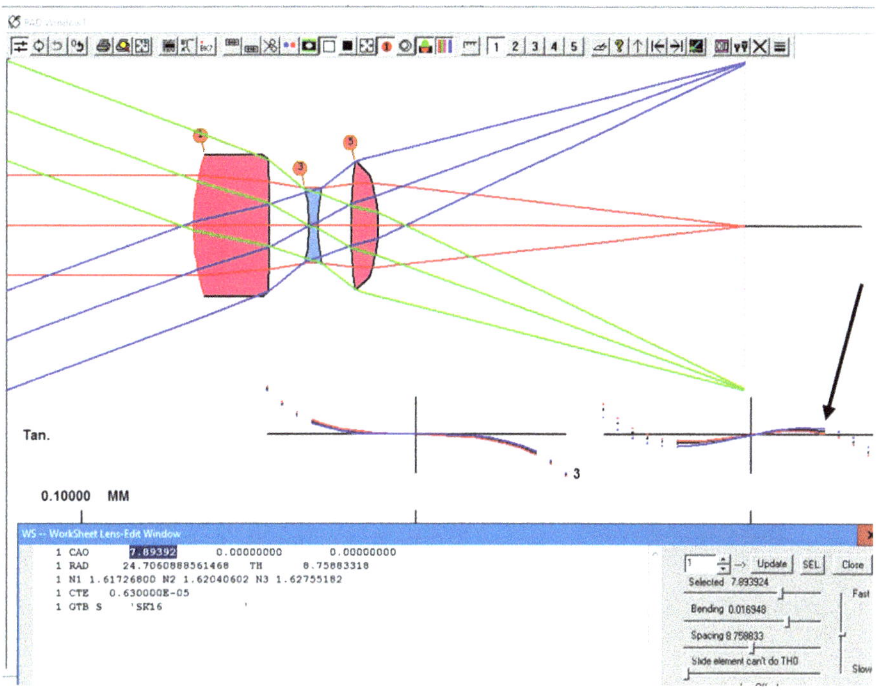

Figure 11.7. Lens with aperture on 1 adjusted for vignetting on the right side of the TFAN.

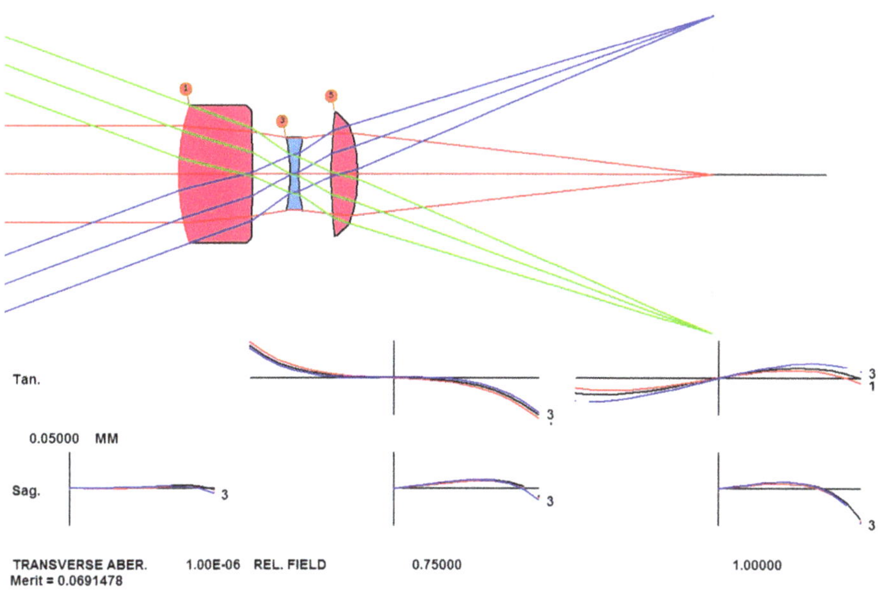

Figure 11.8. Vignetting accounted for via reduced apertures and VFIELD.

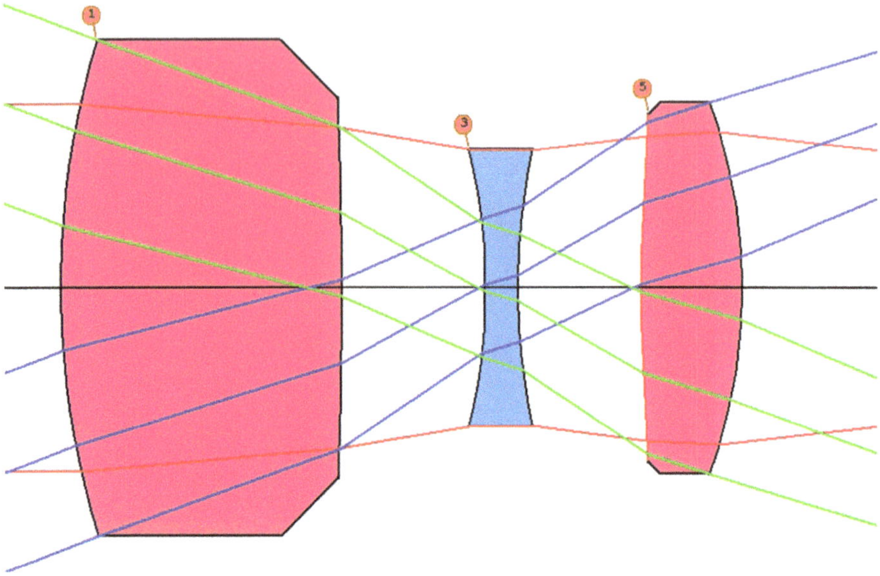

Figure 11.9. Default apertures assigned to honor the vignetting applied with VFIELD.

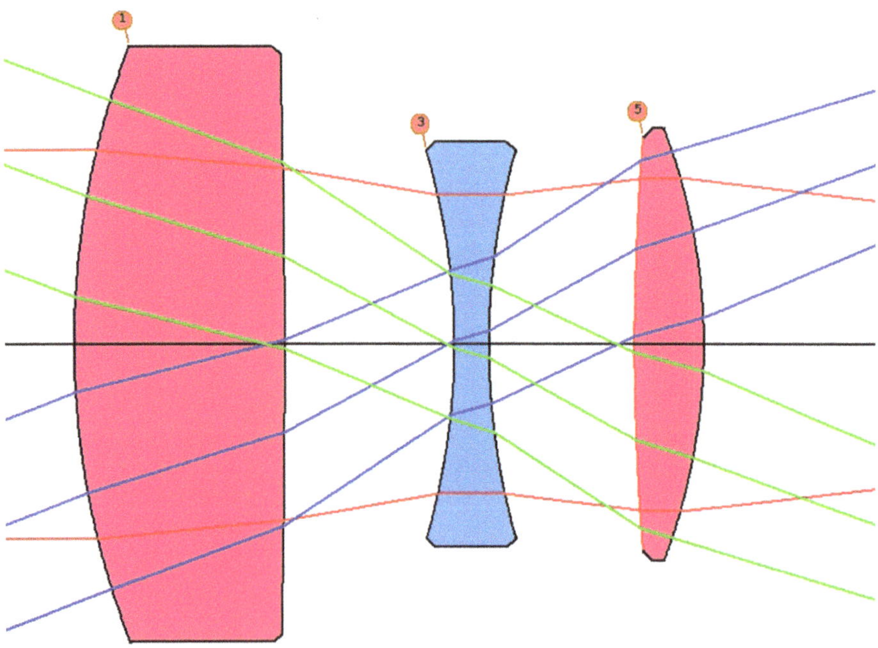

Figure 11.10. Final triplet with vignetting and apertures properly assigned. One must now design the lens cell so that vignetting occurs where the default CAOs were. The elements themselves are larger than this to accommodate the cell and retainer rings, which are usually used with lenses like this one.

You see the pattern. If you now go back to your optimization MACro, remove the VSET directive, and reoptimize, the edge-control monitor will look at the rays in the fans shown above, not the nominal rays. It will therefore reoptimize the lens, reducing the thicknesses if the merit function no longer benefits from the larger values. Then you can adjust the edge geometry with the Edge Wizard (**MEW**) just as you like them, as shown in figure 11.10 (**C11L2**).

What's the difference between the WAP 3 option and VFIELD, since they look much the same? I'm glad you asked. WAP 3 requires aiming five rays every time you do something that requires ray tracing. That is a rather slow option. VFIELD, on the other hand, has already done that calculation and only needs to aim the chief ray afterwards, interpolating at the requested field. That is faster.

And that's what vignetting is all about.

IOP Publishing

Lens Design (Second Edition)
Automatic and quasi-autonomous computational methods and techniques
Donald C Dilworth

Chapter 12

The apochromat

Correcting a lens at three wavelengths

This chapter shows how to design a lens with better color correction than one can get with a simple doublet. A concise description of how one can proceed is given in Rutten & van Venrooij's book *Telescope Optics*. The gist of it is, one must use three different kinds of glass that satisfy certain properties. They may easily be selected by inspecting the glass-table display. To illustrate, we will start with a design using glass types N-SK4, N-KZFS4, and N-BALF10 from the Schott catalog. (Those glasses are sometimes recommended for the purpose.) Here is the starting lens file, shown in figure 12.1 (**C12L1**).

```
RLE
ID F10 APO
 WAVL .6500000 .5500000 .4500000
 APS              3
 UNITS INCH
 OBB  0.00    0.5    2.00000   -0.01194    0.00000    0.00000    2.00000
   0 AIR
   1 RAD   -300.4494760791975   TH      0.58187611
   1 N1 1.60978880 N2 1.61494395 N3 1.62386887
   1 GTB S     'N-SK4                '
   2 RAD     -7.4819193194388   TH      0.31629961 AIR
   2 AIR
   3 RAD     -6.8555018049530   TH      0.26355283
   3 N1 1.60953772 N2 1.61628830 N3 1.62823445
   3 GTB S     'N-KZFS4              '
   4 RAD      5.5272935517214   TH      0.04305983 AIR
   4 AIR
   5 RAD      5.6098999521052   TH      0.53300999
   5 N1 1.66610392 N2 1.67304720 N3 1.68543133
   5 GTB S     'N-BAF10              '
   6 RAD    -27.9819596092866   TH     39.24611007 AIR
   6 AIR
   6 CV      -0.03573731
   6 UMC     -0.05000000
   6 TH      39.24611007
   6 YMT      0.00000000
   7 RAD    -11.2104527948015   TH      0.00000000 AIR
 END
```


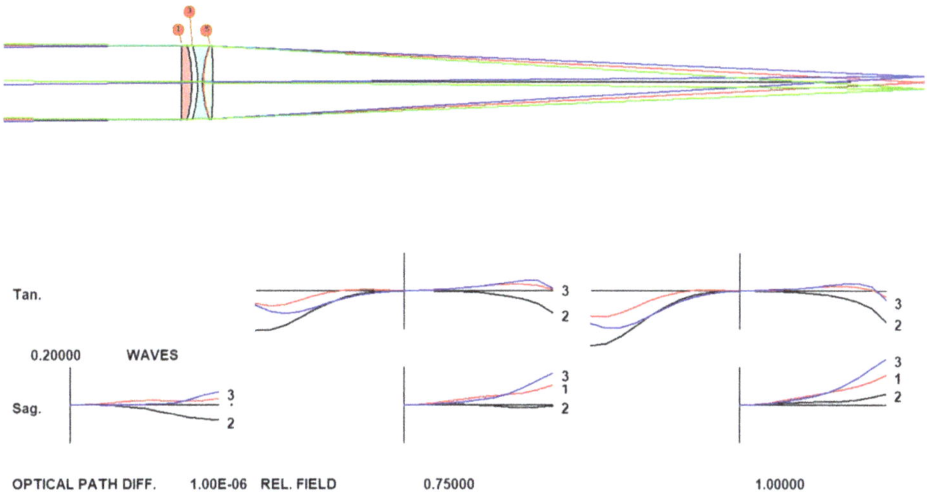

Figure 12.1. Starting point for apochromat design.

Open this file, open PAD, and then click the **GlassTable** button and select the Schott table from the selection that opens. The display is shown in figure 12.2.

This shows the Schott glass map, but it is not what we want for this exercise. Click the Graph button, shown in figure 12.3 and then select Plot P(F,e) vs. Ve, shown in figure 12.4.

The display changes, and now the abscissa is the V-number at the e line (0.546 07 μm) and the ordinate is the quantity $(N_F - N_e)/(N_F - N_C)$.

The theory of the apochromat says that you must select three glasses that *do not* lie on a straight line on this diagram. They must form a triangle, and the larger the area the better. The green circles in figure 12.5 show the present glasses in the triplet. They work quite well—but we can do somewhat better.

Click on the green circle with the number '1' beside it. That is the glass currently on surface 1, N-SK4. Now click the **Properties** button, to see the properties of this glass, shown in figure 12.6.

Hmmm … this glass is not all that stable: humidity rating of 3 and acid sensitivity of 5. Let's see if we can find a better glass for the first element. (This is exposed to the environment, so it's important.) Close the properties window and click the Graph button again; then click the radio button for Acid Sensitivity and OK. Zoom in with the mouse wheel as needed near the green circle at N-SK4 so things get bigger, and then click the Full Name button. You see figure 12.7.

Now you see a red vertical line through the glass locations, showing the acid sensitivity. The line for glass N-SK4 is rather long, since that glass is not very resistant. To the left you see N-BAK2, with no line at all (it's in the best category). Click that glass symbol, and when the name appears in the window on the right, as shown in figure 12.8 click the Properties button again. Aha! This glass has an acid rating of 1, better humidity tolerance, and a lower price as well. There's no reason we can't use it instead of the previous N-SK4, since it makes just as good a triangle with the other glasses. Type the surface number 1 into the Surface box and click \Apply/. Glass N-BAK2 is now assigned to surface 1.

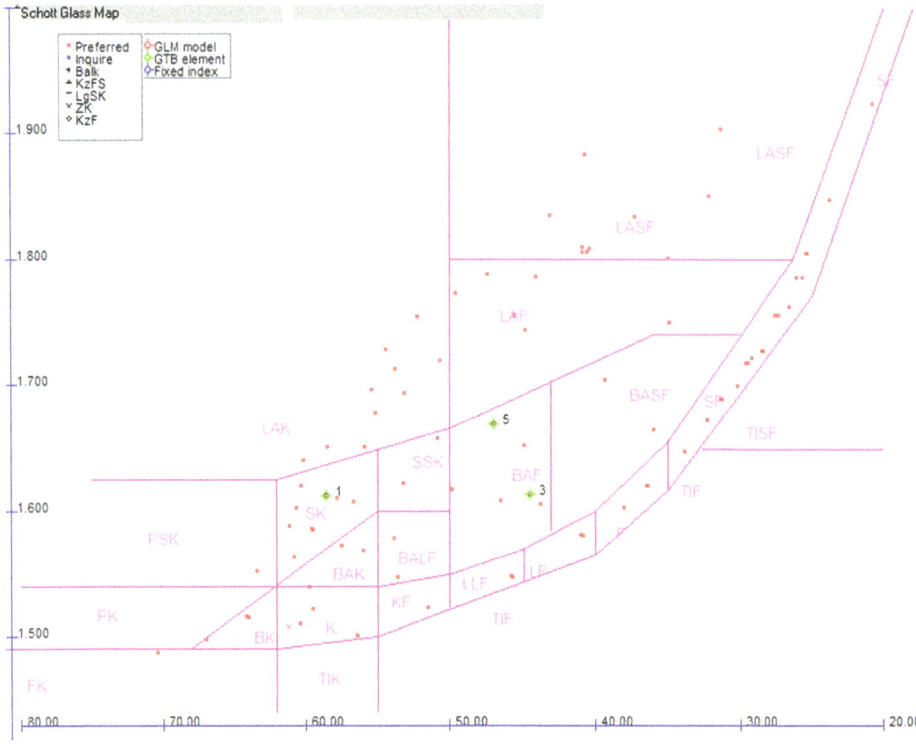

Figure 12.2. The glass map display, showing Nd and Vd.

Now clean up the display by deleting the names (click Spots Only), and then click Graph and select No Graph and OK. The triangle is just as nice as before.

Of course, the lens is not optimized for this glass, so we have to run the optimization program. Here is a MACro that will do the job (**C12M1**).

```
PANT
VLIST RAD 1 2 3 4 5 7
VLIST TH 2 4
END

AANT
AEC
ACC
GSO 0 1 4 M 0 0
GNO 0 .2 3 M .75 0
GNO 0 .1 3 M 1.0 0
END

SNAP
SYNO 30
```

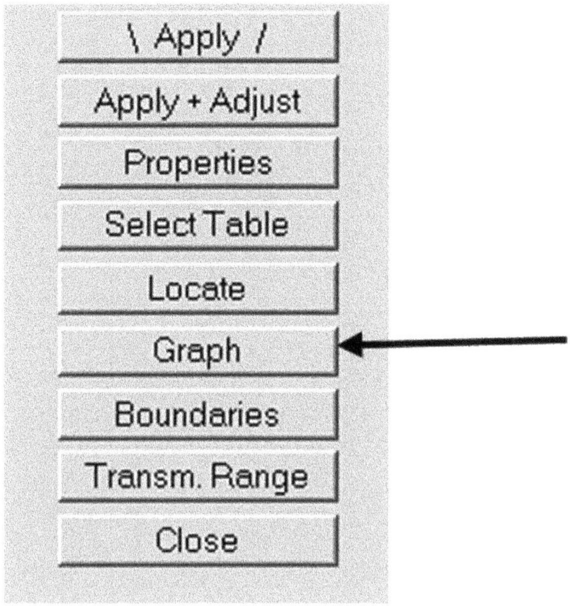

Figure 12.3. Button to show a selected graph in the glass map.

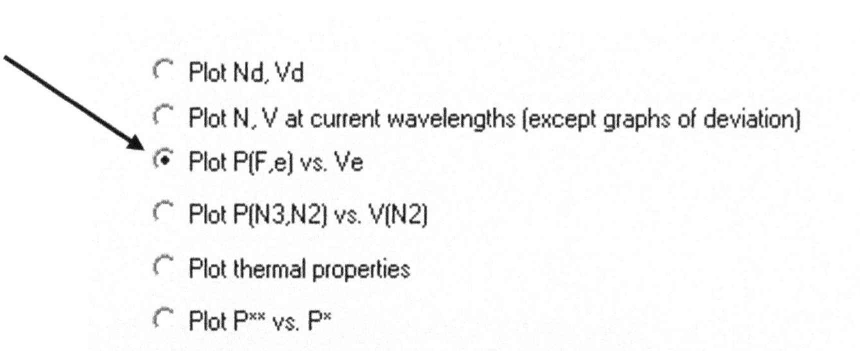

Figure 12.4. Selecting a graph of partial dispersions.

Run this MACro, and now the correction is better than ¼ wave on axis. We have a better design that is cheaper to make, more resistant to the elements, and corrected over the range 0.45–0.65 μm. Here is the RLE file for that design, shown in figure 12.9 (**C12L2**).

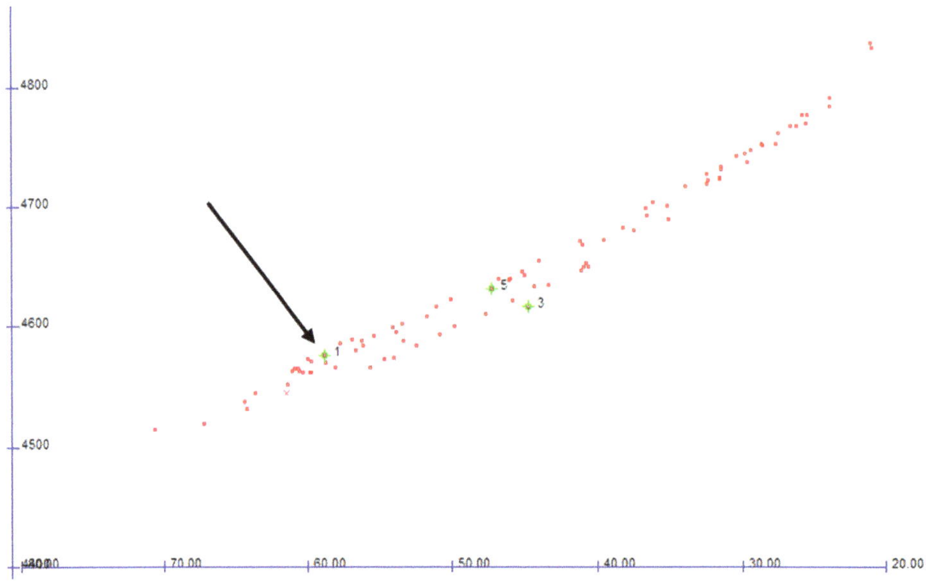

Figure 12.5. The glass map showing a plot of P(F,e) versus Ve.

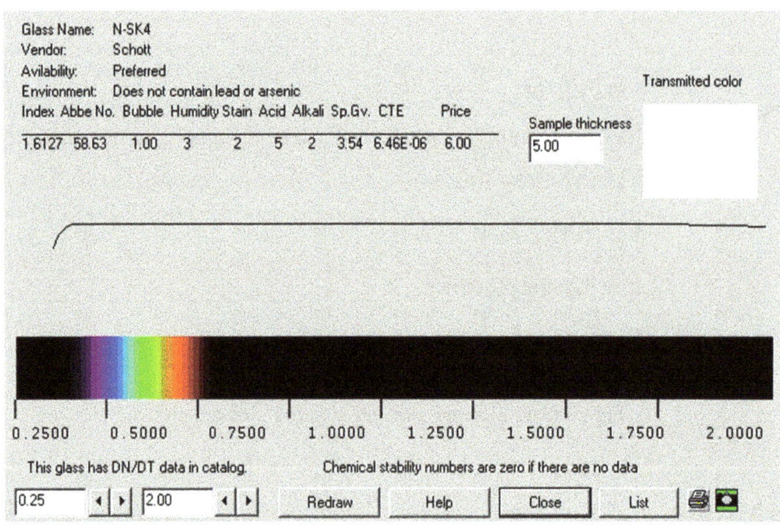

Figure 12.6. Properties of glass N-SK4.

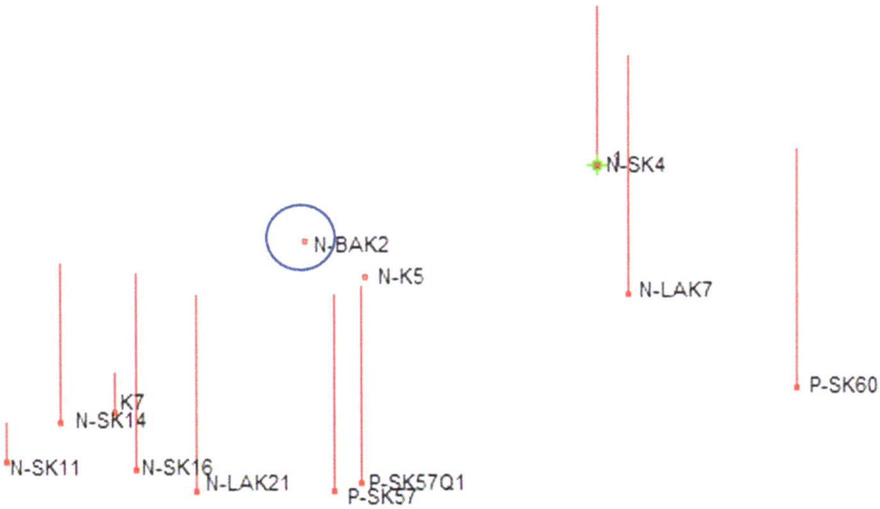

Figure 12.7. Glass map showing acid sensitivity. Glass N-BAK2 is a better choice than N-SK4.

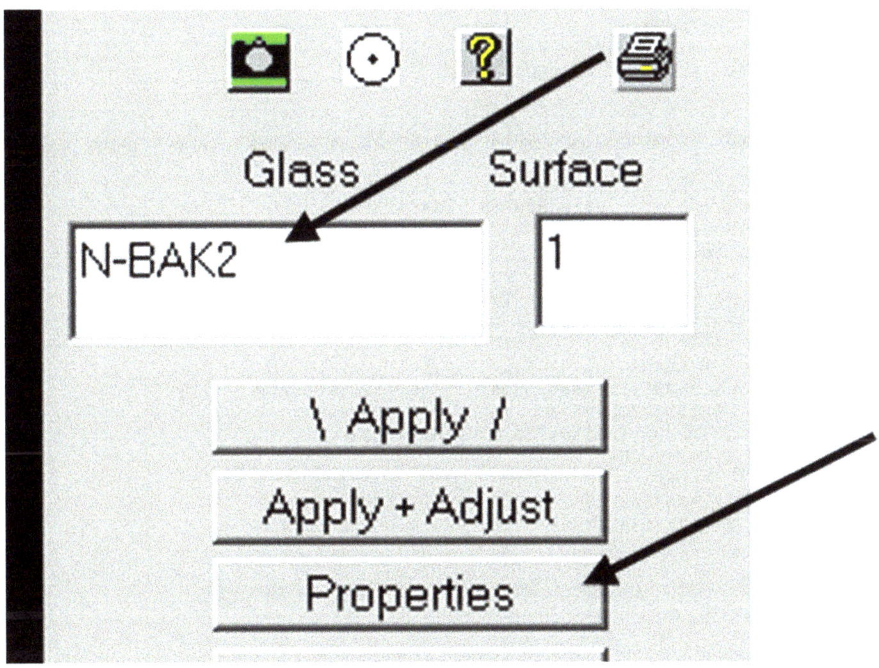

Figure 12.8. Applying a selected glass to surface 1.

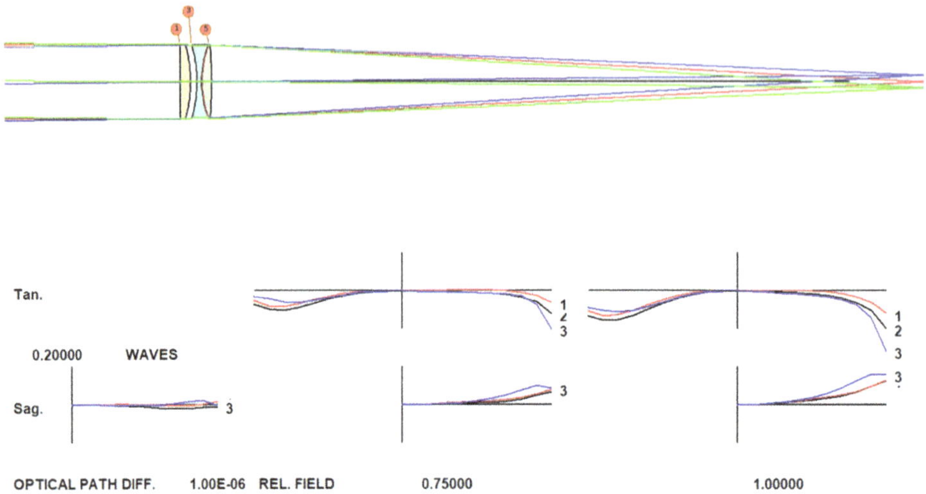

Figure 12.9. Lens reoptimized with improved glass selection

```
RLE
ID F10 APO                                          55567
 FNAME 'C12L2.RLE
 MERIT    0.637627E-02
 LOG    55567
 WAVL .6500000 .5500000 .4500000
 APS                 3
 UNITS INCH
 OBB  0.000000  0.5000000  2.0000000 -0.00652  0.0000000  0.0000000   2.0000000
   0 AIR
   1 RAD    -169.2935630392332   TH      0.58187611
   1 N1 1.53742490 N2 1.54188880 N3 1.54960358
   1 CTE    0.800000E-05
   1 GTB S     'N-BAK2          '
   2 RAD      -7.0670416441280   TH      0.36131941 AIR
   3 RAD      -6.5556994058248   TH      0.26355283
   3 N1 1.60953772 N2 1.61628830 N3 1.62823445
   3 CTE    0.730000E-05
   3 GTB S     'N-KZFS4         '
   4 RAD       5.3108511182960   TH      0.03937000 AIR
   5 RAD       5.4054270177678   TH      0.53300999
   5 N1 1.66610392 N2 1.67304720 N3 1.68543133
   5 CTE    0.618000E-05
   5 GTB S     'N-BAF10         '
   6 RAD     -19.4383633248526   TH     39.42616962 AIR
   7 RAD     -11.1847134357355   TH      0.00000000 AIR
 END
```

Let’s see how the defocus varies with color in the new design.

```
CHG
NOP
END
PLOT DELF FOR WAVL = .45 TO .65
```

The result is in figure 12.10.

This analysis shows a defocus of about 0.0026 inches over the design range, and a perfect Airy disk, shown in figure 12.11. (The latter was calculated by Image Tools (**MIT**) with 10 wavelengths assigned to the lens to make a good white in the center, and with the coherent effect.) The defocus is not zero because the program has balanced a small shift against the change of spherical aberration with wavelength. Both are small. How did we get 10 wavelengths? With the Spectrum Wizard (**MSW**).

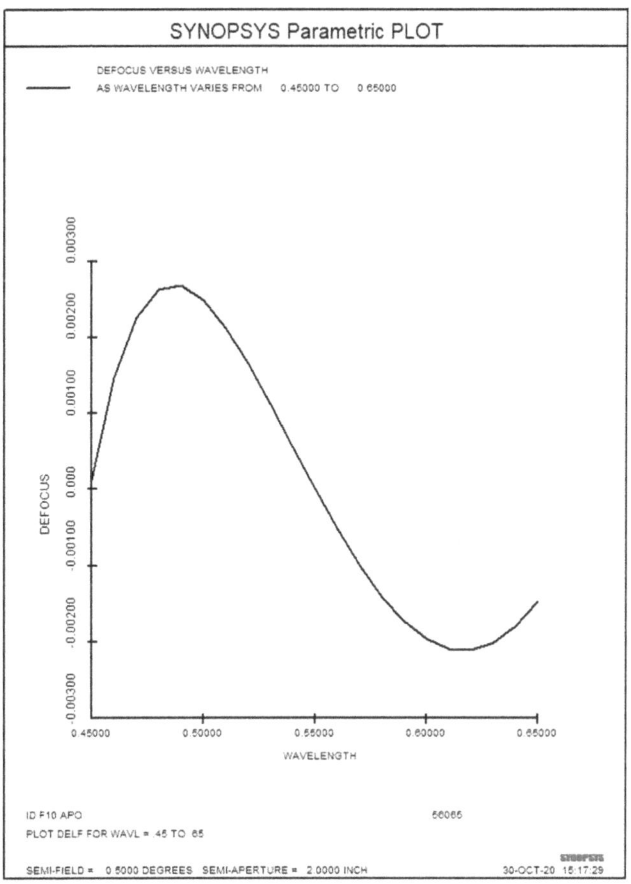

Figure 12.10. Color correction curve for the new design.

Figure 12.11. Diffraction image calculated by the Image Tools (MIT).

Note how we plotted **DELF** in this case instead of BACK, as we did previously. Why? Well, we eliminated all solves with the NOP directive, so the back focus is now fixed. But the paraxial defocus, DELF, now changes with wavelength.

Chapter 13

Tolerancing the apochromatic objective

Tolerance budget calculation; fabrication adjustments

In chapter 12 you designed an apochromatic telescope objective with an extremely good image. In this chapter, you will calculate a tolerance budget for that lens.

Before you can send drawings of lens elements to the shop, you have to determine the largest manufacturing errors you can tolerate and still achieve adequate performance. The result is called a *tolerance budget*. Since this example is a very well-corrected astronomical objective—the fussiest lenses you are ever likely to meet unless you design microlithography objectives—the budget will probably be very tight. That is exactly what we need for this exercise: most lenses are very easy to tolerance with SYNOPSYS, but that would not be an interesting lesson. Use lens **C12L2** from the previous lesson.

Before you can get tolerances, however, you have to remove the curvature solve on surface 6.

```
CHG
6 NCOP
END
```

The on-axis image is the most important for this kind of objective, which is often used for planetary astronomy, and there is some field curvature (which is why we varied the image surface radius in the previous lesson) and some astigmatism, which you cannot correct with a compact lens group at the stop.

The principle behind a tolerance budget is that, if you make all the elements according to that budget, holding radii, thicknesses, wedge, and so on to the tolerances assigned to those parameters, and then assemble the lens while holding all airspaces, and element tilts and decenters to their tolerances, the lens will then yield the requested image quality to the requested statistical confidence level. If that level is, say one sigma, then the quality will be equal or better than requested for 84.27% of a large batch of lenses. If the criterion is two sigma, that confidence rises to 99.53%, and so on.

First, try a simple **BTOL** evaluation. BTOL has many options, but there is a menu for simple cases: **MSB**, for **M**enu, **S**imple **B**TOL. Type **MSB** in the Command Window, and then fill in the boxes as follows: (Most of them are already filled in for you; but select the TOLERANCE and WAVE radio buttons (instead of DEGRADE SPOT), and click the Prepare MC box to select that option. Everything else can be left as it is, as shown in figure 13.1. Then click the **GO** button. (The logic behind BTOL is explained in appendix C.)

When the calculations are finished, look up a little from the bottom line in the CW and you will see the expected performance:

```
SUMMARY OF OPTICAL PERFORMANCE
   REL. Y-HEIGHT  REL. X-HEIGHT              ANTICIPATED STATISTICS
                                             OF QUALITY DESCRIPTOR
        (HBAR)          (GBAR)       MEAN VALUE   MULTIPLE DEV      EXPECTED ZOOM
   ______________________________________________________________________________
         0.000           0.000        0.03472       0.01563         0.05035    1
         1.000           0.000        0.03940       0.00858         0.04798    1
         0.500           0.000        0.03520       0.01482         0.05002    1
```

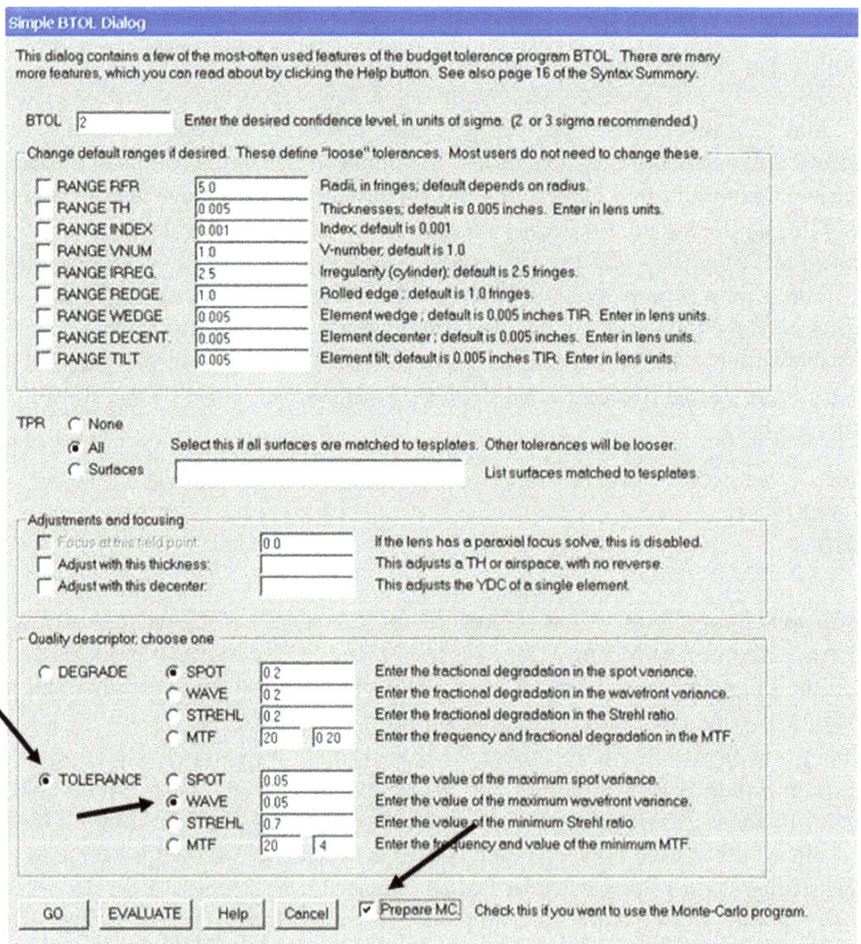

Figure 13.1. Input selection for BTOL via MSB.

This says that the on-axis image will get a variance of 0.050, as specified, and the edge of the field is slightly better. What does the budget look like? Scroll the display up a bit more until you come to the budget results:

BUDGET TOLERANCE ANALYSIS

EL.	SURF	RADIUS	RADIUS TOLERANCE (RADIUS)	(FRINGES)	THICKNESS	-----B----- THICKNESS TOL
1	1	-169.29356	0.85585	4.82632	0.58188	0.00500
1	2	-7.06704	0.00091	2.95742	0.36132	0.00157
2	3	-6.55570	7.71235E-04	2.72720	0.26355	0.00487
2	4	5.31085	2.93360E-04	1.62121	0.03937	4.25336E-04
3	5	5.40543	2.91133E-04	1.57239	0.53301	0.00496
3	6	-19.43836	0.01035	4.32620	39.42617	0.00000
	7	-11.18471	0.00000	0.00000	0.00000	0.00000

ELE	SURF	GLASS NAME	BASE INDEX	INDEX TOL	V-NUMBER	V-NUMBER TOL
1	1	N-BAK2	1.53996d	4.28522E-04	59.70771d	0.15531
2	3	N-KZFS4	1.61336d	1.68886E-04	44.49298d	0.05357
3	5	N-BAF10	1.67003d	1.92707E-04	47.11137d	0.06828

Note: The symbol "d" indicates that the quantity is estimated at 0.58756 uM.
The symbol "F" indicates that the quantity is taken at the primary color.

ELE	SURF	WEDGE TOLERANCE (ARC MIN)	(TIR)	IRREG. TOL (FRINGES)	ROLLED EDGE TOL (FRINGES)
1	1	0.00000	0.00000	0.37528	0.23092
1	2	0.44078	0.00051	0.37250	0.22642
2	3	0.00000	0.00000	0.34578	0.20624
2	4	0.23309	0.00027	0.34266	0.20094
3	5	0.00000	0.00000	0.32697	0.19086
3	6	0.51763	0.00060	0.33254	0.19868
	7	16.68593	0.00339	0.00000	0.00000

ELE	SURF	ELEMENT TILT TOLERANCE (ARC MIN)	(TIR)	Y-DECENT TOL	X-DECENT TOL
1	1	0.53379	0.00062	0.00066	0.00000
1	2	0.00000	0.00000	0.00000	0.00000
2	3	0.30556	0.00035	0.00024	0.00000
2	4	0.00000	0.00000	0.00000	0.00000
3	5	0.24560	0.00028	0.00027	0.00000
3	6	0.00000	0.00000	0.00000	0.00000
	7	0.00000	0.00000	0.00000	0.00000

This is scary. The lens has a 0.0016-inch airspace tolerance between elements 1 and 2, and 0.000 43 between 2 and 3. The V-number tolerance is 0.053 on the center element, and you need to hold centration of 0.000 24 on that element. *Nobody* can make this lens by just following a budget.

We have to loosen these tolerances, but how? Well, one reason tolerances can be tight is because the aberrations of individual elements are large. While third-order aberrations are no longer as useful for the lens designer as they were a generation ago, this is a case where they actually can make a difference, as we saw in chapter 10—but we will not control them directly. Type the command **THIRD SENS**.

```
SYNOPSYS AI>THIRD SENS

 ID F10 APO

 NORMALIZED 3RD-ORDER ANALYSIS OF TOLERANCE SENSITIVITY

 SS OF SA3 BY SURFACE (SAT) =                   8.376289
 SS OF CO3 BY SURFACE (COT) =                   0.018302
 SS OF CO3/YDC BY SURFACE (ACD) =               0.133163
 SS OF CO3/TILT BY SURFACE (ACT) =              4.161976
 SS OF CO3/YDC BY ELEMENT (ECD)  =              0.038178
 SS OF CO3/TILT BY ELEMENT (ECT) =              1.185957
 SS OF SA3 BY ELEMENT (ESA) =                   0.043012
 SS OF CO3 BY ELEMENT (ECO) =                   0.000094
```

This listing shows the sum of the squares of the individual surface contributions to various aberrations and their derivatives. The idea is that if those are large, even if they are compensated by other surface contributions, the system becomes sensitive to small errors since the cancellation will then not be as good. The value of SAT, the spherical aberration sum, is 8.376. Let us modify the merit function to bring that sum down. Here is the new MACro (**C13M1**).

```
PANT
VLIST RAD 1 2 3 4 5 7
VLIST TH 2 4
END
AANT
AEC
ACC
M 4 1 A SAT
GSO 0 1 5 M 0 0
GNO 0 .2 4 M .75 0
GNO 0 .1 4 M 1.0 0
END
SNAP
SYNO 30
```

Here we ask for a value of 4 for the quantity SAT, which was previously over 8. After running this, the lens (**C13L1**) is slightly changed, and SAT now has a value of 4, as requested. Now prepare a new BTOL run: (**C13M2**)

```
CHG
NOP
END
BTOL 2

EXACT INDEX 1 3 5
EXACT VNO 1 3 5
```

```
TPR ALL
TOL WAVE 0.1
ADJUST 6 TH 100 100

PREPARE MC

GO
STORE 4
```

This increases the wavefront variance tolerance to 0.1 and specifies an adjustment of thickness 6. (The first BTOL run utilized the *paraxial* thickness solve on 6, but sometimes the tolerances are looser if you let the program depart slightly. The adjustment will take care of that. The **NOP** directive removes all paraxial solves, so this thickness will be free to vary.) Also, specify that the index and Abbe number of the three glasses are exact, which removes them from the budget. In systems as fussy as this, one always requests melt data from the glass vendor, which give the measured index, and then one tweaks up the design with those values, entered as melt data. So errors in those values do not have to be part of the budget anymore. (Look up MELT in the help file to see how those data are entered.)

Run this, and the tolerances are somewhat looser.

BUDGET TOLERANCE ANALYSIS -----B-----

EL.	SURF	RADIUS	RADIUS TOLERANCE (RADIUS)	RADIUS TOLERANCE (FRINGES)	THICKNESS	THICKNESS TOL
1	1	-169.29356	0.87776	4.94985	0.58188	0.00500
1	2	-7.06704	0.00141	4.56750	0.36132	0.00332
2	3	-6.55570	0.00127	4.49205	0.26355	0.00499
2	4	5.31085	5.77901E-04	3.19368	0.03937	8.65592E-04
3	5	5.40543	5.78255E-04	3.12311	0.53301	0.00499
3	6	-19.43836	0.01177	4.91933	39.42617	0.00000
	7	-11.18471	0.00000	0.00000	0.00000	0.00000

ELE	SURF	GLASS NAME	BASE INDEX	INDEX TOL	V-NUMBER	V-NUMBER TOL

Note: The symbol "d" indicates that the quantity is estimated at 0.58756 uM.
The symbol "F" indicates that the quantity is taken at the primary color.

ELE	SURF	WEDGE TOLERANCE (ARC MIN)	WEDGE TOLERANCE (TIR)	IRREG. TOL (FRINGES)	ROLLED EDGE TOL (FRINGES)
1	1	0.00000	0.00000	0.75817	0.26098
1	2	0.72708	0.00085	0.75025	0.25606
2	3	0.00000	0.00000	0.69583	0.23444
2	4	0.38476	0.00044	0.68283	0.22883
3	5	0.00000	0.00000	0.65092	0.21749
3	6	0.84925	0.00098	0.66899	0.22703
	7	22.01870	0.00447	0.00000	0.00000

ELE	SURF	ELEMENT TILT TOLERANCE (ARC MIN)	ELEMENT TILT TOLERANCE (TIR)	Y-DECENT TOL	X-DECENT TOL
1	1	0.86200	0.00101	0.00108	0.00000
1	2	0.00000	0.00000	0.00000	0.00000
2	3	0.51283	0.00058	0.00039	0.00000
2	4	0.00000	0.00000	0.00000	0.00000
3	5	0.40480	0.00047	0.00045	0.00000
3	6	0.00000	0.00000	0.00000	0.00000
	7	0.00000	0.00000	0.00000	0.00000

Now the lens will get a variance of 0.1 over most of the field, at the two-sigma level. Is that too large? To find out, run the Monte-Carlo routine to see what the as-made lenses will look like. The starting lens is now in the library at location 4, and we want a worst-case example to be placed into location 5.

Type in the CW:

```
MC 50 4 QUIET -1 ALL 5.
```

This will run a batch of 50 lenses, made according to the budget above, compile statistics for the batch, and then save the worst-case example. If you do not know the arguments of the command, just type the letters **MC** and look at the tray. The format of the command is displayed for you, as shown in figure 13.2, and if you need more information, just press the F2 key while the command is shown in the tray to open the Help File at that topic (or type **HELP MC**). You have to run BTOL before MC will work, since it uses the budget from BTOL.

MC runs its 50 cases and prints some statistics. You can look at a histogram of the results by typing **MC PLOT** after the run is finished. The on-axis image now has a variance of 0.1 at a two-sigma confidence.

But this looks too easy. We need to examine the worst-case example. Switch to ACON 2 (type **ACON 2** or click the button 2) and type **GET 5**. This is where MC put that example. Now look at the PAD display, shown in figure 13.3. (Your results will be somewhat different; the Monte-Carlo analysis has to employ different random changes each iteration, ignoring switch 98, which you turned on earlier.)

Select the OPD Fan Plots option for the bottom display, and you see that the lens has more than a quarter-wave of aberration at the on-axis image. We told you this would not be an easy problem! And this budget is still very tight, with centration tolerances of less than a micrometer.

Anyone with experience with astronomical optics could have predicted all of this, of course. One can rarely just build such a lens according to the print and expect it to work. We have a great budget that nobody can hold, and a slightly more practical budget that makes a bad lens. What's next?

13.1 Fabrication adjustment

Clearly, we need some fabrication adjustments. In this scenario, one makes an element, measures it, puts those data into the lens prescription, and reoptimizes, varying the other elements. Then one makes another element to the new design, measures and adjusts all over again, continuing in this way until everything is made. At assembly, one then adjusts centration and possibly tilts to get the best possible image. Go back to ACON 1 again and store the lens in library location 4. We will now analyze the

Figure 13.2. The Trayprompt display for MC.

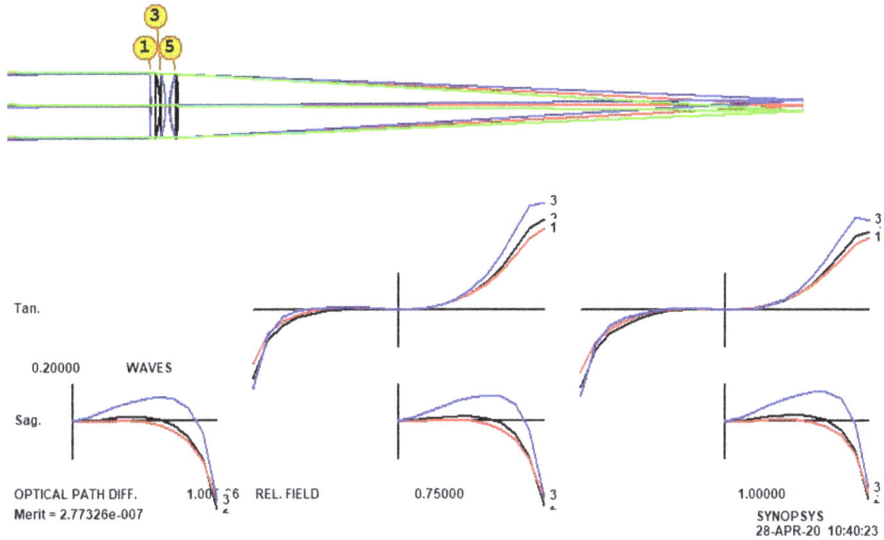

Figure 13.3. Worst-case example from MC. Fabrication adjustment is needed.

statistics with FAMC. (This analysis will use the BTOL budget you prepared above, with a tolerance of 0.1 on the wavefront.)

You really should read about this topic before you use it. Type **HELP FAMC.** (FAMC is Fab Adjust MC.) Here is our MACro (**C13M3**):

```
FAMC 50 4 QUIET -1 ALL 5
PASSES 20
FAORDER 5 3 1

PHASE 1
PANT
VLIST RAD 1 2 3 4 5 6
VLIST TH 2 4 6
END

AANT
GSO 0 1 5 M 0
GNO 0 1 5 M 1
END

SNAP
EVAL

PHASE 2
PANT
VY 3 YDC 2 100 -100
```

```
VY 3 XDC 2 100 -100
VY 5 YDC 2 100 -100
VY 5 XDC 2 100 -100
VY 6 TH
END
AANT
GNO 0 1 4 M 0 0 0 F
GNO 0 1 4 M 1 0 0 F
END
SNAP
SYNO 30

PHASE 3
```

Here is what is going on in this MACro:

1. Request **FAMC**, with arguments the same as for the MC analysis that was run above.
2. In Phase 1, the program will alter the elements in the order given in the **FAORDER** line, changing the parameters by a random amount within the BTOL budget. This simulates making the hardest element first, and so on. It will optimize the lens with the variables and merit function listed in the **PHASE 1** portion, as each element is manufactured, deleting those variables that apply to elements that are already finished.
3. When the simulated elements are all manufactured, it simulates mounting in the cell according to the tilt and decenter tolerances. Then it optimizes once again, varying the decenter of elements 2 and 3 in both X and Y (errors are simulated in both directions, so the compensation must be as well) in accordance with the **PHASE 2** parameters. It also varies thickness 6 once again, since a large centration change produces a small focus shift as well. The merit function also corrects rays over both halves of the pupil (via the '**F**' in the GNO lines) because there is no bilateral symmetry anymore once the errors are simulated.

Run this MACro and look at the worst case again, shown in figure 13.4.

This lens has wavefront errors of just over a quarter wave on axis—but remember this is the worst case. Most of the 50 in this batch are much better. Type MC PLOT, and look at the histogram that describes the on-axis image, which is on the lower left of the plot, shown in figure 13.5

The worst-case example is that lonely box way out at the end, and it is much worse than the others. There is always a small chance of getting a lemon, after all, and this is it. Nonetheless, there is a high probability that the lens will work just fine if built according to this rather tight budget.

Note that the budget itself was not altered, or recalculated, when we decided to use FAMC. What we did was to employ a budget that did not work very well and made it work much better. As a fringe benefit, we don't have to worry about the

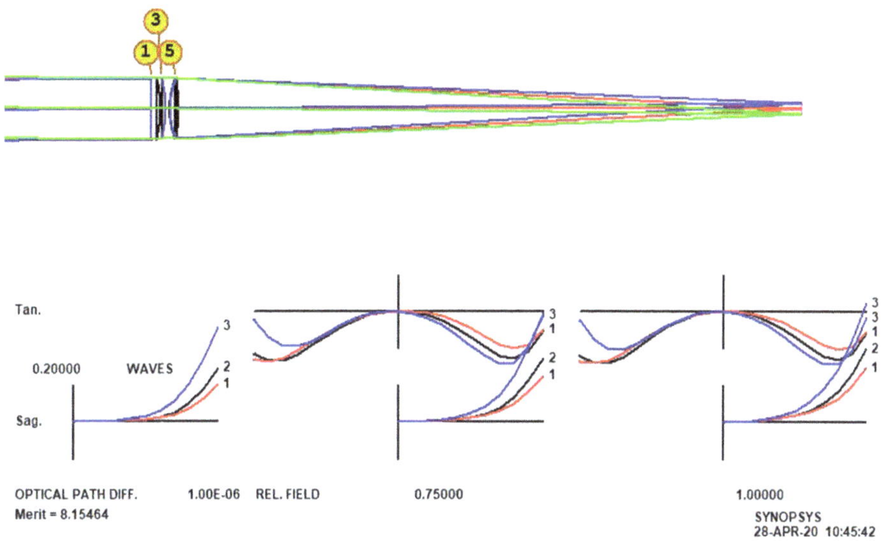

Figure 13.4. Worst-case example with fabrication adjustments.

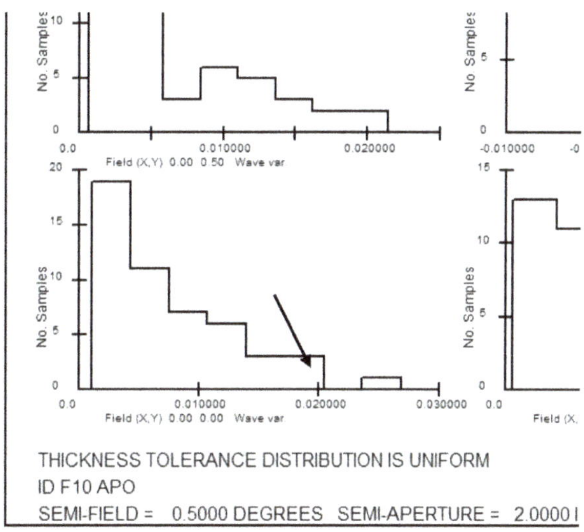

Figure 13.5. Statistics of the MC analysis with fabrication adjustments.

tight centration errors anymore—which are expensive to hold—since the elements are to be adjusted at assembly anyway. So things have gotten much easier.

But there is a price to pay: you have to get melt data when the glass is delivered, tweak up the design with those data, the shop must make the elements in the order given, measure them carefully, and then send those data back to the designer, who will reoptimize everything at each step. And provision must be made for adjusting the centration of elements 2 and 3 on the test bench, and then locking everything down once the image is tuned up. But that's what precision optics is all about.

And about those adjustments: It's easier to adjust the centration of only one element, not two as in this example. What happens if you delete the adjustments on, say surface 3 and redo the whole procedure? If it works, the assembly technician will thank you. Try it and find out!

13.2 Transferring tolerances to element drawings

Now that you have a tolerance budget, you can prepare lens drawings to send to the shop. That can be done automatically too. Open the **MPL** dialog and enter data as shown in figure 13.6.

The **USE BTOL** checkbox tells the program to pick up the tolerances from the BTOL budget and add them to the element drawing, as shown in figure 13.7. (For this to work, the current configuration must be the same one that was active when the BTOL analysis was run.)

These tolerances are all added as *annotations*, rather than as plain text, which means you can edit them with the Annotation editor and change any values if you need to. And if you select the first **DWG** button, the program creates an assembly drawing, also with tolerances added. Figure 13.8 shows an example, where we entered a scale factor of –0.8. (The minus sign tells the program to draw just the

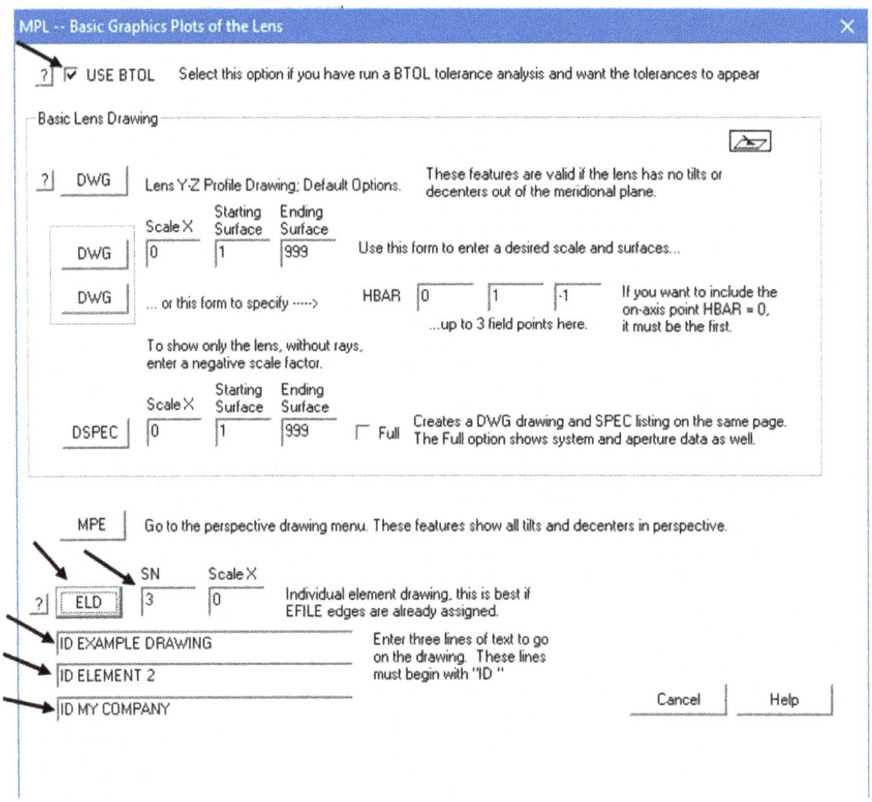

Figure 13.6. The MPL dialog for preparing input for an element drawing.

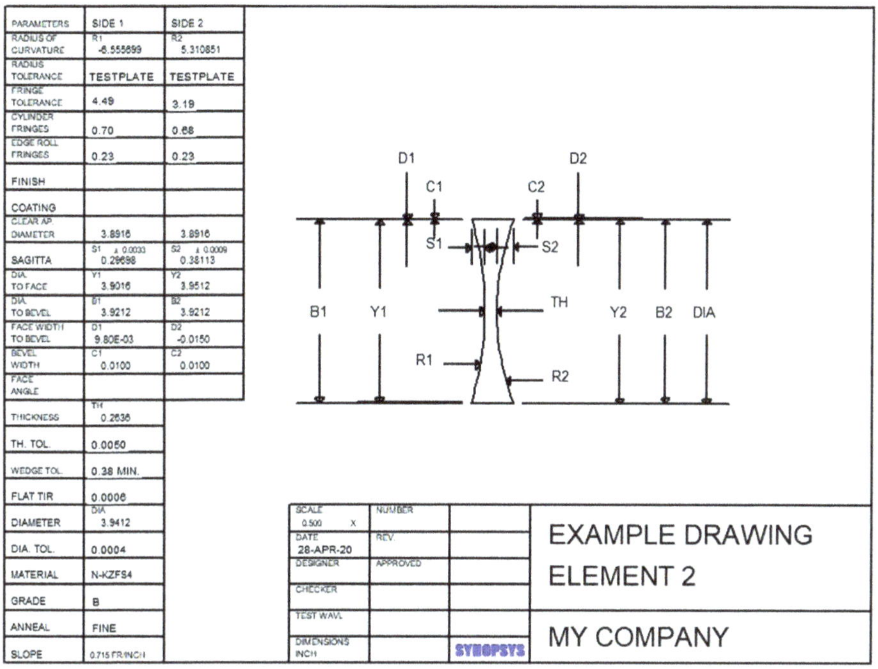

Figure 13.7. Element drawing, annotated with BTOL tolerances.

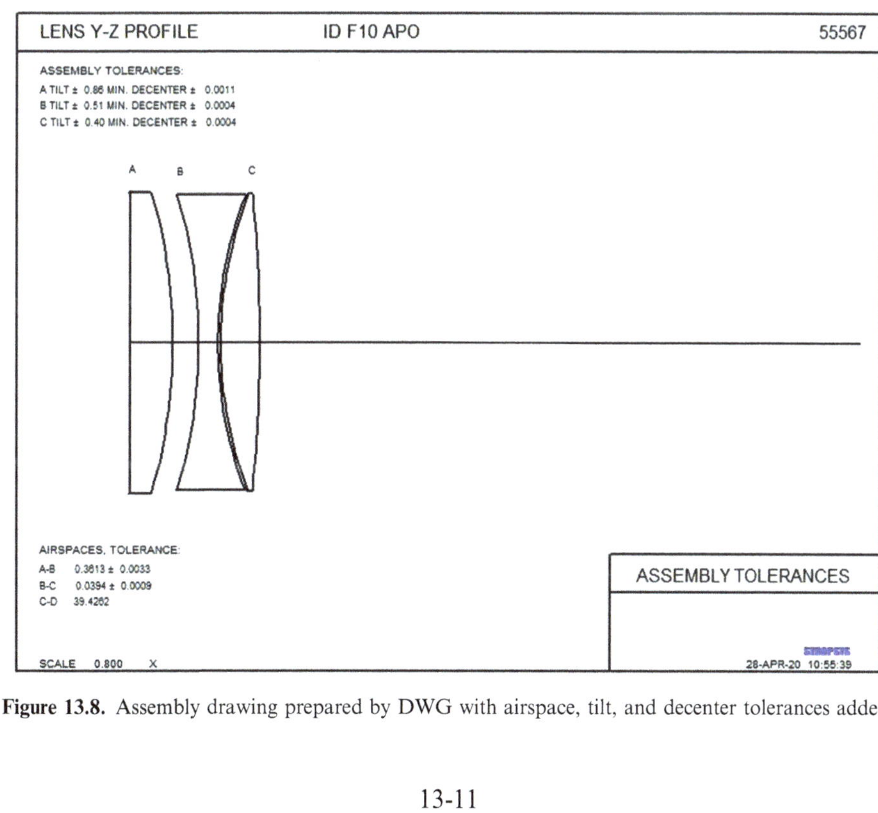

Figure 13.8. Assembly drawing prepared by DWG with airspace, tilt, and decenter tolerances added.

elements and not add rays to the drawing.) This drawing shows the decenter tolerances as calculated by BTOL, and in this case you will want to edit those annotations and add a note that the centration is to be adjusted at assembly.

The program has also added an annotation box in the lower right corner, where you may enter your company name, project, name, or whatever identification you wish, using the Annotation Editor in the Graphics Window. Just click the toolbar button Ab, and select the type of annotation you want with the toolbar that opens.

Chapter 14

A near-IR lens example

Glass selection for the near IR spectrum

Chapter 12 showed how to design an apochromatic objective for use in the visible spectrum. Now we will design one over the wavelength range from 1.06 to 1.97 um, in the near infrared, which is more difficult.

The challenge when designing a lens for the infrared is finding optical materials that are useful over the spectral range and whose cost and chemical properties are attractive. The task in this lesson is to redesign an existing lens, replacing some undesirable materials with ordinary optical glass. The reference system is bundled as 1.RLE, with the ID MIT 1 TO 2 UM LENS. You can FETCH that lens and examine its performance. Set the scale of the OPD curves to 0.5 waves, as shown in figure 14.1.

This lens has three elements of ZNS and one of AS2S3, making four in all. Those names refer to zinc sulphide and arsenic trisulphide glass, and we would like to avoid those materials if possible. The first-order properties we need to match are as follows (dimensions are in mm):

- Entering beam radius 17.5
- Chief-ray angle 0.935 degrees
- Back focus distance 16.3
- Cell length 50
- Wavelengths 1.9701, 1.5296, and 1.06 μm

14.1 Design approach

Rather than try to change the materials in the present lens, all of which have an index greater than 2.0, let us start from scratch. For this we will use the design search program—but first we have to decide which glass types to use: if we just run **DSEARCH** and let it find model glasses, it will not come back with any of the unusual glasses that make a big difference in the NIR. (The model represents an *average* of selected glasses in the visible region.) So we have to steer it.

doi:10.1088/978-0-7503-3695-6ch14

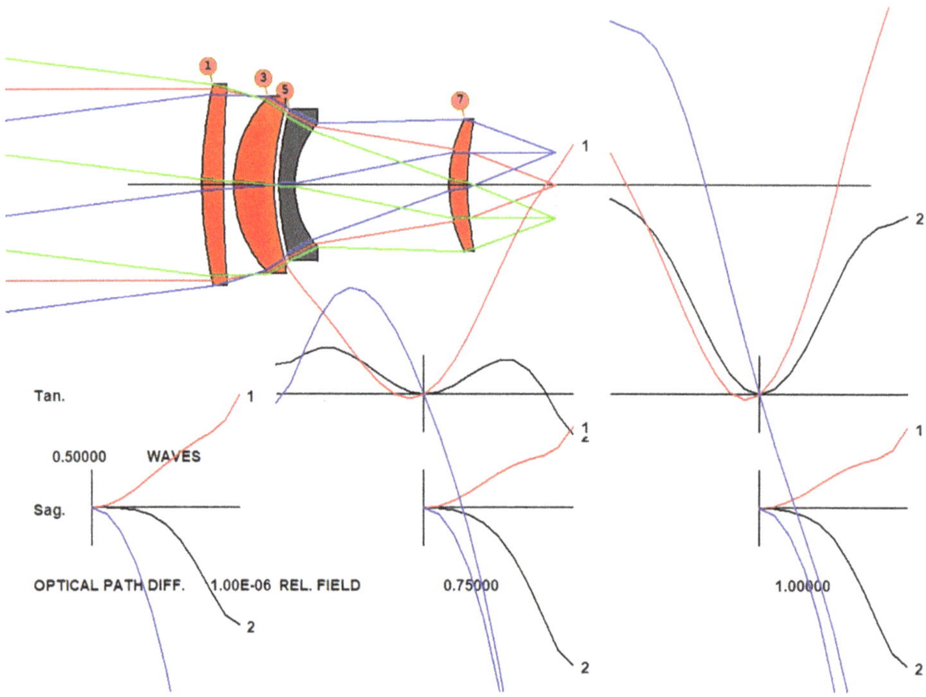

Figure 14.1. An NIR lens, to be redesigned.

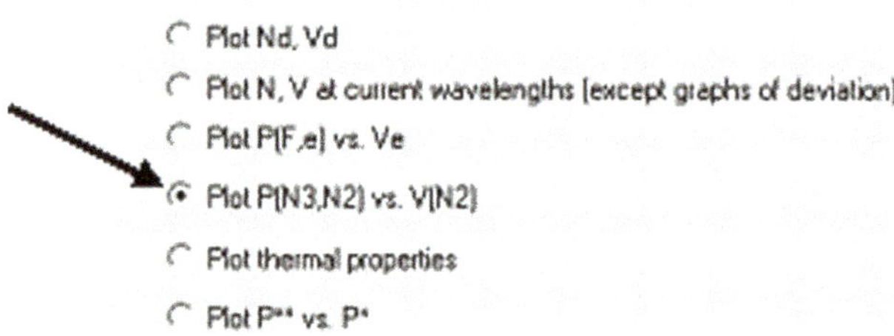

Figure 14.2. Selecting a graph option in the glass map utility.

Open the glass table display (**MGT**), select the Guangming table, and then click the Graph button and select the option shown in figure 14.2.

The data are now off the screen, so click in the display and zoom out with the mouse wheel until you see a collection of red dots. Then pan with the right mouse button to center things and zoom in again. You should see the display shown in figure 14.3. Click the Full Name button, then click on each of the spots circled and write down the name of the glass.

The four glass names circled, D-FK61, G-ZF52, H-ZH88, and H-F51 definitely do not behave the same as all the others. We will direct DSEARCH to only use two of them, and later use all four for a comprehensive glass search.

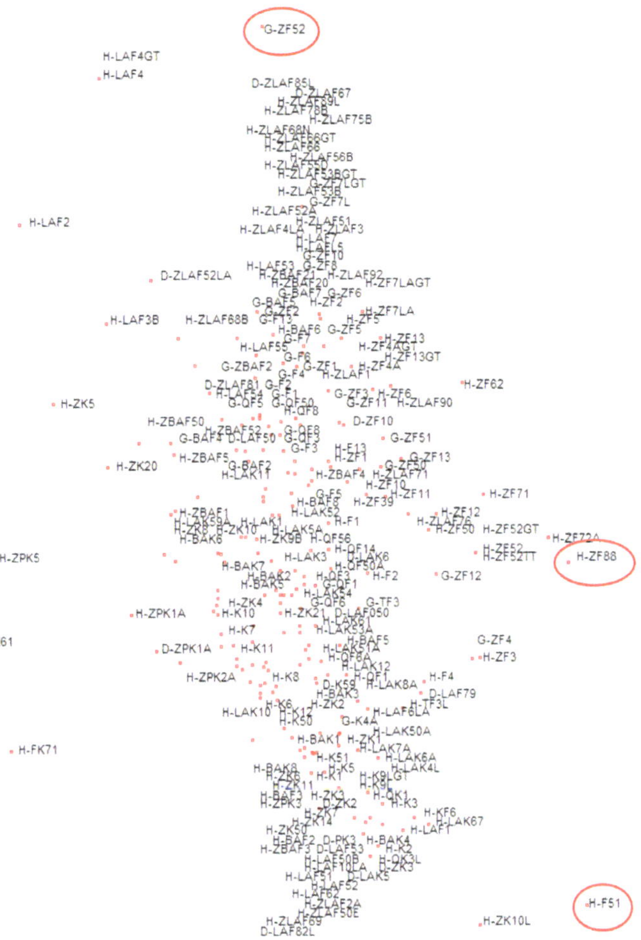

Figure 14.3. Selecting four promising glass types for the NIR design.

Here is the DSEARCH input (**C14M1**):

```
CORE 32
ON 1
TIME
DSEARCH 3  QUIET        ! the best lens will show up in library location 3 (and also in PAD)

 SYSTEM                 ! system requirements follow
 ID NIR EXAMPLE         ! lens ident
 OBB 0 .935 17.5        ! specify the object
 WAVL 1.97 1.53 1.06    ! and the wavelength range
 UNITS MM
 END
 GOALS             ! here we set the goals
 ELEMENTS 5        ! since glass has a lower index, we'll ask for 5.
 FNUM 1.428
 BACK 16 .1
 TOTL 50 .1
```

```
 STOP FIRST        ! there seems to be no reason to let the stop position vary
 STOP FIX          ! so we put it in front and keep it there
 NPASS 100
 ANNEAL 200 10 100
 RSTART 300           ! a useful starting radius,
 TSTART 1             ! and this thickness on each element to start with
 QUICK 50 90
FOV 0 .5 1
FWT 2 1 1
GLASS POS           ! positive elements will use this glass type
G D-FK61
GLASS NEG          ! and negative this type.
G H-ZF88
END

 SPECIAL        ! here we give requirements that are not defaults
 ACC 10 .1 1    ! maximum thickness 10 mm
 ACM 3 .1 1     ! auto edge and center control (AEC), (ACC) are defaults
 ACA            ! but we add to these ACM, so thicknesses do not get too thin, ACA,
 ASC            ! to avoid the critical angle, and ASC so surfaces do not get too steep
 END

 GO             ! this starts the process.
TIME
```

In 25 s, DSEARCH produces a picture of the 10 best configurations it found, shown in figure 14.4.

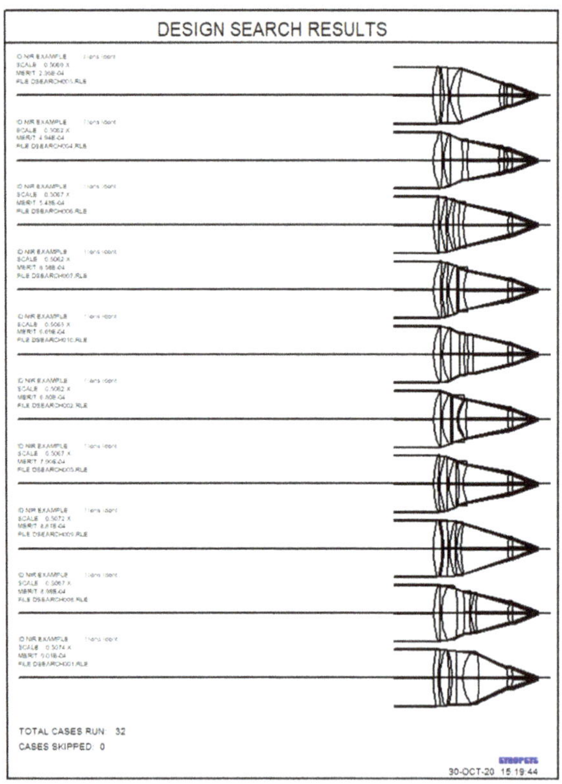

Figure 14.4. Display of the 10 best lenses returned by DSEARCH.

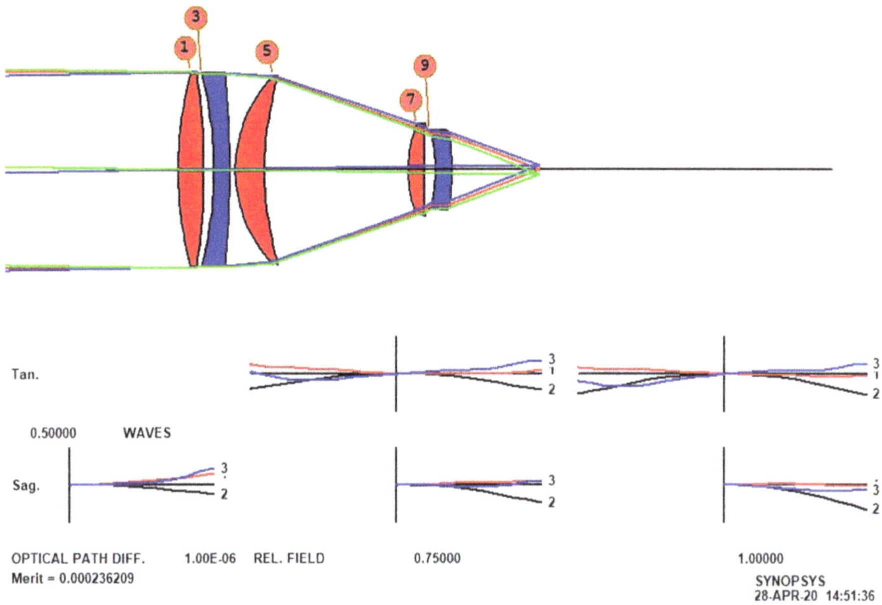

Figure 14.5. Best lens returned by DSEARCH for a NIR design.

We now have a very good five-element lens, shown in figure 14.5, but it has only the two glass types we specified. It's time to do a more comprehensive search.

Look at the MACro DSEARCH_OPT .MAC, which DSEARCH has constructed and should be open in a new editor window.

```
PANT
VLIST RD ALL
VLIST TH ALL
END
AANT P
AEC
ACC
GSR      0.000000      2.000000      4  M      0.000000
GNR      0.000000      1.000000      4  M      0.500000
GNR      0.000000      1.000000      4  M      1.000000
M   0.160000E+02  0.100000E+00 A BACK
M   0.500000E+02  0.100000E+00 A TOTL
 ACC 10 .1 1
 ACM 3 .1 1    ! auto edge and center control (AEC), (ACC) are defaults
 ACA           ! but we add to these ACM, so thicknesses do not get too thin, ACA,
 ASC        ! to avoid the critical angle, and ASC so surfaces do not get too steep
END
SNAP   0/DAMP    1.00000
SYNOPSYS   100
```

Save this MACro with the name NIR_OPT.MAC. This is the optimization MACro that will be run over and over when you execute **GSEARCH**. That program will determine which glass should go on which elements.

Now make a new MACro (type **AEE** to open a new editor, and type the data **(C14M2)** below.)

```
GSEARCH 3 QUIET LOG

SURF
1 3 5 7 9
END

OFILE NIR_OPT.MAC
NAMES
G G-ZF52
G D-FK61
G H-ZF88
G H-F51

END
USE 2
GO
```

Run this MACro, and the lens has improved even more, as shown in figure 14.6. The performance is now just over 0.25 waves of aberration in color 3 at full field.

It looks like we have a solution. (**C14L1**) There is almost no primary or secondary chromatic aberration. We have succeeded in replacing the undesirable materials with ordinary glass, and the performance became much better than the original at the same time. Here is a SPEC listing of this lens:

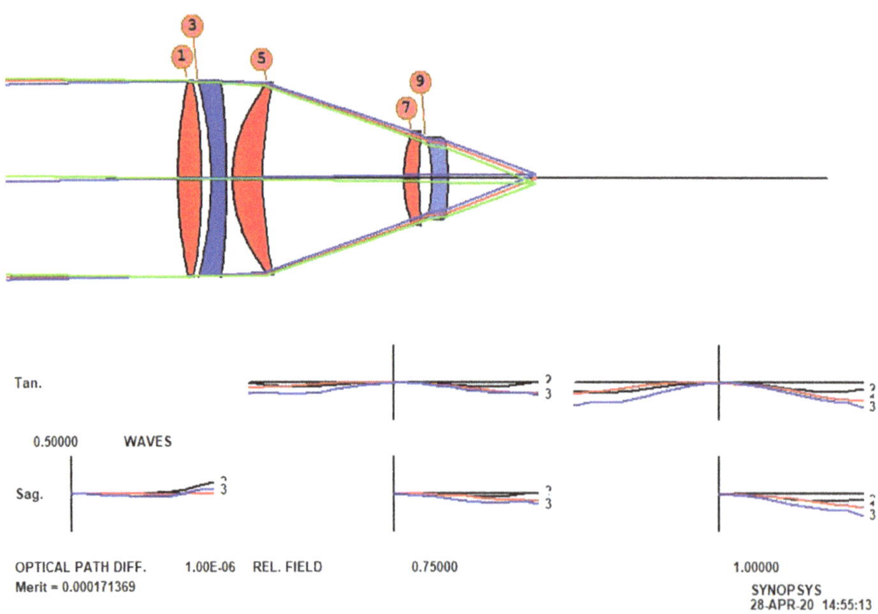

Figure 14.6. Lens returned by GSEARCH.

```
SYNOPSYS AI>SPE

ID NIR EXAMPLE          ! lens identifi   55564            04-AUG-20   10:52:16
ID1 DSEARCH CASE WAS 0000000000000000000010110     22
LENS SPECIFICATIONS:

SYSTEM SPECIFICATIONS
--------------------------------------------------------------------------------
OBJECT DISTANCE     (THO)        INFINITE  FOCAL LENGTH      (FOCL)       49.9800
OBJECT HEIGHT       (YPPO)       INFINITE  PARAXIAL FOCAL POINT           15.9991
MARG RAY HEIGHT     (YMP1)        17.5000  IMAGE DISTANCE    (BACK)       15.9991
MARG RAY ANGLE      (UMPO)         0.0000  CELL LENGTH       (TOTL)       50.0024
CHIEF RAY HEIGHT    (YPP1)         0.0000  F/NUMBER          (FNUM)        1.4280
CHIEF RAY ANGLE     (UPPO)         0.9350  GAUSSIAN IMAGE HT(GIHT)         0.8157
ENTR PUPIL SEMI-APERTURE          17.5000  EXIT PUPIL SEMI-APERTURE       24.7780
ENTR PUPIL LOCATION                0.0000  EXIT PUPIL LOCATION           -54.7668

WAVL (uM) 1.970000 1.530000 1.060000
WEIGHTS   1.000000 1.000000 1.000000
COLOR ORDER    2    1    3
UNITS                                 MM
APERTURE STOP SURFACE (APS)            1     SEMI-APERTURE      17.53073
FOCAL MODE                            ON
MAGNIFICATION                -4.99800E-11
POLARIZATION AND COATINGS ARE IGNORED.
SURFACE DATA
--------------------------------------------------------------------------------
SURF        RADIUS       THICKNESS    MEDIUM            INDEX       V-NUMBER
--------------------------------------------------------------------------------
  0        INFINITE       INFINITE    AIR
  1        82.54737        4.57087    D-FK61           1.48647      78.02 GUANGMIN
  2       -90.10221        1.75340    AIR
  3       -61.34858        2.89125    H-ZF88           1.87811      26.89 GUANGMIN
  4      -137.03664        1.00000    AIR
  5        25.82112        5.68677    D-FK61           1.48647      78.02 GUANGMIN
  6        79.75284       25.92645    AIR
  7        23.78844        2.91256    D-FK61           1.48647      78.02 GUANGMIN
  8       111.32895        2.38658    AIR
  9       -24.11535        2.87455    H-F51            1.60755      25.46 GUANGMIN
 10       -40.17000S      15.99909S   AIR
IMG        INFINITE

KEY TO SYMBOLS
--------------
A  SURFACE HAS TILTS AND DECENTERS    B   TAG ON SURFACE
G  SURFACE IS IN GLOBAL COORDINATES   L   SURFACE IS IN LOCAL COORDINATES
O  SPECIAL SURFACE TYPE               P   ITEM IS SUBJECT TO PICKUP
S  ITEM IS SUBJECT TO SOLVE           M   SURFACE HAS MELT INDEX DATA
T  ITEM IS TARGET OF A PICKUP
THIS LENS HAS NO SPECIAL SURFACE TYPES
THIS LENS HAS NO TILTS OR DECENTERS
SYNOPSYS AI>
```

If these lenses are okay mechanically, the problem is solved.

Except … what is the transmission at 1.97 μm? Type **FIND TRANS IN COLOR 1**. It comes back 98.18%. (Coatings and reflection losses are ignored here because the lens is not in polarization mode.) Very good!

But what if the value had come back too low? Well, then go back to the glass map and display the absorption at 1.97 μm—and select glasses with shorter data bars. Lens design is all about tradeoffs, after all, and with these tools you can get the best one rather easily.

Lastly, we must explain the setting ON 1 in the DSEARCH file. That switch causes the optimization program to delete any variables that reach the boundary of their allowed region when the program tests new solutions. This can sometimes speed things up, since the program has to iterate to find a solution that does not violate that boundary otherwise, and it likely will continue to try to violate it in future passes. But it is not a default setting, since once a thickness had been dropped, it no longer is available to satisfy the edge monitor AEC. In some cases, when running DSEARCH, the process will climb different branches of the chaotic tree, depending on how this switch is set, so it is another item you can experiment with. For this lesson, we got better results with it turned on.

IOP Publishing

Lens Design (Second Edition)
Automatic and quasi-autonomous computational methods and techniques
Donald C Dilworth

Chapter 15

A laser beam shaper, all spherical

Reshaping a laser beam from Gaussian to uniform with spherical elements

The output from a laser has a Gaussian intensity profile, which is nonuniform, and for some applications one would like to make it uniform. That is the job of a laser beam shaper.

The job can be done in several ways. With simple lenses, with spherical surfaces, it requires balancing rather large amounts of spherical aberration in a way that redistributes the light, reducing the energy density at the center of the beam while increasing it near the edges, at the same time keeping the wavefront aberrations under control. It is easier to do with aspheric surfaces, where one has more control over the amount of aberration to be introduced, and also easier if one uses diffractive optical elements. The problem is that the latter two are more expensive than spherical lenses, so we want to see what we can do with those first.

With this in mind, we will try an all-spherical approach to start with, to determine how uniform we can make the beam and how many elements we will need. Let us start with a very simple setup, one that we will modify as we go along.

The problem is to convert a HeNe laser with a waist radius of 0.35 mm to a beam that is 10 mm in diameter and uniform to within 10%. (There is no reason to use DSEARCH for this problem, since the only practical starting design is very simple.)

Here is the input file for our very primitive starting point (**C15M1**), shown in figure 15.1.

doi:10.1088/978-0-7503-3695-6ch15

Figure 15.1. Rough-guess initial system for a laser beam shaper.

```
RLE                      ! Beginning of lens input file.
ID LASER BEAM SHAPER
WA1 .6328                ! Single wavelength
UNI MM                   ! Lens is in millimeters
OBG .35 2                ! Gaussian object; waist radius -.35 mm; define full
                         ! aperture as twice the 1/e**2 point.
1 TH 22                  ! Surface 2 is 22 mm from the waist .
2 RD -5 TH 2 GTB S       ! Guess some reasonable lens parameters; use glass type SF6
                         ! from Schott catalog
SF6
3 UMC 0.3 YMT 5          ! Solve for the curvature of surface 3 so the marginal ray
                         ! has an angle of 0.3;
                         ! find spacing so ray height is 5 mm on next surface
4 RD 20 TH 4 PIN 2       ! Guesses for surface 4
5 UMC 0 TH 50            ! Solve for curvature of 5 so beam is collimated.
7                        ! Surfaces 6 and 7 exist
AFOCAL                   ! because they are required for AFOCAL output.
END
```

(This system is in **AFOCAL** mode, which means that the output will be collimated. Since transverse aberrations make no sense if the image is at infinity, the program translates them into *angular* aberrations, no 'perfect lens' required. The program needs two dummy surfaces at the end, where it does the translation, here surfaces 6 and 7, which are coincident at the end.)

First, let us examine how the energy density falls off from the center to the edge of the aperture. There are three ways to do this. The simplest is with the FLUX command:

```
FLUX 100 P 3
```

This input will show the flux on surface 3, which shows the expected falloff, as illustrated in figure 15.2.

Another way to visualize this uses the FLUX aberration. This form gives you more flexibility, since you can specify the aperture and field points yourself. Make a MACro (**C15M2**) with the following lines. (Note the symbol definition for DD):

```
DD: DO MACRO FOR AIP = -1 TO 1
COMPOSITE               ! Ready a composite definition.
CD1 P FLUX 0 0 AIP 0 3  ! Composite data number 1 is flux at Y- coordinate of AIP
                        ! (defined later) on surface 3.
= CD1

 Z1 = FILE 1
 = 1 + Z1
 ORD = FILE 1
```

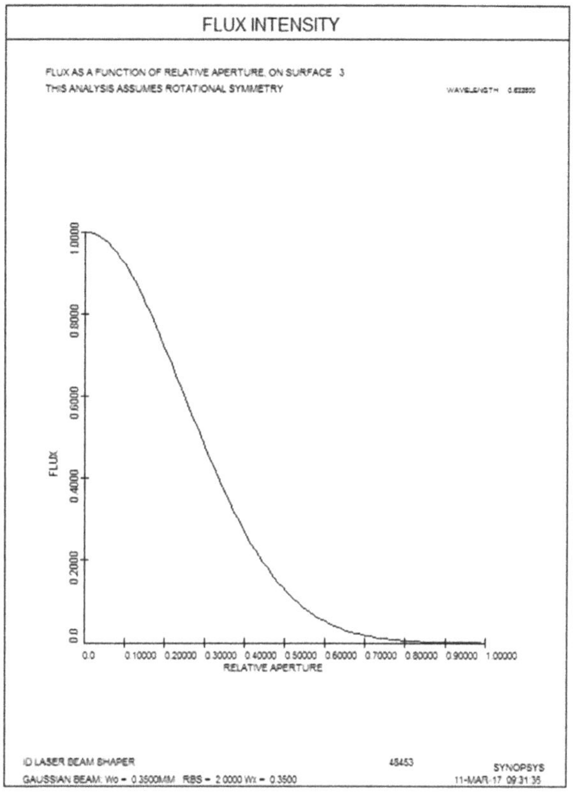

Figure 15.2. Flux falloff due to Gaussian intensity profile, calculated with FLUX command.

Run this MACro once, and then type **STEPS = 100** and then **DD**. The program loops over values of AIP from −1 to 1 and plots the flux density. Let me explain the logic:

- **CD1 P FLUX** … calculates the flux falloff at zone AIP (which is a loop variable) at surface 3.
- **= CD1** is an equation, the result of which is automatically placed in FILE location 1.
- **Z1 = FILE 1** picks up that value and puts it into variable Z1.
- **= 1 + Z1** adds 1.0 to the result. This is then the total flux, since Z1 is the falloff.
- **ORD = FILE 1** picks up this value and uses it for the ordinate of a plot, the abscissa of which is the loop variable, AIP.

Here is the plot, in figure 15.3.

Again, you see the Gaussian flux profile, evaluated out to twice the $1/e^2$ point according to the OBG definition. (The third way uses the **DPROP** diffraction propagation feature. That is more complicated to set up and run, but can account for diffraction along the beam, which the other two cannot. An example will be shown later.)

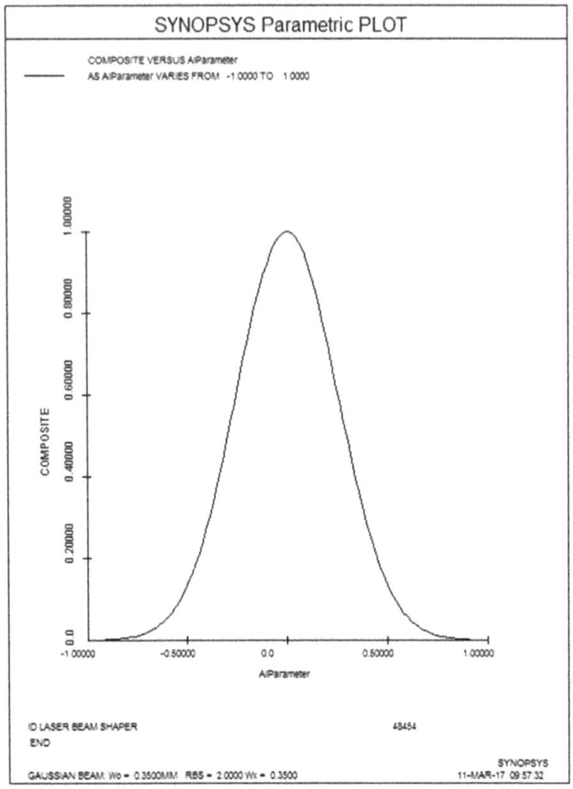

Figure 15.3. Flux intensity as calculated by an AI loop.

The object of this lesson is to make the flux as uniform as possible, with a target of 10% change over the aperture.

Here, we have simply guessed some starting lens dimensions, but already we see a hint of a solution: Notice how the marginal rays converge toward the axis while the center rays are more collimated. The energy will indeed be more concentrated at the edges than before, which is a step in the right direction—but we also want the entire beam to be collimated, so we need a way to straighten out the rays afterwards. That is more difficult.

We suspect that we need more than just two elements, so let us add two more. Using the WorkSheet, as you did in chapter 6, click the button, and click the Insert Element button in the WorkSheet toolbar, . Then click on the axis in the PAD display, to add an element to the right of surface 5. Do the same thing again, a little further to the right. Now the system should appear as shown in figure 15.4.

We will try to optimize this system, but first let us make a Checkpoint, so we can instantly go back if things do not work as we hope. Click the Checkpoint button, .

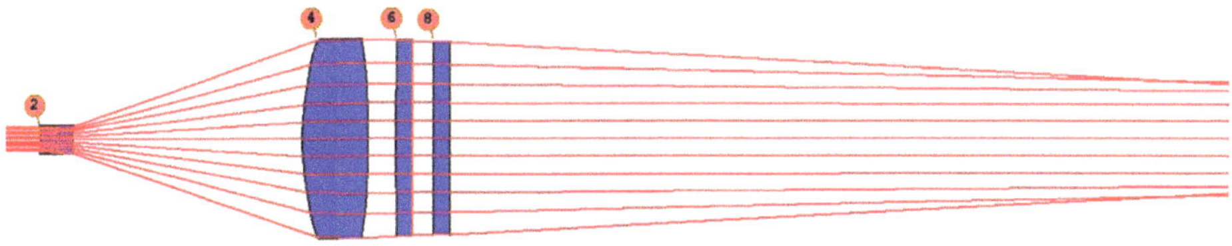

Figure 15.4. Lens with two added elements.

Now we need to set up an optimization MACro to see if we can even things out. Here is a start (**C15M3**):

```
CHG
NOP
9 UMC
END
OFF 1

PANT                    ! Start of variable parameter definitions.
VLIST RAD ALL           ! Vary all radii.
VLIST TH 3 5 6 7 8      ! Vary the airspaces and five thicknesses (so AEC !works or
                        ! those elements).
END

AANT                    ! Start of merit function definition
AEC 1 1 1
ACC 4 1 1
ACA 60 10 1             ! stay away from the critical angle
LUL 100 1 1 A TOTL      ! Prevent the system from growing too large; assign upper-
                        ! limit of 150 on TOTL.
M 5 10 A P YA 0 0 1 0 9   ! Ask for a beam radius of 5 mm on surfaces 9 and 10
M 5 10 A P YA 0 0 1 0 10
M 0 1 A P FLUX 0 0 1 0 10 ! Ask for a flux falloff of zero at several zones on 10
M 0 1 A P FLUX 0 0 .99 0 10
M 0 1 A P FLUX 0 0 .985 0 10
M 0 1 A P FLUX 0 0 .98 0 10
M 0 1 A P FLUX 0 0 .97 0 10
M 0 1 A P FLUX 0 0 .96 0 10
M 0 1 A P FLUX 0 0 .95 0 10
M 0 1 A P FLUX 0 0 .94 0 10
M 0 1 A P FLUX 0 0 .93 0 10
M 0 1 A P FLUX 0 0 .92 0 10
M 0 1 A P FLUX 0 0 .91 0 10
M 0 1 A P FLUX 0 0 .85 0 10
M 0 1 A P FLUX 0 0 .8 0 10
M 0 1 A P FLUX 0 0 .7 0 10
M 0 1 A P FLUX 0 0 .5 0 10
M 0 1 A P FLUX 0 0 .3 0 10
C 0 1 A P HH 0 0 1
GSO 0 1 5 P             ! Control the output ray OPD over an SFAN of 5 rays.
END                     ! End of merit function definition.
SNAP
SYNO 100
```

Run this MACro, and the new lens is shown in figure 15.5. (Your results may differ, since the exact place you clicked to insert the elements is unpredictable.)

To see if things improved, we need to evaluate the flux uniformity again. Type

`FLUX 100 P 10.`

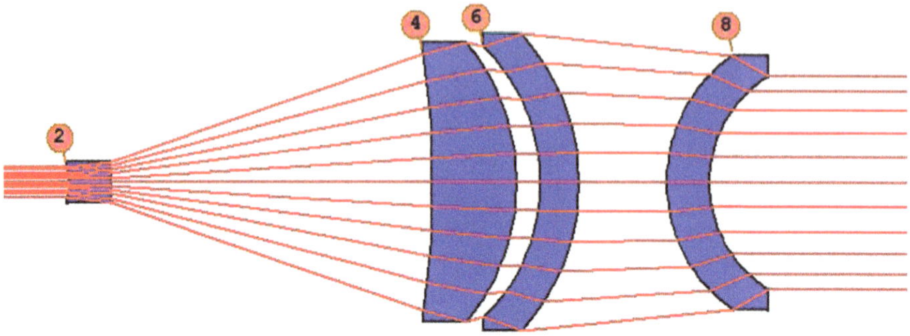

Figure 15.5. Lens optimized with FLUX aberrations.

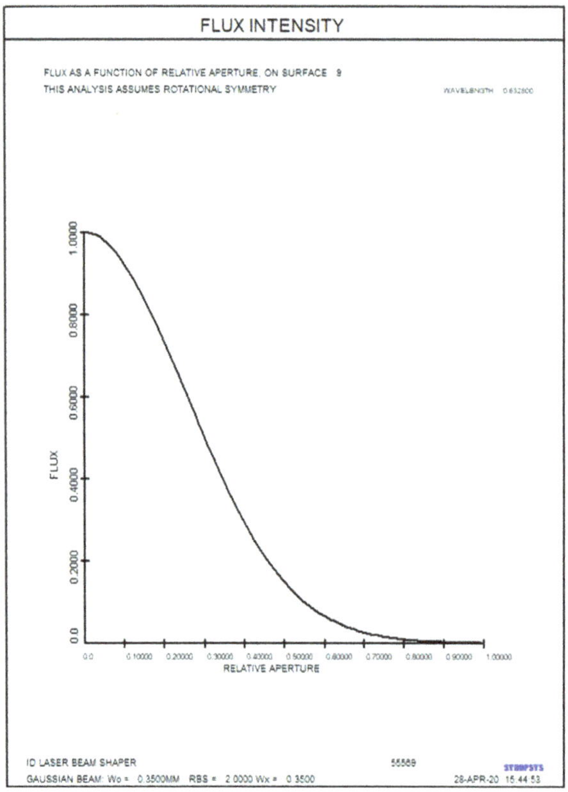

Figure 15.6. Flux falloff after reoptimizing.

The flux is not improved, as shown in figure 15.6.

It looks like we're not there yet. What to do?

Well, remember that we defined our Gaussian object so that 'full aperture' is twice the $1/e^2$ point. That means we are trying to redistribute the energy so the very faint outer edge is as bright as the center. Maybe that's asking too much. Let's restore the Checkpoint and edit the object specification. In the WorkSheet, enter the number 0 in the surface box and click Update. This shows the current system specifications in the edit pane, including the object selection, as seen in figure 15.7.

The object is currently defined as **`OBG 0.35 2.000000.`** Change this to **`OBG .35 1`** and click Update. Now, 'full aperture' is at the $1/e^2$ point, not twice as far out as before. Save this as a new Checkpoint, run the last MACro again, and anneal for a few cycles (**22, 1, 50**). We get the results in figure 15.8 and the flux uniformity shown in figure 15.9.

Now the flux is somewhat better, but still not uniform enough. We mentioned that it was not so easy to keep the ray angles under control at the same time one flattens the intensity profile. The OPDs are at about 0.5 waves.

This seems to be about as good a balance as we can achieve with four elements. What happens if we add some more? Let's start with this design and add two more elements, shown in figure 15.10.

Now we need to add the new variables to the PANT file; also, since the merit function specifies the surface number where certain quantities are to be evaluated, and this number changes whenever we add an element, we can make things easier by changing the surface number to a special symbol, 'LB1', which means 'last but one'. Since the highest surface in the lens is currently number 15, this symbol automatically turns into the number 14. Now, if we decide to add or delete elements as we search for a solution, we will not have to edit that number every

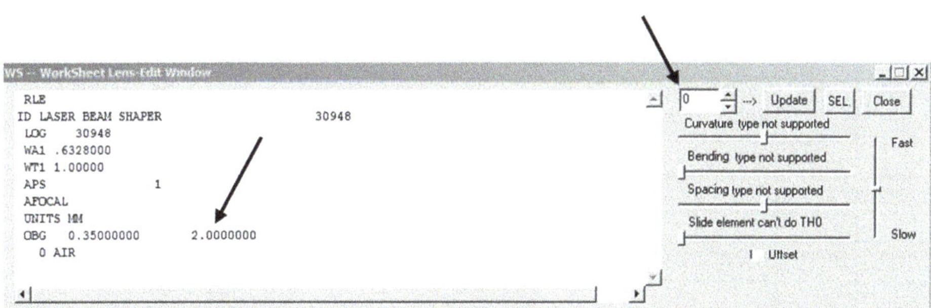

Figure 15.7. Edit pane in WS, showing the object specifications.

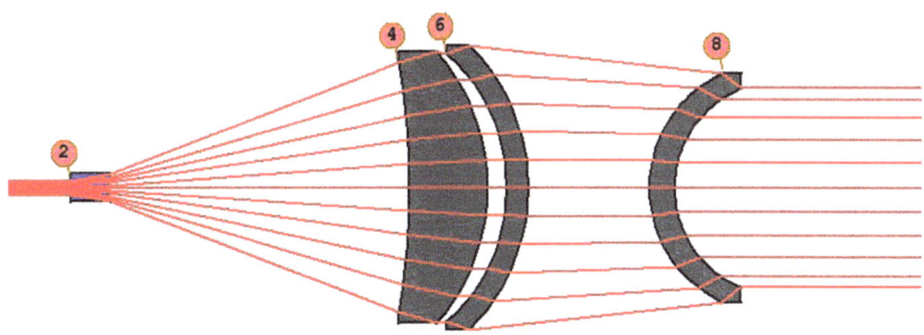

Figure 15.8. Lens reoptimized.

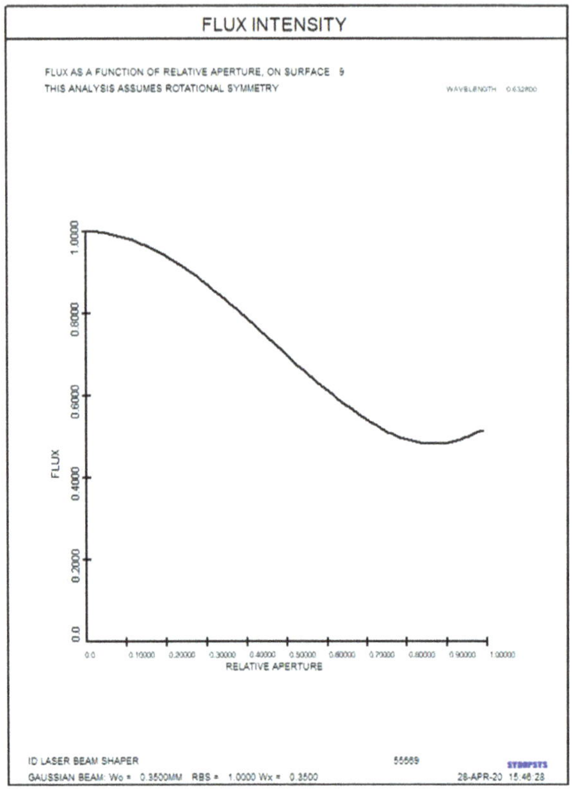

Figure 15.9. Flux uniformity after optimizing with added elements.

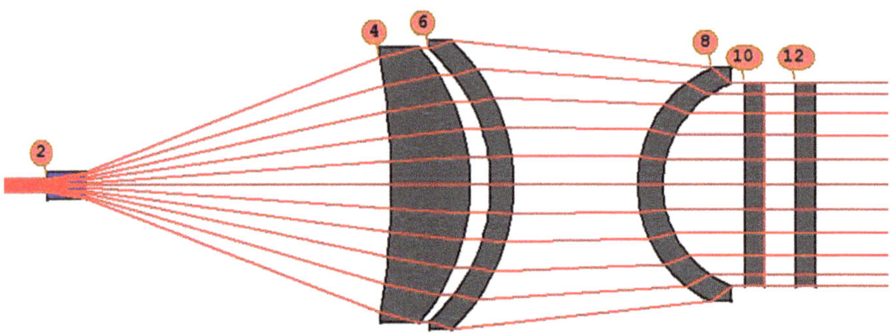

Figure 15.10. Lens with six elements, before optimizing.

time. We also add a GSR directive after the GSO, to better keep the ray angles under control, and we lower the weight on the OPD fan to better balance things. We also assign the UMC solve to surface 13 instead of 9 and declare all thicknesses variables (**C15M4**).

```
CHG
NOP
13 UMC
END

PANT              ! Start of variable parameter definitions.
VLIST RAD ALL     ! Vary all radii.
VLIST TH ALL      ! Vary the airspaces and all thicknesses (so AEC works on those
                  ! elements).
END

AANT                     ! Start of merit function definition
AEC 1 1 1
ACC 4 1 1
LUL 100 1 1 A TOTL       ! Prevent the system from growing too large; assign upper-
                         ! limit of 150 on TOTL.
M 5 10 A P YA 0 0 1 0 LB1 ! Ask for a beam radius of 5 mm on surfaces 9 and 10
M 5 10 A P YA 0 0 1 0 LB1
M 0 1 A P FLUX 0 0 1 0 LB1 ! Ask for a flux falloff of zero at several zones on 10
M 0 1 A P FLUX 0 0 .99 0 LB1
M 0 1 A P FLUX 0 0 .98 0 LB1
M 0 1 A P FLUX 0 0 .97 0 LB1
M 0 1 A P FLUX 0 0 .96 0 LB1
M 0 1 A P FLUX 0 0 .95 0 LB1
M 0 1 A P FLUX 0 0 .94 0 LB1
M 0 1 A P FLUX 0 0 .93 0 LB1
M 0 1 A P FLUX 0 0 .92 0 LB1
M 0 1 A P FLUX 0 0 .91 0 LB1
M 0 1 A P FLUX 0 0 .85 0 LB1
M 0 1 A P FLUX 0 0 .8 0 LB1
M 0 1 A P FLUX 0 0 .7 0 LB1
M 0 1 A P FLUX 0 0 .5 0 LB1
M 0 1 A P FLUX 0 0 .3 0 LB1
C 0 1 A P HH 0 0 1       ! Correct the marginal ray angle
GSO 0 .1 5 P             ! Control the output ray OPD over an SFAN of 5 rays.
GSR 0 1 5 P
END                      ! End of merit function definition.

SNAP
SYNO 100
```

Run this MACro, and then anneal (**22, 1, 50**). After many iterations of annealing, the lens (**C15L1**) looks like figure 15.11.

And the flux is well within the target of 10% uniformity, as shown in figure 15.12.

```
FLUX 100 P 14
```

The OPD errors are now under 0.3 waves. It appears that doing this job with all-spherical lenses is possible but requires six elements.

Now we understand why people generally use aspherics or diffractive elements for this kind of job. Then there are fewer elements, so it can be worth the extra manufacturing trouble.

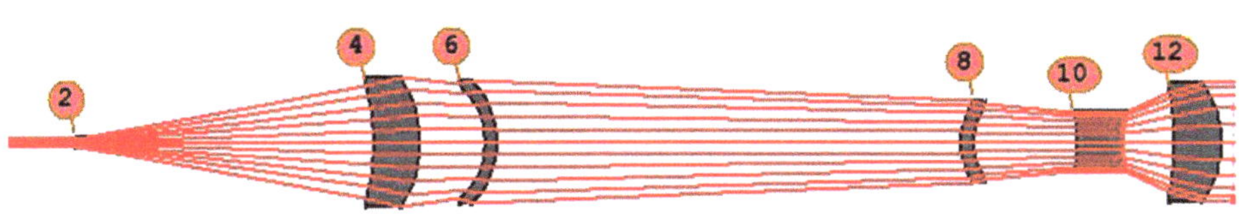

Figure 15.11. Lens optimized with six elements.

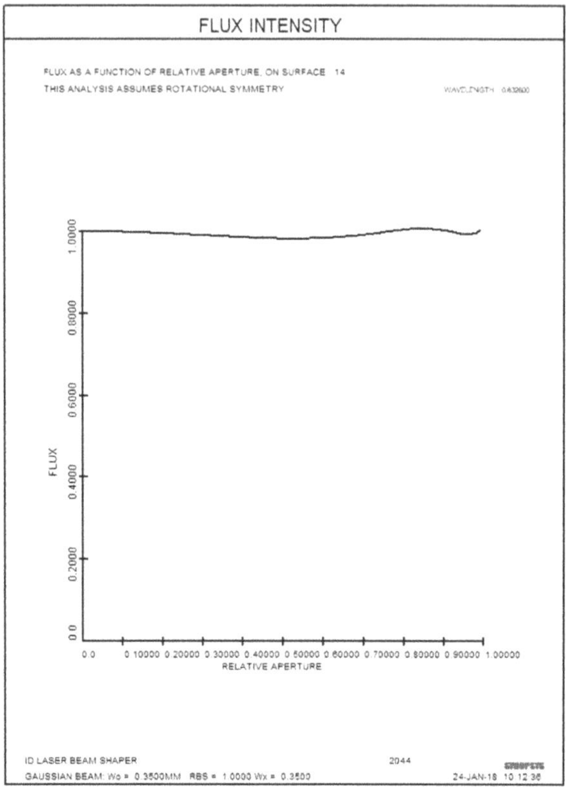

Figure 15.12. Flux uniformity with six-element design.

To finish this lesson, let us look at the ray pattern coming out of the system. Go to the footprint dialog (**MFP**), turn off switch 27, select surface 13, 10x scale, and ask for 600 rays. Here is the ray distribution, shown in figure 15.13.

This is pretty good for a beam shaper; rays are more spread out near the center and compressed near the edge, which is exactly the right way to make the beam more uniform.

Now we will use **DPROP** to evaluate the final envelope. In this design, the beam is expanded early on, from surface 3, and diffraction will not play a significant role thereafter.

Here is the input needed to run DPROP:

```
CHG
CFIX
1 TH 0
END
DPROP P 0 0 13 SURF 3 R RESAMPLE
```

Here, we changed all apertures to fixed values with **CFIX**, which is always recommended when running DPROP, since that program will often examine a larger area

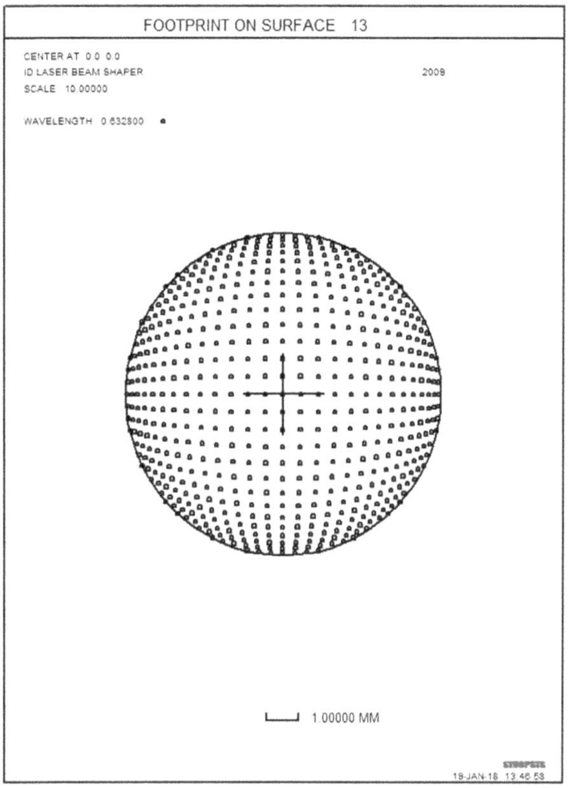

Figure 15.13. Footprint of final lens; rays are concentrated near the edges, compensating for the Gaussian falloff of the entering beam.

than the lenses allow, if diffraction can send even a small amount of energy there. The fixed apertures prevent this. The wavefront on surface 13 now looks like figure 15.14. It is not too far from flat, but still not as good as we'd like.

In this lesson we have introduced a new concept. In the optimization files you saw the entry

```
C 0 1 A P HH 0 0 1,
```

which is a way of controlling (C) a quantity that is especially important. Instead of adding it to the merit function as something to be minimized (M) to a target, which is the usual case, the program controls it via a technique known as 'Lagrange multipliers'. The effect is to apply an infinite weight, but without altering the value of the merit function otherwise. In this case, it is vitally important since we want to keep the output ray angle very close to zero at the edge of the beam. Why? Because we want to avoid a *caustic* at the output surface, which would make the flux infinite at that point (and numerically uncontrollable). A condition like that is never a good idea. A caustic shows up anytime a portion of the beam crosses over a lower part, and this is a simple way to avoid it.

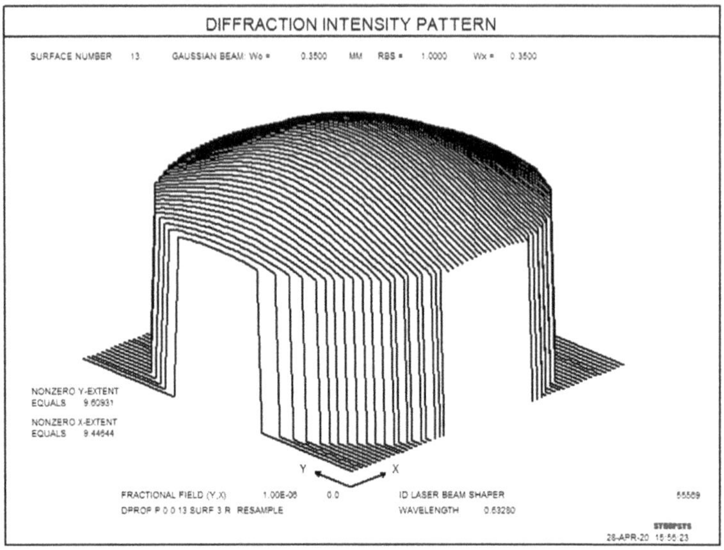

Figure 15.14. DPROP analysis of the final design.

In the next chapter we will develop a beam shaper with aspheric surfaces to illustrate the effect of those very useful variables.

This exercise was intended to show you some of the tools you have available and illustrate what to look out for and how to deal with the inevitable surprises and tradeoffs that turn up when you face a difficult and unfamiliar challenge.

IOP Publishing

Lens Design (Second Edition)

Automatic and quasi-autonomous computational methods and techniques

Donald C Dilworth

Chapter 16

A laser beam shaper with aspherics

With aspherics one can design a laser beam shaper with only two elements

In chapter 15 you designed a laser beam shaper to flatten out the Gaussian beam profile of a small HeNe laser. To keep manufacturing costs down, you tried in that lesson to do the job with spherical surfaces, since those are much easier to make than aspherics. You came up with a six-element design that came close to meeting the specs. Perhaps that design could be further improved, but we also must ask whether six spherical lenses will be cheaper to make than two aspherics. If not, then an aspheric design looks more attractive.

Let us start with the same two-element configuration that we used in chapter 15, modified so that we only flatten the flux out to the $1/e^2$ point. Going out to twice that aperture appeared to be impractical before, and we suspect that it will again. Here is our starting point (**C16M1**), shown in figure 16.1.

```
RLE                          ! Beginning of lens input file.
ID LASER BEAM SHAPER
WA1 .6328                    ! Single wavelength
UNI MM              ! Lens is in millimeters
OBG .35 1           ! Gaussian object; waist radius -.35 mm; define full aperture at the 1/e**2 point.
1 TH 22             ! Surface 2 is 22 mm from the waist .
2 RD -5 TH 2 GTB S  ! Guess some reasonable lens parameters; use glass type SF6 from Schott catalog
SF6
3 UMC 0.3 YMT 5     ! Solve for the curvature of surface 3 so the marginal ray has an angle of 0.3; find
                    ! spacing so ray height is 5 mm on next surface
4 RD 20 TH 4 PIN 2           ! Guesses for surface 4
5 UMC 0 TH 50                ! Solve for curvature of 5 so beam is collimated.
7                            ! Surfaces 6 and 7 exist
AFOCAL                       ! because they are required for AFOCAL output.
END                          ! End of lens input file.
```

Since we learned a great deal in chapter 15, we can start this problem with this merit function (**C16M2**).

doi:10.1088/978-0-7503-3695-6ch16

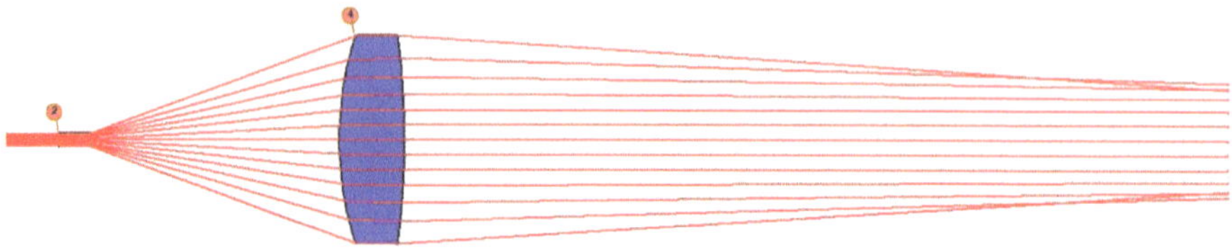

Figure 16.1. Starting system for beam shaper.

```
CHG
NOP               ! Be sure there are no pickups or solves.
4 PIN 2
5 TH 10 UMC 0    ! move surface 6 before the caustic
END
PANT                    ! Start of variable parameter definition.
VLIST RAD 2 3 4 5       ! Vary four radii.
VLIST TH 3              ! Vary the central airspace.
VY 3 CC           ! Vary the conic constant on surface 3.
VY 4 CC           ! And on surface 4.
VY 3 G 3          ! Add three aspheric terms to surface 3.
VY 3 G 6
VY 3 G 10
VY 4 G 3          ! And three to surface 4.
VY 4 G 6
VY 4 G 10
END

AANT              ! Start of merit function definition.
AEC 1 1 1         ! Enable automatic edge feathering control.
ACC 4 1 1         ! Enable automatic center thickness monitoring
ASC               ! Enable automatic slope control, so curves don't get too steep.
LUL 100 1 1 A TOTL              ! Limit the paraxial total length to no more than
                                ! 150 mm.
M 5 100 A P YA 0 0 1 0 LB1
M 5 100 A P YA 0 0 1 0 LB2      ! Assign a target of 5 mm to the marginal ray on
                                ! surfaces 5, 6.

M 0 1 A P FLUX 0 0 1 0 LB1      ! Target the flux difference between the marginal
                                ! ray point and the on-axis point to 0 on 6.
M 0 1 A P FLUX 0 0 .99 0 LB1    ! Target the flux at the 0.99 aperture point.
M 0 1 A P FLUX 0 0 .98 0 LB1    ! And so on, for a set of zones.
M 0 1 A P FLUX 0 0 .97 0 LB1
M 0 1 A P FLUX 0 0 .96 0 LB1
M 0 1 A P FLUX 0 0 .95 0 LB1
M 0 1 A P FLUX 0 0 .94 0 LB1
M 0 1 A P FLUX 0 0 .93 0 LB1
M 0 1 A P FLUX 0 0 .92 0 LB1
M 0 1 A P FLUX 0 0 .91 0 LB1
M 0 1 A P FLUX 0 0 .9 0 LB1
M 0 1 A P FLUX 0 0 .89 0 LB1
M 0 1 A P FLUX 0 0 .88 0 LB1
M 0 1 A P FLUX 0 0 .86 0 LB1
M 0 1 A P FLUX 0 0 .84 0 LB1
M 0 1 A P FLUX 0 0 .82 0 LB1
M 0 1 A P FLUX 0 0 .8 0 LB1
M 0 1 A P FLUX 0 0 .7 0 LB1
M 0 1 A P FLUX 0 0 .5 0 LB1
M 0 1 A P FLUX 0 0 .3 0 LB1
GSO 0 .01 10 P          ! Target the OPD of an SFAN of 10 rays to zero, with a
                        ! weight of .01
GSR 0 50 10 P           ! And also target the ray angles to zero.
END

SNAP
SYNO 50
```

While this is pretty straightforward, we should point out some things: Why is GSR used to target the ray angles? Usually, GSR controls the actual X-coordinate of each ray relative to the chief ray—but since this system is in AFOCAL mode, where the output is collimated, this entry targets the output *angles* instead. We elect to use this method of controlling ray angles instead of the 'C' operand of the previous chapter.

Where did we specify that the ray and flux targets are to be taken on surface 6? Well, this system has a total of seven surfaces, counting the two dummies at the end that are required for AFOCAL angle conversion. In the previous chapter you learned that the mnemonic 'LB1' means 'last but one', and here it gets replaced by the number 6 when the input is processed. This form of input is valid within PANT and AANT files and is a real time-saver when you want to use the same number in several places.

In this exercise, we have chosen to vary the conic constant and three aspheric terms on two surfaces. There are also higher-order terms available, using a variety of surface specifications—but this is a simple task and we want to see how close we can come with just those terms. This form of aspheric has 22 terms available, but only terms G 3, 6, 10, 16, 18, 19, 20, 21, and 22 are rotationally symmetric. They vary the 4th, 6th, 8th, and 10th—through 20th-order aspheric terms, and here we did not even use the last six.

Okay, let's run this MACro. Things might still get better, so anneal (**22, 1, 50**).

That brings the merit function down to 3.7E-5, which is a sign that we have reached a good solution, shown in figure 16.2.

Here is the FLUX plot for our final design (**C16L1**), in figure 16.3.

```
FLUX 100 P 5
```

Wow! The flux is almost perfectly uniform! How about the OPD errors?

```
SYNOPSYS AI>OPD

 SYNOPSYS AI>TFAN 5 P

 ID LASER BEAM SHAPER                              55571            29-APR-20   08:48:57
 TANGENTIAL RAY FAN ANALYSIS

 FRACT. OBJECT HEIGHT                       HBAR       0.000000    GBAR        0.000000
 COLOR NUMBER                                  1
  REL ENT PUPIL   WAVEFRONT ABERR
       YEN              OPD (WAVES)
 ________________________________
      -1.000           -0.000559
      -0.800           -0.001216
      -0.600            0.000922
      -0.400           -0.000552
      -0.200           -0.000721
       0.200           -0.000721
       0.400           -0.000552
       0.600            0.000922
       0.800           -0.001216
       1.000           -0.000559
 SYNOPSYS AI>
```

Okay, this design is essentially perfect, with just over .001 wave of error, and it only required two elements. We can be pleased at that. Looks like we don't need the six-element design of chapter 15 after all.

Just to be sure, let's also examine the output wavefront with DPROP, shown in figure 16.4.

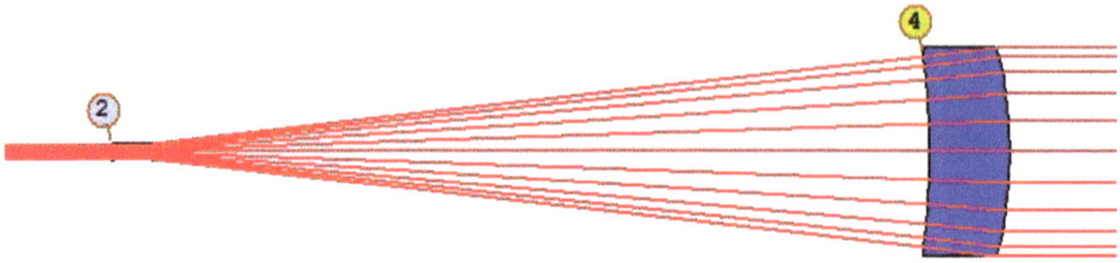

Figure 16.2. Two-lens design with aspherics.

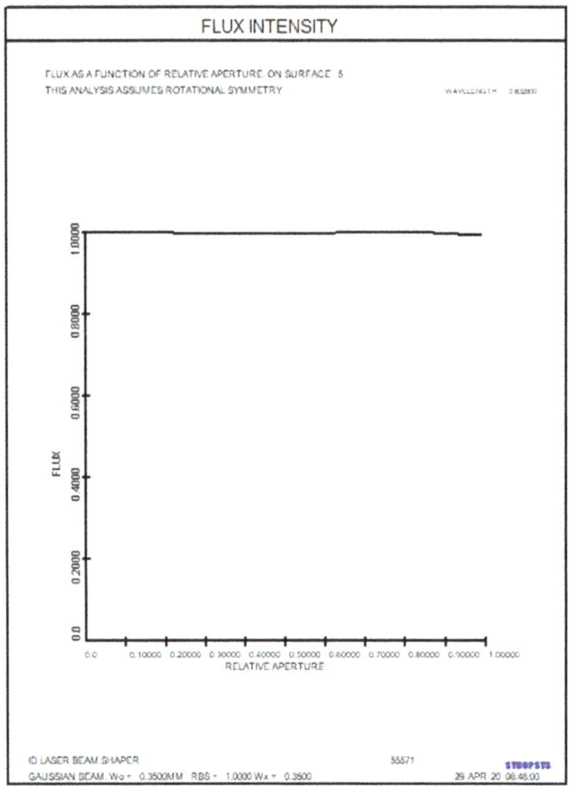

Figure 16.3. Flux uniformity of the aspheric design.

```
STORE 9
CHG
CFIX
END
DPROP P 0 0 5 SURF 2.5 R RESAMPLE
GET 9
```

Yes, this is very close to what we are after.

Now the only question is how difficult the aspherics will be to manufacture. To see how far the aspherics are from the closest fitting sphere (CFS), enter

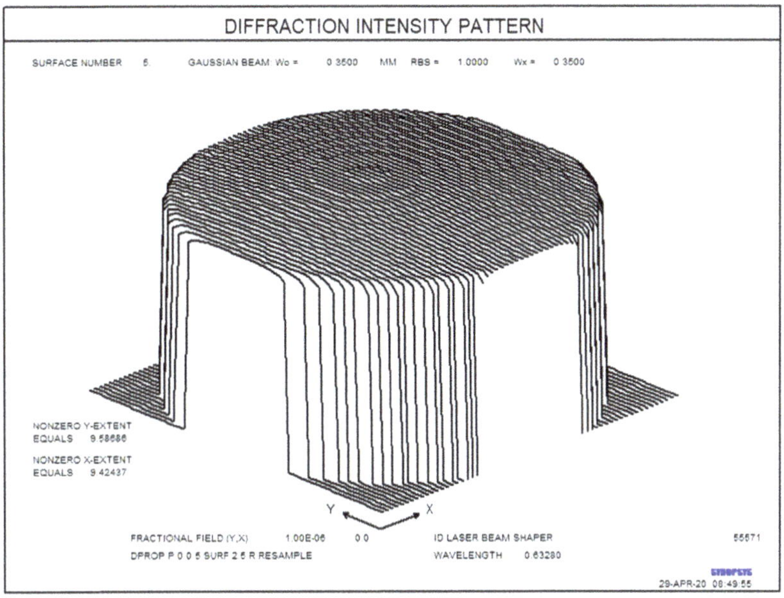

Figure 16.4. Output wavefront of aspheric design.

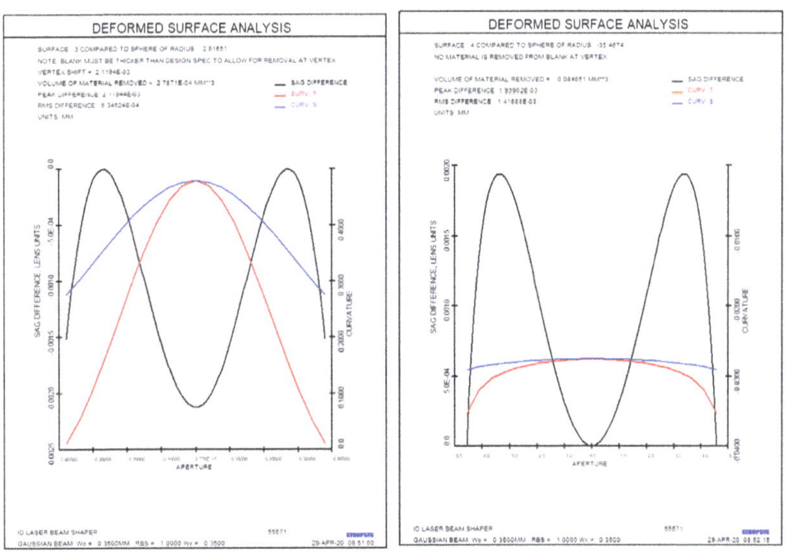

Figure 16.5. Analysis of aspheric surfaces.

```
ADEF 3 PLOT
ADEF 4 PLOT
```

and get the results in figure 16.5.

Both of these aspherics are only a few micrometers away from the CFS. It appears that this is manageable. Let us look at the fringe pattern, relative to the CFS, shown in figures 16.6 and 16.7.

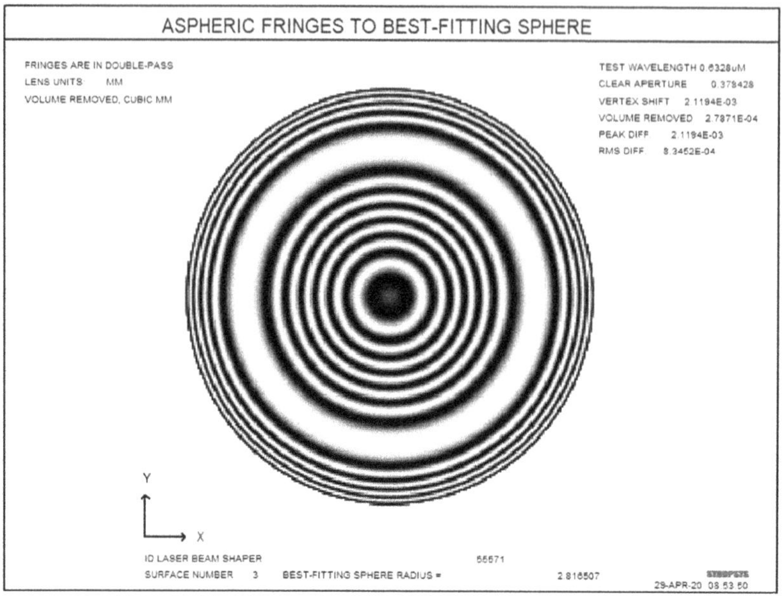

Figure 16.6. Fringe pattern relative to the closest-fitting sphere to surface 3.

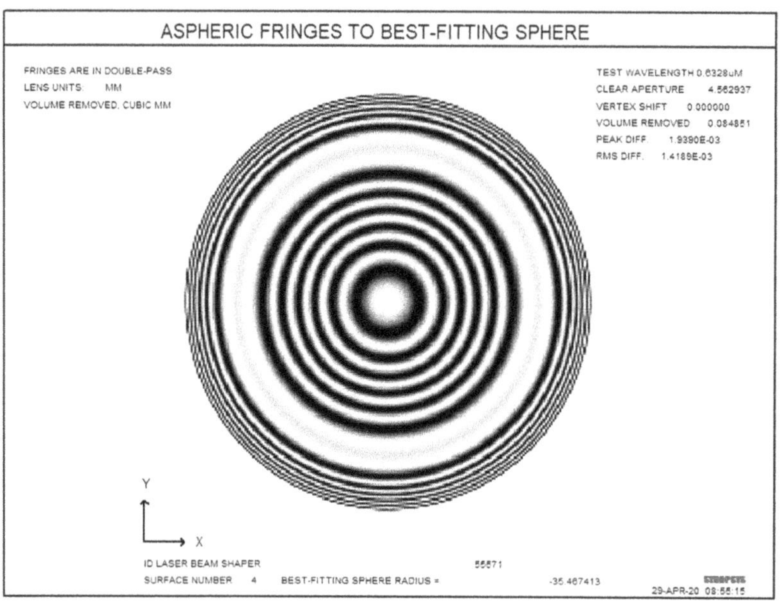

Figure 16.7. Fringe pattern for surface 4.

```
ADEF 3 FRINGES
ADEF 4 FRINGES
```

This may or may not be a challenge to the folks in the shop. Depending on how the aspherics are made and measured, one might try to reduce the aspheric departure somewhat, trading off that departure against the performance. See chapter 24 to learn how one can use the CLINK optimization feature to do just that.

IOP Publishing

Lens Design (Second Edition)

Automatic and quasi-autonomous computational methods and techniques

Donald C Dilworth

Chapter 17

A laser beam expander with kinoform lenses

A laser beam shaper with kinoform lenses requires only two elements

In Chapter 15 you saw how a laser beam expander could be designed with ordinary spherical lenses and learned that one would need six elements to achieve good performance that way. Chapter 16 did the same thing with only two aspheric elements, with excellent results. This lesson will show that you can do as well with diffractive optical elements (DOEs), also known as kinoform lenses.

The problem is to convert a HeNe laser with a waist radius of 0.35 mm to a beam that is 10 mm in diameter and uniform to within 10%.

Here is the input file for our starting point (**C17M1**):

```
RLE                          ! Beginning of lens input file.
ID KINOFORM BEAM SHAPER
WA1 .6328                    ! Single wavelength
UNI MM                ! Lens is in millimeters
OBG .35 1             ! Gaussian object; waist radius -.35 mm; define full
                      ! aperture = 1/e**2 point.
1 TH 22               ! Surface 2 is 22 mm from the waist.
2 RD -2 TH 2 GTB S    ! Guess some reasonable lens parameters; use glass
                      ! type SF6 from Schott catalog
SF6

3 TH 20               ! Surface 3 is a kinoform on side 2 of the first
                      ! element
3 USS 16              ! Defined as Unusual Surface Shape 16 (simple DOE)
CWAV .6328            ! Zones are defined as one wave phase change at this
                      ! wavelength
HIN 1.7988 55         ! Assume the zones are machined into the lens.  You
                      ! can also apply a film of a different index.
RNORM 1

4 TH 2 GTB S          ! The first side of the second element is also a DOE
SF6
4 USS 16
CWAV .6328
HIN 1.7988 55
RNORM 1
```

doi:10.1088/978-0-7503-3695-6ch17

```
5 CV 0 TH 50          ! Start with a flat surface
7                     ! Surfaces 6 and 7 exist
AFOCAL                ! because they are required for AFOCAL output.
END                   ! End of lens input file.
```

We guessed a value for RD number 2, and we came close enough to start with. Here is the system at this stage, with no aspheric DOE terms yet, shown in figure 17.1. If you are not yet familiar with the USS surface shapes, open the help file by typing

```
HELP USS
```

in the Command Window and selecting type USS 16, a simple DOE.

The beam is expanded but not collimated, and the intensity profile is still that of the Gaussian input beam. The task is to find the DOE OPD terms that will accomplish both of our goals. To start with, we will keep both sides of the second element flat but add aspheric terms to it to define the DOE. Here is an optimization MACro that might do the job (**C17M2**):

```
PANT             ! Start of variable parameter definitions.
RDR .001         ! This is a very small beam, so use smaller derivative
                 ! increments to start with
VY 2 RAD
VLIST TH 3       ! Vary the airspace
VY 3 G 26        ! Vary term Y**2,
VY 3 G 27        ! Y**4,
VY 3 G 28        ! and Y**6
VY 3 G 29        ! and Y**8

VY 4 G 26        ! Do the same at surface 4
VY 4 G 27
VY 4 G 28
VY 4 G 29

END

AANT             ! Start of merit function definition
AEC 1 1 1
ACC 4 1 1
LUL 150 1 1 A TOTL            ! Prevent the system from growing too large
M 5 1 A P YA 0 0 1 0 5        ! Ask for a beam radius of 5 mm on surface 5
```

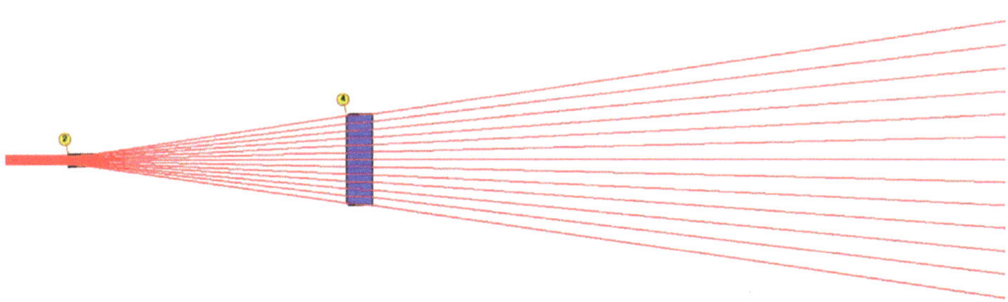

Figure 17.1. Starting point for DOE beam shaper.

```
M 0 1 A P FLUX 0 0 1 0 6          ! Ask for a flux falloff of zero at several zones
M 0 1 A P FLUX 0 0 .98 0 6
M 0 1 A P FLUX 0 0 .97 0 6
M 0 1 A P FLUX 0 0 .96 0 6
M 0 1 A P FLUX 0 0 .95 0 6
M 0 1 A P FLUX 0 0 .94 0 6
M 0 1 A P FLUX 0 0 .93 0 6
M 0 1 A P FLUX 0 0 .92 0 6
M 0 1 A P FLUX 0 0 .91 0 6
M 0 1 A P FLUX 0 0 .85 0 6
M 0 1 A P FLUX 0 0 .8 0 6
M 0 1 A P FLUX 0 0 .7 0 6
M 0 1 A P FLUX 0 0 .5 0 6
M 0 1 A P FLUX 0 0 .3 0 6
GSO 0 .1 10 P                  ! Control the output ray OPD over an SFAN of 10 rays,
GSR 0 100 10 P                 ! and some transverse aberrations too.
END                            ! End of merit function definition.

SNAP
SYNO 40
```

This PANT file varies some of the general-purpose G variables, which we used in the previous lesson to vary the aspheric terms on the lens elements. But in this case, surfaces 3 and 4 are already defined as USS type 16, which is a simple DOE surface, and those terms therefore alter selected coefficients defining that shape. (The help file describes how the G terms are applied to each of the USS surface types.)

Run this, and the lens looks promising. So run it again, and then anneal (**22, 1, 50**). You get the lens in figure 17.2.

This is not bad at all. Let's try varying some of the higher-order coefficients. Add new terms on both DOEs in the AANT file, up to G 31, which is the Y**12 term. After reoptimizing, the lens (**C17L1**) looks much the same, but the merit function drops to 2.16E-7. Looks like the run converged!

How does the flux vary over the aperture now? Type the command

FLUX 100 P 6

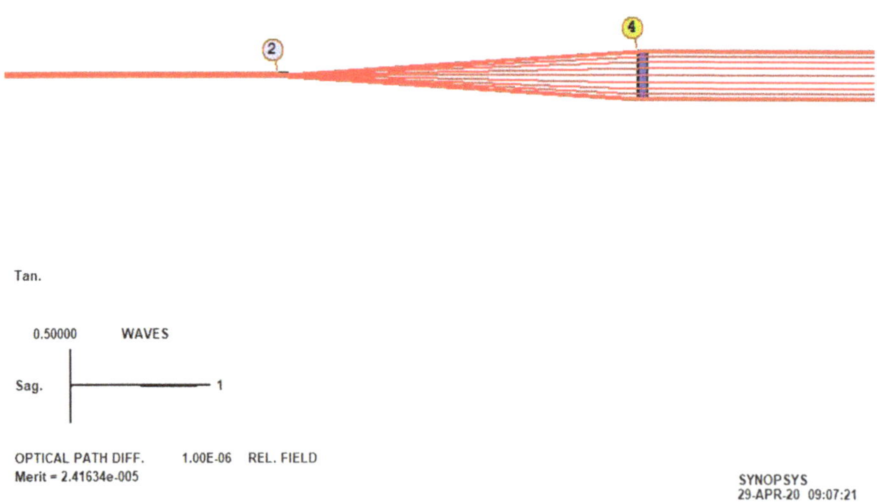

Figure 17.2. DOE-based beam shaper, after optimization.

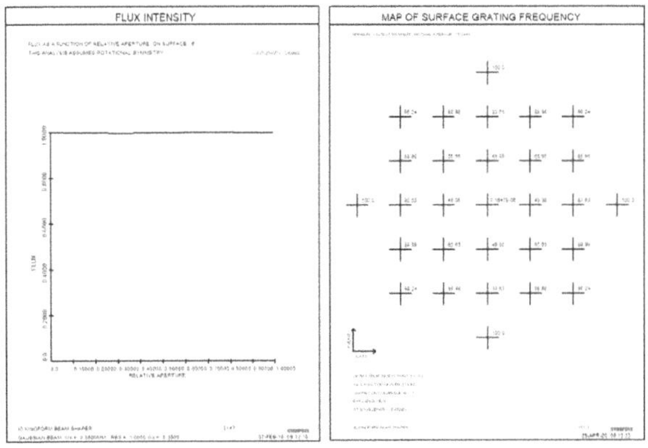

Figure 17.3. Flux uniformity of DOE-based beamshaper (left) and MAP of surface spatial frequency on surface 4.

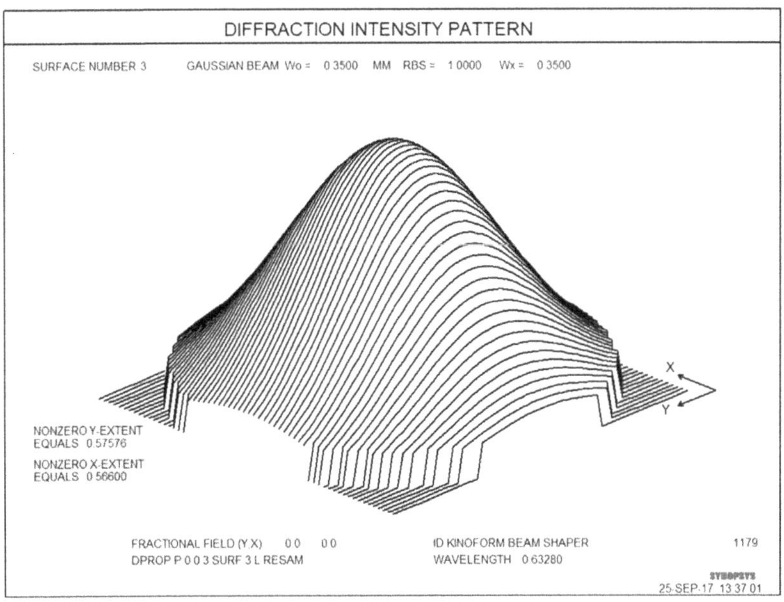

Figure 17.4. Beam intensity plot at surface 3.

and you get a beautiful curve, almost straight, shown on the left in figure 17.3.

This is indeed an excellent design. The question now is, can anyone make it? What is the spatial frequency on surface 4? If it is too high, fabrication technology may have trouble with it. Open the **MMA** dialog to select the input for a **MAP** command, select a map of HSFREQ over PUPIL with object POINT 0 and raygrid CREC with a grid of 7, DIGITAL output, and PLOT. The result shows a frequency of 100.0 c/mm at the edge of the lens, on the right in figure 17.3.

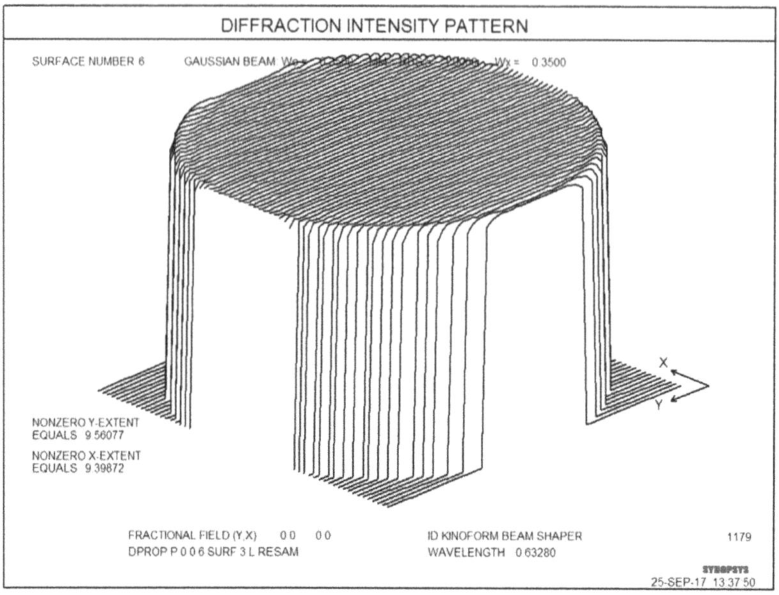

Figure 17.5. Beam intensity plot at surface 6.

That works out to just 10 um/cycle, which is possible but perhaps not easy. Can we reduce that to, say, 50 c/mm? Add the variable 5 RAD to the variable list and add a new aberration to the AANT file:

```
M 50 .01 A P HSFREQ 0 0 1 0 4
```

The program now controls the spatial frequency at the marginal ray intercept on surface 4. Reoptimize, and now surface 5 is slightly convex and the spatial frequency on 4 is right at 50 c/mm. The flux uniformity is as good as before. Mission accomplished!

How well did we do? Run the **DPROP** command, asking for the profile at surface 3, before the beam has been restructured. This shows the Gaussian profile of the beam at that point, as in figure 17.4.

```
DPROP P 0 0 3 SURF 3 L RESAMPLE
```

And now we do the same on surface 6, results shown in figure 17.5. Essentially perfect!

```
DPROP P 0 0 6 SURF 3 L RESAMPLE
```

IOP Publishing

Lens Design (Second Edition)
Automatic and quasi-autonomous computational methods and techniques
Donald C Dilworth

Chapter 18

A more challenging optimization challenge

Designing a lens with plane-parallel plates; inserting real glass types for the glass models

In chapter 7 you designed a seven-element lens starting from plane-parallel surfaces, which is about as close to starting from scratch as you can get in this business. That lesson was intended to demonstrate the speed of the PSD III optimization algorithm, which is one of the factors that make modern number crunching so effective.

In this lesson, you will start with the same system—but in this case, you want to achieve a high modulation transfer function (MTF) at four field points and substitute catalog glass types for the glass models of the earlier lesson. To do the latter, you will use the automatic real-glass insertion program, **ARGLASS**[1].

Here is the input and optimization MACro (**C18M1**):

```
RLE                     ! The starting system.
ID TEST PSD III
OBB 0 20 12.7
WAVL CDF
UNITS MM
1 TH 5 GLM 1.6 50
2 TH 5
3 TH 5 GLM 1.6 50
4 TH 5
5 TH 5 GLM 1.6 50
6 TH 5
7 TH 5 GLM 1.6 50
8 TH 5
9 TH 5 GLM 1.6 50
10 TH 5
11 TH 5 GLM 1.6 50
12 TH 5
13 TH 5 GLM 1.6 50
14 TH 50
15
APS 7
END
PAD/U     ! Show the initial system.
TIME      ! Start a timer, then define a symbol, AWT, for the ap. weight

AWT: 0.5     ! almost equal weight over aperture
QUIET        ! not showing everything on the monitor speeds things up
```

[1] ARGLASS™ is a trademark of Optical Systems Design LLC, an Arizona, USA corporation.


```
PANT          ! Define variables.
VY 1 YP1      ! Vary the paraxial stop position.
VLIST RAD 1 2 3 4 5 6 7 8 9 10 11 12 13 14
VLIST TH ALL
VLIST GLM ALL
END

AANT            ! Start of merit function definition.
AEC
ACC
M 33 2 A GIHT
GSR AWT 5 5 M 0     ! Note how weights are assigned to the several fields.
GNR AWT 5 4 M .3    ! This creates a ray grid at the .3 field point
GNR AWT 5 4 M .5    ! These for the 0.6 field point
GNR AWT 5 4 M .65   ! These for the 0.75 field point
GNR AWT 4 4 M .8    ! These for the 0.8 field point
GNR AWT 4 4 M 1     ! Full field
END

SNAP 100
DAMP 1
SYNOPSYS 10
SYNOPSYS 50
SYNOPSYS 50
SYNOPSYS 100
ANNEAL 50 10
ANNEAL 50 10

LOUD            ! Restore output to the monitor
MERIT?

STORE 3         ! Store the results in the library.

TIME                        ! See how long the job took
MOF M 0 40 80 0 Q 30 20 10  ! Calculate the MTF over field.
```

The job runs for about 60 s and produces the lens in figure 18.1 (**C18L1**) and the MTF in figure 18.2.

Optically, this lens is superb—but some elements are too thin. We have to fix that. Select the lines in the MACro from PANT to the first SYNOPSYS command. Type <Ctrl>+C to copy them to the clipboard. Then click the NewMACro Window

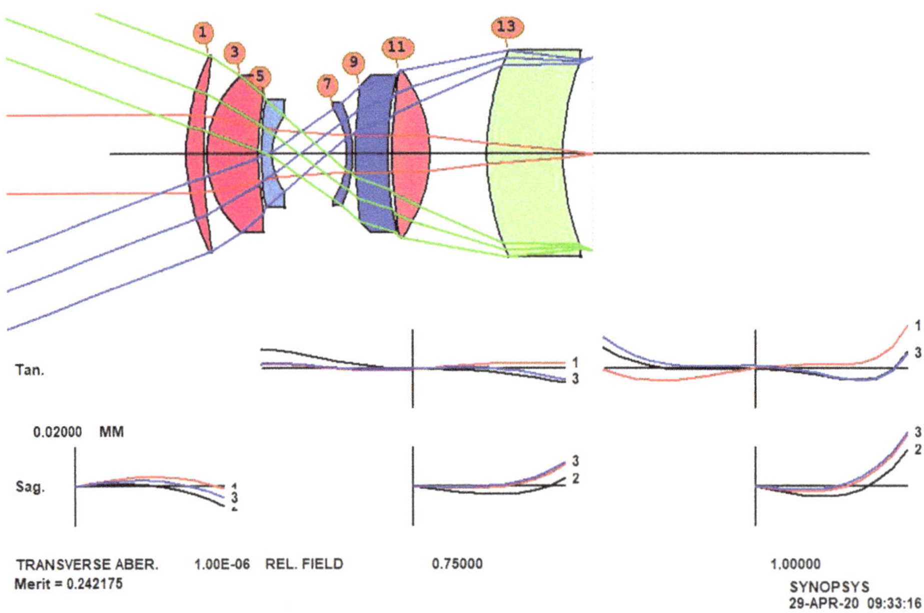

Figure 18.1. Results of first optimization.

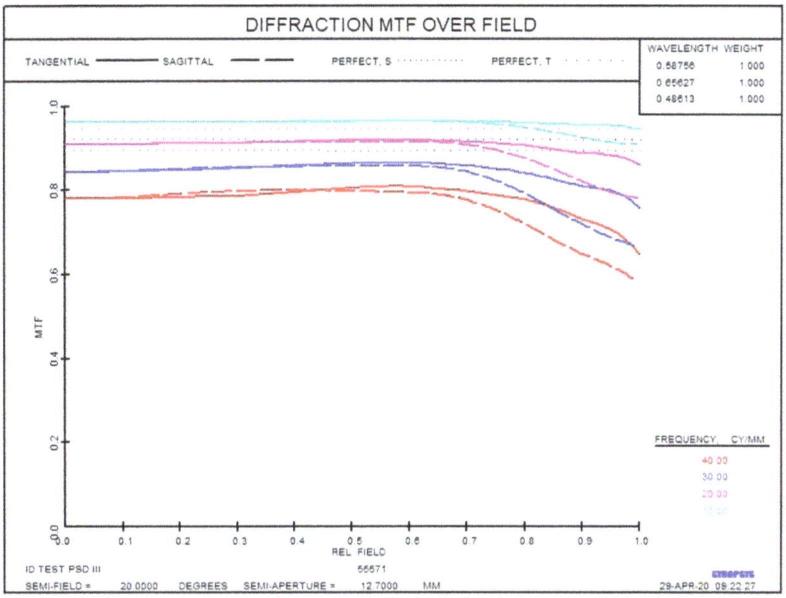

Figure 18.2. MTF over field of the optimized lens.

button and type <Ctrl>+V to paste those lines into the new editor window. Now add a line to the AANT section:

```
ADT 6 1 10
```

and run this version. The elements are thicker, and the MF goes up to 0.259. Anneal (**55, 2, 50**), and the MTF is again very good.

Now open the **MRG** dialog, which prepares the input for **ARGLASS** and lets you specify a number of filters that affect which glasses the program selects. You might only want inexpensive glasses, or those with good acid resistance, for example. Here is what is selected when we run the program, selecting the Schott catalog and Sort (**C18L2**):

```
--- ARGLASS 1 QUIET
Lens number      1 ID TEST PSD III
 GLASS N-SK16            HAS BEEN ASSIGNED TO SURFACE    1; MERIT =    0.257108
 GLASS N-LAK21           HAS BEEN ASSIGNED TO SURFACE    3; MERIT =    0.270393
 GLASS N-LAK14           HAS BEEN ASSIGNED TO SURFACE   11; MERIT =    0.287159
 GLASS F5                HAS BEEN ASSIGNED TO SURFACE    5; MERIT =    0.288196
 GLASS N-LASF46A         HAS BEEN ASSIGNED TO SURFACE   13; MERIT =    0.295070
 GLASS N-SF14            HAS BEEN ASSIGNED TO SURFACE    7; MERIT =    0.275430
 GLASS N-SF66            HAS BEEN ASSIGNED TO SURFACE    9; MERIT =    0.263344
 Type <ENTER> to return to dialog.
  SYNOPSYS AI>
```

To examine the properties of these glasses, enter the command

```
PGA ALL                    !Print Glass Attributes, all glasses
```

And you get a table, part of which is shown here:

```
************************************************************
 GLASS ATTRIBUTE FOR SURFACE NO.  11
 SCHOTT          N-LAK14
 GLASS IS A PREFERRED TYPE.
 GLASS IS ENVIRONMENTALLY SAFE (NO Pb OR As).

 PRICE   BUBBLE   HUMIDITY   STAIN   ACID RESIST   ALKALI RESIST   SP GRAVITY
   3.0       0          3        3             6               1         3.63
 THIS GLASS HAS A LIST OF TRANSMISSION VALUES ATTACHED
 VALID RANGE OF TRANSMISSION DATA:
 LOW     HIGH
  0.300   2.500
 GLASS HAS SELLMEIER INDEX COEFFICIENTS:
  1.5078120E+00 3.1886680E-01 1.1428720E+00 7.4609870E-03 2.4202480E-02
8.0956520E+01
 GLASS HAS 6 DNDT VALUES FROM GLASS TABLE:
  2.6800E-06  1.1500E-08 -1.4400E-11  3.7200E-07  5.5300E-10  2.2600E-01
 THERMAL COEFFICIENT (ALPHA) =  0.550E-05

************************************************************
```

This looks like what you are after. This lens is shown in figure 18.3. To finish up, you can insert a dummy surface where the stop wants to be and then assign a real stop there and reoptimize.

We recommend you run this exercise yourself (you will need a license, since the read-only mode will not allow you to save the lens, and the 12-surface mode will not allow seven elements). Try changing some of the field weights or the aperture weight and running it again. The results are rather sensitive to those changes, and you will need to get a feel for what works and what doesn't as you develop your own lens design skills.

This example started with plane-parallel plates and produced a rather good lens. The final form strongly resembles the classic Double Gauss, with a central stop and elements on each side bent inward. It is always pleasant when SYNOPSYS rediscovers one of the classical lens forms.

What happens if you run this job on DSEARCH? (That program starts with nonzero powers, assigned according to its rules, and finds many more designs.) I tried it on this problem and got an even better solution. Try it yourself and see! Adjust the input variables to see what happens. This is your most powerful tool, so it makes sense to learn how to use it.

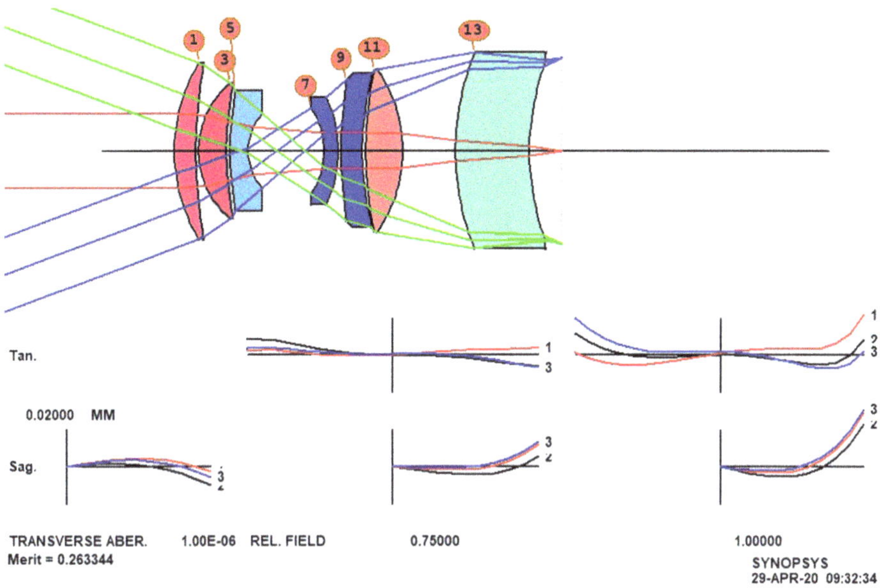

Figure 18.3. The final lens, with real glass.

18.1 Glass absorption

A final question: This lens has two dense flint elements, at surfaces 7 and 9. Glasses in that region of the glass map tend to absorb at the shorter wavelengths—which can affect transmission in the blue. It's a good idea to check. Type the command **XCOLOR**, and you get the plot in figure 18.4. There you see that the transmitted color indeed has a slight yellow tint. If that is a problem for the application, you have to find a glass type with better blue transmission than N-SF66 from the Schott catalog, used on surface 9 in this example. The glassmap display (**MGT**) can clarify the situation, as illustrated in figure 18.5, where the length of the red line is a function of the absorption at 0.4 μm. (Click Graph, and then Absorption at Wavelength 0.4). On the left you see that glass N-SF66 has a long red line, because its absorption at that wavelength is rather high, and there are no glasses nearby that are much better. But sometimes a different glass company can provide a more suitable glass, as shown on the right, where glass S-NPH4 from Ohara looks like a better choice for this lens.

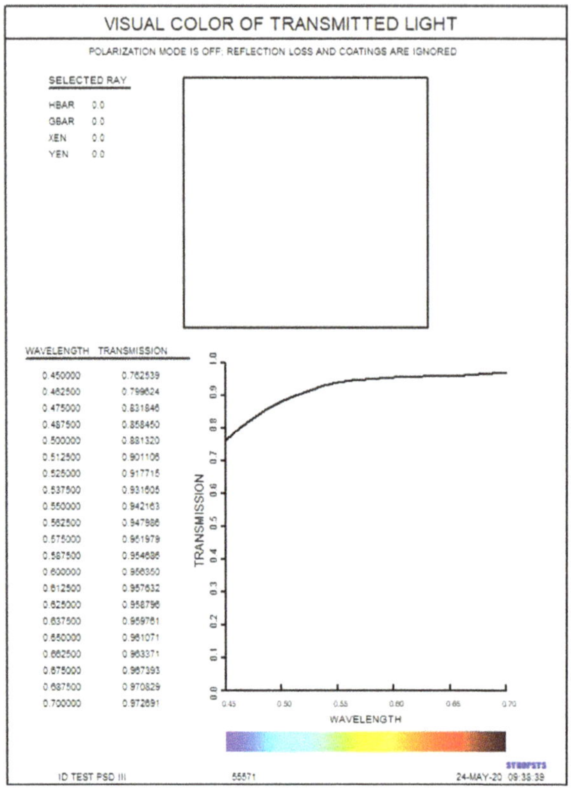

Figure 18.4. The transmitted color through the lens. Absorption in the blue region of the spectrum causes the transmitted beam to appear somewhat yellow, which may be a problem. A change of glass type may be in order.

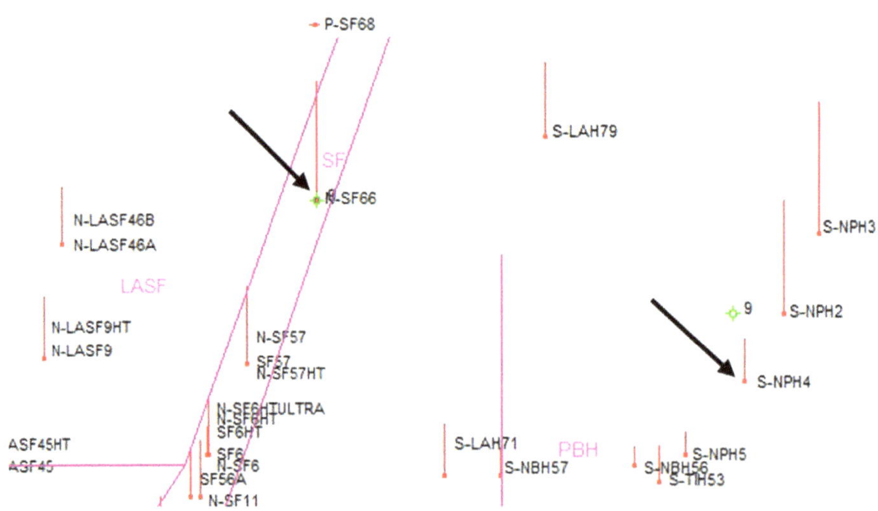

Figure 18.5. Absorption at a wavelength of 4 μm; the Schott catalog on the left, Ohara on the right. For this lens, type S-NPH4 may be a better choice than N-SF66.

In the event you need a certain glass type and cannot find a suitable substitute, the only other option is to reduce the thickness of the offending element as much as physically practical. And, if the lens is intended for aerial reconnaissance, the yellow tint is of no concern, since those systems generally use a yellow filter to cut through atmospheric haze anyway.

IOP Publishing

Lens Design (Second Edition)
Automatic and quasi-autonomous computational methods and techniques
Donald C Dilworth

Chapter 19

Real-world development of a lens

Global search for a seven-element lens; correcting for two object distances

In chapter 18 you designed a seven-element lens starting with nothing but plane-parallel surfaces and had the program fit the design to catalog glass types automatically with the ARGLASS feature. This lesson will develop the lens further, as you would do if you wanted to manufacture it, and describes some additional procedures that would then be appropriate. To make it a real 'real world' lesson, we will show how a designer will follow various clues in order to arrive at a solution, and how not all clues lead to success. That is important too: it is instructive to see how sometimes one wanders into blind alleys. As you develop your skills as a lens designer, you will encounter many of them, and should not be discouraged since it happens to all of us.

You will do this lesson in two ways; first with DSEARCH along with a number of other tools. Then, in chapter 21, we show another approach that is actually quicker and easier. You should know about the tools used in both of those approaches.

First, run DSEARCH to find a good starting point. Here is the input: **(C19M1.MAC)**

```
CORE 64
ON 1
OFF 99
TIME
DSEARCH 6 QUIET
SYSTEM
ID DSEARCH SAMPLE
OBB 0 20 12.7
WAVL 0.6563 0.5876 0.4861

UNITS MM
END
```

doi:10.1088/978-0-7503-3695-6ch19

```
GOALS
ELEMENTS 7
FNUM 3.575
BACK 50 SET
STOP MIDDLE
STOP FREE
RSTART 600
RT 0.5
FOV 0.0 .5 .7 .9 1
FWT 2 2 2 1 1
NPASS 55
ANNEAL 200 20 Q 44
COLORS 3
SNAPSHOT 10
QUICK 55 55
END
SPECIAL PANT

END
SPECIAL AANT
LUL 150 1 1 A TOTL
END
GO
TIME
```

Notice the **RT** parameter in this file. That controls how individual rays are weighted in the merit function, as you saw in chapter 7. A value of zero gives all the rays in a given grid the same weight, while a higher value will weight rays near the center of the pupil more than rays near the edge. That is a useful way to increase the resolution of a lens; the ray fans may fly away strongly right near the edge, but if the central portion is very flat the resolution will be high anyway. This is a parameter you will often want to experiment with. A value of 0.5 is a good initial guess. The initial radius assigned to each surface is also a useful parameter. Here, we set it about 6 times the desired focal length, which is often a good choice too.

We also set the back focus distance to exactly 50 mm in this case, which means the program will not use a YMT solve on the last airspace. This, too, is a parameter that can sometimes explore different branches. We will fix up element thicknesses later, when we have a good configuration, so there Is no custom monitor applied at this stage.

Run this, and then optimize and anneal (**50, 2, 50**), using the file DSEARCH_OPT, which is in a new editor window. The lens, shown in figure 19.1, is quite good.

Suppose you want the lens to work over a range of object distances from one meter to infinity. There are two ways to implement that requirement: with multi-configurations, which is very flexible but complicated, or by declaring it a zoom lens

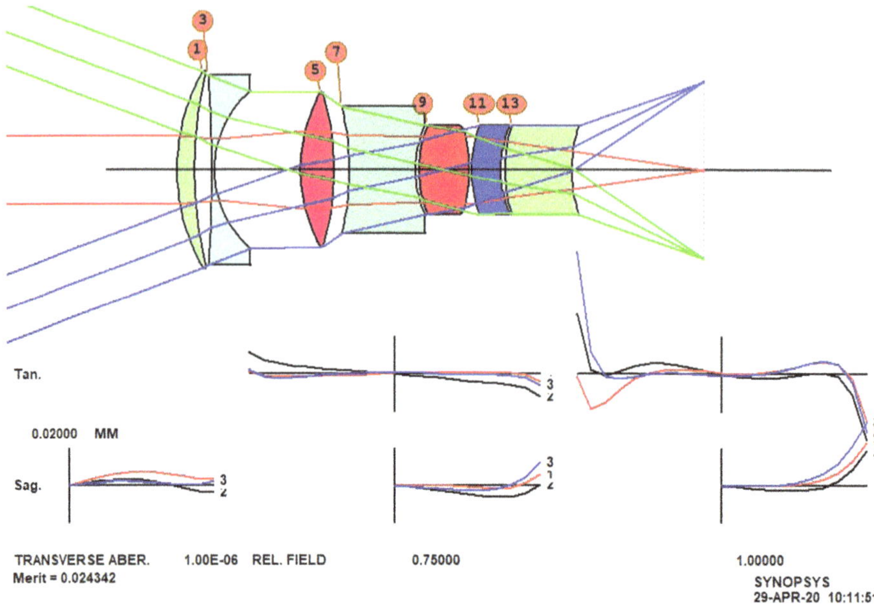

Figure 19.1. Lens as returned by DSEARCH, then optimized and annealed.

in which the object distance zooms. The second approach is better here, since it is simpler, does what you want, and you can examine intermediate object distances very easily. You have to set up this lens as a ZFILE zoom lens. But first, you have to assign a real stop where it looks like it wants to be, at surface 9.

```
CHG
APS -9
END
```

Then optimize and anneal again, after deleting the variable YP1. Now the lens is ready to be turned into a zoom lens.

Enter the following in a new editor:

```
CHG
15 CAO 32        ! fix the CAO on the image (so FFIELD works)
FFIELD           ! adjust the object height so the image fills the CAO there
14 YMT           ! assign a paraxial focus solve to surface 14
ZFILE 1          ! start of the ZFILE section
14 14            ! there is one zooming group, the last thickness
ZOOM 2           ! ZOOM 1 is default; ZOOM 2 gets OBA object on the next line
OBA 1000 -366.554 12.7     ! the object description at this zoom
END              ! end of changes
```

This input sets a hard aperture at the image so the FFIELD directive has a target, puts a thickness solve on 14 so all zooms refocus automatically, and declares a single zooming group, surface 14. Then it defines the object distance for ZOOM 2 at 1000 mm distance, with a negative YPP0 because the value in ZOOM 1 is also negative, and they have to have the same sign.

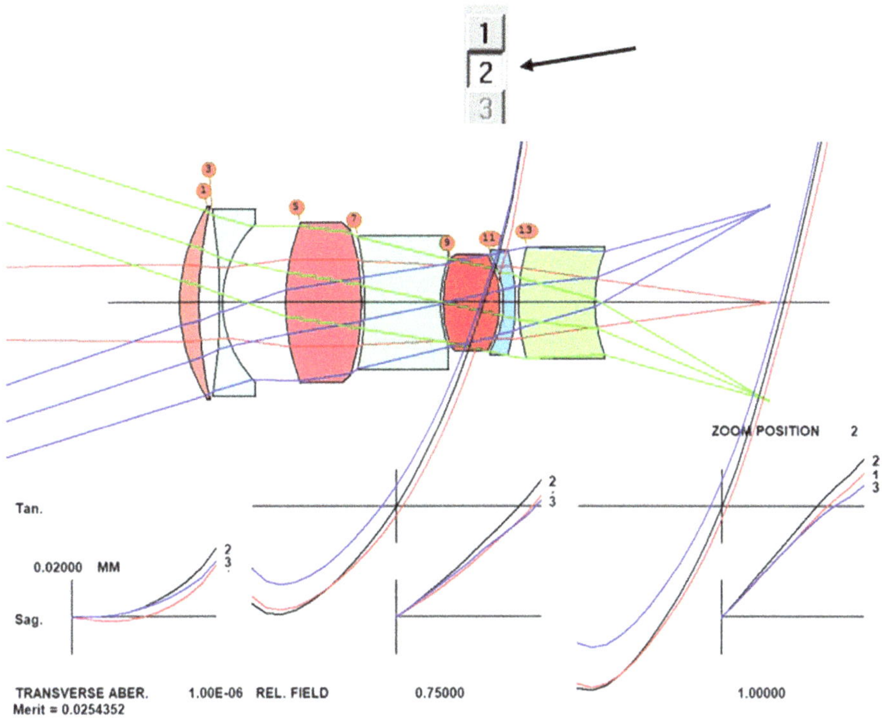

Figure 19.2. Lens with modified object distance.

Run this MACro, and the lens changes to a zoom lens, with only a single airspace zooming in this case. Now you see a new toolbar on the right side of the monitor. What does the image look like in ZOOM 2? Click on button 2, and you see the lens at that zoom setting, shown in figure 19.2.

Pretty awful! The aberrations changed markedly when we changed the object distance. We have to correct the image at both conjugates. Here is a MACro that might do the job (**C19M2**):

```
AWT: 0.5
PANT            ! Define variables.
CUL 1.9         ! Set upper limit of 1.9 on index variables.
FUL 1.9
!VY 1 YP1       ! Don't vary YP1
VLIST RAD ALL ! Varies all radii that are not flat.
VLIST TH ALL  ! varies all thicknesses and airspaces except for the
! back focus, thickness 14, which has a solve in effect
VLIST GLM ALL
END

AANT                    ! Start of merit function definition.
AEC                     ! Activate automatic edge-feathering monitor
ACC                     ! and maximum center thickness monitor.
ADT 6 .1 10             ! Keep diameter/thickness ratio 6 or more
!M 33 2 A GIHT ! Comment this out, since the FFIELD will control scale
LUL 150 1 1 A TOTL
```

```
M 50 .1 A BACK        ! Since the back focus will vary, keep it reasonable
M 90.61 1 A FOCL      ! Add this requirement so the focal length doesn't
                      ! change
GSR AWT 10 5 M 0      ! Note how weights are assigned to the several field
                      ! points,
       ! and the symbol AWT controls the aperture weighting.
GNR AWT 5.5 4 M .5    ! This creates a ray grid at the ½ field point
GNR AWT 5.5 4 M .7    ! These for the 0.7 field point
GNR AWT 3 4 M 1       ! Full field gets the lowest weight.

ZOOM 2        ! Targets for zoom 2 (with the object at one meter)
GSR AWT 10 5 M 0      ! Note how weights are assigned to field points.
GNR AWT 5.5 4 M .5    ! This creates a ray grid at the ½ field point
GNR AWT 5.5 4 M .7    ! These for the 0.7 field point
GNR AWT 3 4 M 1       ! Full field gets the lowest weight.
END

SNAP
SYNO 50
```

Run this and anneal, and the lens is better but still not very good, with about equal and opposite errors at the ends of the zoom range, as shown in figure 19.3.

Some subtleties deserve mention: the **GLM ALL** variable will vary all glass models currently in the lens, which means all elements, since DSEARCH uses the glass model unless told otherwise. We have to control the focal length, since the object height will be continuously adjusted so the image CAO is filled at full field, and the image height will therefore not be available as a target to control it. This is done with the FOCL operand in the AANT file.

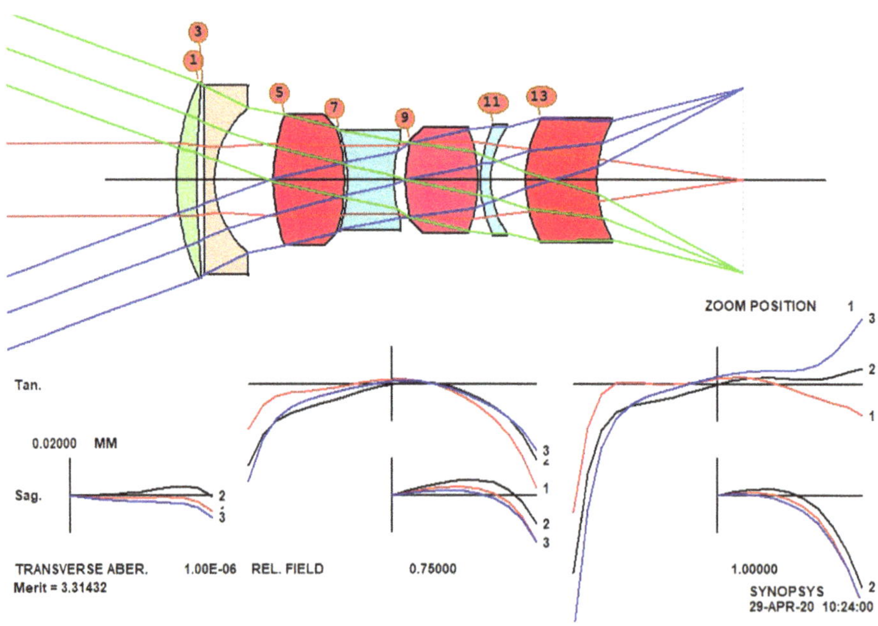

Figure 19.3. Lens reoptimized for two object conjugates; the quality is not good.

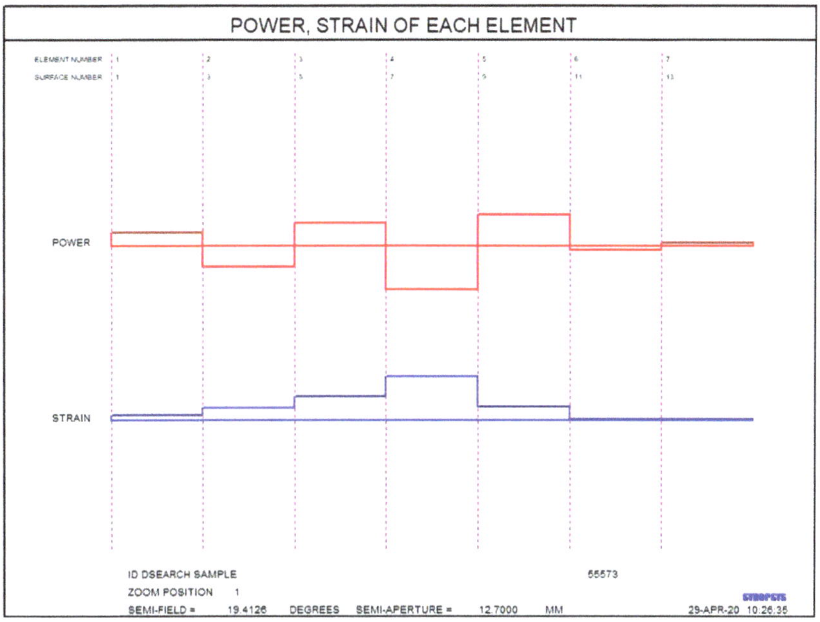

Figure 19.4. Plot produced by STRAIN P command.

This is better than zoom 2 was before, but there is still a loss of resolution. What to do? We need more variables. What should we add?

A classic tool for cases like this is the **STRAIN** calculation. The idea is that the elements with the largest strain are contributing most of the aberrations, and splitting an element there might relieve it. (Strain is defined here as the sum of the squares of the 3rd and 5th-order aberrations of that element.)

Type **`STRAIN P`** in the CW. You see the results in figure 19.4.

Indeed, element 4 has the largest strain. Now you can do one of two things: You can split that element and reoptimize, or you can use a different tool that can figure out the best place to add an element. We will try it both ways. First, save this version, so you can go back if things don't work out. Type

```
STORE 1.
```

Then go to the WorkSheet (type **`WS`** or click on the button). Then click the button , which lets you split an element by clicking in the PAD display on the axis inside that element. Click between surfaces 7 and 8, splitting the element. Your lens now looks like figure 19.5.

When the program splits (or adds) an element, it assigns an index pickup, because at that moment it has no other index data. Change the index pickup on surface 9 to a glass model by typing

```
9 GLM
```

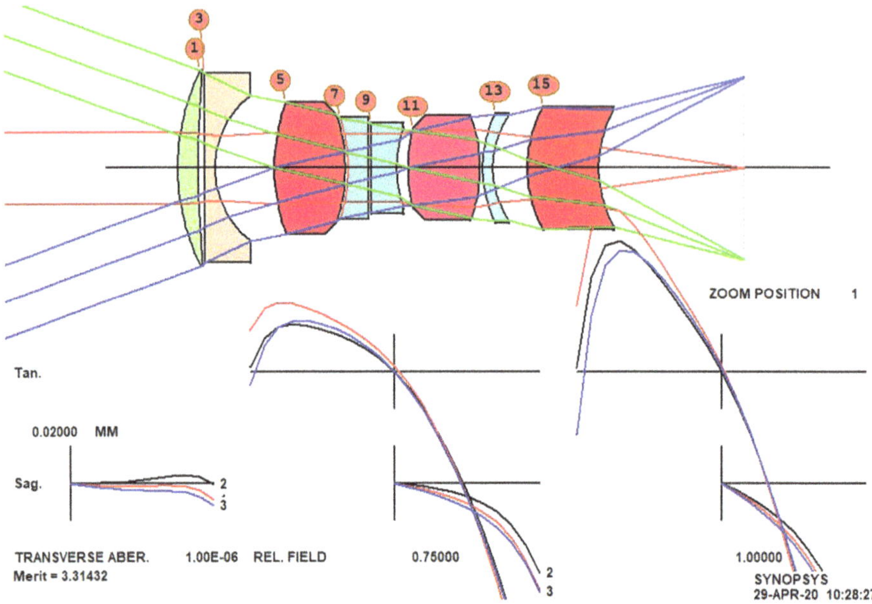

Figure 19.5. Lens with split element, before optimizing.

in the WS edit pane, and click Update. That assigns the element a model glass with properties similar to what were there before.

Make a new checkpoint, close WS, and run the optimization again and anneal. The MF goes to 1.88. This is better, but we wonder if it is good enough. Now what?

This is the way lens design has long been done, using classic tools, and it was a slow and arduous process. One can modify the lens and then try optimizing again. And again. And again....

Today we have better tools. Go back to the version you stored before splitting the element:

```
GET 1
```

and then add a line before the PANT file:

```
AEI 2 1 14 0 0 0 10 2
```

This will run the Automatic Element Insertion tool (**AEI**). Now the program will search for the best place to insert a new element. Run this, and the lens is better. Comment out the AEI line and run the MACro again, then anneal. The result (**C19L1**) is shown in figure 19.6.

The program has inserted a new element at surface 15, and the merit function came down to 1.54. There's a lesson here: The program can usually figure out how to improve a lens better than you can, so it's better to let AEI do it than to try things yourself, even though you have ideas that seem to make sense. Those things sometimes work, but AEI is better.

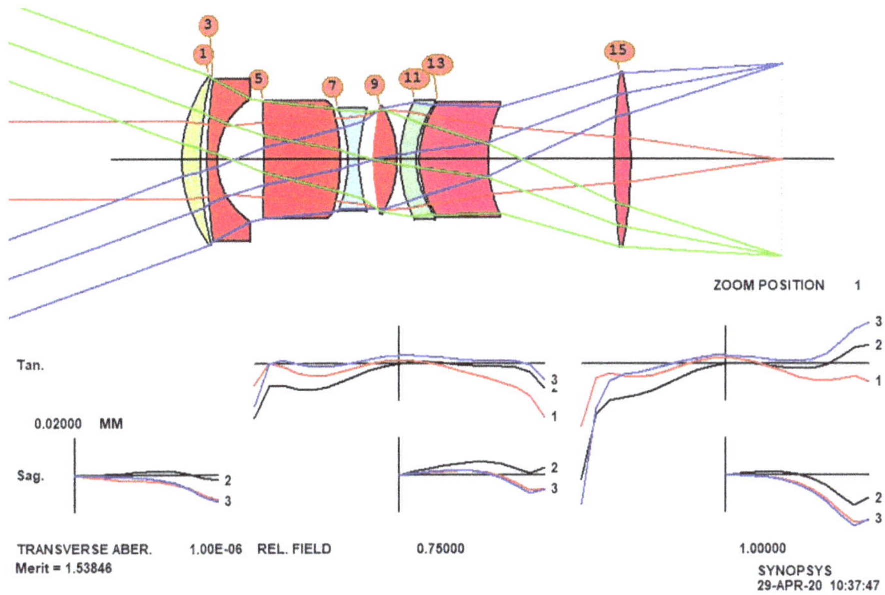

Figure 19.6. Lens reoptimized with new element inserted by AEI.

Here you see an improvement, and the modulation transfer function is also better, as you can check yourself. Now we have a lens that is somewhat corrected for both infinity conjugate and at one meter. But what about in-between distances? It would be a rude surprise if you built the lens and found that intermediate distances yielded a poor image. We have to check.

That is one of the reasons we chose to use the ZFILE zoom feature for this job. One can easily scan over the zoom range and spot any points that perhaps need attention. Click the button at the bottom of the zoom-selection bar: This opens the Zoom Slider[1], shown in figure 19.7, which is fun to watch.

Click the SCAN button and watch the PAD display. The image plane slowly moves back, from the infinity focus to the one-meter focus position, and then forward again. The good news is, the image quality shows little change over the entire range. (If it had changed, you could use the CAM command to create an intermediate focus position, making a total of three zooms, and then add some more targets for the new ZOOM 3 position in the AANT file. You can create and target up to 20 zooms this way, as you will learn if you type **HELP CAM** to read about that feature.)

This lesson shows how asking a lens to do something it was not selected for is generally a bad idea. We got a lens that compromises many of the goals we set and then balanced things as best we could. Chapter 21 shows how a different approach works better in cases like this.

But let us continue. Now we need to assign real glasses again.

---

[1] Zoom Slider™ is a trademark of Optical Systems Design LLC, an Arizona, USA, corporation.

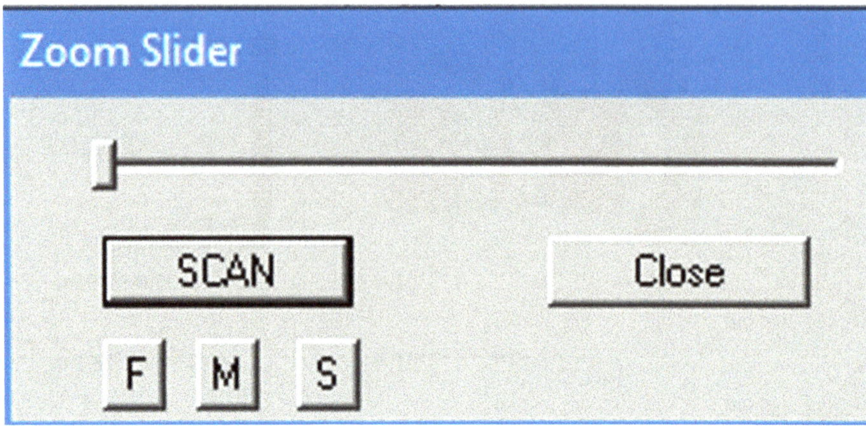

Figure 19.7. The Zoom Slider.

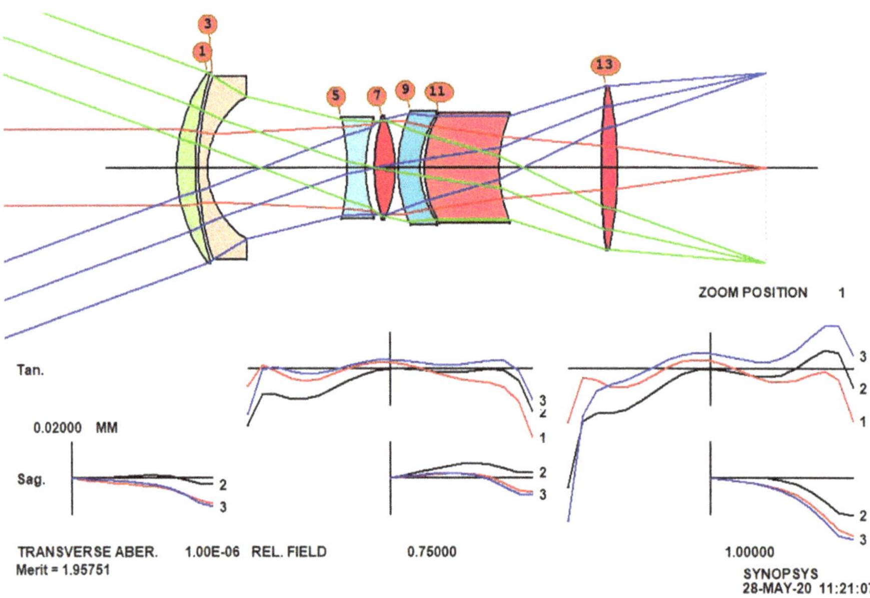

Figure 19.8. Lens with element removed by AED.

But … wait a minute. The sixth element shown in figure 19.6 bothers us. What is it doing? Can we remove it? We have to try! Remove the AEI directive and replace it with

```
AED 6 QUIET 1 15
```

And run it again, and—the program says the *third* element can be removed! Allow it to do this, then comment out the AED directive and optimize and anneal some more (**C19L2**). The merit function goes to 2.08. And you have eliminated an element, as shown in figure 19.8. See how AED can make better decisions than you can? This design is not quite as good as before, however, so we would probably go back to that version.

So that's how it's done: Figure out what's wrong and use the tools in SYNOPSYS to fix it. Sometimes it's quick and sometimes not. That's what lens design is all about, blind alleys and all. If this lens is still not good enough, another run on AEI might do the trick. Chapter 21 shows a better way to approach this kind of problem, letting the software handle the whole job.

But that is enough for this lesson.

Oh, I almost forgot: Why did we enter the surface number (14) for the zooming group, since the YMT solve will override it anyway? Well, the program requires a group definition, and it won't work otherwise. That's to save you from a serious mistake if you ever leave those data out for a real zoom lens.

Chapter 20

A practical camera lens

Global search for a camera lens design

Here are the goals for this chapter:

1. Focal length 90 mm
2. Semi field angle 20 degrees
3. Semi aperture 25.4 mm
4. Cell length approximately 100 mm
5. Back focus distance 50 mm or greater

In this chapter we will let **DSEARCH** find a starting point. Type **MDS** in the Command Window, to open the design search menu, shown in figure 20.1.

Enter the data shown by the arrows and click OK. We will modify this input later, when we see the results, and we guess the lens will need seven elements. The program asks you for a file name, so type a name of your choice. An editor window opens, containing the input required to run DSEARCH (**C20M1**).

```
CORE 64
TIME
OFF 1
ON 99
DSEARCH 1  QUIET
SYSTEM
ID DSEARCH SAMPLE
OBB 0 20 12.7
WAVL 0.6563 0.5876 0.4861

UNITS MM
END
GOALS
ELEMENTS 7
FNUM 3.54
BACK 0 0
TOTL 100 0.1
STOP MIDDLE
```



```
STOP FREE
RSTART 900
THSTART 12
ASTART 12
RT 0.75
FOV 0.0 0.75 1.0 0.0 0.0
FWT 5.0 3.0 3
NPASS 50          ! this gives the number of passes in the final MACro
ANNEAL 200 20 Q
COLORS 3
SNAPSHOT 10
QUICK 50 50      ! 50 passes in quick mode, 50 in real mode
END
SPECIAL PANT

END
SPECIAL AANT
LLL 50 .1 1 A BACK
END
GO
TIME
```

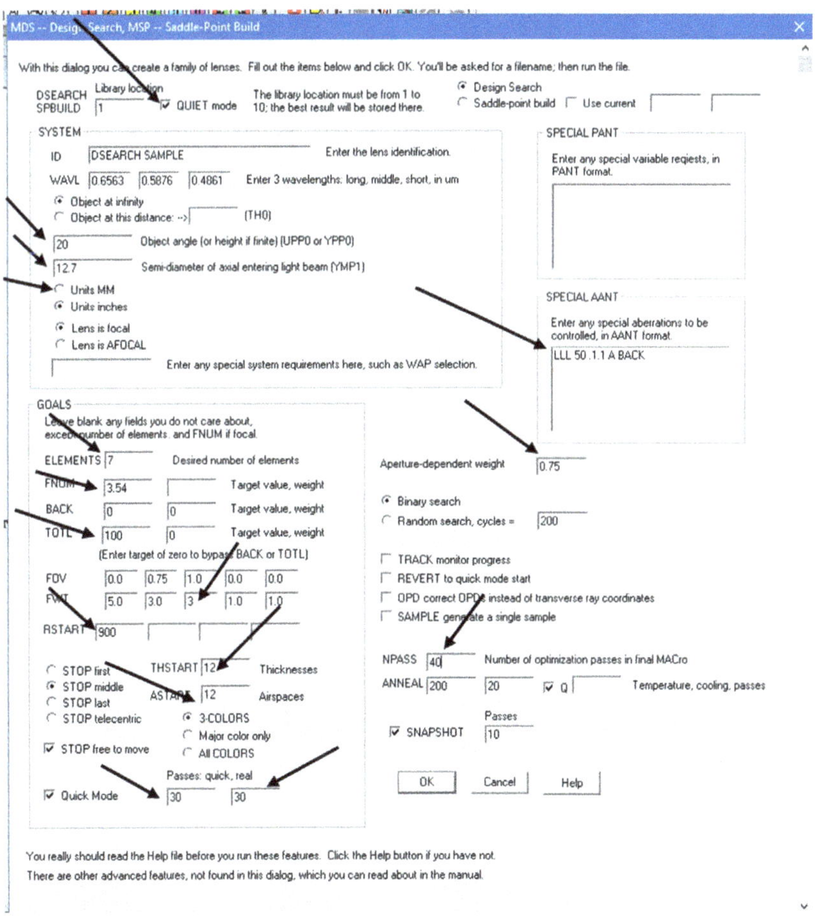

Figure 20.1. The DSEARCH dialog, opened with the command **MDS**.

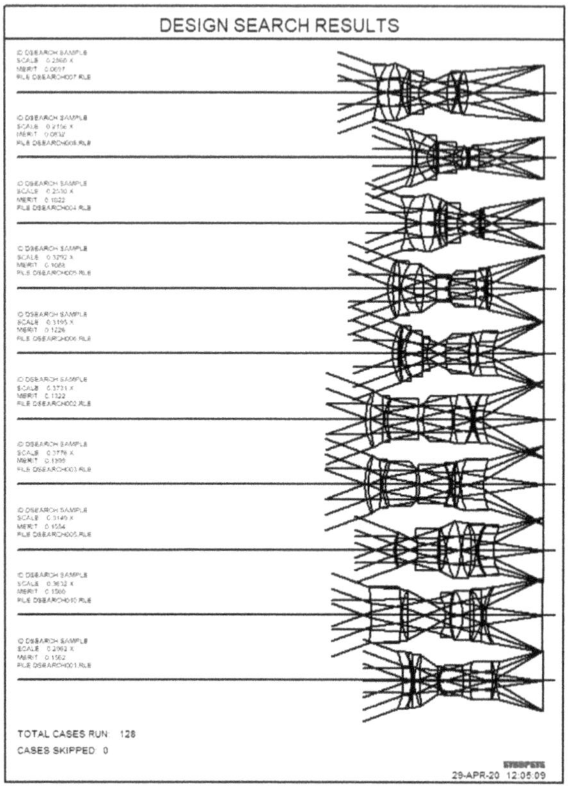

Figure 20.2. Ten lenses returned by DSEARCH.

We elected not to assign a weight to the back focus distance in the dialog, preferring to put that requirement in the SPECIAL section, where we gave it a one-sided requirement. **LLL** means Limit, Lower Limit, and this lets the back focus become larger than 50 without any penalty—but not smaller.

Run this macro, and you get a collection of potential starting points, shown in figure 20.2. Some of the lenses returned by DSEARCH have a negative element in front. We term such lenses '*inverse telephoto*'. A telephoto lens has a negative group at the *image* end, which serves to increase the size of the image, thereby getting the focal length of a longer lens in a small package. The inverse telephoto configuration, with a negative group in front, is often used for wide-angle lenses, where one wants just the opposite: the focal length is shorter than the physical size of the lens.

Let's examine these lenses more closely. Type, in the CW,

```
EM DSS
```

This command loads and runs the file DSS.MAC, which DSEARCH has created. That macro will open each of the lenses returned by DSEARCH, show it on the PAD display, and then wait for you to press the <Enter> key before opening the next. You will see that many of them have similar quality. When you see one you

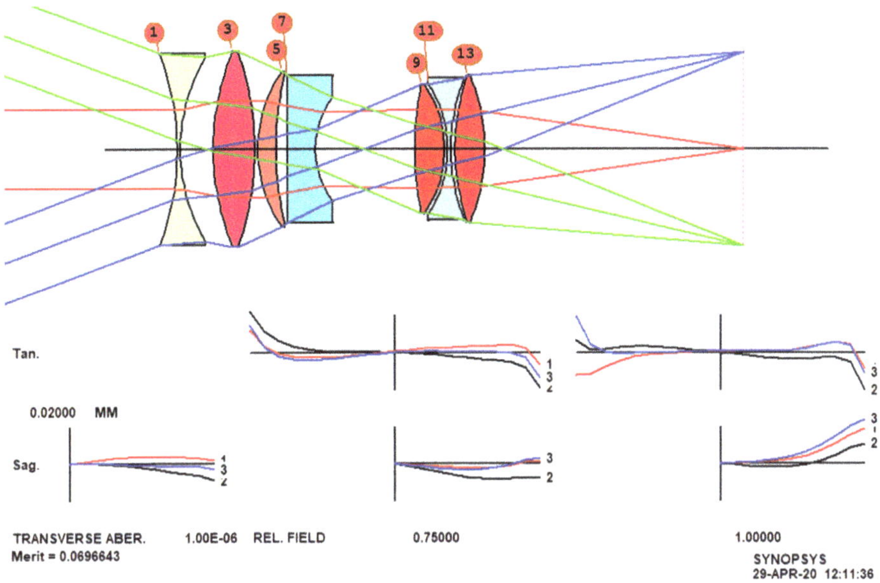

Figure 20.3. A good candidate from the DSEARCH results.

like, just press the <Esc> key to stop the macro at that point. We like the one saved by DSEARCH with the name DSEARCH07.RLE, which was also the top one in this example. (The names in your results may be different, since the order depends on which cores finished when.) The lens is shown in figure 20.3.

Run the optimization MACro DSEARCH_OPT, which DSEARCH has opened in a new window; then anneal (**50, 2, 50**). The lens changes slightly.

How good is this lens? Open the dialog **MOP** and enter the data shown in figure 20.4. Click the MOF button, and you get the modulation transfer function (MTF) curves in figure 20.5. Call this lens version 1.

Not too good—and we learned some things. The MTF varies with field, so we might need more field points in the DSEARCH input. We also decide that, since this lens is not far from the diffraction limit and we want to improve the MTF, we should try the **TOSHEAR** directive instead of the default transverse aberrations.

To understand this input, you have to understand how the MTF is calculated. A common method is to evaluate a *convolution integral*, which combines two copies of the exit pupil, one of them sheared in X or Y by an amount that depends on the frequency to be calculated. If the OPD errors at a given point are the same as those at the sheared point, the MTF is perfect, as far as those points go. Calculated over the whole pupil, the result is the MTF of the lens, in the scalar approximation. These calculations are explained in more detail in chapter 59.

The requests below combine a set of transverse ray targets with a set of GSHEAR requests when it makes the MF. The latter target the difference in OPD over a set of sheared pupil locations. We also increase the weight on fields 0.5 and 0.7, since those were more poorly corrected than the others.

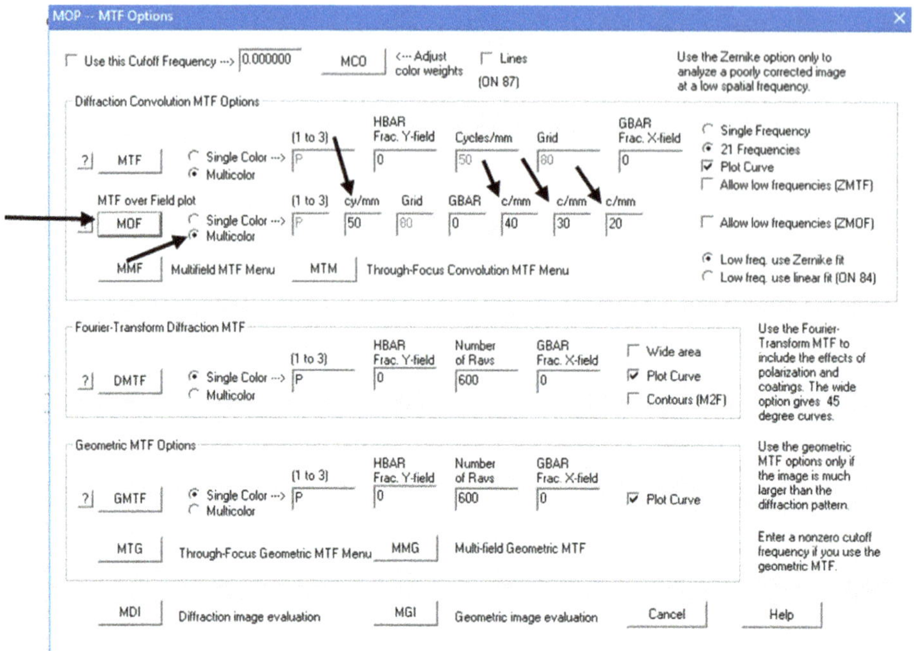

Figure 20.4. Dialog to run MOF, with four spatial frequencies requested.

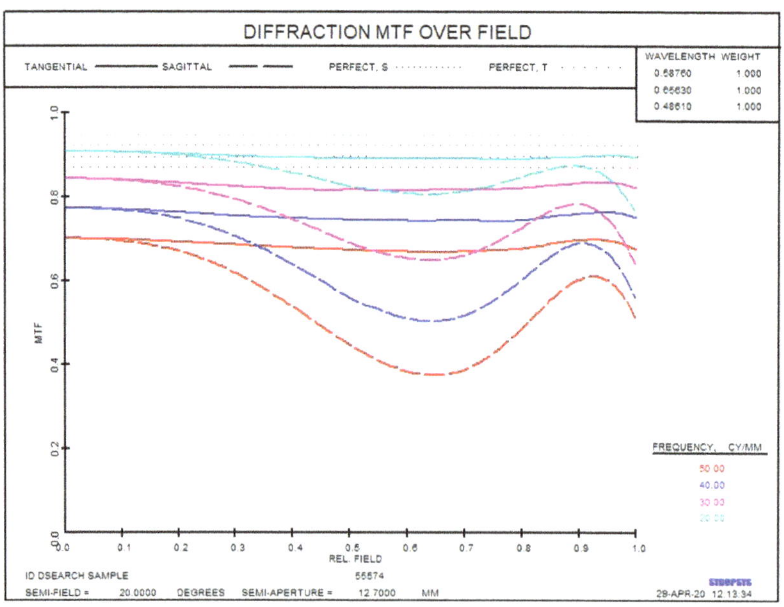

Figure 20.5. MTF curves for Version 1.

Here is the revised input for the GOALS section:

```
GOALS
 ELEMENTS 7
 FNUM 3.54
 BACK 0 0
 TOTL 100 0.1
 STOP MIDDLE
 STOP FREE
 RSTART 900
 THSTART 12
 ASTART 12
 RT 0.75
 FOV 0.0 .5 .7 .9 1
 FWT 1 1.5 1.5 1 1
 TOSHEAR
 NGRID 6
 NPASS 40
 ANNEAL 200 20 Q
 COLORS 3
 SNAPSHOT 10
 QUICK 80 80
 END
```

Run this DSEARCH file, and then optimize and anneal. You get a very different lens (**C20L1**), shown in figure 20.6, and higher MTF curves, shown in figure 20.7.

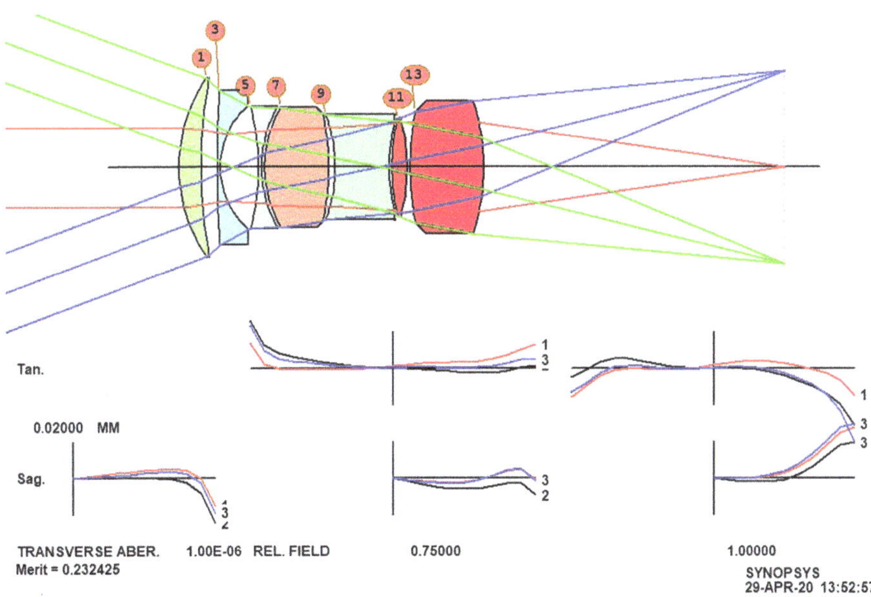

Figure 20.6. Version 2 lens, optimized and annealed.

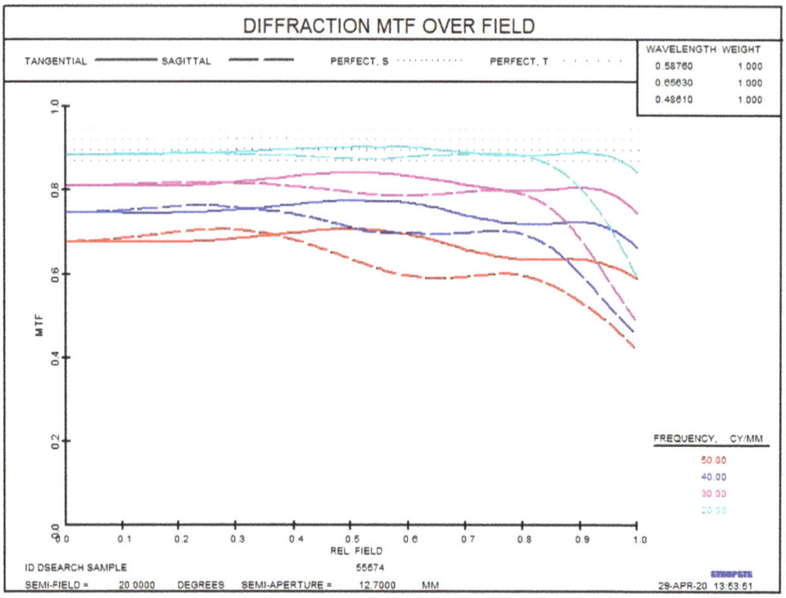

Figure 20.7. MTF of version 2 lens.

Perhaps we cannot do better with only seven elements. Add a line before the PANT command,

```
AEI 4 1 123 0 0 0 50 10
```

and run the MACro again. The program adds an element at surface 13. Then comment out the AEI line, optimize, and anneal once more. The MF comes down to 0.15.

This is Version 3, shown in figure 20.8, with the MTF in figure 20.9. This is an excellent lens.

This lens works very well, but some elements are too thin. It's time to gently push thicknesses up.

Add a new monitor to the AANT file:

```
ADT 7 .1 1
```

Then run it again and anneal. The elements are more reasonable. This is the version 4 design (**C20L2**) in figure 20.10; MTF curves are in figure 20.11. Can we do even better? Perhaps we can by relaxing the requirements somewhat.

We ran DSEARCH again, but with a target of 150 mm for the total length instead of 100. The lens form that came back was very different, and MTFs were higher (after running AEI, optimizing, and annealing), as seen in figures 20.12 and 20.13 (**C20L3**).

Now it is easy to tell a customer that, with *this* length you can get *this* MTF, and with *that* length you get *that*. What happens if you allow the back focus distance to be shorter? Again, tradeoffs are simple with this kind of autonomous lens design

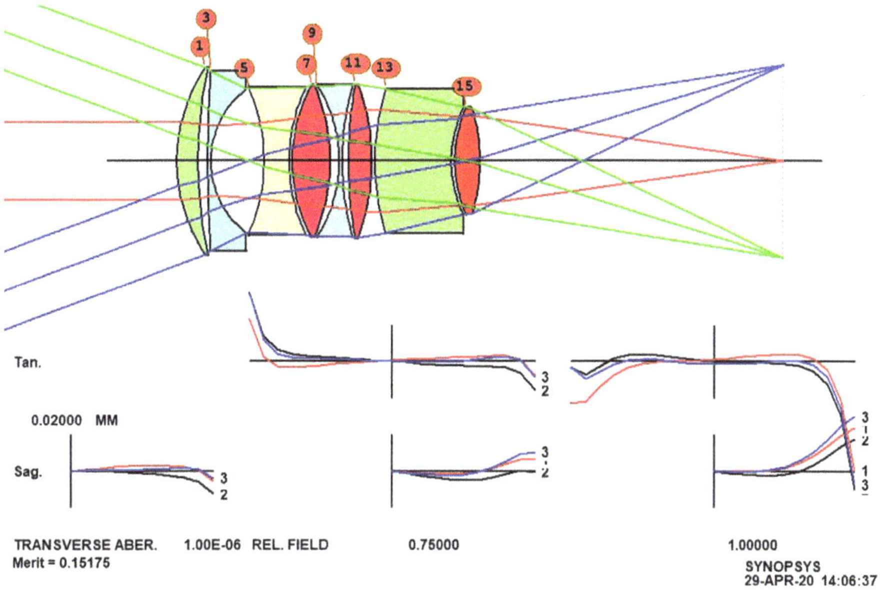

Figure 20.8. Version 3 lens, with element 7 inserted by AEI, optimized and annealed.

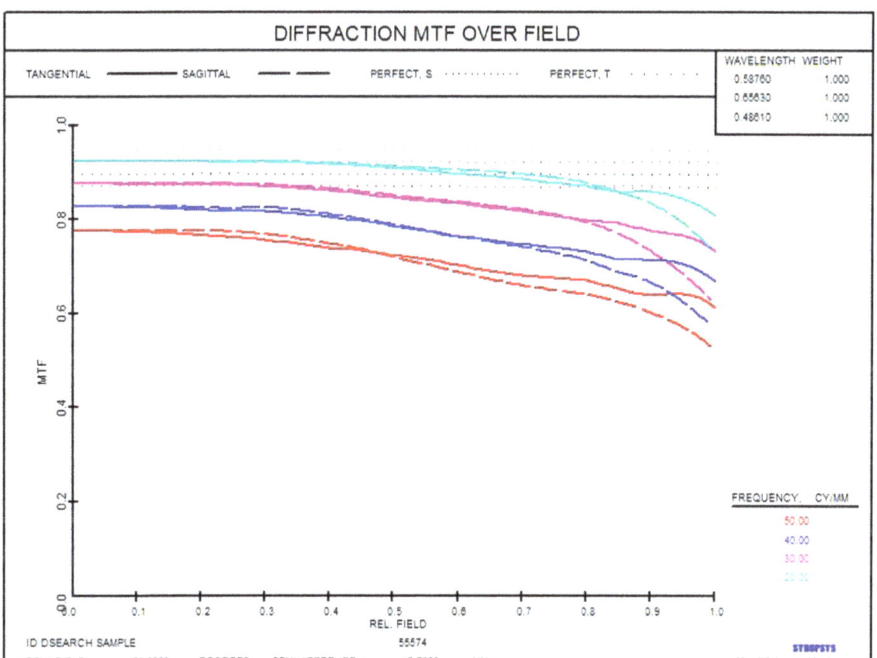

Figure 20.9. MTF of version 3 lens.

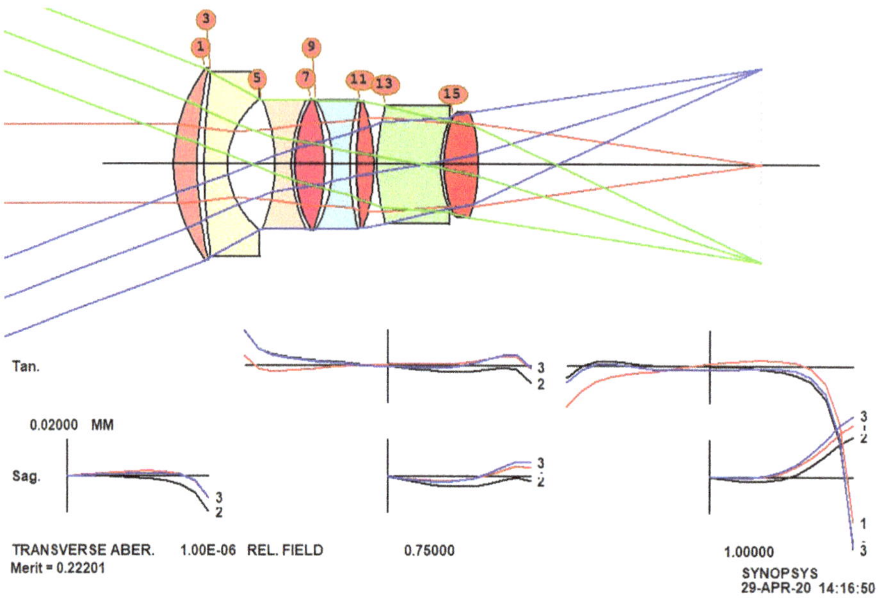

Figure 20.10. The version 4 lens.

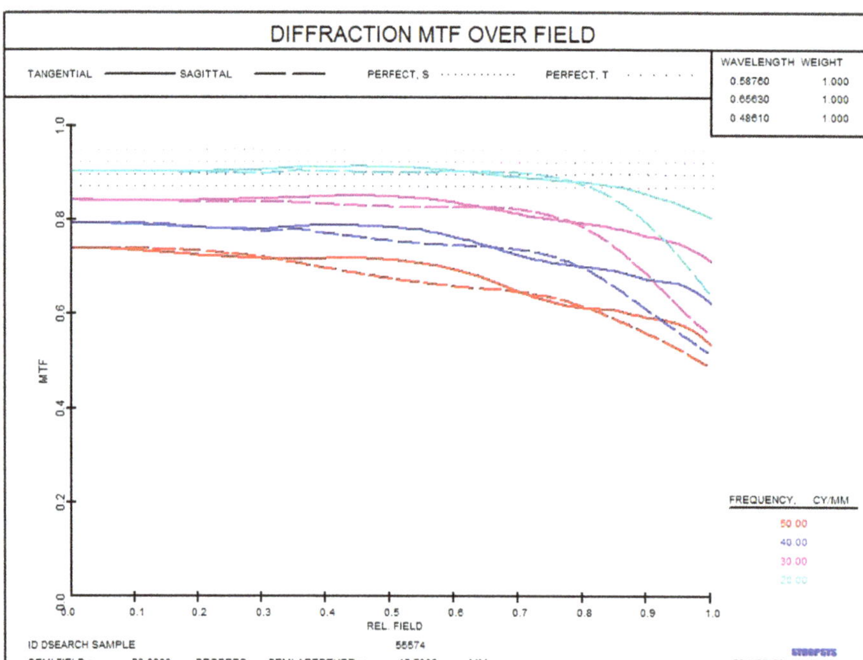

Figure 20.11. MTF of the version 4 lens.

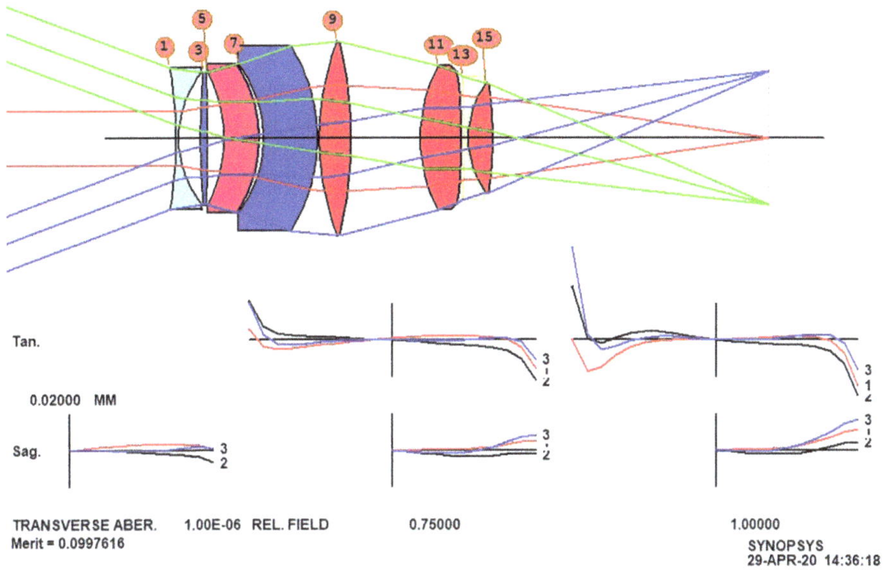

Figure 20.12. Lens form returned by DSEARCH when a longer lens was permitted, then improved with AEI.

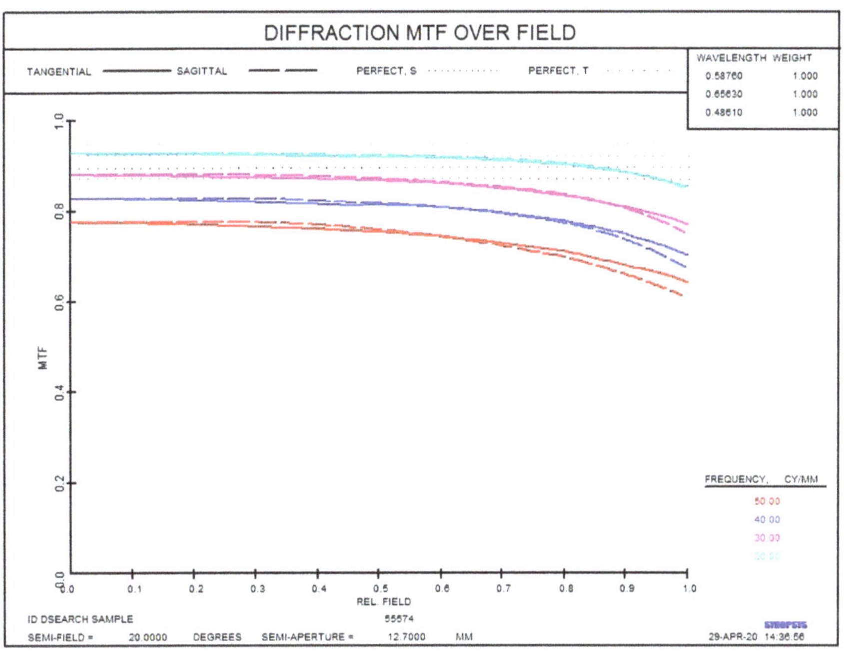

Figure 20.13. MTF of the lens in figure 20.12.

tool. Asking for a lower limit of 10 mm, we got a design with only seven elements that was even better than the eight-element lens we got in version 4. What would have taken weeks to develop in a former period can now be evaluated in a few minutes with these powerful tools.

This lens appears to have a very weak second element, and we wonder if AED could remove it with little loss of quality. Try it and see!

With so many tools available, one naturally wants to know which ones to use. If you want an 8-element lens, for example, should you ask DSEARCH for eight, or ask it for, say, six—and then use AEI twice to get to eight elements? Can we predict what Nature will favor? Guidance, please.

Experience alone will tell. We ran this job again doing just that, asking DSEARCH for six elements, and then running AEI twice. Testing all 10 of the DSEARCH results in this way, we found that nine of them had quality similar to the previous designs. Figure 20.14 shows the best of the lot (**C20L4**), with MTF curves in figure 20.15. It seems this might be a good strategy, but owing to the chaotic nature of the process, one cannot draw a firm conclusion; a different problem might work best the other way. So put this idea in your toolbox and use it when it makes sense.

This lens does not resemble any of the classic forms derived from a triplet, which have the stop in the middle near a negative lens. One of the other cases, with essentially identical performance, is shown in figure 20.16; that lens does somewhat resemble a classic form. The ancient designers found constructions such as the Double Gauss that worked well and did not explore many other configurations thereafter, giving us another case when automatic methods prove superior. The search tools are not biased and can find solutions a human would likely never have imagined.

Back to our question; we ran DSEARCH again, asking for eight elements this time instead of using AEI to get to that number. The lens that came back was, in this example, not quite as good as the above, but close. It seems that AEI is a powerful tool you will want to use often.

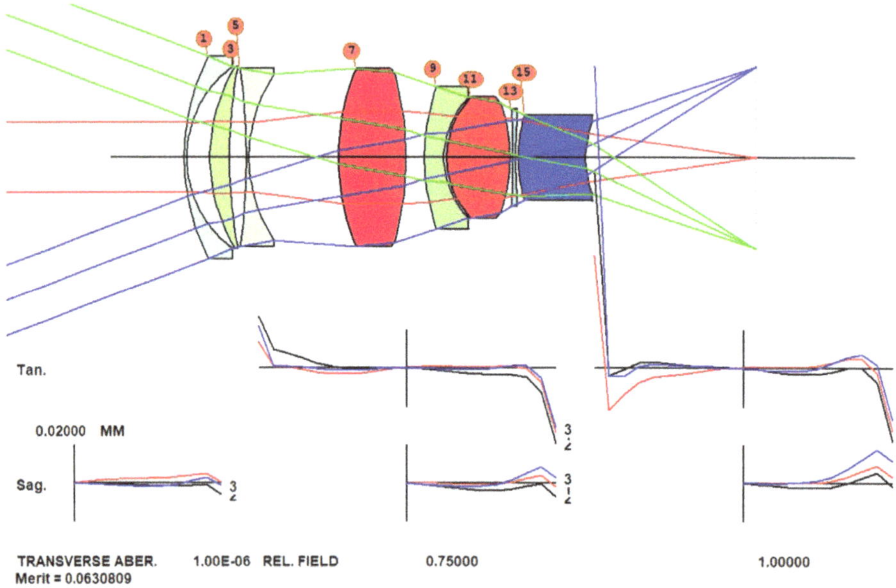

Figure 20.14. Lens found by asking DSEARCH for six elements, running AEI twice on all 10 cases, then optimizing the best one.

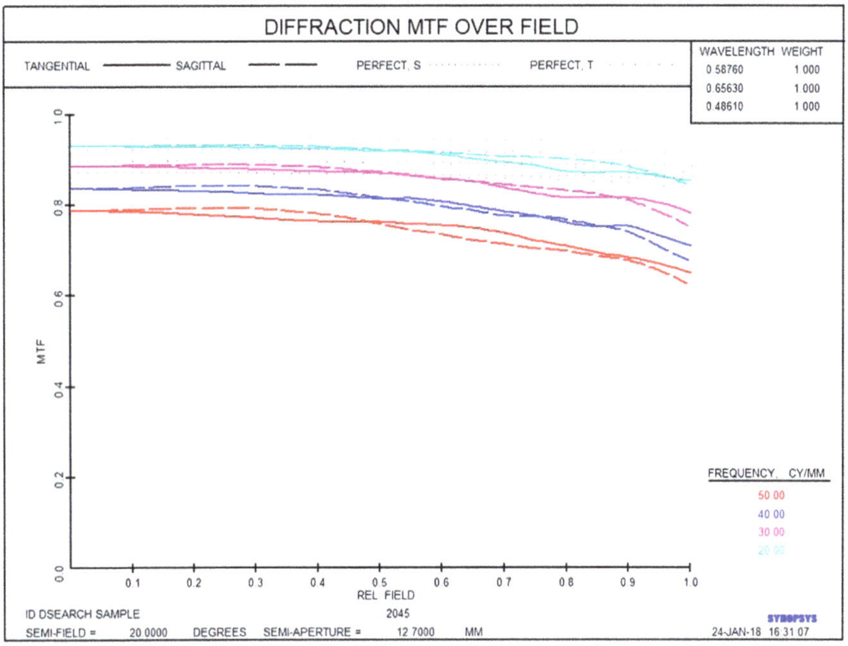

Figure 20.15. MTF of the lens in figure 20.15. This was found when a DSEARCH lens of six elements was increased to eight with AEI.

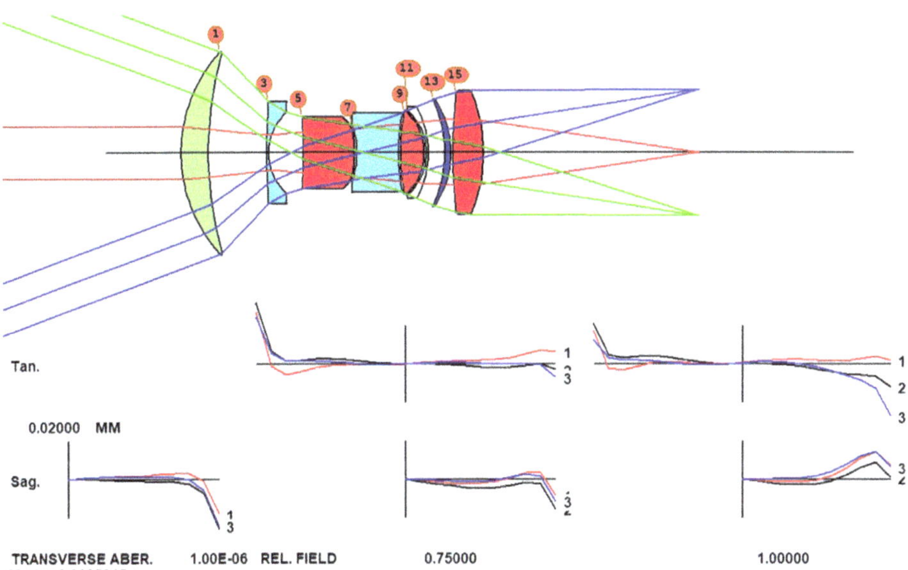

Figure 20.16. A different configuration found by increasing a 6-element lens to eight elements.

This lesson illustrates quite graphically that, in lens design, you are dealing with a landscape with a great many solutions. Unless you already have a very good starting lens, your chance of finding a great one improves as you experiment with the parameters of DSEARCH.

The art of lens design has changed significantly with the advent of new tools like DSEARCH. Whereas in former ages an expert designer would work for days or weeks on a single design, using sophisticated knowledge to steer the process, today one can produce a great many designs in a matter of minutes and then select the most promising for further work. Some of those designs often turn out to be superior to what an expert was able to come up with in days past.

In a practical situation, I would next finalize the lens, see how the image correction changes with conjugate (if that is a requirement, and reoptimize the lens as shown in chapter 19 if so), look at the distortion with GDIS, and add a requirement to the AANT file to control it if necessary, insert real glass types with ARGLASS or GSEARCH, assign a real stop where implied, decrease the thickness of some elements and increase others, reoptimize, define edge geometry with the Edge Wizard, match the curves to a vendor's testplate list with TPM, prepare tolerances with BTOL, make element drawings with ELD, system drawings with DWG, and so on. There is much to do when designing a lens, and this has been a practical lesson that shows how to use a few of the many tools that are available. We started with nothing more than a list of requirements and in a fairly brief time came up with some rather good lenses.

Feel free to experiment with this problem, and try various values for the starting radius, thickness, airspaces, and other parameters for DSEARCH, with switch 98 turned off.

20.1 Reusing dialog commands

Before leaving this chapter, I would like to point out some very handy features. You have run the **MOF** (MTF **O**ver **F**ield) analysis several times from the dialog **MOP** (MTF **OP**tions) when analyzing these lenses, and of course you can always type **MOP** to go back there when you want. But after you have run a command from a dialog such as this, notice the prompt in the CW:

```
Type <ENTER> to return to dialog
```

If you now just press the <Enter> key, you go right back there, which is simpler. And if you have run some other command in the meantime, and that prompt is no longer there, just click the **Last Menu** button on the top toolbar and you go back to the most recent dialog, which also saves some typing.

And here's another handy tool: After you run a command from a dialog, if you type **LMM** (**L**oad **M**enu **MA**Cro) in the CW, the program opens a new editor window with the command form already filled in, in this case

```
MOF M 0 50 80 0 Q 40 30 20
```

Now you can save this as a new file, giving it a name like MOF.MAC, and you can run it with the command

```
EM MOF
```

which will save the labor of typing the data next time if you want to do the same analysis. And there's more. If you define a symbol, say

```
U9: EM MOF
```

Then just clicking the U9 button on the left toolbar will execute that MACro. Put that symbol definition in your CUSTOM.MAC MACro, and it comes back whenever you restart SYNOPSYS. You have just defined a whole new command. Just hit the U9 button and you get that MTF analysis. No typing needed.

IOP Publishing

Lens Design (Second Edition)

Automatic and quasi-autonomous computational methods and techniques

Donald C Dilworth

Chapter 21

An automatic real-world lens

Using ZSEARCH to design a lens from scratch with two conjugates

In chapter 19 you developed a seven-element lens with DSEARCH and then changed it into a zoom lens so you could correct at two different object conjugates. You found you needed to add an additional element with AEI to improve performance and ended up with eight elements. That exercise was not completely successful, and was tedious besides—and illustrates an important concept: ask for *exactly* what you want, if possible. In that case, you did not.

Since DSEARCH does not know about zoom lenses, what you got was a very good *fixed-focus* lens—and then you added a new requirement to focus at a shorter distance. So the lenses that came back from DSEARCH were not selected for that property. That's always a risky process. Perhaps they were not well suited to the job.

Wouldn't it be nice if the search routines could monitor the performance at two different conjugates? Then the lenses that come back would naturally tend to perform well on that score. Well, they can. We have said nothing so far about **ZSEARCH**[1], which is similar to DSEARCH except it works on zoom lenses. Let me summarize the requirements for this same job by listing the input for ZSEARCH (**C21M1**). (You should also review ZSEARCH in the UM at this point, to better follow along.)

```
CORE 64
ON 1
ON 99

TIME
ZSEARCH 4 QUIET
SYSTEM
ID CHAPTER 21 ZOOM
OBB 0 20 12.7        ! OBJECT AT ZOOM 1, AT INFINITY
WAVL CDF
UNI MM
END
```

[1] ZSEARCH™ is a trademark of Optical Systems Design LLC, an Arizona, USA corporation.

doi:10.1088/978-0-7503-3695-6ch21

```
GOALS
ZOOMS 3                 ! CORRECT AT THREE ZOOMS
GROUPS 8 0 0            ! ONE GROUP, EIGHT ELEMENTS
ZGROUPS Z 0             ! ZOOM THE FIRST GROUP
GIHT 32 32    .1        ! KEEP THE IMAGE HEIGHT CONSTANT
BACK 50 .01             ! BACK FOCUS DISTANCE IN ZOOM 1
RSTART 400
THSTART 10
ASTART 20
RT 0.25

FINAL
OBA 1000 -388 12.7  ! OBJECT DEFINITION FOR ZOOM 3
APS 10        ! PUT THE STOP NEAR THE MIDDLE OF THE LENS
NPASS 50

COLOR M

FOV 0 .5 .7 1
FWT 2 1 1 1
NGRID 5
SNAP 10
ANNEAL 30 5 Q 50
QUICK 50 100
AGROUP
END

SPECIAL AANT
ADT 6 .01 10
LUL 350 .1 1 A TOTL
END

GO
TIME
```

This input declares two different object distances, infinity and 1000 mm, specifies that the lens will have three zoom positions (the minimum allowed by ZSEARCH), declares a single zooming group of eight elements (the entire lens), and gives a maximum allowed total length of 350 mm.

Run this job, and then optimize and anneal **(55, 2, 50)** the top lens that comes back. The result (**C21L1**) is shown in figures 21.1 and 21.2.

Compare this to the results from chapter 19 and you see the importance of our observation. When you ask ZSEARCH for what you *really* want, it has no trouble finding a configuration that meets those requirements. The lens looks nothing like the previous results and works better besides. This kind of tool is a real time-saver, for sure.

But, like all tools, it has limitations. ZSEARCH does not support a curvature or thickness solve, since too many things might be zooming, and it does not support the variable YP1, which we have used before to find the best place for the stop. So we assigned the stop to surface 10 so it would be near the middle of the lens. We also put an upper limit on the total cell length.

If you look closely at the lens at both conjugates, you'll notice that the size of the beam at the stop, on surface 10, changes slightly with conjugate. It's a fine point, but easy to address. Open WS, and type in the edit pane

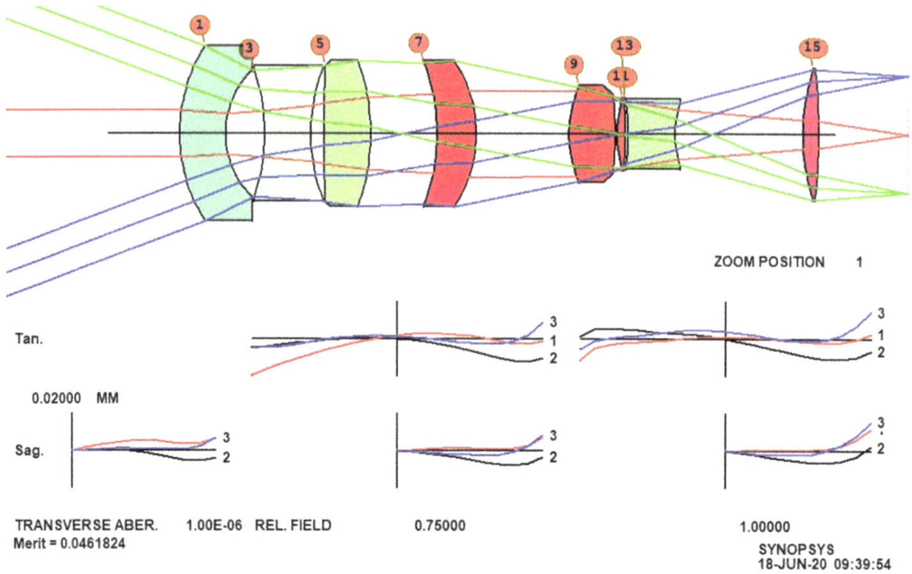

Figure 21.1. Lens corrected at two conjugates returned by ZSEARCH, optimized and annealed, at infinity conjugate.

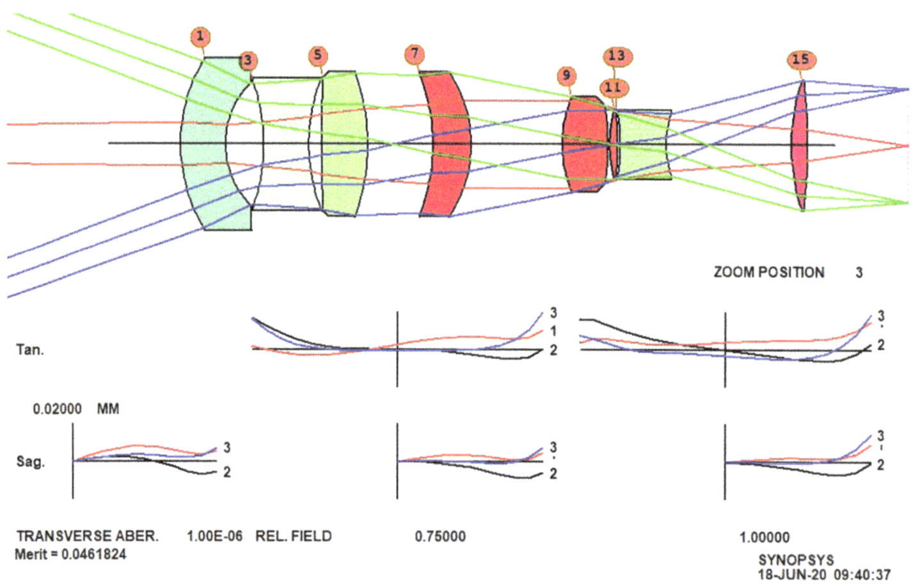

Figure 21.2. Lens at near conjugate.

```
APS -10
CSTOP
WAP 2
```

and click Update. Then optimize and anneal. This will fix the size of the clear aperture on 10 to whatever the paraxial raytrace needs for the marginal ray, and then adjust the size of the entrance pupil at each zoom and field point so the real beam just clears that aperture. All of this is discussed more fully in the next chapter. And the ray fans look even better with the corrected pupil.

The input to ZSEARCH included the monitor

```
ADT 6 .01 10
```

to prevent lenses from getting too thin. This monitor must be used with caution, since you don't want to miss a great configuration because the program penalized it early on, never mind it could be fixed up later. But here we have specified a low weight, 0.01, which means that it will not have much influence in the early stages when the other aberrations are large, but will steer the design when it approaches optimum and those become much smaller. We often try our searches both ways.

Now, we ask, do we really need all those elements? Add the line

```
AED 4 QUIET 1 123
```

before the PANT file and optimize again. The program says element number 6 can come out and asks if you want to delete it. Select that option, and the element is removed.

Now delete the AED line, reoptimize and anneal again. The new lens is shown in figure 21.3. This is not as good as before, so it looks like we really need eight elements.

Or do we? What happens if we run ZSEARCH again, asking for seven elements at the start, instead of asking for eight and then deleting one? We did just that, also moving the stop to surface 8 so it would remain near the center of the lens. Here is the result, after optimizing and adjusting the pupil as we did above, in figure 21.4. This lens is much different from and about the same quality as before. Both methods can be very effective, and you simply have to try them in any given case to see which works best for that lens.

What would happen if you assigned the stop to a different surface? Try it and see! And experiment with the value of RSTART. Every change you make sends the program up a different branch of the tree. You can quickly identify a great many promising configurations with these tools, and you can always run AEI and AED to improve the lens further or to change the number of elements.

Lastly, you'll notice that we turned on switch 99 for this session. When you authorize many cores, as we did here, the search programs automatically utilize all of them. But the basic optimization feature does not—for the simple reason that for simple jobs, starting and stopping all those cores uses so much overhead that it can

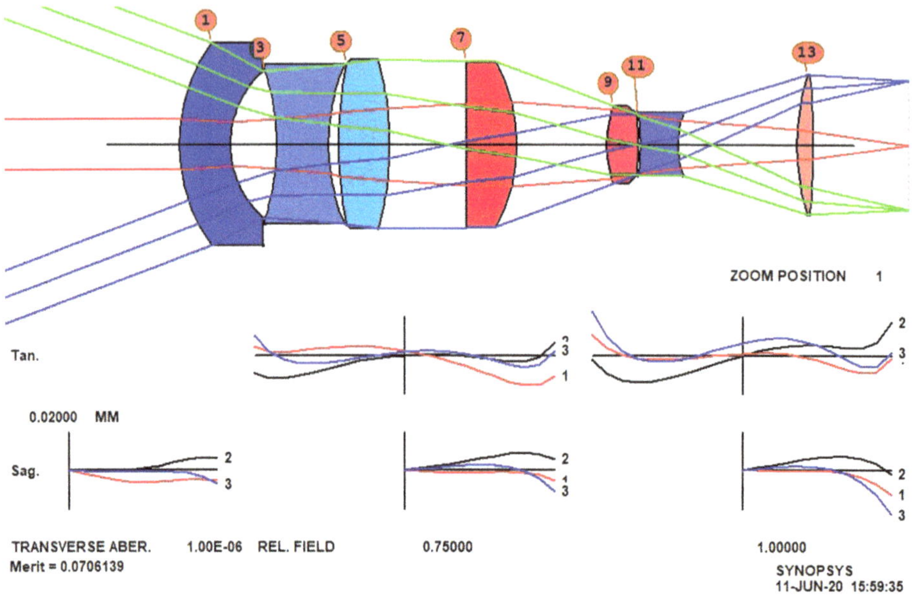

Figure 21.3. Zoom lens after AED has removed an element.

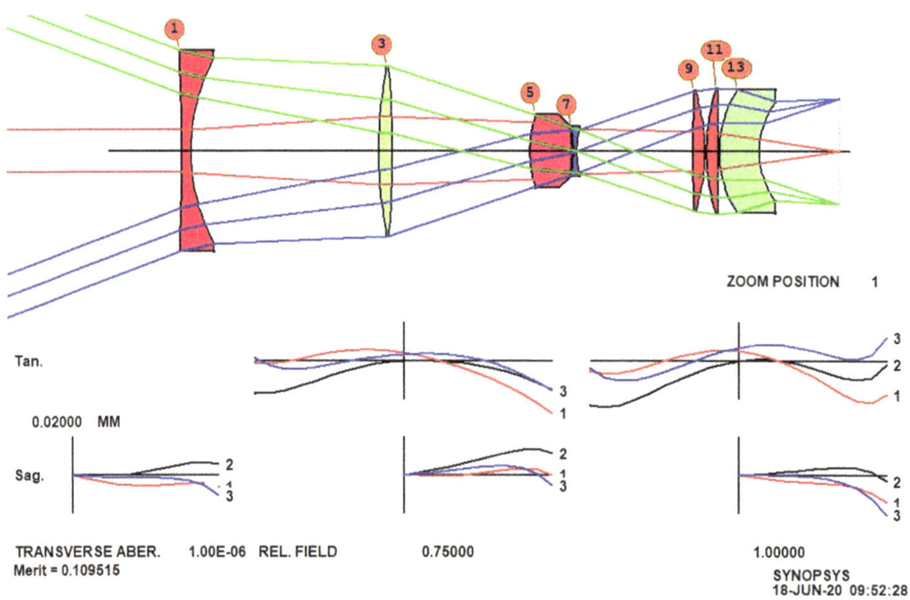

Figure 21.4. Lens returned when seven elements were requested.

slow things down instead of speeding things up. But for complicated tasks, such as optimizing a zoom lenses, a single cycle usually takes more time than does the overhead, so the process actually runs faster then. So put this in your toolbox: if optimization runs very slowly, try turning on switch 99 and authorizing all your cores.

Chapter 22

What is a good pupil?

Pupil definition; vignetting; ray aiming; wide-angle pupil options

Those familiar with other lens design codes already know about two common pupil definitions: a paraxial pupil that only applies to simple systems or where the stop is out in front, and, for more complex systems, 'ray aiming', which is intended to model a real stop somewhere inside the system.

For those not familiar with these concepts, let me explain. Start this lesson with the lens (**C22L1**) shown in figure 22.1. Surface 7 has been declared the stop surface—but it is a *paraxial* stop at the moment.

This lens has two problems: the chief ray does not go through the center of surface 7, and the upper and lower rim rays do not hit that surface at the edge, as they should. Figure 22.2 shows the lens after those problems have been corrected. Rays from every field point now fill the aperture of surface 7. (And the EFILE edges will vignette some rays at the edge at full field.)

When you tell the program to trace a ray, it first has to know where to aim the ray so it hits the stop at the desired point. For example, the ray at HBAR = 1 and YEN = 1 (the full-field marginal ray) should hit surface 7 (the stop) at the edge of the aperture. How does it know where to aim? That is the whole issue with pupil definitions.

There are two possible stop definitions: **paraxial** and **real**. The definition in the first example was *paraxial*, declared with

```
RLE
...
APS 7
...
END
```

doi:10.1088/978-0-7503-3695-6ch22

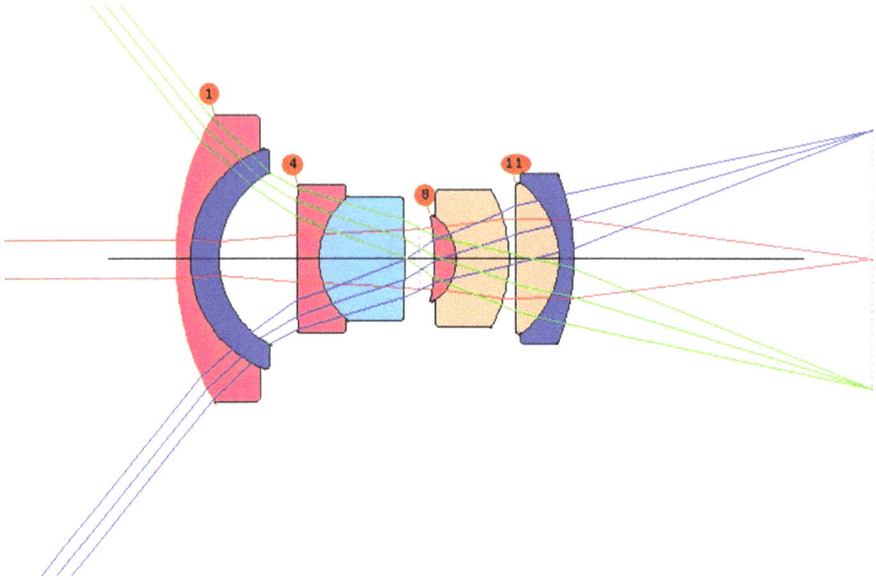

Figure 22.1. Wide-angle lens with paraxial pupil, which is not adequate in this case.

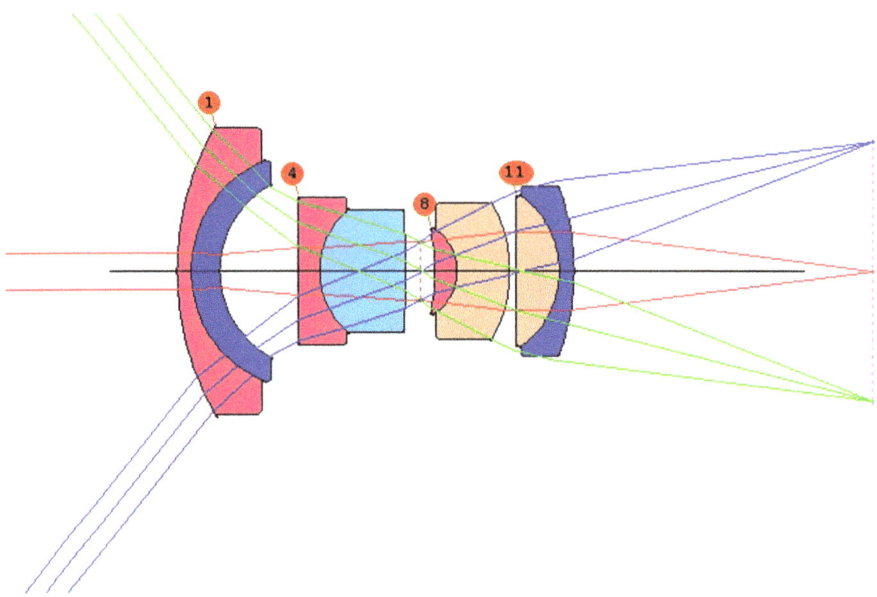

Figure 22.2. Wide-angle lens with correct pupil definition.

To fix these problems, we first declare surface 7 to be a *real* stop, with

```
CHG
APS -7
END
```

Now the chief ray is traced properly, as shown in figure 22.3. The minus sign indicates that this is a real stop and the chief ray will be found by iteration.

Although the chief ray is now okay, the marginal rays are not. You need another common declaration, which will adjust the size of the pupil so the stop is nicely filled. This is the **WAP 2** option (there are three Wide-Angle-Pupil possibilities, illustrated in figure 4.7). It finds the shape of the entrance pupil by iterating a few rays at the edge of the stop. But this option requires a hard aperture on the stop surface so it knows where to aim. Let us assume for the moment that the aperture is not defined. You can do a **CAP** listing—to see the values of all current apertures—and then assign a 'hard aperture' to surface 7. The value in this case turns out to be 3.9937, so you could enter that value either in a CHG file or with the WorkSheet. Here's how to do it with a CHG file (CAO means Clear Aperture Outside):

```
CHG
7 CAO 3.9937
END
```

An easier way is simply to type, in a CHG file or in the WorkSheet edit pane, **7 CFIX**. That fixes the current value, whatever it is at the moment, so you don't need

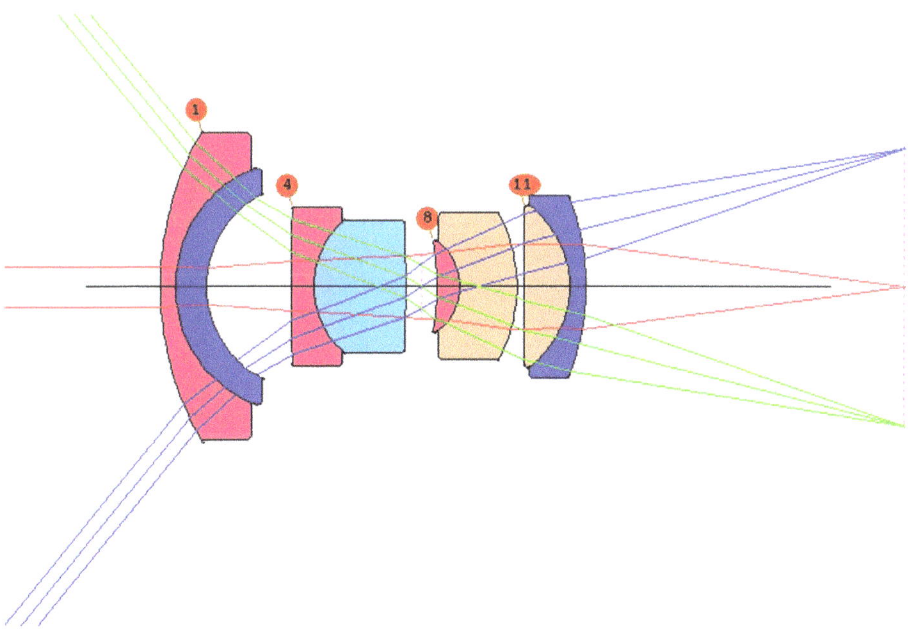

Figure 22.3. Lens with real pupil assigned but without aperture aiming.

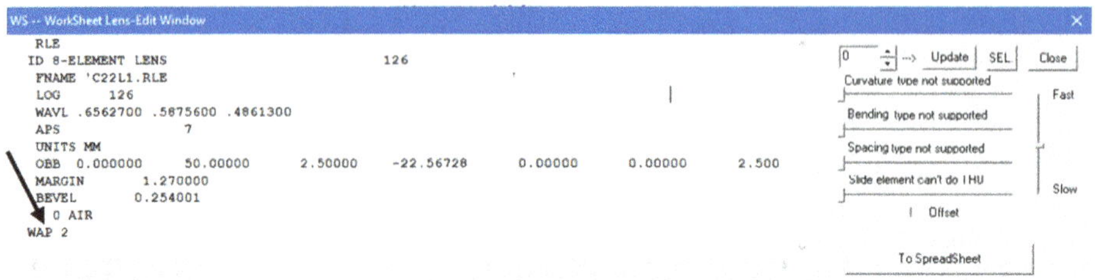

Figure 22.4. Worksheet edit panel, where a wide-angle pupil (WAP) number 2 has been declared.

to type it yourself. Now change to WAP 2 with the Worksheet as shown in figure 22.4 and click the Update button. You get the lens in figure 22.2. Now both the chief ray and the marginal rays go to the right place on surface 7. You have turned on ray aiming for a total of five rays.

So far, this is not too complicated, and many users will not need anything else. But suppose you are optimizing the lens and the required aperture on surface 7 keeps changing. In that case, the hard aperture you assigned will be incorrect almost immediately.

No problem. You can activate an option to *recalculate* that aperture every time the lens is changed. This is done by adding the directive **CSTOP** to the lens input file. Then the program will alter the CAO on 7 so it always equals the paraxial marginal ray height there.

(And if the lens has pupil aberrations so large that a *real* axial marginal ray requires a different aperture than the *paraxial* ray, change this to **CSTOP REAL**. You can even specify *which* real ray is used to define this aperture, as explained in the UM.)

But what is the point of all this? Isn't it easier just to use the kind of 'ray aiming', that some other codes do?

Actually, to do it right, is not easier at all. As usually implemented, when those programs trace a grid of rays for any kind of image analysis, they create a square grid at the stop and then iterate *every ray* so it goes through the required grid point. All that iteration takes time. Lots of time.

Here is an example of a very wide-angle design (**C22L2**), in figure 22.5.

The stop is on surface 9, and it is nicely filled by the WAP 2 option. Look at a footprint on that surface showing rays from the full-field point, in figure 22.6.

Wow! This is certainly *not* a uniform square grid! Those programs that employ 'ray aiming' fill this aperture with the *wrong* distribution, and then, to account for this wrong distribution, they alter the effective energy of each ray according to the actual ray density at that point. That requires tracing neighboring rays, finding the local density, making the adjustment—and doing all this for every ray in the pupil. It is not so simple at all.

While this can indeed produce a correct evaluation of the image, one has to ask why they spend so much time with all the ray iterations in the first place.

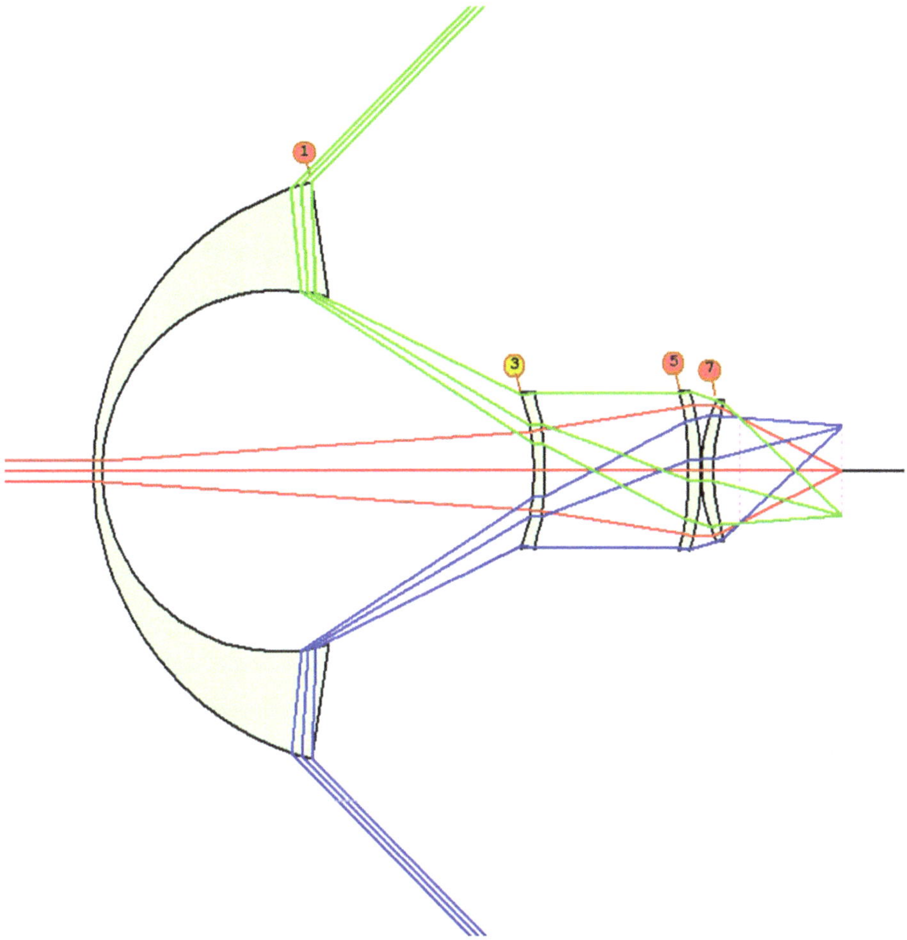

Figure 22.5. A very wide-angle lens using the WAP 2 option.

Instead, SYNOPSYS finds the size and shape of the *entrance pupil*, and then fills *that* with a uniform grid. Here is that pupil, on surface 1, for the above lens. It is regular, as it should be, shown in figure 22.7:

The WAP pupil options in SYNOPSYS model the *outline* of this distribution, so a regular grid can fill it as it should. It is not necessary to iterate each ray, so it is much faster, and the distribution at the stop is correctly modeled. For this extreme example, a simple outline is a bit too small, however (normally it is modeled by an elliptical shape). A better pupil, in this case, is found by declaring **RPUPIL** in the lens file. Now it starts with a rectangle that encloses that ellipse, and deletes any rays that fall outside the stop aperture. Figure 22.8 shows that shape as it enters the lens, and figure 22.9 shows an outline of what actually gets through.

We like this better than the slow 'ray aiming' method.

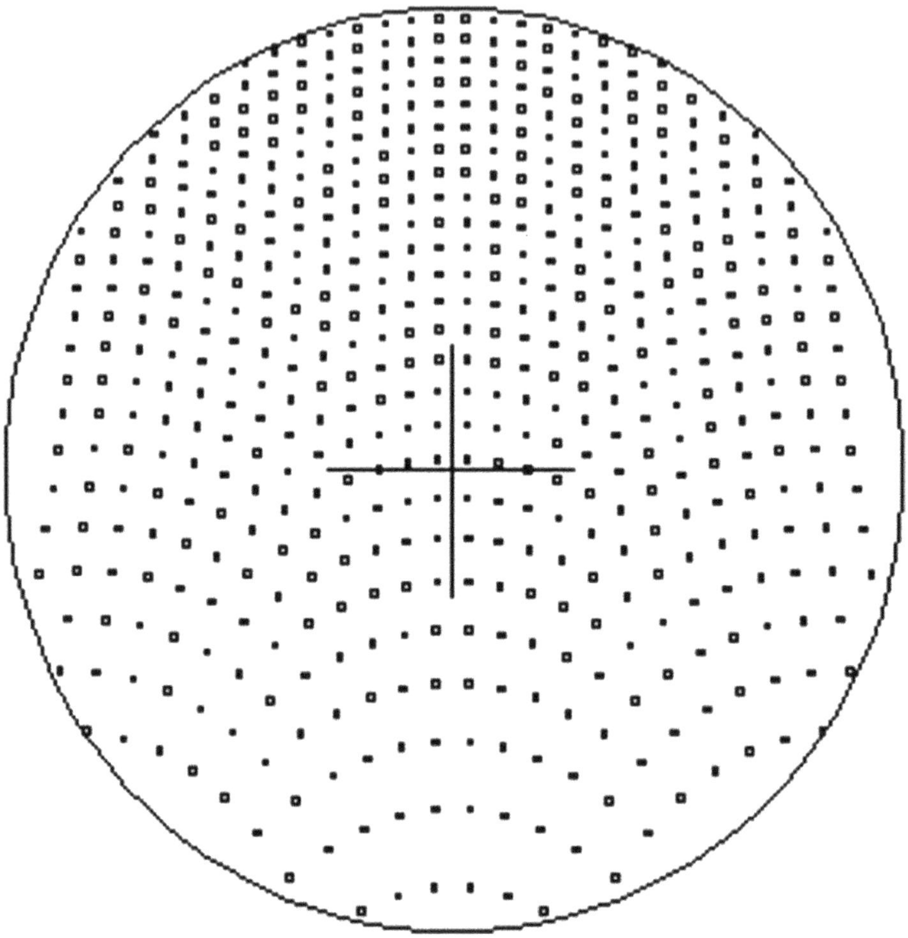

Figure 22.6. Ray distribution at the stop for the lens in figure 22.5. This is not a uniform distribution.

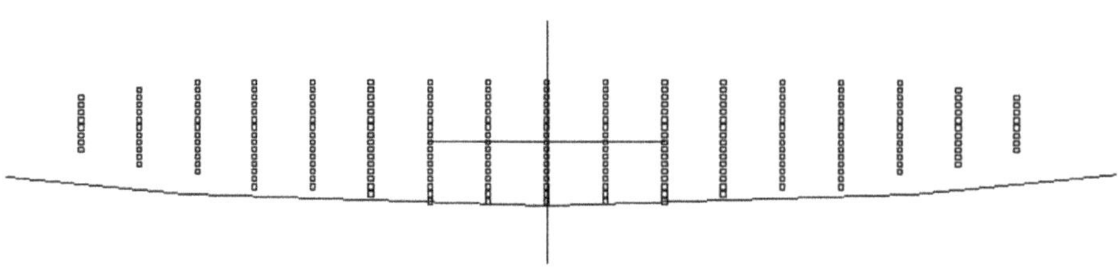

Figure 22.7. Ray grid at entrance pupil for the wide-angle lens.

Don't forget to check out the dialogs **MPW** (Menu, Pupil Wizard) and **MOW** (Menu, Object Wizard), where you can define the kind of pupil you need by checking boxes and selecting from various options. Both of those dialogs do much the same thing, but they are organized differently, so you can choose which you like best.

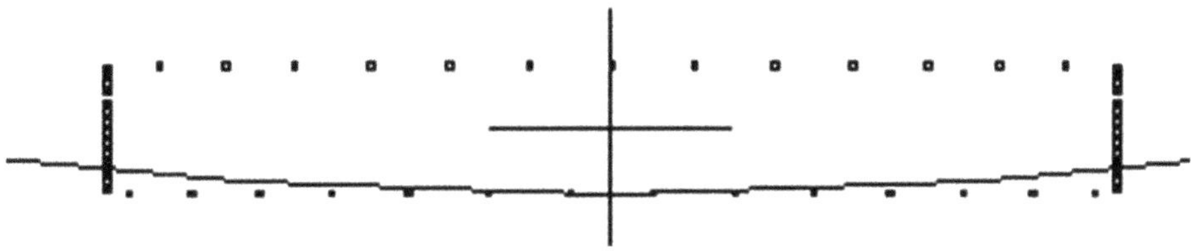

Figure 22.8. Outline of input ray pattern.

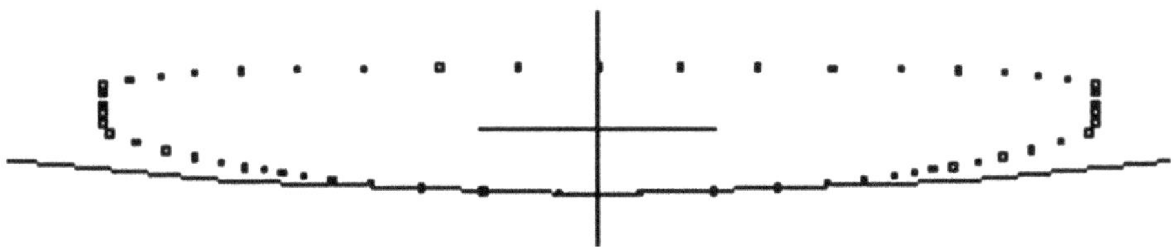

Figure 22.9. Outline of pattern that gets through the lens.

22.1 Which way is op? Orientation of pupil

The unique pupil definition in SYNOPSYS offers an interesting possibility—which is handy but takes some getting used to.

Let me illustrate. First, I will show you some rays that don't go where you expect them too, and then I'll describe a simple way to keep things straight. Fetch the lens bundled as 1.RLE.

```
FETCH 1
```

Now look at it in PAD, as shown in figure 22.10.

The lens at the moment has been assigned a *paraxial* stop on surface 4. In PAD, click the PAD Top button, and select the option to draw a single ray. Click OK, and a small box opens where you can select which ray to draw with two sliders. Move the top slider to full field (HBAR = 1) and the bottom slider to full aperture (YEN = 1). This object has been defined with a positive *angle* coming in, which means that the 'full field' rays start from an object *below* the axis, as shown in figure 22.11.

You see the full-field marginal ray, as expected. Now move the top slider to the bottom of the field (HBAR = −1), as shown in figure 22.12.

Again, the ray enters at the top of the pupil. This is the basic idea of the paraxial pupil. Simple—but not always adequate. Close the ray display dialog and change the stop designation to

```
APS -4
```

in the WorkSheet. Remember, the full-field object is located at a *negative* Y-coordinate, far to the left of the lens.

Now it's time to get interesting. Open the single-ray dialog again, and set it to full aperture and full field once more. Now you see the ray in figure 22.13.

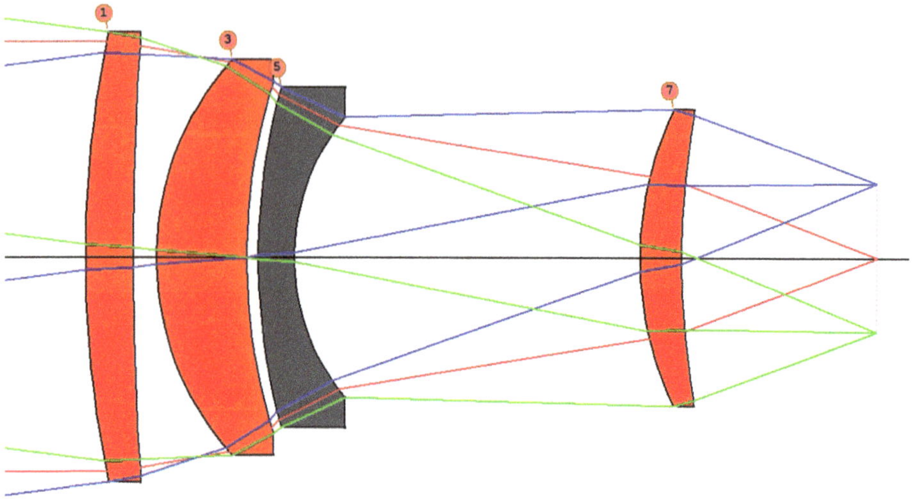

Figure 22.10. PAD display of lens for demonstrating pupil options.

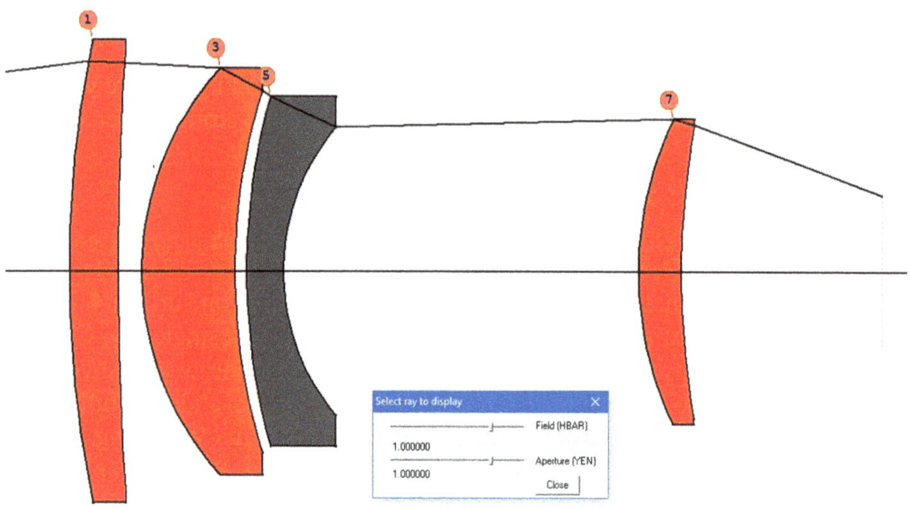

Figure 22.11. PAD display showing a single user-selected ray at the top of the field and aperture.

Wow! What happened? The 'full aperture' ray is now at the *bottom* of the pupil. What's going on?

Simple. This feature is designed to make it easy to correct feathered edges, no matter where in the field you are looking. (We have used the AEC monitor in many of these lessons to control edge feathering, a feature that usually works very well. But there is also a provision in the program to control feathering along a selected ray at a given surface, and to use that effectively, you have to know which ray to trace. Look up ECP and ECN in the help file to read about that feature.)

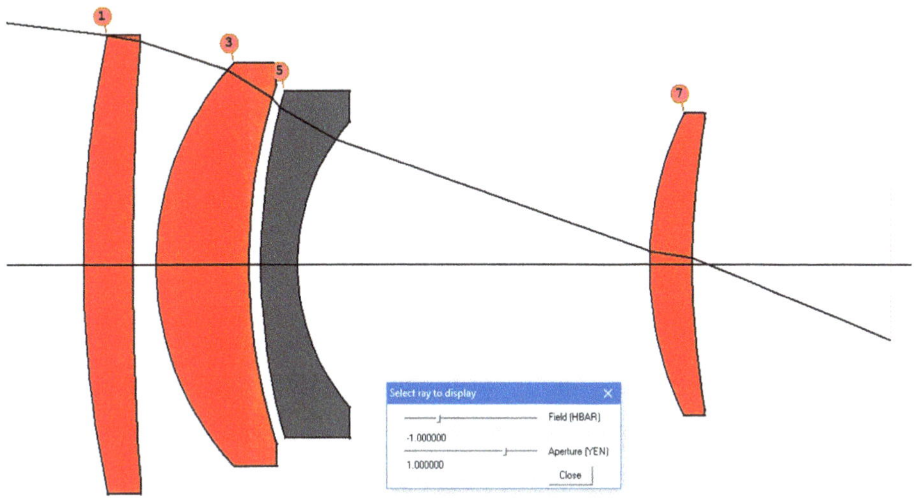

Figure 22.12. PAD display showing a single ray at the bottom of the field.

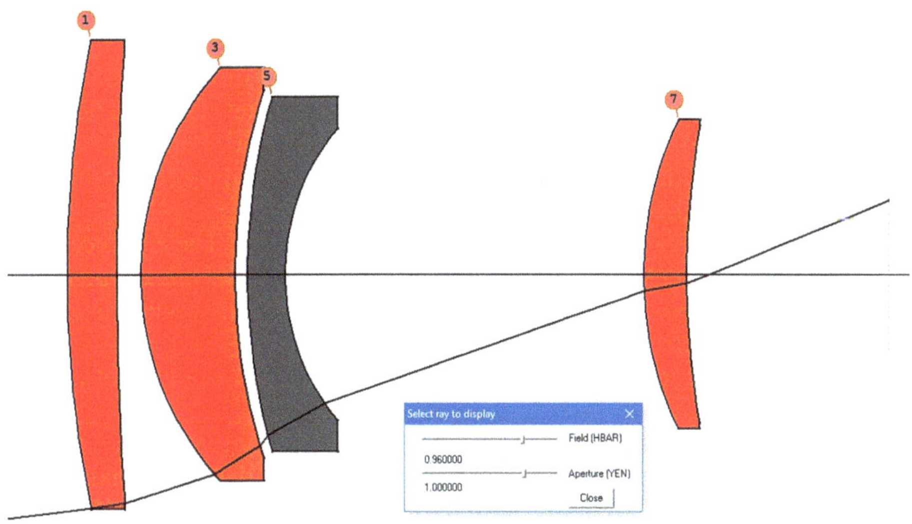

Figure 22.13. Example lens with a real pupil defined, at the top of the field.

In the lens in figure 22.13, you would correct along the 'upper' rim ray (the ray shown) if feathering were a problem. Now go to the lower field point, HBAR = −1, as in figure 22.14.

Aha! The ray to correct is *still* the upper rim ray! The program *rotates* the entire entrance pupil according to the orientation of the field point from which you are tracing. If you trace a point in the skew field, the 'upper marginal ray' would turn out to be the extreme *skew* ray, so again you can easily control feathered edges. If the program left the definition of the upper and lower marginal rays the same for all field points (as it is with a paraxial pupil), it would not be so easy to do; you'd have to

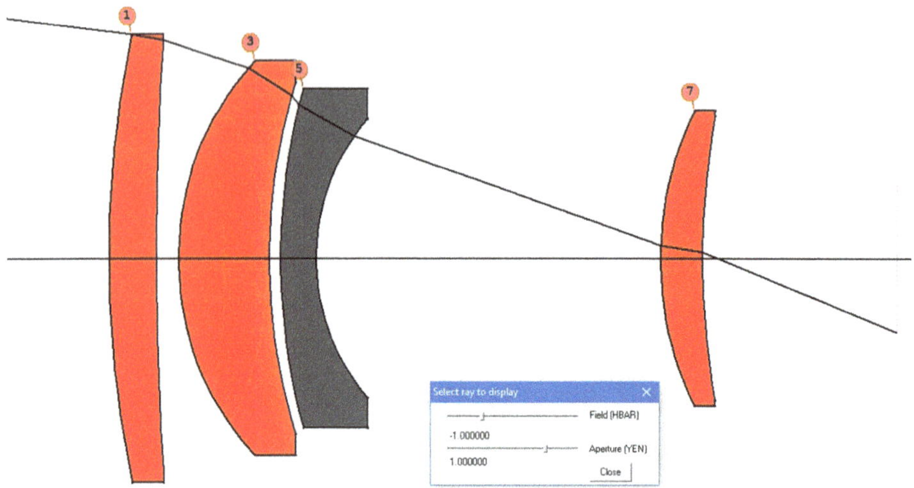

Figure 22.14. Example lens at the bottom of the field with a real pupil.

figure out which skew ray to fix and then create an aberration for it. This is much simpler—once you get used to it.

So how can you easily figure out which ray to examine or correct for feathered edges? Simple. While the PAD display is open, press the **F7** key. Only the 'lower' rim ray at full field shows up. The **F8** key shows only the 'upper'. Presto! You can tell which ray is where with a single keystroke, and if you hold down the <shift> key, rays are traced from the lower field point instead.

There is another advantage to this kind of pupil definition: The entrance pupil is usually modeled as an ellipse, as illustrated in the first part of this chapter, and it turns out that the ellipse also rotates with field point. So it can model the vignetted pupil at all points in the field. Neat.

Take a look at section 2.6.1 in the User's Manual for an example of a rotated pupil.

Oh, and one more thing: The program decides which ray to call the 'upper' ray based on the sign of the height of the full-field object. Since that was negative in this example, it flipped the marginal rays for positive HBAR. At negative HBAR, the object comes from a positive Y-coordinate, and the opposite happens.

And at HBAR = 0? Well, to avoid confusion from being neither positive or negative, the program displays a very small but nonzero field point there.

Confused? Try the F7 and F8 keys. Simple.

IOP Publishing

Lens Design (Second Edition)

Automatic and quasi-autonomous computational methods and techniques

Donald C Dilworth

Chapter 23

Using DOEs in modern lens design

Improving a lens with a kinoform surface

In this chapter, we will start from scratch, design a five-element lens, and then see if adding a diffractive optical element (DOE) to the lens can improve its performance.

Here is the problem, as defined by the entries in the Design Search dialog **MDS**. This dialog will create a MACro that runs the DSEARCH command, and is shown with all the data filled in in figure 23.1.

This input will design a lens at F/3.5 with a semi-field angle of 25 degrees and an aperture radius of 12 mm. We elect to control the back focus with the SPECIAL AANT entry, which lets the distance grow but will not let it become less than about 22 mm. We also ask for the angles of the upper and lower rim rays at full field with respect to the surface normal at every surface to be no more than 60 degrees, with a low weight, with the ACA request, so we don't get solutions with wild angles at the image and to avoid steep ray refraction, which would introduce high-order aberrations that are better avoided.

When you click the OK button, the program loads the MACRO. The **CORE 32** directive at the top will speed things up, and we specify a grid number of 6, because aspherics and DOEs can cause high-order aperture aberrations and we will likely need more than the default grid of 4 (**C23M1.MAC**).

```
CORE 32
OFF 1
OFF 99
DSEARCH 1  QUIET
SYSTEM
ID DSEARCH SAMPLE
OBB 0 25 12
WAVL 0.6563 0.5876 0.4861

UNITS MM
END
GOALS
```

doi:10.1088/978-0-7503-3695-6ch23

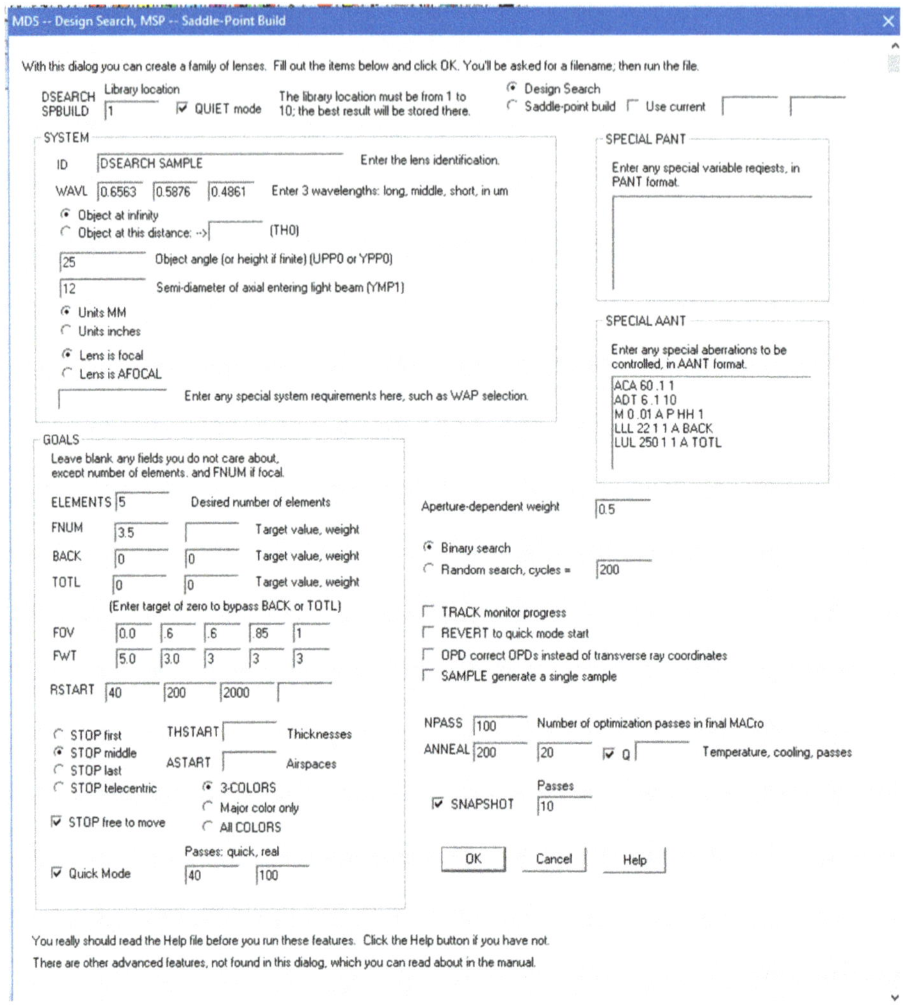

Figure 23.1. MDS dialog with selections for an example lens.

```
ELEMENTS 5
FNUM 3.5
BACK 0 0
TOTL 0 0
STOP MIDDLE
STOP FREE
RSTART 40 200
RT 0.5
FOV 0.0 .4 .6 .85 1
FWT 5.0 3.0 3 3 3
NPASS 100
NGRID 6
ANNEAL 200 20 Q
COLORS 3
SNAPSHOT 10
```

```
QUICK 40 100
END
SPECIAL PANT

END
SPECIAL AANT
 ACA 60 .1 1
 ADT 6 .1 10
 M 0 .01 A P HH 1
 LLL 22 1 1 A BACK
 LUL 250 1 1 A TOTL
END
GO
```

Since we are going to implement DOE surfaces, we also elect to specify five field points for correction. This is a good idea when using any kind of aspheric surfaces, since otherwise one might get great correction where specified and poor correction at intermediate fields.

We also specify two different starting values for the radius of curvature of each case, to be investigated in turn. Remember, even a small change to the initial conditions can send DSEARCH to a different branch of the design tree, and this will increase the number of cases searched by a factor of two.

Run this MACro, and you see that the best lens that comes back from DSEARCH is not too good—but what can you expect with only five elements at this field and speed? Optimize with the MACro prepared by DSEARCH, and then anneal (**50, 2, 50**). The lens is shown in figure 23.2, a classic inverse telephoto.

We can probably get better results by requesting a greater number of elements—but here we want instead to see how much improvement we can get by changing one

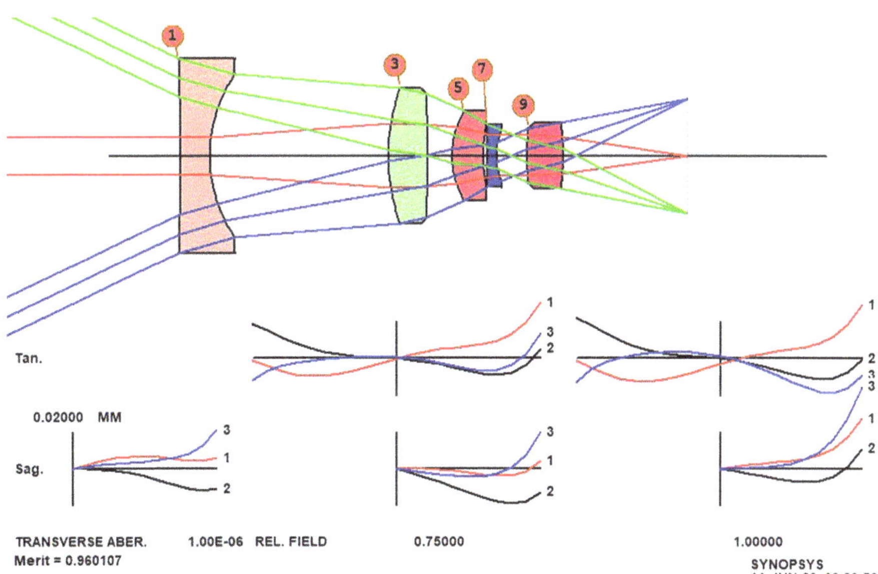

Figure 23.2. Five-element lens, all spherical, returned by DSEARCH, optimized, and annealed.

of the lenses to a DOE. Let's try it. Add to the optimization MACro another line at the top. ('**ADA**' means **A**utomatic **DOE A**ssignment.)

```
ADA 5 QUIET 1 123
PANT
VY 0 YP1
VLIST RD ALL
VLIST TH ALL
VLIST GLM ALL
END
AANT P
AEC
ACC
GSR      0.500000     5.000000     6  2    0.000000
GSR      0.500000     5.000000     6  1    0.000000
GSR      0.500000     5.000000     6  3    0.000000
GNR      0.500000     3.000000     6  2    0.400000
GNR      0.500000     3.000000     6  1    0.400000
GNR      0.500000     3.000000     6  3    0.400000
GNR      0.500000     3.000000     6  2    0.600000
GNR      0.500000     3.000000     6  1    0.600000
GNR      0.500000     3.000000     6  3    0.600000
GNR      0.500000     3.000000     6  2    0.850000
GNR      0.500000     3.000000     6  1    0.850000
GNR      0.500000     3.000000     6  3    0.850000
GNR      0.500000     3.000000     6  2    1.000000
GNR      0.500000     3.000000     6  1    1.000000
GNR      0.500000     3.000000     6  3    1.000000
  ACA 60 .1 1
  ADT 6 .1 1
  M 0 .01 A P HH 1
  LLL 22 1 1 A BACK
  LUL 250 1 1 A TOTL
END
SNAP/DAMP 1
SYNOPSYS  100
```

Run this version, and the program finds that a DOE at surface 9 works best, as shown in figure 23.3. The lens is much better.

The command **ASY** shows the data for this DOE.

```
SYNOPSYS AI>ASY

 SPECIAL SURFACE DATA
 ____________________________________________________________________________
  SURFACE NO.   9 -- UNUSUAL SURF TYPE 16 (SIMPLE DOE)
 WAVELENGTH OF OPD DEFINITION:         0.587600
 INDEX OF DOE MATERIAL: PICKUP SUBSTRATE
 NORMALIZING RADIUS:        17.532700
 DIFFRACTION ORDER:         -1
 XD  1  0.007901242 (CV)    XD 11   7.875870E+01 (p**2)   XD 12 -1.287067E+01 (p**4)
 XD 13  6.007757130 (p**6)  XD 14  -1.185123280 (p**8)

 THIS LENS HAS NO TILTS OR DECENTERS
```

We are curious what would happen if we added a *second* DOE. That's simple to test. Add variables to the PANT file for the DOE terms that ADA just added,

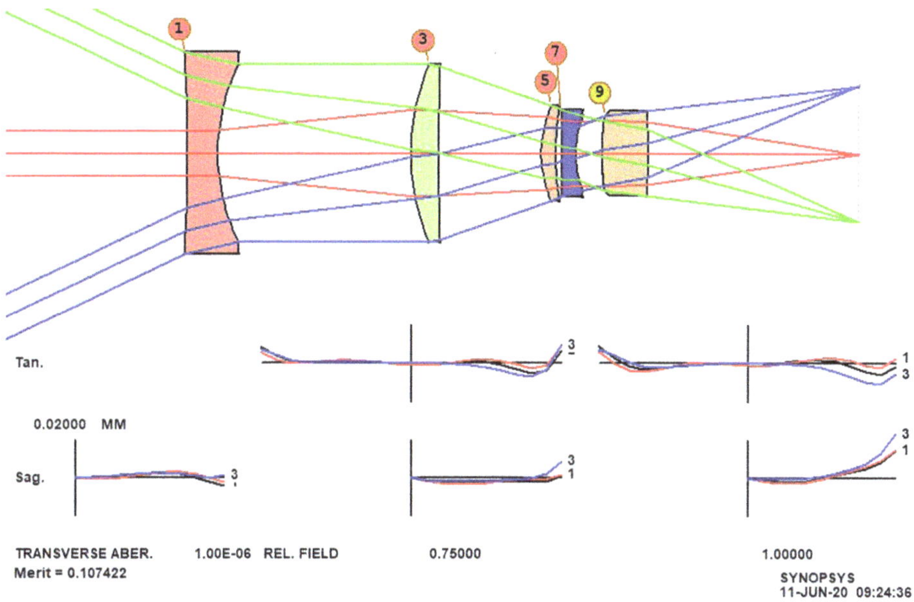

Figure 23.3. Lens with a DOE added by ADA.

```
VY 9 G 16
VY 9 G 26
VY 9 G 27
VY 9 G 28
VY 9 G 29
```

and run the MACro again. How can you tell which G-variables to vary? Look in the UM under **USS** (Unusual Surface Shapes), select type 16, and you see that those terms will vary the base curvature and OPD terms from second to eighth order.

XD number	Function	XDD index	G-variable
1	Base curvature	1	16
2	Base conic constant	1	17
3	Rho**4 term of base curve	1	18
4	Rho**6	1	19
5	Rho**8	1	20
6	Rho**10	2	21
7	Rho**12	2	22
8	Rho**14	2	23
9	Rho**16	2	24
10	Rho**18	2	25

(*Continued*)

(Continued)

XD number	Function	XDD index	G-variable
11	Rho**2 term of OPD expansion	3	26
12	Rho**4	3	27
13	Rho**6	3	28
14	Rho**8	3	29
15	Rho**10	3	30
16	Rho**12	4	31
17	Rho**14	4	32
18	Rho**16	4	33
19	Rho**18	4	34
	D0: Blaze depth constant term		54
	D1: linear term		55
	D2: quadratic term		56

This time it wants a DOE at surface 3, shown in figure 23.4, and the merit function was reduced to 0.0435.

It's time to assign a real stop to surface 7. Then modify the PANT file so it will vary the coefficients on both DOEs, and include some higher-order terms as well. Term G 32 is the 12th-power coefficient, while the default from ADA only goes to the 8th power. (And be sure to comment out the ADA command, so you don't get a third DOE!) Also, comment out the YP1 variable so the stop stays on 7.

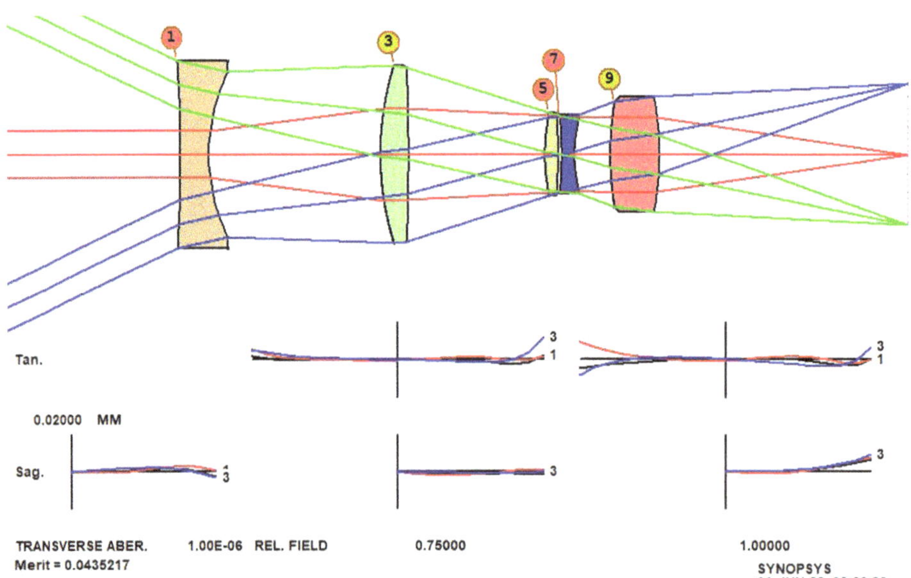

Figure 23.4. Lens with DOEs on two surfaces, assigned by ADA.

```
!ADA 5 QUIET 1 123

PANT
!VY 0 YP1
VLIST RD ALL
VLIST TH ALL
VLIST GLM ALL

VY 3 G 16
VY 3 G 26
VY 3 G 27
VY 3 G 28
VY 3 G 29
VY 3 G 30
VY 3 G 31
VY 3 G 32

VY 9 G 16
VY 9 G 26
VY 9 G 27
VY 9 G 28
VY 9 G 29
VY 9 G 30
VY 9 G 31
VY 9 G 32

END
...
```

Run this again, and then anneal **(50, 2, 50)**. You get the design in figure 23.5. (**C23L1**)

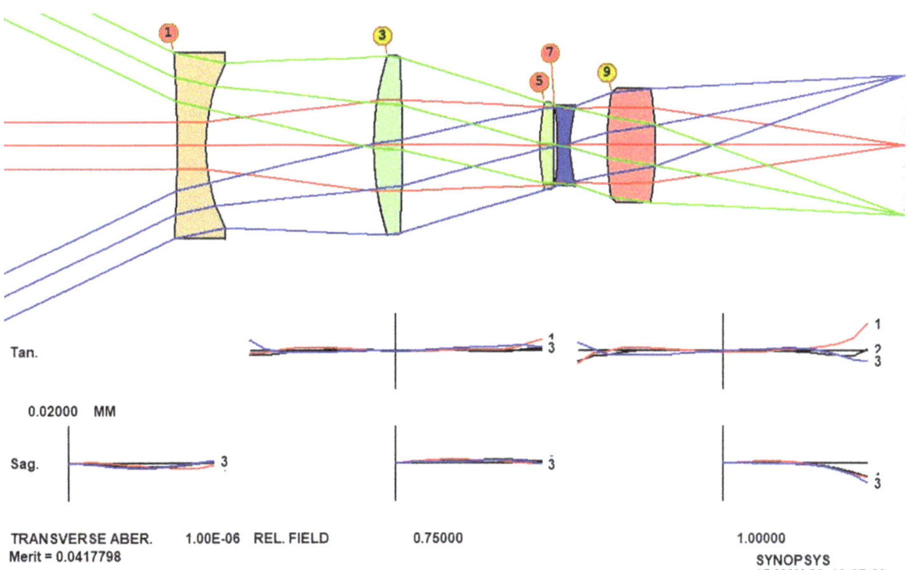

Figure 23.5. Lens optimized with two DOEs, with higher-order terms.

This lens is rather close to the diffraction limit, and it might make sense to do this exercise again, this time with OPD targets. But we'll continue as it is.

One naturally asks, how much light gets through? A DOE has to *diffract* light into a desired direction, and of course diffraction involves multiple orders. It would be wise to check. The program assumes that the zones on the DOEs are *blazed* in such a way that the diffracted direction coincides with the *refracted* direction, but that assumption cannot be exact for all rays. Open the **MMA** dialog, select the transmission of a ray (not the transmission of a *beam*), Map over PUPIL, Y-point 1.0 in the object field, ray pattern CREC with a grid number of 51, EXPLODED, and Show color scale. Click Execute, and you get a picture showing the lens transmission mapped over the aperture, shown in figure 23.6. The assumption mentioned above, also called the *Bragg condition*, seems to be valid enough for practical purposes. Transmission is very high. (The system is not in polarization mode, so reflection losses are ignored, and the materials at the moment are all glass models, which do not come with absorption coefficients. Those effects can be analyzed later.)

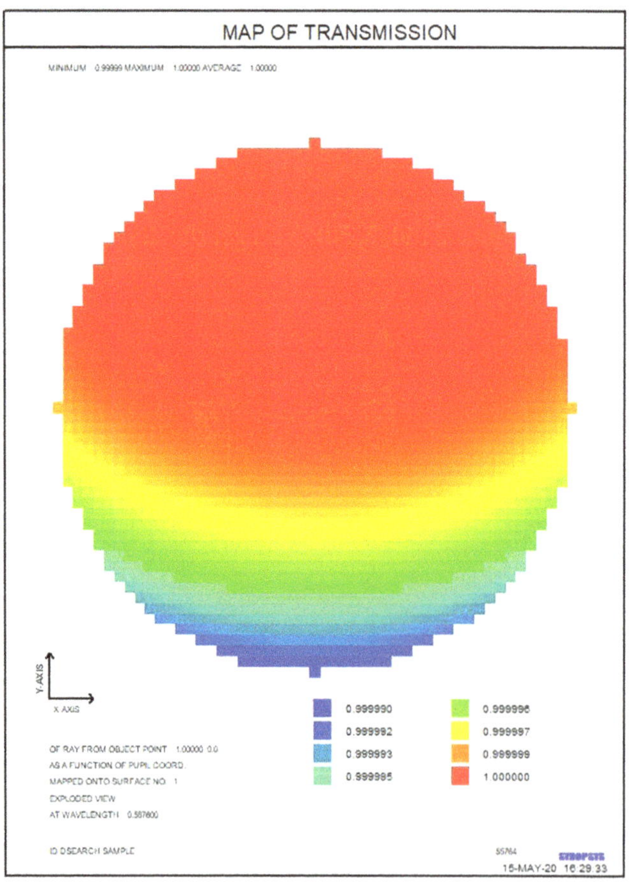

Figure 23.6. MAP of ray transmission over the pupil at the edge of the field.

It would be interesting to see how many spherical elements one would need to get this kind of quality. For sure it will be more than five. And the next step would be to replace the glass models with real glasses—but we'll leave that exercise for the student.

This lesson has shown how converting a lens surface to a DOE can significantly improve image quality—or let you get the quality you need with fewer elements. Of course, it all depends now on whether the lens vendor can *make* the DOEs. These

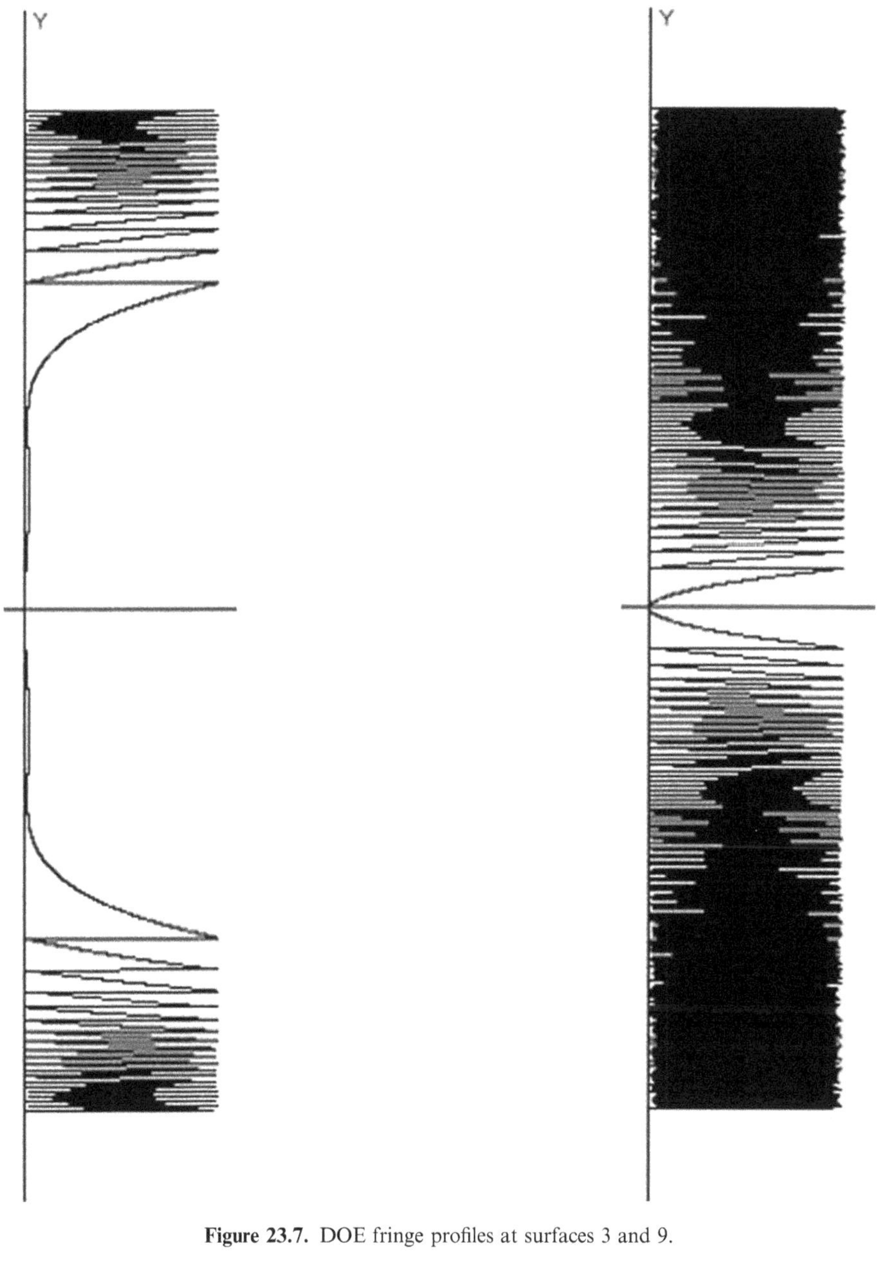

Figure 23.7. DOE fringe profiles at surfaces 3 and 9.

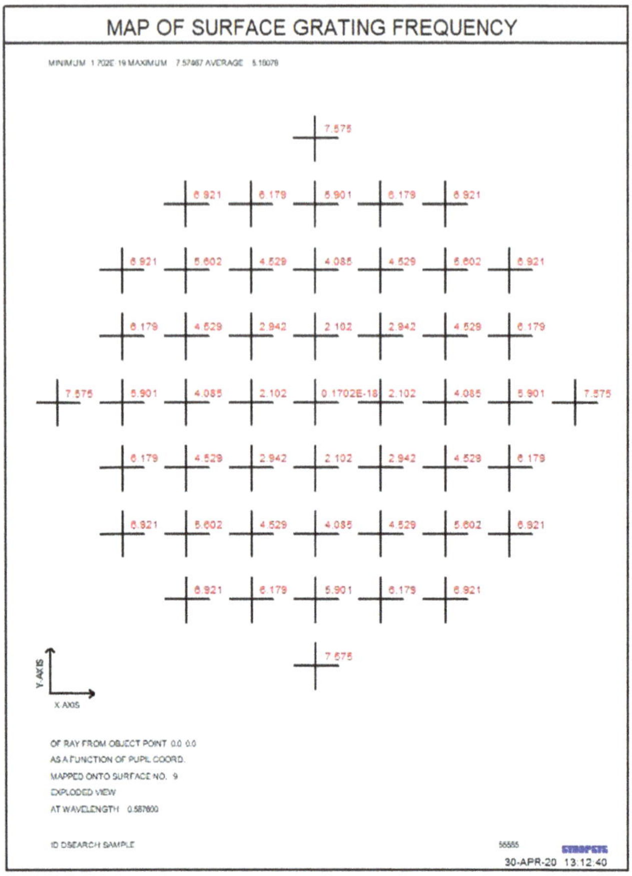

Figure 23.8. Surface grating frequency of the DOE on surface 9.

may not be too easy. Here is the DMASK profile at surfaces 3, on the left, and 9, on the right, shown in figure 23.7.

```
DMASK 3 PROFILE
DMASK 9 PROFILE
```

The second might be a challenge for the shop. Let's examine the spatial frequency. Open the MAP dialog again with **MMA**, select a map of HSFREQ over PUPIL, on surface 9, object point 0, Ray Pattern CREC 9, DIGITAL, and Execute. The highest frequency is over 7 c/mm at the edge. This is looking pretty good, but that of course depends on the capability and technology of the shop that will make them. This plot is in figure 23.8.

We expect that, as this technology matures, the designs presented here will become more and more practical. In any event, it is better to be ahead of the technology rocket than running behind, trying to keep up. We invite lens vendors with DOE capability to comment on this lesson and perhaps offer insights and design tradeoffs as they understand them today.

IOP Publishing

Lens Design (Second Edition)

Automatic and quasi-autonomous computational methods and techniques

Donald C Dilworth

Chapter 24

Designing aspheres for manufacturing

Improving a lens with aspherics; controlling departure from best-fitting sphere; CLINK

This lesson shows how to add aspheric terms to a spherical surface, thereby improving the image. Then the optimization controls the RMS departure of the aspheric surface from the closest fitting sphere (CFS) in order to make it easier to manufacture.

This is the starting lens (**6.RLE**), a poorly corrected triplet, shown in figure 24.1. Type `FETCH 6` to open this lens.

Let's do a simple optimization run, to see how things improve. First, we will use only spherical surfaces. Here is the MACro (**C24M1**):

```
PANT
VLIST RAD 1 2 3 4 6
VY 1 TH 20 3
VY 2 TH
VY 3 TH 20 3
VY 5 TH
VLIST GLM 1 3
END

AANT
AEC ; AUTOMATIC EDGE CORRECTION
ACC ; AUTOMATIC CENTER THICKNESS CONTROL

GNR .5 1 3 2 0
GNR .5 1 3 2 .5
GNR .5 1 3 2 .7
GNR .5 1 3 2 1.
GNR .5 1 2 1 0
```

doi:10.1088/978-0-7503-3695-6ch24

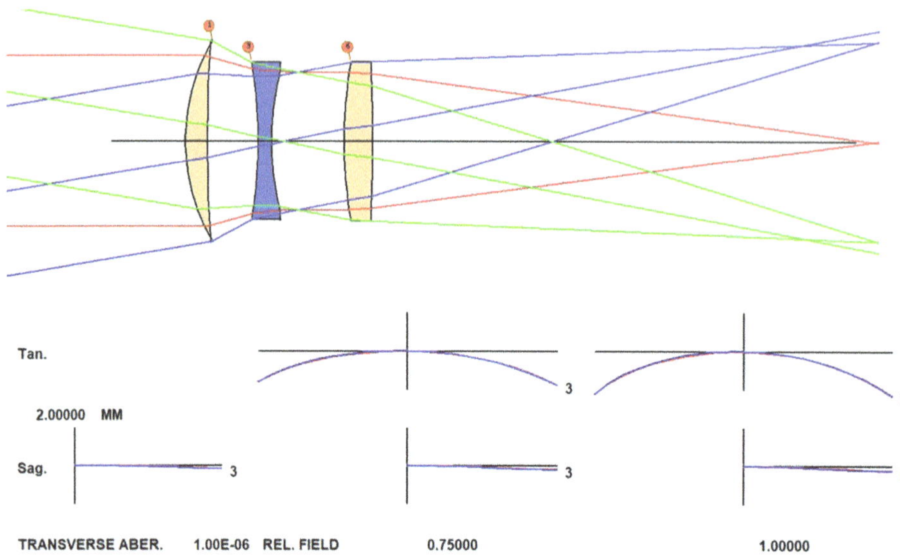

Figure 24.1. Starting triplet lens.

```
GNR .5 1 2 3 0
GNR .5 1 2 1 1
GNR .5 1 2 3 1
END

SNAP ; REQUEST SNAPSHOTS AS OPTIMIZATION RUNS
SYNO 50 ; REQUEST OPTIMIZATION FOR 50 PASSES
```

Run this, and the lens is improved; but the fifth-order spherical aberration is balanced by under-corrected third-order spherical, as shown in figure 24.2.

Now we will use the Automatic G-term Testing feature **AGT** to see if adding some general aspheric terms will improve things. Add a line before the PANT file in your MACro:

```
AGT 5 QUIET 1 .01 3 6 10 16
```

Here, we ask the program to test terms G 3, 6, 10, and 16 on surface 1, and keep any that reduce the merit function by 1% or more. Since surface 1 is not currently aspheric, those terms will apply to the default power-series asphere and will vary terms of power 4, 6, 8, and 10 in the aperture.

Run this, and the lens is better, as one would expect, as shown in figure 24.3. The program reports that only term G 3 was useful.

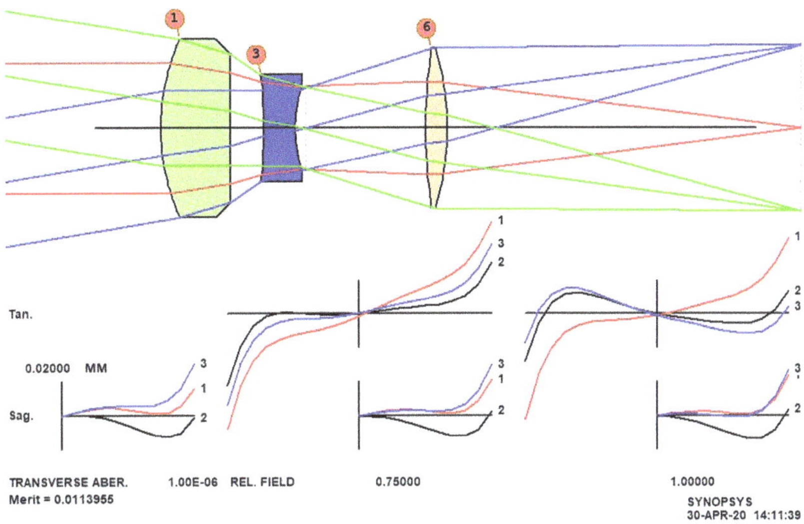

Figure 24.2. PAD display of the improved lens.

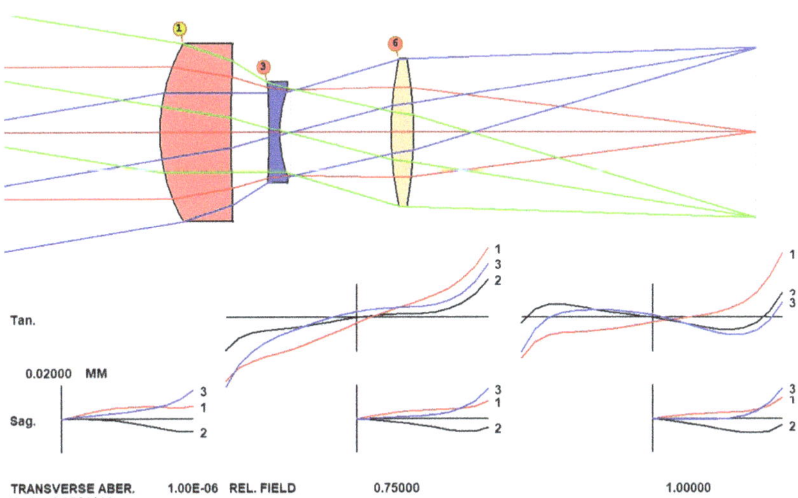

Figure 24.3. Lens optimized with one aspheric term.

What is the value of those aspheric terms? The **ASY** listing gives the coefficients:

```
SYNOPSYS AI>ASY

SPECIAL SURFACE DATA
------------------------------------------------------------
 SURFACE NO.   1 -- RD + POWER-SERIES ASPHERE

G 3   -8.491556E-08 (R**4)

THIS LENS HAS NO TILTS OR DECENTERS
```

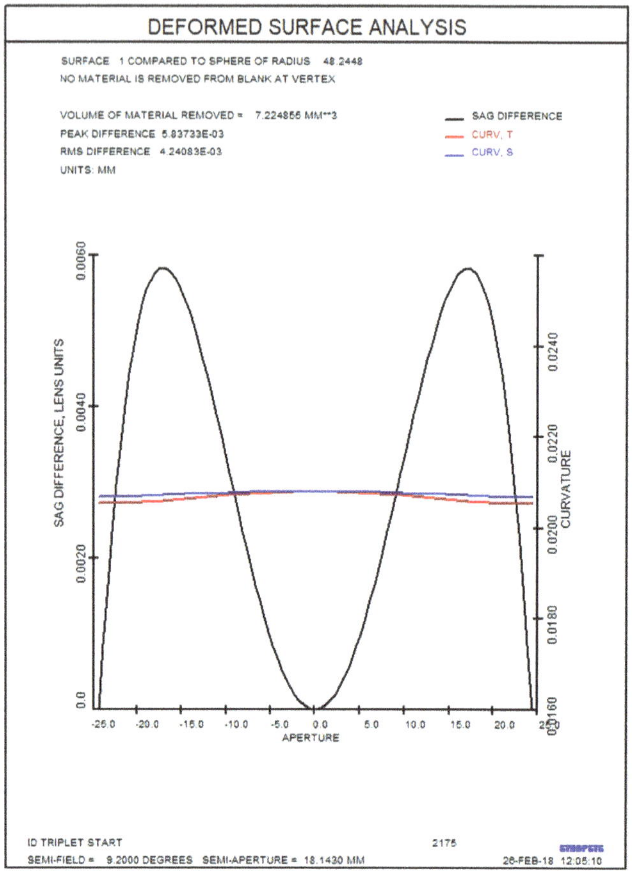

Figure 24.4. Analysis of aspheric surface 1.

We have used only one of the 22 possible G-series aspheric terms so far. How close is this surface to the closest-fitting sphere?

Type the command **ADEF 1 PLOT**, and you get the plot in figure 24.4. The maximum sag difference is about 5.8 micrometers.

24.1 Adding unusual requirements to the merit function with CLINK

The plot in figure 24.4 shows a large departure from the CFS, almost 6 micrometers. That might be hard to control accurately. Let's see if we can get similar performance with an asphere that has a smaller departure. Add variables to the PANT file as shown below, to see if we can get similar performance with a smaller RMS departure:

```
VY 1 G 3
VY 1 G 6
VY 1 G 10
VY 1 G 16
```

Comment out the AGT line, and add to the AANT file the lines

```
M 0 5 CLINK
ADEF 1
CD1 FILE 6
= CD1
```

Here we use the CLINK option, which causes the optimization program to run the next command (**ADEF** in this case) and then pick up the desired quantity from the FILE BUFFER. (To learn about this useful feature, type **HELP CLINK** in the CW.)

SYNOPSYS already has a command for this kind of analysis, described in Section 10.3.3 of the User's Manual, but what would you do if it did not? This lesson shows how other features can be used to do the same thing, and it is a good idea to know how to use these other features in case you want to do something for which there is no command. (To use the intrinsic form, add to the AANT file the line M 0 50 A ADIFF sn, instead of the CLINK section above, where sn is the surface number.)

Where did the weighting factor of 5 come from? The RMS difference between the asphere and the CFS is 0.0037 mm, and the largest of the other aberrations is 0.015. (To see those values, you can use the handy FINAL nb command. Enter FINAL 5, to see the five largest aberrations, and you get the following table):

```
FINAL 5
 ABERRATION LIST
        NAME            TARGET          WEIGHT              RAW VAL.  FINAL ERROR   R. EFFECT

   80                0.0000000    0.9362033 SR     0.0144   0.134860E-01 0.029585
     A    2  YC    0.70000    0.50000    0.83333   0.00000       ACON 2

   90                0.0000000    0.9362033 SR    -0.0131  -0.122242E-01 0.024307
     A    2  YC    0.70000    0.50000   -0.83333   0.00000       ACON 2

  102                0.0000000    1.7262735 SR    -0.0068  -0.118239E-01 0.022742
     A    2  YC    1.00000    0.16667    0.50000   0.00000       ACON 2

  156                0.0000000    1.1508490 SR     0.0133   0.152882E-01 0.038020
     A    1  YC    1.00000    0.25000    0.75000   0.00000       ACON 2

  160                0.0000000    2.5733766 SR    -0.0057  -0.147917E-01 0.035591
     A    1  YC    1.00000    0.25000   -0.25000   0.00000       ACON 2
 SYNOPSYS AI>
```

So this weighting will give the RMS departure an error comparable to the error of the largest ray aberration. Of course you can adjust it later, when you see how things work out.

And how can one figure out the other lines to be added to the AANT file? Simple. When the ADEF command runs, it puts a copy of some of its output into the AI buffer. Run the command **ADEF 1**, and then ask the AI question

BUFFER?

to see what got put where.

```
SYNOPSYS AI>BUFF?

The current FILE BUFFER contains
   1         0.02072660  BEST FIT CV
   2        48.24717805  BEST FIT RD
   3  -4.52970994E-13  VERTEX SHIFT
   4         7.22272358  VOL. REMOVED
   5         0.00583572  PEAK DIFF.
   6         0.00423966  RMS DIFF.
```

Location number 6 has the desired RMS difference, the quantity we want to reduce.

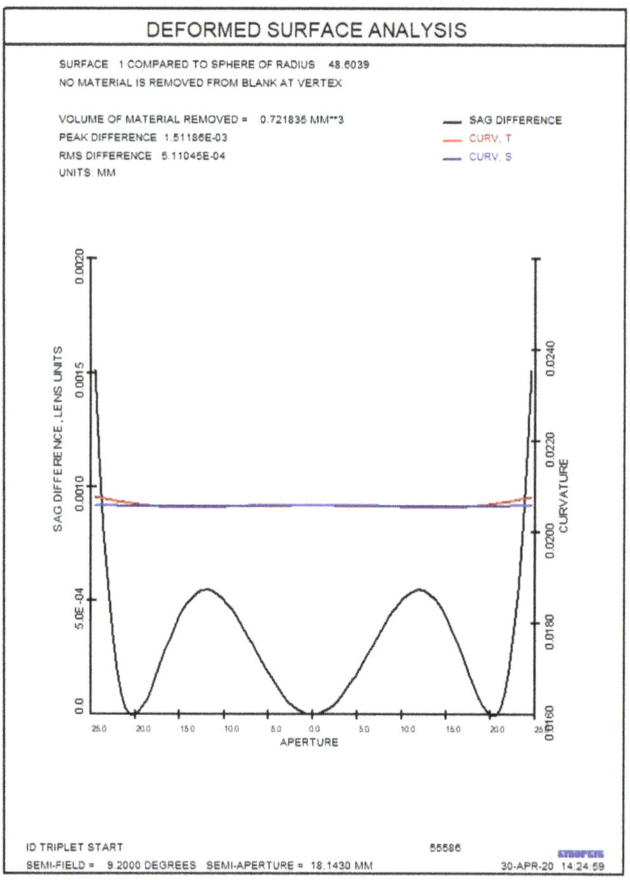

Figure 24.5. Aspheric analysis shows a smaller departure from the CFS.

Run this new optimization, and the ray fans are almost unchanged. What does the aspheric look like now? The result is in figure 24.5.

```
SYNOPSYS AI>ASY

SPECIAL SURFACE DATA
--------------------------------------------------------------------------------
  SURFACE NO.    1 -- RD + POWER-SERIES ASPHERE
 G 3  -3.459206E-08 (R**4)   G 6  5.725129E-12 (R**6)   G 10 6.886139E-14 (R**8)
 G 16 -2.957288E-17 (R**10)
```

Now the RMS difference is only 0.0005. Here we have a lens with only 12% as much aspheric departure as before, and essentially identical performance. (**C24L1**)

Now it might make sense to adjust that weighting factor, increasing it in stages until the performance starts to degrade. That might yield a lens that is even easier to make than the one above.

One can also control the *peak* difference instead of the RMS, by simply picking up FILE 5 instead of 6. Let's try that idea. Change the extra AANT entries you added above to

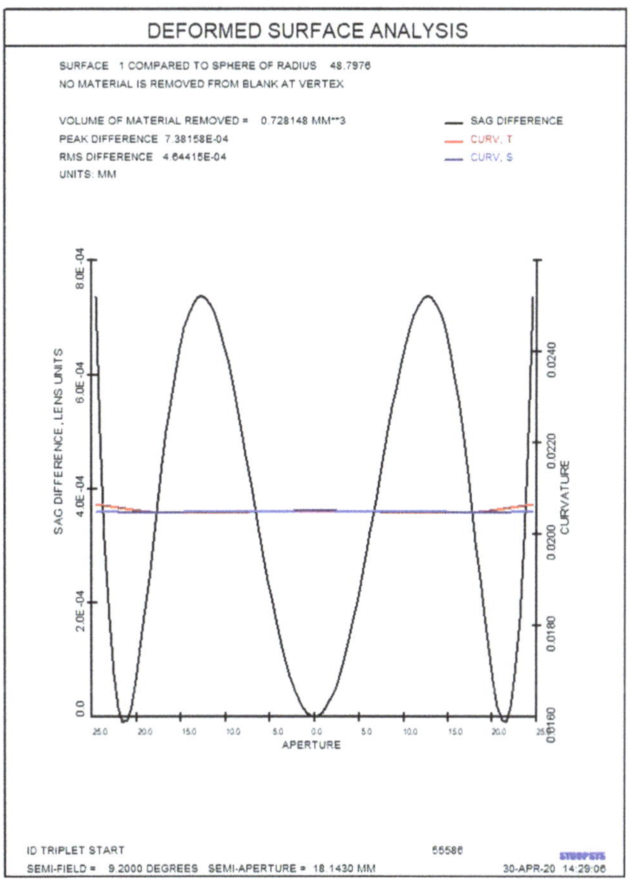

Figure 24.6. Analysis of aspheric surface after controlling the peak difference with the CFS.

```
M 0 5 CLINK
ADEF 1
CD1 FILE 5
= CD1
```

and reoptimize. Now the peak departure is 0.000 74 mm, shown in figure 24.6.

How hard would it be to make this aspheric? Let's see:

```
ADEF 1 FRINGES
```

The result is in figure 24.7.

This is a very nice gentle asphere, with only a few fringes to the best-fitting spherical surface. The fringes are seen in double-pass; the actual difference is only half that shown.

With these tools, you can utilize aspheric elements while taking care that they are easily manufactured.

24.2 Defining an aberration with COMPOSITE

Yet another way to create a custom aberration definition is with the COMPOSITE format, which lets you combine quantities with an algebraic equation. For example, consider the following AANT file (which Is not related to the lens above):

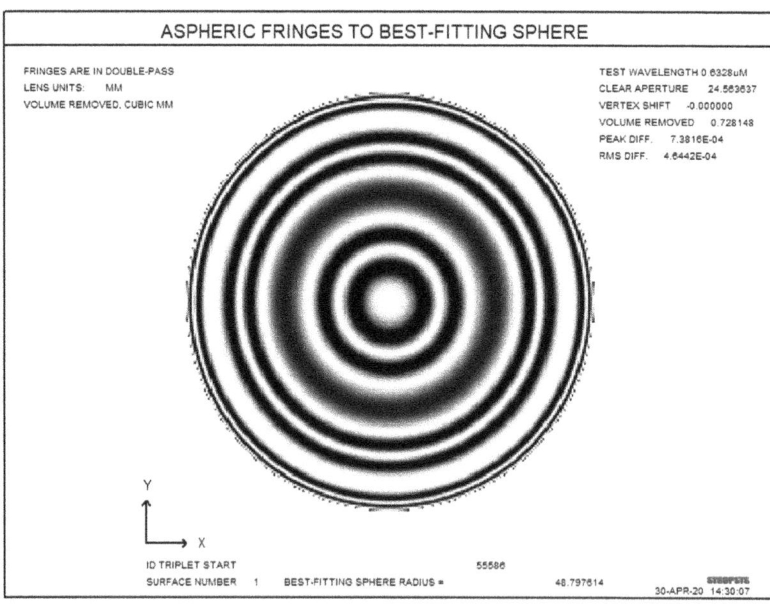

Figure 24.7. Analysis of aspheric surface with reduced peak difference.

```
AANT
M 0 1 COMPOSITE
CD1 PYA 10
CD2 PYB 10
CD3 GBD
CD4 RAD 10
= ATAN(CD3) + ASIN((SQRT(CD1**2 + CD2**2))/CD4)
END

SYNO 5
```

Here we use four of the nine CDn variables, which are defined in the AANT file as shown above. Variable CD1 is set equal to the paraxial Y coordinate of the marginal ray on surface 10, CD2 is the Y coordinate of the chief ray there, CD3 is the Gaussian beam divergence at the last surface, and CD4 is the radius of curvature of surface 10. When these are all defined, the aberration consists of the entered equation, which involves the arctan, arcsine, and square root. The result of that calculation is then targeted to the value 0 with a weight of 1.0. Chapter 10 of the User's Manual describes this feature in more detail. (This example is shown only as an illustration; we would likely never actually use that definition for a real lens!)

Chapter 25

Designing an athermal lens

Selecting glass types and housing material to correct both chromatic aberration and thermal effects

This chapter will show how to design a lens that must stay in focus over a wide temperature range.

First, we must discuss the interesting concept of the 'achrotherm', which applies to lenses that are corrected for both chromatic aberration and temperature changes at the same time. The theory is actually quite simple, and you can read the original paper at Applied Optics Vol. 33, No. 34, 8009-8013 (1 December 1994).

To design this kind of system, you select two glass types that obey a special requirement. They may be found by using the glass table display (**MGT**), where you click on the **Graph** button and then select **Thermal properties**.

The graph in this case shows the quantity 1/Vd on the abscissa (Vd is the Abbe number) and the quantity β on the ordinate, defined as

$$\beta = \alpha_g - \frac{1}{n-1}\frac{dn}{dT}$$

where α_g is the coefficient of thermal expansion of the glass, n is the index of refraction, and dn/dt is taken from the glass table. The idea is to select two glasses on the diagram and draw a line connecting them. Extend this line to the right to where 1/V equals zero. The height of the intercept, which can be read from the ordinate axis, is the CTE (coefficient of thermal expansion) of the housing required to athermalize the lens.

This feature needs to know the thermal coefficient of the housing material. Open any lens file and specify aluminum 6061:

```
CHG
ALPHA A6061
END
```

doi:10.1088/978-0-7503-3695-6ch25

Now you can use the glass table to select some likely glass pairs. In **MGT**, select the Ohara catalog and look at the thermal properties, as explained above. The program draws a green symbol over to the right, which is a function of the **ALPHA** of the lens housing that you just entered. Click on the left-most glass, and it draws a line between that glass and the symbol, as shown in figure 25.1.

The line starts at glass S-NPH3. Now you need another glass that falls close to the same line. Type S-FPM3 is quite close, so those two look like a good pair. It might be useful to have a second set, so click the glass S-TIM35 and then see glass type S-PHM53 close to the new line, as shown in figure 25.2.

Okay, we have selected four glass types. Now let's design a lens with those glasses and hope the thermal properties can be controlled.

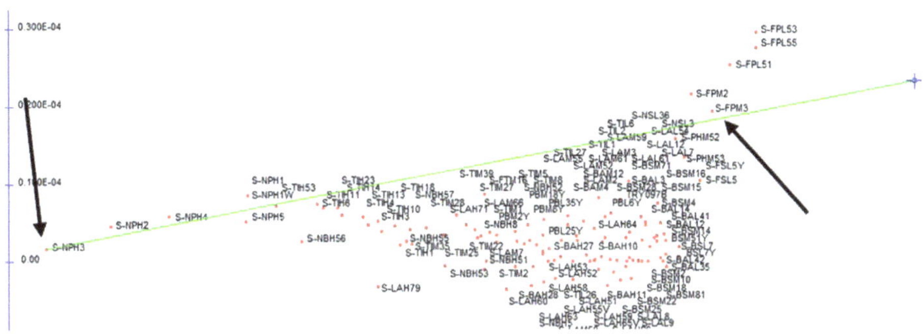

Figure 25.1. Glass-table display showing thermal properties.

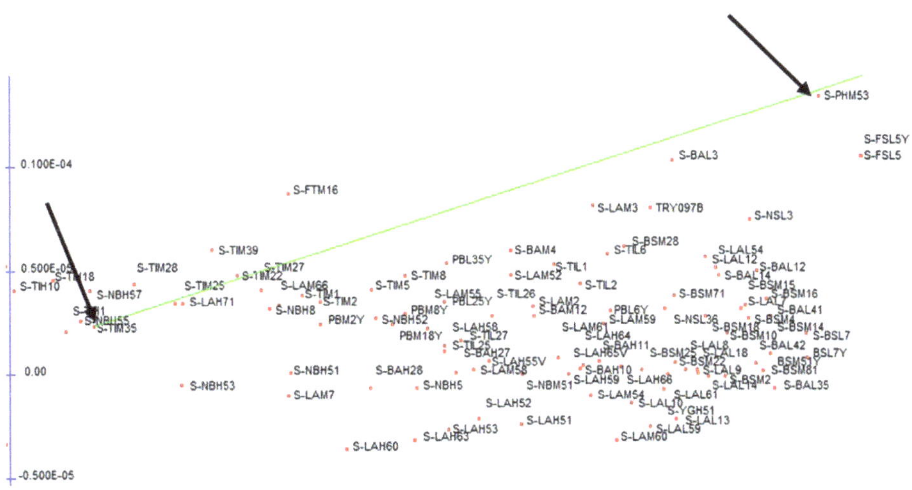

Figure 25.2. Display when a second glass is selected.

Create a DSEARCH MACro, specifying only the first two glasses found above. The dialog **MDS** is a good place to enter the requirements. Give it a file name when you click OK, and it creates a MACro, which you can further edit as you need. Here is a good example (**C25M1**):

```
CORE 64
OFF 1
TIME
DSEARCH 1 QUIET
SYSTEM
ID DSEARCH ATHERMAL
OBB 0 25 2.5
WAVL 0.6563 0.5876 0.4861

UNITS MM
END
GOALS
ELEMENTS 5
FNUM 2
BACK 0 0
TOTL 10 0.1
STOP MIDDLE

STOP FREE
RT 0.5
FOV 0.0 0.75 1.0 0.0 0.0
FWT 5.0 3.0 1.0 1.0 1.0
NPASS 40
GLASS POSITIVE
O S-NPH3
GLASS NEGATIVE
O S-FPM3
ANNEAL 200 20 Q
COLORS 3
SNAPSHOT 10
QUICK 33 40
END
SPECIAL PANT

END
SPECIAL AANT
END
GO
TIME
```

Run this, and in a minute you see the best lens it found. Optimize the lens some more, using the MACro that DSEARCH created. The image is pretty good, at least at a temperature of 20 °C (**C25L1**).

Now you have to check the thermal characteristics. In the WorkSheet, declare the housing to be aluminum 6061, as you did above. Then delete the thickness solve, so the lens won't automatically get refocused as the temperature changes. In WS, type **NTOP** to remove the solve. (The shadowed lenses will automatically remove the curvature solve.)

Now prepare another MACro to initiate *thermal shadowing* of this lens.

```
THERM
ATS 50 2
ATS 100 3
END
```

Run this, and the program puts a copy of the lens in ACON 2, recalculated for 50 degrees, and in ACON 3 at 100 degrees. The ACON 1 display is in figure 25.3.

Now the critical test: Click the button for ACON 2,

and … nothing changes! The rayfan curves look almost identical to those in ACON 1. How about ACON 3, at 100 degrees? Again, almost identical! Our lens is satisfactorily athermal.

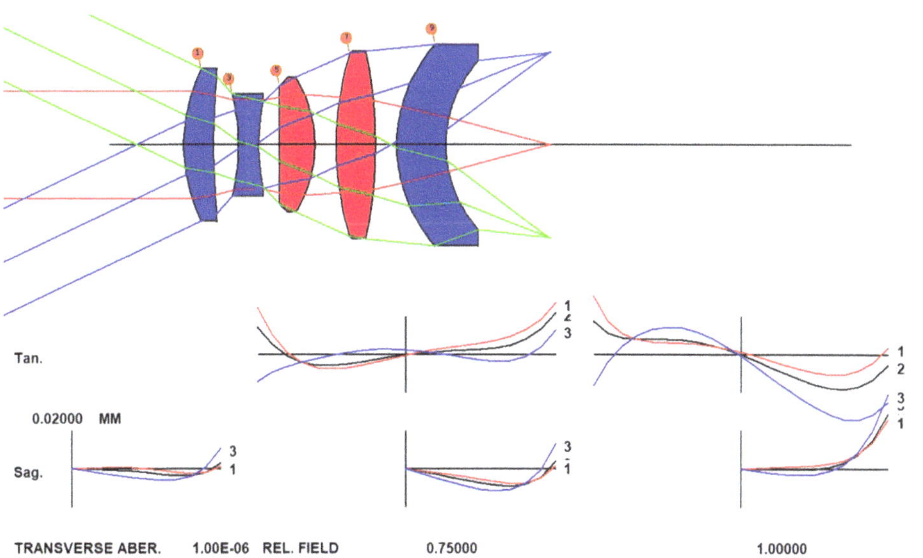

Figure 25.3. Athermalized lens at 20 degrees.

What if it had not been so good? Well, then you could correct the image in all three ACONs in the optimization file. That should tweak up any remaining problems.

And you didn't even need the second pair of glasses we found. What happens if you use those instead of the first pair? Try it and see!

This exercise shows how designing an achrotherm lens is not difficult, using the proper tools. When you have finished, be sure to type

```
THERM OFF
```

In the command window. Then you can work on other projects and they won't be shadowed—which you very likely wouldn't want.

IOP Publishing

Lens Design (Second Edition)

Automatic and quasi-autonomous computational methods and techniques

Donald C Dilworth

Chapter 26

Using the SYNOPSYS glass model

Glass variables; boundary conditions

When you vary the properties of an optical glass in SYNOPSYS, you are asking the program to find values of the index Nd and Abbe number Vd that will correct aberrations and that are within the boundaries of the commercial glass maps. That's pretty straightforward—but you need more than just those two parameters. The program must calculate the index of refraction at each wavelength in the lens, and you hope the value so found will closely resemble the behavior of real glasses in that section of the map. That is the purpose of the glass model. The area on the glass map over which glass models can roam is shown in figure 26.1, which also shows the preferred glasses from the Schott glass company. This display is shown in response to the command **MGT**. The boundaries are given by the lines on the left and right side of the glass display.

A model glass can be inserted into the lens via the SpreadSheet, or more quickly with the keyboard or WorkSheet. For example, the input

```
CHG
1 GLM 1.6 55
END
```

assigns a model glass to surface 1 with the Nd and Vd values specified. You can declare glass variables in the PANT file with input such as

```
VY 1 GLM
VY 3 GBC
VY 5 GBF
VLIST GLM 1 5 8
VLIST GLM ALL
```

The **VLIST GLM ALL** form varies all glasses that are already declared model glasses, while the form **VY sn GLM** forces the material to be a glass model, if it is not

doi:10.1088/978-0-7503-3695-6ch26

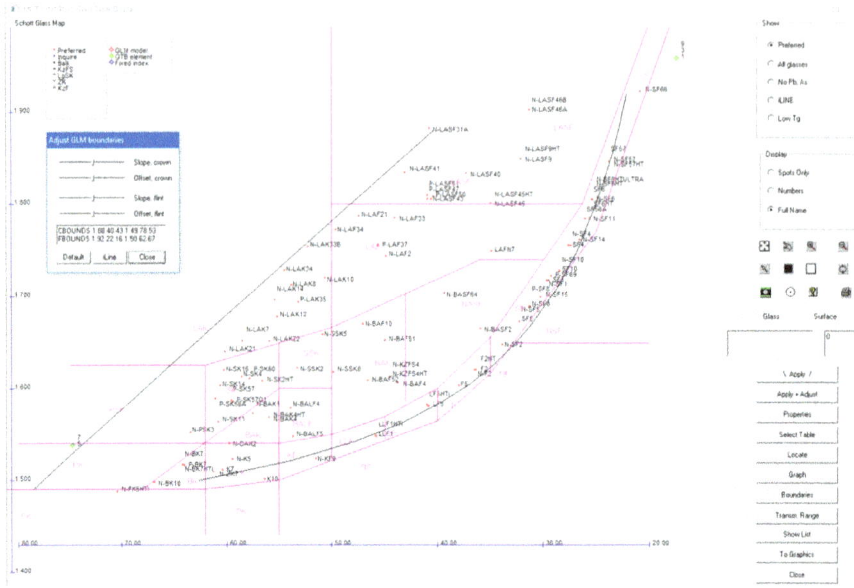

Figure 26.1. MGT display showing glasses from Schott and the boundaries of GLM variables.

already, as does the **VLIST** form with surface numbers added. In that case, the program first finds a model that closely resembles the current glass and assigns it to the lens. **GBC** and **GBF** are used to vary a glass along the crown or flint boundary. (This is an example of the input formats; you could not actually submit this list to the PANT file, since it will not accept duplicate variables.)

Glass boundaries are tricky to implement. During optimization, the index often wants to become very high, and of course the Vd of many elements would like to be infinite. That would be great mathematically, but such materials do not exist—so the program has to constrain the glass model to the usable part of the glass map. To do so, it does something clever: when any of the models tries to go over the boundary on the left or right, the program first restricts the change so it goes exactly to that boundary, and then redefines that variable, changing the **GLM** variable to a **GBC** (glass bounded, crown) or **GBF** (glass bounded, flint) instead. Then the glass model will move up or down along that boundary. As a result, the glass model remains within the glass map, and you are left with a single variable where previously there were two. If the glass tries to go over the upper or lower limit on the index, the program reduces the change so it goes exactly to that boundary. In this way, the glass model variable always remains inside the region of the glass map where actual glasses are to be found.

Once a glass has become pinned to the crown or flint boundary, however, it remains there for the duration of that run. It sometimes happens that after a design has been much improved, one of the glasses would work better if it left the boundary. This is simple to test: just run the optimization again. The glasses will start out free to move anywhere, and they can immediately leave the boundary if it improves the lens.

The simulated annealing program also frees up all glass models that have become pinned to a boundary before it reoptimizes the lens the first time, and can free them every iteration if you select that option on the Simulated Annealing dialog.

Of course, you do not expect the model glass found during optimization to coincide exactly with any real glass in a selected vendor's catalog, but that is not a problem since you usually can find one whose properties resemble the model closely enough. Then you just substitute that glass and reoptimize. ARGLASS, which you have used in previous chapters, makes that job easy. And there is a search program for glasses, GSEARCH, that can often find just the right combination. Chapter 35 gives an example.

Many high-quality designs have to compensate for secondary color to some degree, and in order for the program to optimize glasses while accounting for that aberration, the partial dispersion of the model must be reasonably close to that of nearby real glasses so the correction is maintained when a real glass is substituted later. 'Partial dispersion' refers to the curvature of the index curve, which is different for different glass types.

But now it gets tricky. SYNOPSYS uses a polynomial expression for its glass models that yields the index at any wavelength in the visible region, given only the glass-map coordinates (Nd, Vd), the coefficients having been found via a least-squares fit to selected glasses from the Schott table. Figure 26.3 shows the Schott glass map, where the Graph option has been selected to show the partials P(F,e) vs. Ve. (Open the glass map with MGT or the PAD button , select Schott, click the button

and select that option, as shown in figure 26.2.)

For the above example, we prepared a lens of eight elements with glass models assigned as shown by the red circles in figure 26.3. The goal is for the model to approximate the same distribution as the real glasses, which it does well enough to be useful.

How close is the glass model to a real glass? We prepared a lens where element 1 uses glass SK6 and surface 3 is assigned a GLM with the same Nd and Vd as that glass. Then we prepared an AI plot showing how the two compare, as shown in figure 26.4.

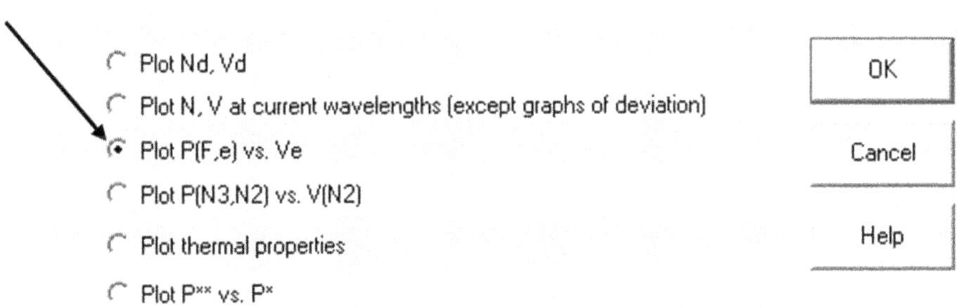

Figure 26.2. Selecting the partials map.

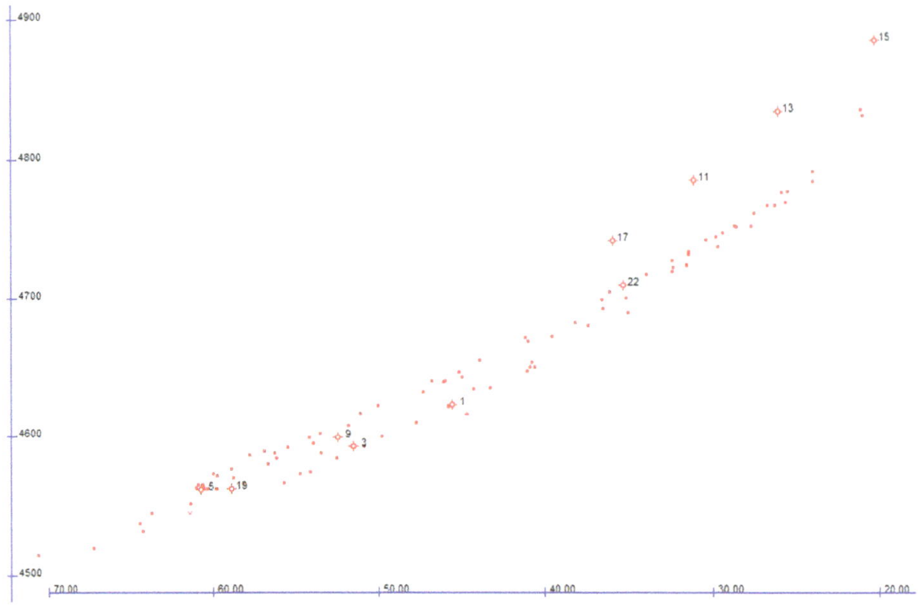

Figure 26.3. Glass map display showing selected partials along with glass model equivalents.

```
STEPS = 50
MULTI PLOT INDEX OF 1 FOR WAVL = .3 TO 2
ADD PLOT INDEX OF 3 FOR WAVL = .3 TO 2
END
```

The fit is close enough for design purposes, especially over the visual range of about 0.4–0.7, and is useful over a wider range, although slightly less accurate, as shown on the plot. (The model coefficients were calculated over the range 0.35–0.9 um.) For wavelengths far outside this range, such as for the NIR design in chapter 14, it is best to proceed as outlined in that chapter.

Now we will show how to adapt the glass model for special conditions. A good example is when designing for the UV spectrum, where one is restricted to iLine glasses from the Ohara glass company. How can one vary the glass model while staying in the region where those glasses can be found? Simple. Here is the glass map, in figure 26.5, showing just the iLine glasses, selected by that radio button. The vertical lines show the relative prices; the glasses in red are preferred, while black denotes an inquiry glass.

If we just varied the GLM variables as usual, we would probably wind up with very high-index materials that are not very close to one of the iLine glasses. We can prevent that by changing the boundaries. Click the button

and the program displays the current (in this case the default) boundaries, as in figure 26.6.

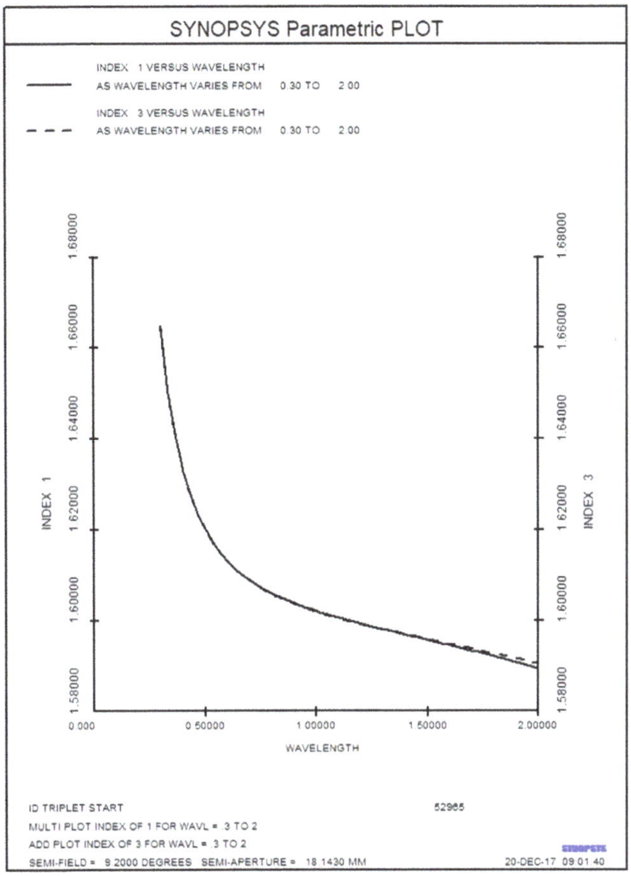

Figure 26.4. Comparison of real and model glass index over the spectral range of 0.3–2 um.

Now, click the iLine button on the boundaries dialog. You see the region where iLine glasses are to be found, shown in figure 26.7. You can also adjust the boundary lines with the sliders in this dialog, should you want to.

There are four parameters that can be specified in the PANT file for controlling the glass boundaries, and the edit box shown in figure 26.7 gives data for the CBOUNDS and FBOUNDS directives. Select those lines and then copy-paste them into the PANT file, near the top. Then add another line, giving an upper limit to the GLM index variables of 1.6, with the CUL (crown, upper limit) line. The PANT file now starts with

```
PANT
CBOUNDS 1.88 9.43 1.49 82.55
FBOUNDS 1.92 22.16 1.50 62.67
CUL 1.6

...
```

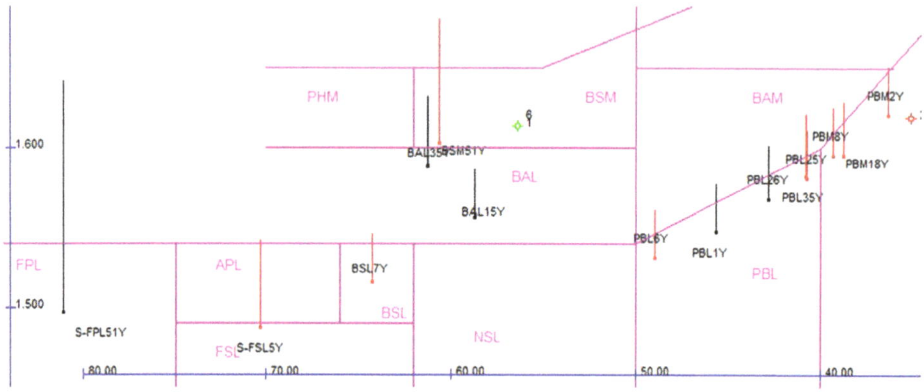

Figure 26.5. Glass map display showing only iLine glasses from Ohara.

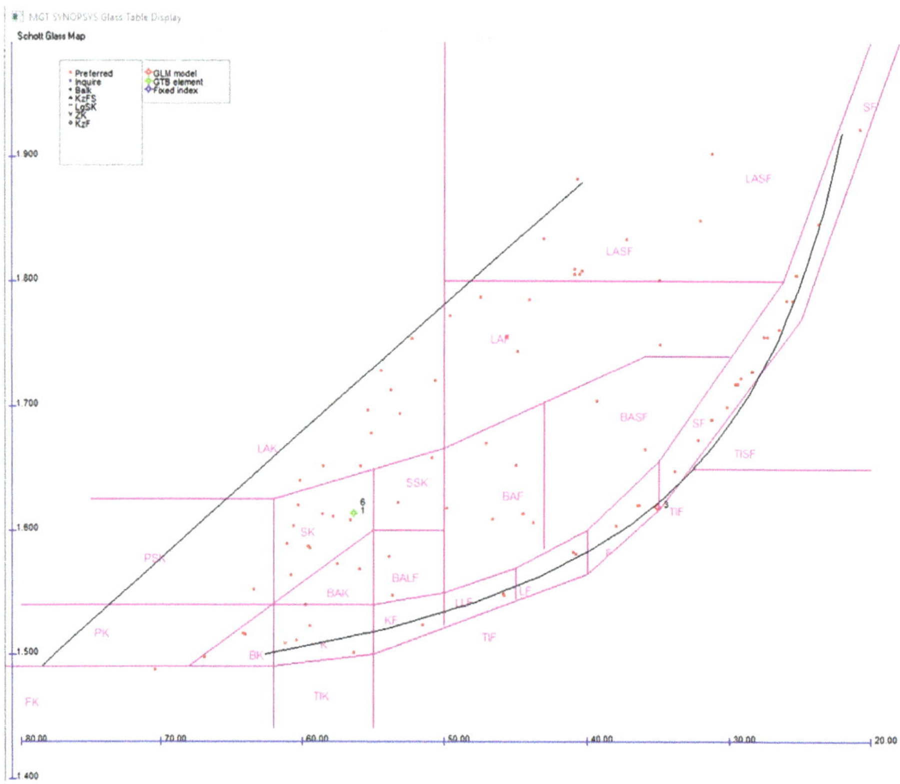

Figure 26.6. Glass map display showing default model boundaries.

Now, when the glasses are varied, they will remain in the area shown in figure 26.7, and one would expect no trouble finding an iLine glass to match the model.

One final note: When you give the program a glass model, you are specifying the *input* to the polynomial. The actual index that comes back at each wavelength is the

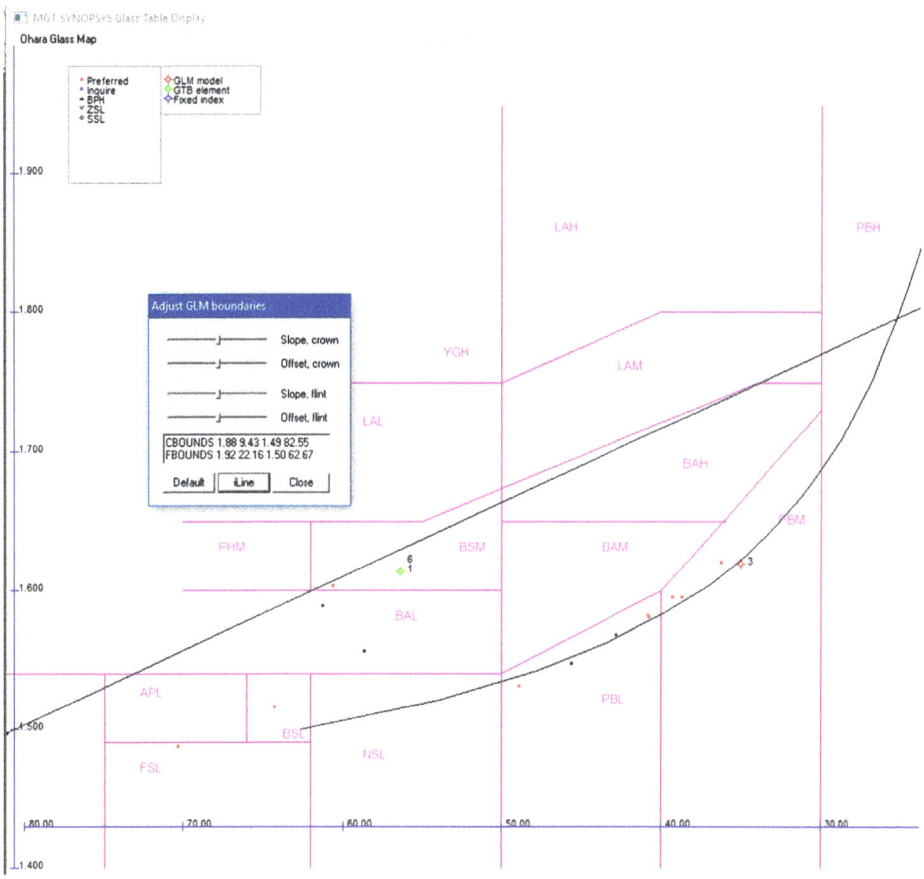

Figure 26.7. Glass map display with boundaries appropriate for iLine glasses.

output from the model, and the two can be slightly different. If the lens is assigned the CDF spectral lines, they will be very close—but if your spectrum is anything else, then you can expect the index listing from SPEC (which gives the model *input*) to be a little different from the output of PRT (which lists the *output* indexes).

This glass model has proven to be invaluable for finding where on the glass map the lenses work best. In some cases, the program has even managed to correct secondary color by a good choice of glasses, all by itself. And if the **ARG** dialog cannot find glasses that maintain the excellent correction provided by the glass models, the search program **GSEARCH** will usually do the job. An example is given in chapter 47.

IOP Publishing

Lens Design (Second Edition)
Automatic and quasi-autonomous computational methods and techniques
Donald C Dilworth

Chapter 27

Chaos in lens optimization

Effect of PSD optimization; three-parameter plots; automatic ray-failure correction

In this chapter we will explore a powerful feature of SYNOPSYS: it can do parametric studies showing the effects of two variables on a third. In this case, we want to see how the end point of a lens optimization run depends on the starting point. In a perfect world, every starting point would go to the best of all possible optima, but the world is not perfect. There are usually many local minima for any given problem, and the best we can expect is that a good optimization algorithm should go reliably to the closest one. (Of course, *global* optimization algorithms such as DSEARCH can find a variety of solutions, but that is a different topic. Here we will analyze the process that simply minimizes the merit function, starting with a single configuration.)

One would therefore expect that two starting points that are almost exactly the same would go to the same local minimum, even if it is not global. How well do current algorithms perform on this score? Some interesting results were discovered by Dr. Florian Bociort of TU Delft. He ran a simple case, the doublet shown in figure 27.1.

To make the job very simple, he corrected rays at three field points in the major color only, ignoring edge violations. Then he varied the starting value of radii 2 and 3 in a raster fashion and made a plot where the color of each pixel on the grid encodes the final value of the merit function. He found that there are several local minima, which is not surprising, even for so simple a problem—but what was completely unexpected was how the merit function varied in a chaotic manner in many places. Thus, nearby starting points often go to very different end points. (He did this analysis on a different program, with a different optimization algorithm than the PSD method used in SYNOPSYS.) Here is a figure from his article at http://homepage.tudelft.nl/q1d90/FBweb/fractals.html, shown on the right in figure 27.1.

(We have turned this picture on its side so it will line up with the SYNOPSYS analysis below.)

doi:10.1088/978-0-7503-3695-6ch27

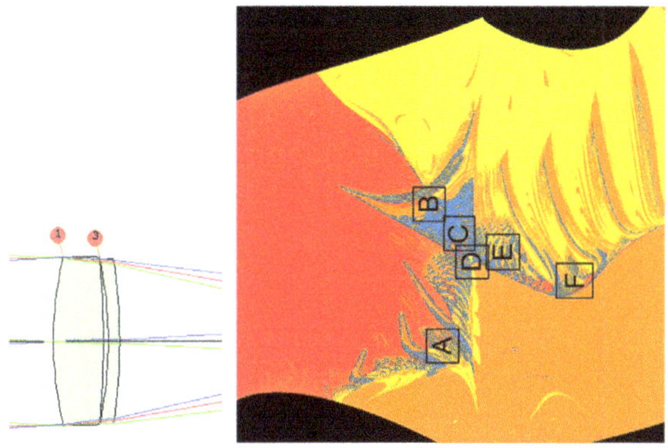

Figure 27.1. Doublet lens and a map of the merit function as a function of starting radii. Right-hand panel reproduced with permission from van Turnhout M and Bociort F 2009 Instabilities and fractal basins of attraction in optical system optimization *Opt. Express* **17** 314–28. Copyright The Optical Society.

Notice how the results near the boundaries of the zones of attraction are complex and chaotic. The black areas show starting points that gave ray failures and were not analyzed.

It would be interesting to see whether the PSD III algorithm in SYNOPSYS is more reliable and stable than the method used for the above picture. Here is the input to run the three-parameter evaluation feature **PA3**, to find out.

This is the starting doublet (**C27M1**):

```
RLE
ID FLORIAN STARTING DOUBLET
WA1 .5876000
 WT1 1.00000
 APS 1
 UNITS MM
 OBB 0.000000 3.00000 16.66670 0.00000 0.00000 0.00000
16.66670
 0 AIR
 1 CV 0.0146498673770 TH 10.34600000
 1 N1 1.61800000
 1 GID âGLASS â
 2 RAD -174.6512432672814 TH 1.00000000 AIR
 2 AIR
 3 RAD -80.2251653581521 TH 2.35100000
 3 N1 1.71700000
 3 GID âGLASS â
 4 RAD -111.8857786363961 TH 92.41206276 AIR
 4 AIR
 4 CV -0.00893769
```

```
  4 UMC -0.16667000
  4 TH 92.41206276
  4 YMT 0.00000000
  5 CV 0.0000000000000 TH 0.00000000 AIR
  5 AIR
  END
STORE 5
```

And this is the input for the PA3 program (**C27M2**):

```
ON 78                   ! use finer grid (118x118 points)
PA3 LOOP COLOR          ! initialize PA3, request color boxes for output
RZ1 -.025 .04           ! set the range of variable Z1
RZ2 -.045 .075          ! set the range of Z2
RZ3 0 3.7               ! display results over this range of merit function
                        ! values
NOSMOOTH                ! there will be steps in the output; do not smooth
XLAB "2 CV -.025 .04"        ! define the label for the X-axis, which is
                             ! variable Z1
YLAB "3 CV -.045 .075"       ! label for Y-axis, Z2
ZLAB "MERIT"            ! label for Z-axis, the final merit function
LOOP                    ! tell PA3 to loop over the above raster of data

GET 5                   ! get the starting lens each time
2 CV = Z1               ! set curvature 2 to the value of variable Z1
3 CV = Z2               ! and CV 3 to Z2, using the artificial-intelligence
                        ! parser

PANT                    ! initialize the variable list
VLIST RAD 2 3           ! and vary two radii
END                     ! end of the variable list

AANT                    ! initialize the merit function definition
GSR .5 10 3 P 0         ! correct a sagittal fan, three rays, on axis
GNR .5 1 3 P .75        ! correct a full grid of rays, primary color, 0.75
                        ! field point
GNR .5 1 3 P 1          ! same, at full field.
END                     ! end of merit function definition

DAMP 10000              ! initial damping (see below)
SNAP 50                 ! watch what happens, but not too often, in order to
                        ! keep it fast
SYNOPSYS 100            ! optimize until it converges

Z3 = MERIT              ! assign the current merit function value to Z3
PA3                     ! tell PA3 to cycle to the next case.
```

Why the high damping? (The default is 1.0 or 0.01, depending on mode switches.) The first iteration in SYNOPSYS is a DLS (Damped-Least-Squares) cycle[1], and we want to avoid any chaos that results on the first pass from that algorithm; the high damping will ensure that the lens changes very little on that pass. The more powerful

[1] This is true if switch 67 is off. Somewhat better results sometimes obtain if that switch is on, causing the first iteration to use calculated second derivatives. The PSD III method is employed from the second pass in either case.

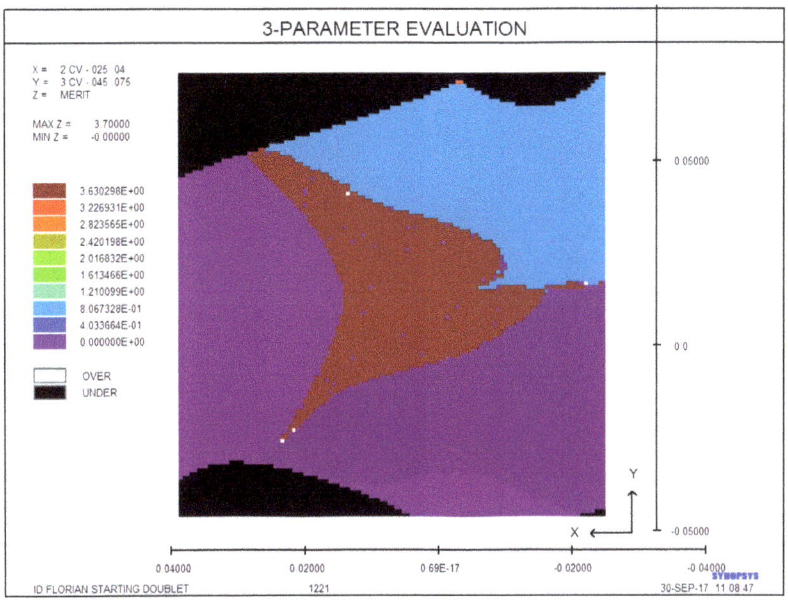

Figure 27.2. Evaluation of merit function with raster scan of two radii.

PSD algorithm keeps track of changes in the first derivatives from pass to pass and deduces information about higher-order derivatives. That is the magic behind the PSD method, as described in detail in appendix B—but it only works starting at the second pass.

The results of this study are shown in figure 27.2. The violet areas on the left and near the bottom show that the program went to the same minimum for very different starting points, whereas in Florian's study those areas went to different minima. There is no evident chaos at the boundaries of the zones of attraction, as we expected would be the case with the PSD method, although scattered poles do show up in the central red region. We attribute the latter to the nonzero changes made by the DLS method on the first pass. Indeed, if we run this again with different initial damping, those random spots appear in different places.

The black areas at top and bottom show where the starting points yielded ray failures, in the same place as they did in Florian's study. We are curious what would happen if we activate the automatic ray-failure correction feature. Change the SYNOPSYS command to

```
SYNOPSYS 100 0 FIX
```

and rerun this job. The result is in figure 27.3.

Now the program has corrected the ray failures over every point where they occurred before. The starting lenses that were not optimized by Florian now all yield respectable solutions. There is some very slight chaos evident at the boundaries in areas that were previously all black, however, and we attribute this to changes the

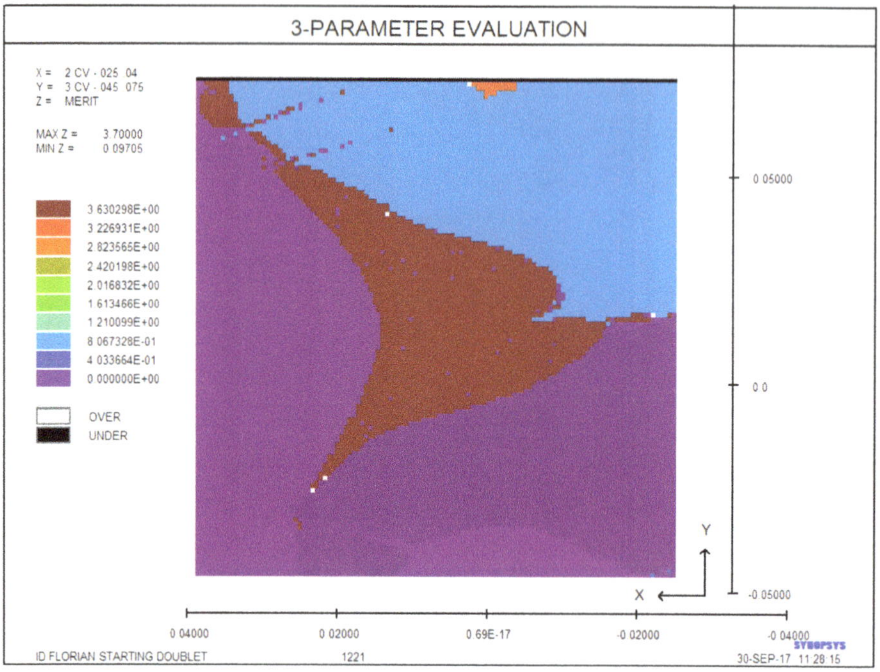

Figure 27.3. Evaluation with automatic ray-failure correction activated.

ray-failure correction program made to that starting point. Those changes sometimes moved the lens closer to another zone of attraction.

This very simple study involved optimizing with only two variables. What happens if you add CV 1 to the variable list? Try it and see! (The boundaries shift about somewhat, and the scattered spots no longer show up.) For those interested in looking further into the subject of chaos in lens design, I quote here an analysis by Florian Bociort, who has studied the effect.

> I use the word ‘Chaos’ in the strict sense of Nonlinear Systems Theory. In that theory it is shown that when the basins of attraction are fractal, then the succession of iterations of the optimization algorithm that leads to the color of a point is always a ‘temporary’ chaotic attractor (the technical term is ‘chaotic saddle’). The iteration process is first attracted to something that seems to be a chaotic attractor (which is something like what you find in the popular Lorenz Butterfly Effect metaphor). But it turns out that this chaotic attractor has ‘escape’ holes, so when the iteration finds such a ‘hole’ and ‘escapes’, it finally converges as expected to some minimum. The finely interwoven structure of the basins results when iterations starting at neighboring points find different escape holes and land therefore in different minima.

IOP Publishing

Lens Design (Second Edition)

Automatic and quasi-autonomous computational methods and techniques

Donald C Dilworth

Chapter 28

Tolerance example with clocking of element wedge errors and AI analysis of an image error

Statistics of tolerance budget with wedge clocking; adding AI commands to a Monte-Carlo simulation

This chapter applies some of the features discussed earlier and adds some interesting and powerful new options. Here, we will use BTOL to calculate a tolerance budget for an eight-element lens, then look at the image quality statistics for the case where the wedge errors are compensated by clocking the elements in the cell. Finally, we will examine the statistics of the lateral color of a set of 100 lenses subject to the budget, after refocusing the lens and clocking the elements.

Here is a MACro that will create the tolerance budget (**C28M1**). The lens (**C28L1**) is shown in figure 28.1.

```
FETCH C28L1          ! Get out the starting lens.
STORE 5
BTOL 90              ! Ask for 90% confidence level.
TPR ALL              ! All surfaces are matched to testplates.
EXACT ALL INDEX      ! Assume melt data are received,
EXACT ALL VNO        ! so the index and dispersion tolerances are zero.
TOL WAF .18 .32 .18 ! Ask for this wavefront variance at three fields.
FOCUS REAL           ! Focus the on-axis image point
ADJUST 14 TH 100     ! with thickness 14 (the last airspace).
PREP MC              ! Prepare the input for Monte-Carlo evaluation.
GO                   ! Start BTOL.
```

This MACro will get the lens out with the **FETCH** command and store a copy in library location 5. Then it creates a BTOL budget, which is listed on the monitor. Now we need to use the Monte-Carlo program MC. An adjustment MACro was prepared by BTOL, called **MCFILE.MAC**, that will be part of the MC analysis. Let's see what it contains (you don't have to open this file to make it work; here we show it for instruction purposes). Type **LM MCFILE** to load that MACro:

doi:10.1088/978-0-7503-3695-6ch28

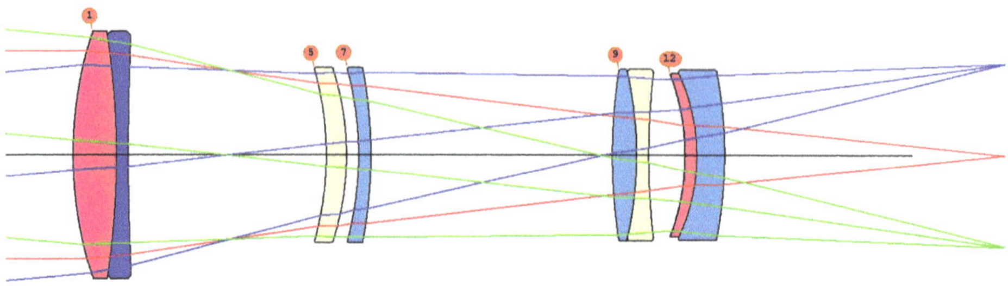

Figure 28.1. Example lens to demonstrate wedge clocking.

```
PANT
VY  14 TH
END
AANT
M    0.000000E+00 0.3333     A  2 XC  0.000 0   .1       0.000
M    0.297888E-05 0.3333 SR A  2 YC  0.000 0   .1       0.000
M    0.000000E+00 0.3333     A  2 XC  0.000 0 -.1        0.000
M   -0.297888E-05 0.3333 SR A  2 YC  0.000 0 -.1        0.000
M    0.297888E-05 0.3333     A  2 XC  0.000 .1 0         0.000
M    0.000000E+00 0.3333 SR A  2 YC  0.000 .1 0         0.000
M   -0.297888E-05 0.3333     A  2 XC  0.000 -.1 0        0.000
M    0.000000E+00 0.3333 SR A  2 YC  0.000 -.1 0        0.000
M   -0.177179E-02 0.3333     A  2 XC  0.000 -.64 .64     0.000
M    0.177179E-02 0.3333 SR A  2 YC  0.000 -.64 .64     0.000
M    0.177179E-02 0.3333     A  2 XC  0.000 .64 .64      0.000
M    0.177179E-02 0.3333 SR A  2 YC  0.000 .64 .64      0.000
M    0.177179E-02 0.3333     A  2 XC  0.000 .64 -.64     0.000
M   -0.177179E-02 0.3333 SR A  2 YC  0.000 .64 -.64     0.000
M   -0.177179E-02 0.3333     A  2 XC  0.000 -.64 -.64    0.000
M   -0.177179E-02 0.3333 SR A  2 YC  0.000 -.64 -.64    0.000
M    0.000000E+00 0.6667     A  3 XC  0.000 0  0.        0.000
M    0.000000E+00 0.6667     A  3 YC  0.000 0  0.        0.000
M    0.000000E+00 0.6667     A  3 XC  0.000 0   .1       0.000
M    0.149917E-03 0.6667     A  3 YC  0.000 0   .1       0.000
M    0.000000E+00 0.6667     A  3 XC  0.000 0 -.1        0.000
M   -0.149917E-03 0.6667     A  3 YC  0.000 0 -.1        0.000
M    0.149917E-03 0.6667     A  3 XC  0.000 .1 0.        0.000
M    0.000000E+00 0.6667     A  3 YC  0.000 .1 0.        0.000
M   -0.149917E-03 0.6667     A  3 XC  0.000 -.1 0        0.000
M    0.000000E+00 0.6667     A  3 YC  0.000 -.1 0        0.000
END
SYNOPSYS 10
MC
```

As requested, the last airspace is varied in the PANT file, and the AANT file defines a merit function that will converge to zero if the adjustment is able to restore exactly the same ray pattern as the nominal design. (Note that we are not trying to *improve* the image here, just trying to keep it as it is.) Now we need to prepare an MC

MACro. (This is the file where we specify what kind of Monte-Carlo analysis we want; the file MCFILE.MAC, shown above, specifies the *adjustments* we want to run on each case. They are separate files.)

To start with, we will run MC with random wedge orientation. Here is the MACro (**C28M2**):

```
MC ITEMIZE
SAMPLES 1              ! One case, please.
LIBRARY 5              ! We saved the initial lens in library location 5.
WORST ALL 1            ! Later we may want to see a worst case.
THSTAT UNIFORM         ! Uniform thickness statistics.
WEDGES RANDOM          ! Wedges have random orientation.
TEST                   ! Let's just look at a perturbed example.
GO                     ! Run MC.
```

Here, we do not optimize anything, just prepare a single perturbed example so we can examine it. (The elements all have a wedge error now, so the PAD display does not color the elements as before.) This is shown in figure 28.2.

Okay, let's run a set of 100 lenses and look at the statistics. First, we **GET 5** (to restore the nominal lens), and then comment out the TEST directive and change the samples number.

```
MC ITEMIZE
SAMPLES 100 ! Ask for a set of 100 lenses.
LIBRARY 5
QUIET
WORST ALL 1
THSTAT UNIFORM
WEDGES RANDOM
! TEST
GO
```

When MC finishes, make a plot of the statistics with **MC PLOT**, as shown in figure 28.3.

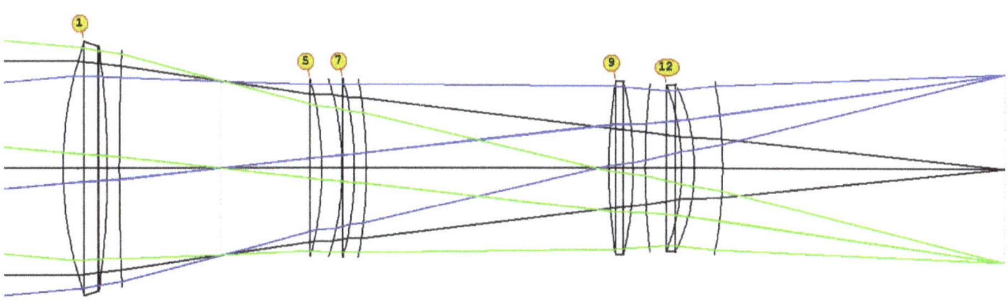

Figure 28.2. Lens perturbed by MC, with wedges on all elements.

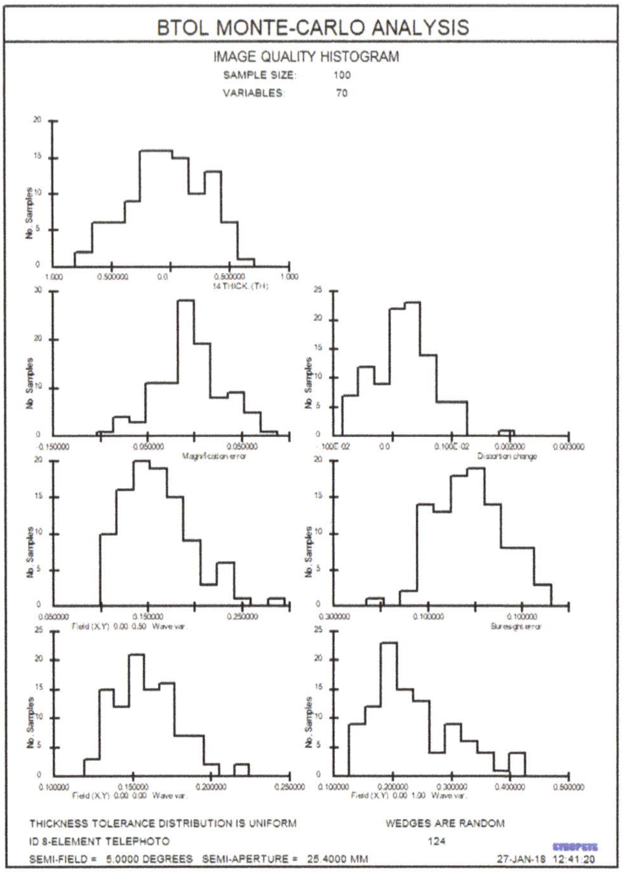

Figure 28.3. Monte-Carlo statistics of lens without wedge clocking.

All of this is pretty simple, so now we make things more interesting. Change the MACro as follows:

```
MC ITEMIZE
SAMPLES 100
LIBRARY 5

QUIET
WORST ALL 1
THSTAT UNIFORM
WEDGES CLOCK        ! Clock the wedge errors for each case.
TEST                ! Again make a single TEST case.
GO
```

Now the program will model the element tilts with a different protocol, using GROUP instead of RELATIVE tilts. This frees up the gamma tilt on each element to be used for clocking the wedge errors. ('Clocking' refers to the practice of rotating each element in the cell, aligning them so that the effects of the wedge error of a given element are compensated by the orientation of the other elements.) Here, we ask for a test case so we can examine how the errors have been defined. When this has executed, look at the **ASY** listing for the perturbed lens:

```
TILT AND DECENTER DATA
 LEFT-HANDED COORDINATES
```

SURF	TYPE	X	Y	Z	ALPHA	BETA	GAMMA	GROUP
1	GROUP	0.00000	-0.00406	0.00000	0.0065	0.0000	0.0000	3
2	REL	0.00000	0.00000	0.00000	-0.0238	0.0000	0.0000	2
3	REL	0.00000	0.00000	0.00000	0.0000	0.0033	0.0000	1
5	GROUP	0.00000	-0.00969	0.00000	0.0413	0.0000	0.0000	2
6	REL	0.00000	0.00000	0.00000	-0.0035	0.0000	0.0000	1
7	GROUP	0.00000	-0.01296	0.00000	0.0200	0.0000	0.0000	2
8	REL	0.00000	0.00000	0.00000	0.0000	-0.0248	0.0000	1
9	GROUP	0.00000	-0.03168	0.00000	-0.0495	0.0000	0.0000	3
10	REL	0.00000	0.00000	0.00000	-0.0848	0.0000	0.0000	2
11	REL	0.00000	0.00000	0.00000	0.0000	0.0021	0.0000	1
12	GROUP	0.00000	-0.01401	0.00000	-0.0103	0.0000	0.0000	3
13	REL	0.00000	0.00000	0.00000	-0.0189	0.0000	0.0000	2
14	REL	0.00000	0.00000	0.00000	0.0000	0.0569	0.0000	1
15	REL	0.00000	0.00000	0.00000	-0.0134	0.0000	0.0000	1

From this listing, you see that surfaces 1, 5, 7, 9, and 12 have been assigned a group tilt. You will vary the clocking angle (gamma tilt) of all of them except for surface 1, which furnishes a reference direction.

Okay, we need to modify the file MCFILE.MAC, adding the gamma tilt variables. We also elect to do a more sophisticated optimization while we are at it. Then save the new MACro so MC will be able to open it and see the changes. (**Do not run this file yourself; it will be used by MC as it optimizes each case.**) It now looks like this:

```
PANT
VY 14 TH
VY 5 GPG          ! Vary group gamma tilt on surfaces 5, 7, 9, and 12
                  !(but not surface 1).
VY 7 GPG
VY 9 GPG
VY 12 GPG
END

AANT
M 0 1 A P YA          ! Control the boresight error this way.
M 0 1 A P XA
GSR .5 10 5 M 0 0 0 F       ! Correct over the full pupil since the
                            ! lens no longer has
GNR .5 2 3 M .7 0 0 F       ! bilateral symmetry.
GNR .5 1 3 M 1 0 0 F

GNR .5 2 3 M -.7 0 0 F      ! For the same reason, we also control
                            ! the negative field.
GNR .5 1 3 M -1 0 0 F

END
SYNOPSYS 10
MC
```

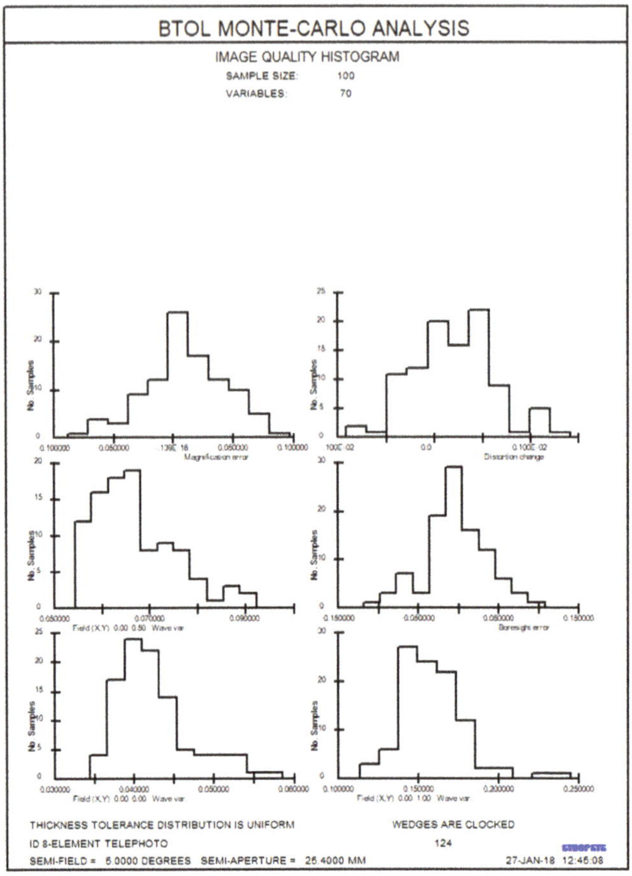

Figure 28.4. Statistics of tolerance budget when wedges are clocked.

Save this version, type **GET 5** so you start with the original lens, and rerun the MC MACro, requesting 100 cases and deleting the TEST directive. When you run it, you get improved statistics, shown in figure 28.4. Indeed; clocking the elements improves performance, as expected.

This exercise is almost finished—but suppose the lens is used in a device where the lateral color must be well controlled. You would like to know the statistics of the resulting aberration as each case is optimized. Add to the file MCFILE.MAC some AI input, and it now reads as follows:

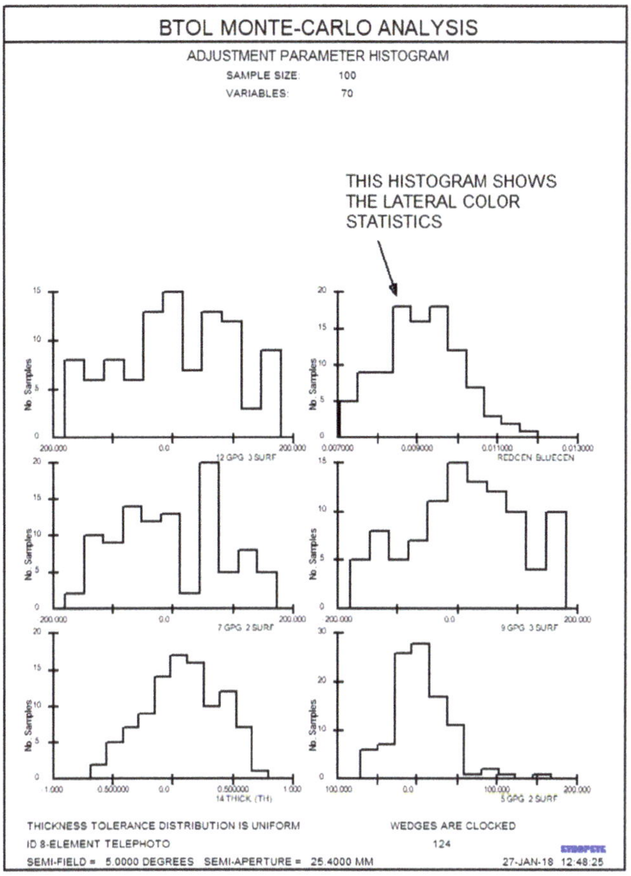

Figure 28.5. MC plot showing statistics of adjustments and lateral color when 100 lenses are manufactured.

```
PANT
VY 14 TH
VY 5 GPG
VY 7 GPG
VY 9 GPG
VY 12 GPG

END
AANT
M 0 1 A P YA
M 0 1 A P XA
GSR .5 10 5 M 0 0 0 F
GNR .5 2 3 M .7 0 0 F
GNR .5 1 3 M 1 0 0 F
```

```
GNR .5 2 3 M -.7 0 0 F
GNR .5 1 3 M -1 0 0 F
END
SYNOPSYS 10

Z1 = XA IN COLOR 1   ! Get the actual X coordinate of the chief ray in
                     ! color 1.
RMS 1 0 555          ! Run the RMS command, which also finds the centroid.
Z2 = FILE 4          ! This is the X-centroid location, relative to the
                     ! chief ray,
Z3 = FILE 5          ! and this is the Y.
Z4 = YA IN COLOR 1   ! Also get the actual Y coordinate.

Z5 = XA IN COLOR 3   ! Do the same thing in color 3.
RMS 3 0 555
Z6 = FILE 4
Z7 = FILE 5
Z8 = YA IN COLOR 3

= SQRT((Z1 + Z2 - Z5 - Z6)**2 + (Z3 + Z4 - Z7 - Z8)**2)
Z9 = FILE 1                  ! Load it into variable Z9, and tell MC
MC IZ9 "RedCen-BlueCen"      ! to gather the statistics and plot Z9 with
                             ! this label.
MC
```

Save this, as before. Now, when you run your MACro, MC adds the statistics of the lateral color to the second plot page, which also shows the adjustment statistics, as seen in figure 28.5.

Your results will differ from these, since MC has to ignore switch 98 and use a different set of random numbers for every case it analyzes. But the statistics should match fairly well.

Chapter 29

Tips and tricks of a power user

Time-saving strategies for effective use of lens optimization software

If you have gone through the previous chapters in this book, you already know the basic procedures for running SYNOPSYS. This chapter goes a little deeper and shows some ways you can get just what you are after, ways that may not be obvious.

Here is an example of a sophisticated optimization MACro. By now, you should be able to read this kind of file fluently.

```
LOG
AWT: 1.0
CHG
NCOP
END

PANT
VY 0 YP1 50 -50
VY 0 BTH
VLIST RD ALL
VLIST TH ALL
END

AANT P
AEC
ACC
M 100 10 A FOCL
LLL 2 1 1 A BACK
AAC 49.5 .5 5

SKIP
GSR    AWT    6.000000    4  1    0.000000
GNR    AWT    3.000000    4  1    0.100000
GNR    AWT    3.000000    4  1    0.300000
```



```
GNR    AWT    3.000000    4  1    0.500000
GNR    AWT    3.000000    4  1    0.70000
GNR    AWT    3.000000    4  1    0.80000
GNR    AWT    3.000000    4  1    0.90000
GNR    AWT    3.000000    4  1    1.000000
EOS

!SKIP
GSO    0    0.8     4  1    0.000000
GNO    0    0.27    4  1    0.100000
GNO    0    0.27    4  1    0.300000
GNO    0    0.27    4  1    0.500000
GNO    0    0.27    5  1    0.70000
GNO    0    0.27    5  1    0.80000
GNO    0    0.37    5  1    0.930000
GNO    0    0.27    5  1    0.950000


GNO    0    0.27    5  1    1.000000
EOS

!SKIP
LUL 29 1 1
A BLTH 8

EOS

END
!EVAL
!EDS

SNAP/DAMP 1
SYNOPSYS 40
```

Tip 1:

In this example, we vary the paraxial quantity YP1. The lens does not have an explicit stop defined, and this variable will let the chief-ray intercept on surface 1 vary—thus sending it into the lens at the current location and *implying* a stop position wherever it happens to cross the axis. This is a powerful way to find out where the stop should go. If the design looks promising, then it is simple to just assign the stop at that location or close to it and reoptimize.

Tip 2:

Note the monitor **AAC 49.5 .5 5**. This lens has to fit inside a tube of 100 mm diameter, and this monitor will penalize the lens if any of the clear apertures exceed the value 49.5 mm in radius. The other two arguments give the relative weight and the monitor window. You can adjust them according to the importance of that control. (And you can always *look up* that topic in the User's Manual if you want to see how it works. Type **HELP AAC** in the Command Window. There are 17 monitors available, and it pays to know how to use them.)

Tip 3:

Notice how the file defined a symbol, **`AWT: 1.0`**. That symbol shows up in the AANT file as the aperture weighting parameter on some ray sets, a parameter we have used in previous chapters. A value of 0 says to weight all rays in the generated grid by the same amount, the value given by the second argument on the raygrid request. A weight of 1.0 weights the central rays rather more heavily than the edge, and for this lens that is a good idea. Uniform weights tend to give an image of high contrast, while a higher weight often gives better resolution. Here, too, you can experiment to see what works best with your lens. I often start with a value of 0.5. And the point of making this a *symbol* is so you can try a different value just by changing that symbol and reoptimizing—so you don't have to change all the weights on every line of the MACro every time you want to try a new value.

Tip 4:

The shop making the lens in figure 29.1 has a curious problem: they already have a lens blank for the element at surface 8, and have measured its thickness at 30 mm. So that lens must be controlled during optimization to be sure it does not require a thicker blank.

That's where the AANT entries below come in.

```
LUL 29 1 1
A BLTH 8
```

Here we specify a maximum value (with LUL) of 29 mm for the thickness of the blank for this element. 'LUL' means Limit, Upper Limit, and you can read about this useful feature by typing **`HELP LUL`**. You can also make your own aberration by combining

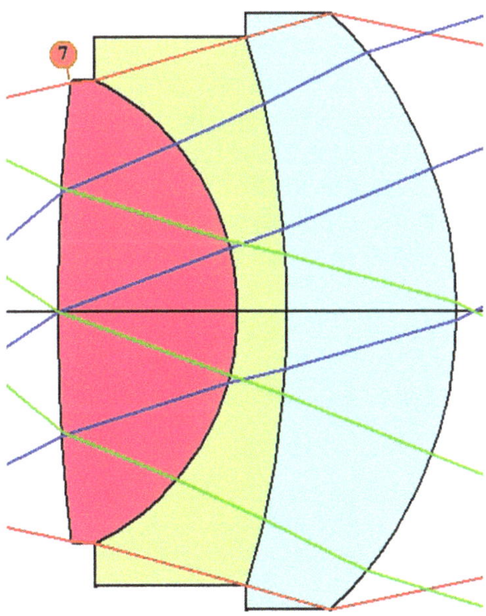

Figure 29.1. Example lens for which the blank thickness must be controlled.

the TH and the sag at any surface that happens to be concave at the current clear aperture. Section 10.3.3 of the User's Manual describes the target SCAO (the sag at the current aperture). It's useful to know how to make aberrations like this.

Tip 5:

Notice the use of the **SKIP** directive in this MACro. Ray grid definitions and weights are easily generated with the buttons on the MACro editor toolbar, which generate rays whose transverse intercepts are to be controlled, or rays with OPD targets. But which should you select? The **SKIP** directive in the above example lets you choose one or the other (or both) just by commenting or uncommenting that directive. As shown above, the first set of ray grids, which target transverse aberrations, will be skipped, because it is within a **SKIP** block. When the program reaches the **EOS** (End Of Skip) line, it stops skipping—so the ray grids targeting OPDs are then in effect. To see the effect of transverse targets instead of OPD, just comment out the first **SKIP** and uncomment the other (delete the '!'). You see how simple it is to switch from one to the other. With a few clicks and keystrokes you can study the effect.

This is one reason why we do not save the merit function along with the lens, as some other programs do. We frequently alter the definition of some of the operands, which is very quick and easy, as we have shown above. So the lens file is just a lens file (and you can save the optimization MACro with a single click of the mouse on the button on the editor toolbar, when you need to.)

Tip 6:

More on the subject of TAP targets *vs* OPD targets: Look at the ray-fan plots in figures 29.2 and 29.3.

You might think this is a terrible image, with the rays flying off like that at the ends of the fans. But look at the OPD fans:

Can't get much better than that! This is a good illustration of why you should switch to OPD targets if your lens is close to the diffraction limit. Someone targeting only ray intercepts would likely look at the top curves and then discard the lens and start over—or give a heavy weight to the marginal rays and keep optimizing. But that would not work very well. Specifying a large value of the aperture weight parameter (which we called **AWT** in the above MACro) might work somewhat better,

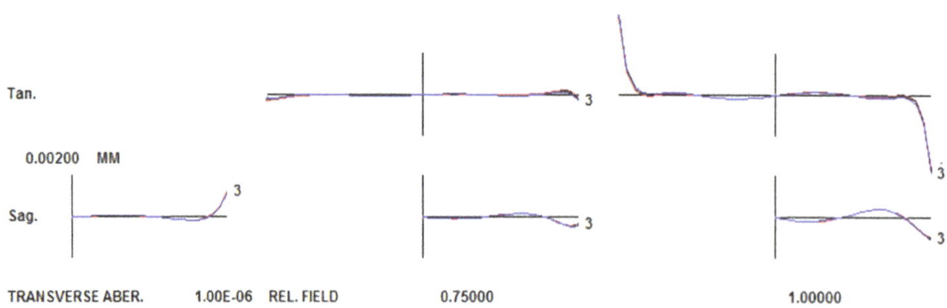

Figure 29.2. Transverse aberration plot example with large aberrations at the ends of the TFAN.

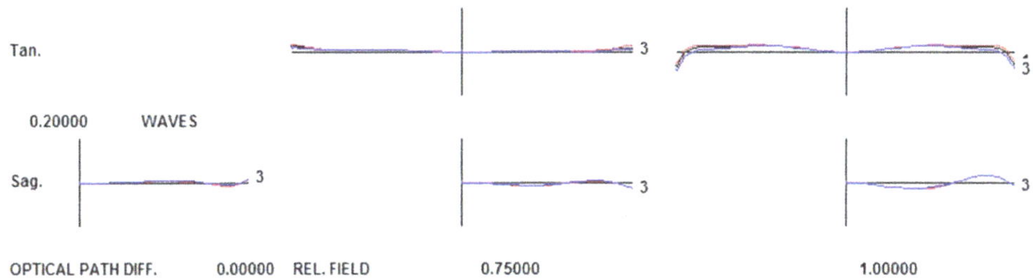

Figure 29.3. OPD plots for the same lens. The large transverse aberrations turn into a large *slope* of the curve, but the OPD errors are small.

but still, OPD targets are better in cases like this. Remember, the OPD fan is the *integral* of the TAP fan. A steep *slope* on the OPD curve means a large transverse aberration—which you don't care about if the OPDs are small enough.

Tip 7:

So why use transverse targets, anyway? Well, for the lenses I have studied, those targets can change a lens faster—and thus arrive at a good configuration—more quickly than can OPD targets. So start with transverse and switch to OPD when you are almost there. DSEARCH and ZSEARCH both have an option to include both kinds of targets (using the **TOPD**, **OPD**, **OPSHEAR**, and **TOSHEAR** directives), and if you want a diffraction-limited image, those options sometimes prove helpful. Try it both ways; you will probably find some interesting configurations.

Tip 8:

Suppose you only want to evaluate the current lens, to spot the largest aberrations—but don't want to change anything just yet. Simple. Just uncomment the lines in the MACro shown above

```
!EVAL
!EDS
```

and run it. The program will evaluate the merit function and then end the data set (the MACro) at the **EDS** line. Later, if you want to optimize the lens, just comment those lines out again and rerun it.

Tip 9:

When you optimize your lens, if you have switch 1 turned on, it is a good idea to run the program twice. Thus,

```
SYNOPSYS 50
SYNOPSYS 50
```

is sometimes better than

```
SYNOPSYS 100.
```

This is because if any of the variables in the first run encounters a boundary, that variable is dropped for the remaining iterations (according to switch 1). The rule saves time, since usually that variable will continue to try to violate the same boundary, and if not dropped it will just slow things down. But if the lens changes its form significantly during the run, the same variable may now want to move in the other direction. The second run frees all variables so it can do so. This is especially true for glass model variables (GLM), which usually hit a boundary early on.

And, by the way, we often see new users asking for, say 500 iterations of optimization. While this may be necessary when running a less powerful program, it is almost never necessary with SYNOPSYS. Ask for a smaller number, and use the simulated annealing button thereafter.

Tip 10:

The MACro given at the start of this lesson includes both transverse aberrations and OPDs. Notice the difference in weights. This reflects the fact that an OPD error of one unit (one wave) usually gives a much better image than a transverse error of one inch (or one mm). Sometimes one can get excellent results from giving targets to *both* kinds of errors—but now the relative weights are important. And if you have, say, a target for a mechanical property, such as an aperture or spacing somewhere, you want to be sure that your very nice balance is not upset when the program thinks the one-wave OPD is just horrible and tries to bring it down at the expense of the other errors. To make finding a suitable weight easier, the program provides two tools. If you click the **Ready Made Raysets** button in the MACro editor and choose option 8, the program will create ray grids for both kinds of targets, assigning weights to the OPD errors that reflect the current wavelength and F/number. These weights ensure that the difference in units is accounted for in a reasonable way, and of course you can adjust things from there after you see the effects. The button also lets you select OPD targets, and in this case, you assign relative weights as usual and then click the box 'Calculate special OPD weights'. The OPD weights will be modified by the same rule when the ray grid requests are added to the MACro. (These calculations are a function of the F/number of the current lens, which must be well defined already.)

Tip 11:

You have seen several pupil definitions in the earlier chapters, from simple paraxial to sophisticated wide-angle options. Some users try to start a new design with the most sophisticated, such as WAP 3—and then wonder why the runs take so long.

We advise them to start instead with the simplest setup that makes sense, a paraxial pupil if possible, and add complications only when clearly necessary. I almost never use the complicated options, and if vignetting is an issue and the paraxial pupil is clearly not adequate, you should switch to a real pupil (with a negative value on the APS input). If the stop is not filled properly, then switch temporarily to WAP 2 with a CAO on the stop (or CSTOP in effect), and then use the **FVF** utility to find a set of vignetted apertures (the VFIELD) that duplicate the WAP 2 results. This removes the WAP option, and everything runs much faster thereafter since all the pupil searching has already been done. If the lens changes form and the current VFIELD is no longer appropriate, just run **FVF** again, and the apertures are updated. If you have not done so, it would be wise to type **HELP VFIELD** and follow the links to the explanation of that very useful feature at this time.

Tip 12:

We mention here an obscure topic that occasionally has proved very mysterious. If you optimize your lens with OPD errors only, in FOCAL mode there are actually two solutions. One is the one you want, and the other is a lens with collimated output, which you probably do not want. The OPD errors are calculated by comparing the path length of a given ray with that of the chief ray—and if the beam enters and leaves the lens collimated and hits a flat image plane, those paths may again be equal. The program will faithfully find this spurious solution if it should head that way. It's just doing its job. The moral is, it is a good idea to add at least one transverse intercept error to the MF. It can have a low weight, but it makes the collimated solution look unattractive. This situation rarely happens, but it is a mystery when it does. Now you know why.

Tip 13:

If your PC has more than a single CPU, you can save much time by authorizing more than a single core when running the search programs—but the gain in speed is not a simple function of that number. It turns out that, if you have N cores, for large values of N the incremental gain in speed by adding yet one more core goes as $1/N^2$, while the overhead required to start, stop, and manage the data from the extra core is a linear function of N. Mathematics tells us those two functions have to cross somewhere, and after that point, adding yet more cores actually makes the run take *longer*.

So the biggest saving of time does not always occur when you activate the maximum number of cores in your PC, if that number is large. Increasing from one to two cores cuts the time by 0.5, while Increasing from 10 to 11 cuts it by only 0.009 09, and so on. Experiment with your PC to find the core number that gives you the fastest performance.

Keep in mind that, if a program requires a single process for half of its calculations, the maximum improvement will be 50%, even with an infinite number of cores.

Tip 14:

Much effort has gone into figuring out how to maximize the modulation transfer function of a lens, and some designers routinely target the wavefront variance in the MF for this purpose. While that works, and can be accomplished with the GNV ray set option in the AANT file, it is not necessarily the best approach since it often converges very slowly. Instead, read about the **GSHEAR** ray sets in the UM. That feature usually works much better and faster. The DSEARCH examples in chapter 20 show how that program can utilize that technique.

Tip 15:

Changing a curvature usually has a much larger effect on the MF than changing a thickness or airspace. In most of these examples, we have not put any monitors on lens thicknesses at all, and the results often came back with elements that were too thick or too thin to be practical.

If an element is too thin, it is hard to keep it from flexing under the pressure of polishing, which ruins the figure. If too thick, it is more expensive and can absorb light. The program has default limits and monitors, but those are often too forgiving and you should reduce the target values in those cases. A lens maker looking at the preliminary results would likely exclaim at how impractical they are—but he would not understand that one approaches the goals in steps, and that issue can be corrected later. Also, be aware that, when the vendor first makes an element, it always has a larger diameter than the print calls for. That excess is ground away later, after the lens is carefully centered on a precision lathe and both sides running true—to remove the residual wedge—and at that time the diameter is cut to the desired dimension. But that does not work unless there is some glass outside of the desired clear aperture to cut away, so be sure the edge thickness on the print is not too thin, or there might not be. Some lens makers greatly prefer elements with a rather thick edge, to facilitate mounting on the polishing machine. Of course, that is not often a good idea for other reasons, so here you must engage the shop in constructive dialog.

When the design is in good shape, it is then time to gently corral those properties. The ADT monitor is very helpful, but should be used gingerly: assign a low weight and large window at first, such as

```
ADT 7 .01 10
```

and increase the weight gradually if you need to. That targets the ratio of the lens diameter to thickness to the entered value. Adjust the target value and weight to see which combination works best. Also useful are the ACM and ACC monitors, which simply penalize any element thickness that is outside the limits on those entries; a minimum thickness for ACM and maximum for ACC. You should apply these monitors gently, as you need them. The goal is not to perturb the design so strongly that it leaves the very nice solution region you are in at the moment.

In rare instances, a lens returned by DSEARCH simply does not work as well when edges are modified with these tools. That is a good time to explore some of the other results from the search routines. Or you can put an ADT monitor in the SPECIAL AANT section, with a low weight, and expect to get a different set of candidates. Experiment with these settings; there are a great many solutions out there.

IOP Publishing

Lens Design (Second Edition)
Automatic and quasi-autonomous computational methods and techniques
Donald C Dilworth

Chapter 30

FLIR design, the narcissus effect

FLIRS; Retroreflection from a cooled detector produces a dark area on a FLIR display; how to correct it

Night-vision systems can see in total darkness. That works because all matter in the universe radiates energy in the form of photons, following the Planck function in the case of a perfect blackbody radiator or approximating that function to some degree otherwise. Since your skin is close to room temperature at 20 °C or 293 K, you emit radiation according to the curve in figure 30.1, as calculated by the SpectrumWizard[1]. Note the peak at about 10 μm. (Type **MSW** to opcn the Wizard.)

Although the atmosphere absorbs much infrared radiation, it has a window of transparency centered at a wavelength of just over 10 μm, which nicely fits the spectrum shown above.

Night-vision systems sense this radiation by means of a detector that converts IR photons into an electrical current. A common material for this purpose is HgCdTe, which has the spectral sensitivity shown in figure 30.2. (This figure is from *The Infrared Handbook*, by Wolfe and Zissis.) The exact sensitivity range depends on the relative proportions of the ingredients. We are fortunate that the source, atmosphere, and detector all work well over the required spectral window.

In order to get a high signal-to-noise ratio, one must ensure that the optics, and even the detector itself, do not radiate unwanted flux at the same wavelengths. This is accomplished by cooling the detector, often with liquid nitrogen, and using high-quality anti-reflection coatings on the lens surfaces. If those steps are not taken, the situation is like what you would see looking through a telescope if the lenses and housing were all glowing white: it would be hard to distinguish what

[1] SpectrumWizard is a trademark of Optical Systems Design LLC, an Arizona, USA corporation.

doi:10.1088/978-0-7503-3695-6ch30

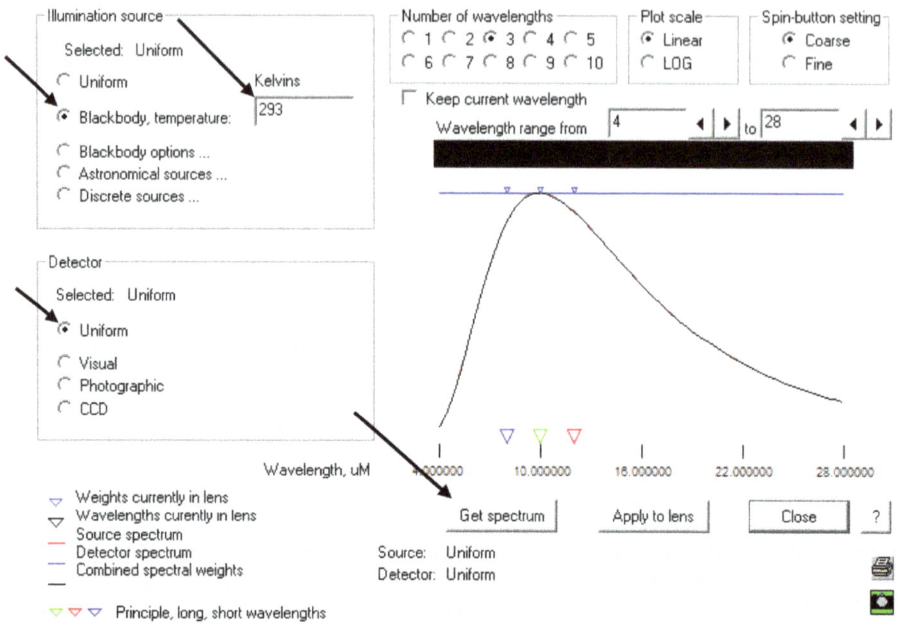

Figure 30.1. Spectrum Wizard showing the blackbody curve at 293 K.

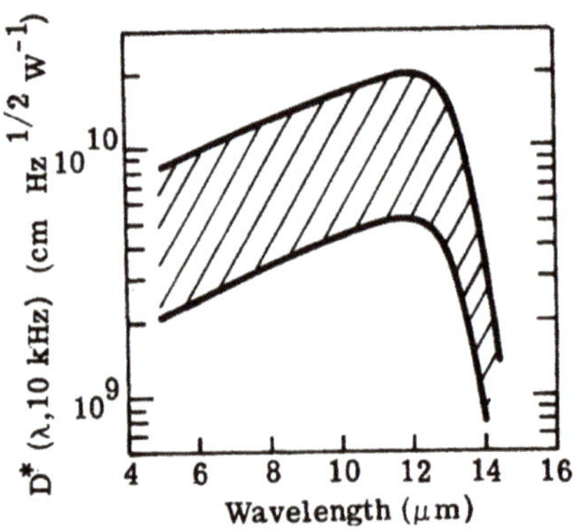

Figure 30.2. Sensitivity of HgCdTe detectors.

you are looking at. In this chapter we will develop a new FLIR design and then analyze the narcissus properties.

Designing Forward-Looking Infrared Radiometers (FLIRS) is not difficult. We will ask DSEARCH to design a five-element lens, using Germanium for the positive elements and ZnSe for the negative. Here is the input file (**C30M1**):

```
CORE 32
OFF 1
 DSEARCH 5  Q
 SYSTEM
 ID FLIR DSEARCH
 AFOCAL
 OBB  0.000000     17.50000      6.35000
 WAVL 12 10 8

 UNITS MM
 END
 GOALS
 ELEMENTS 5

 TOTL 350 .1
 FNUM -25.4 1

 STOP FIRST
 STOP FREE
 RT 0.25
 RSTART 3000

 FOV 0.0 0.75 1.0 0.0 0.0
 FWT 3.0 2.0 2.0 1.0 1.0

 NPASS 100
 ANNEAL 200 20 Q
 TSTART 10 ! This thickness on each element to start with
 QUICK 60 90


GLASS POS ! positive elements will use this glass type
U GE
GLASS NEG ! and negative this type.
U ZNSE
END

SPECIAL AANT
ACC 10 .1 1
AAC 26 .1 1
M 13.37 1 A P YA 1 0 0 0 2
M 0 1 A P YA 1 0 0 0 10
ADT 7 .1 10
END
GO
```

The SPECIAL AANT section is where we request the chief ray at full field to hit surface 2 a height of 13.37 mm (to allow clearance for a scanning prism at surface 1), and hit surface 10 at the axis (since that is the objective lens). This system is set up backwards, and light actually comes through the other way. It can be designed either way. The ACC monitor keeps thicknesses under control, and the AAC keeps the apertures not more than 26 mm in radius. The afocal magnification is controlled by the FNUM requirement, which in this AFOCAL system is interpreted as the radius of the exiting paraxial beam, negative because the interior focal plane flips the beam. (For FOCAL lenses, FNUM of course controls the F/number.)

After this file runs, we select the third one from the top, shown in figure 30.3 (**C30L1**) It is already in the diffraction limit and one might think it requires no more optimization. But we haven't controlled the *narcissus* Yet.

30.1 Narcissus correction

The narcissus effect is too often overlooked, showing up in scanning IR systems as a dark smudge in the center of the displayed image, as in the simulated image in figure 30.4.

This effect occurs because at the center of the field the detector can see a ghost image of itself, reflected from a lens surface somewhere. This ghost is very cold—because the detector is very cold—and the total background signal seen by the detector is therefore lower at the center than it is at other parts of the field, where the ghost image is vignetted by other lens apertures, scanned out of the field, or so aberrated that it cannot form a sharp image. Only at the center are all the ghosts lined up.

Let us start with the lens designed above and get the narcissus under control. It is designed for the 8–12 μm band and uses AFOCAL mode, which means that ray

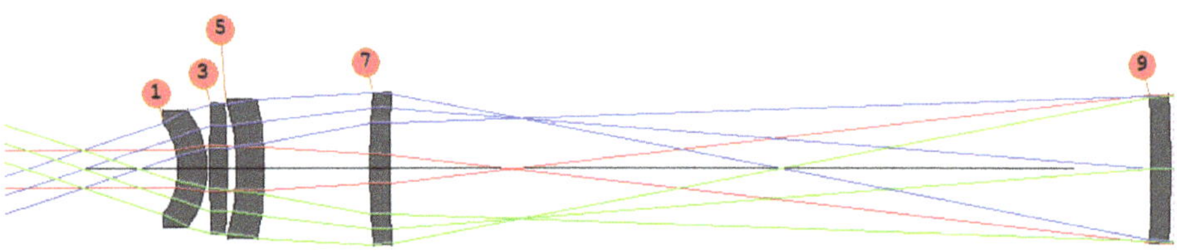

Figure 30.3. FLIR design returned by DSEARCH.

Figure 30.4. Example of the narcissus effect.

output is given in *angles* instead of transverse coordinates. (No 'perfect lens' is required in SYNOPSYS.)

To analyze the narcissus properties of this lens, use the command **NAR.**

```
SYNOPSYS AI>NAR

ID FLIR DSEARCH                          55602           10-NOV-20    07:18:28

 NARCISSUS ANALYSIS

 SURF         YNI          Imarg/Ichief       CYNI         RATIO (M/C)
 ______________________________________________________________________
  1          1.5197          2.8209          0.9007          1.6872
  2          1.7533          1.4301          2.2574          0.7767
  3          0.1769          0.0758          4.3745          0.0404
  4          1.0251        -37.6347          0.0525         19.5353
  5          1.4883          1.4203          2.1474          0.6931
  6          0.9554          9.3410          0.2253          4.2406
  7          0.2329         -0.2301          4.3015          0.0542
  8          0.5113          2.0508          1.0978          0.4657
  9          1.8925          1.0526          0.1941          9.7522
 10          1.8525         -0.8452          0.2243          8.2589
 11          0.0036          0.0018          0.1987          0.0181
 SYNOPSYS AI>
```

The column 'NYI' shows the value of the quantity φ A_0N, where the arguments are defined in figure 30.5. From this we can calculate the approximate size of the retroreflected blur on the detector: Y1 = 2 φ A_0 N/α, where α is the half-angle of the converging beam at the detector, and Y1 is the ray height of the blur at the image after reflection at the given surface.

To control the narcissus, one must ensure that the value of YNI never gets lower than a limiting value, which is a function of the sensitivity of the scanner and user acceptance. A larger value means the ghost image is more out of focus and therefore less intense. According to the above table, the worst narcissus is from surface 3, where the value is 0.1769. That is a little too small. We can use the GHPLOT program to visualize what that means. Type **MGH** in the CW, or navigate to that dialog in **MLI**. Enter the data shown in figure 30.6, and then click the **GHPLOT** button. You get the picture in figure 30.7. Here, we want to look at the beam that reflects from surface 11, then from 3, and then goes to the output.

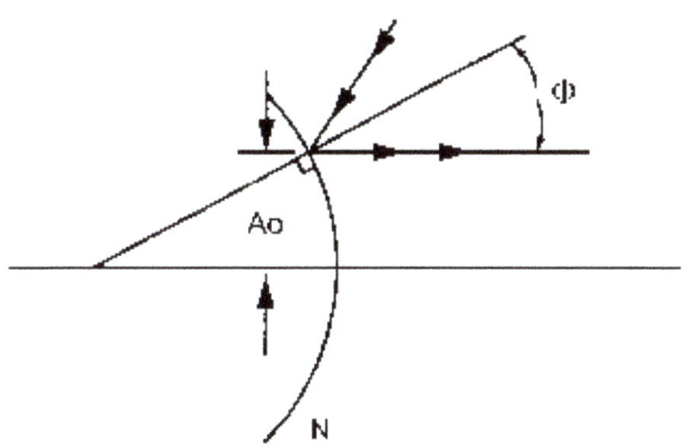

Figure 30.5. Geometry of the narcissus equation.

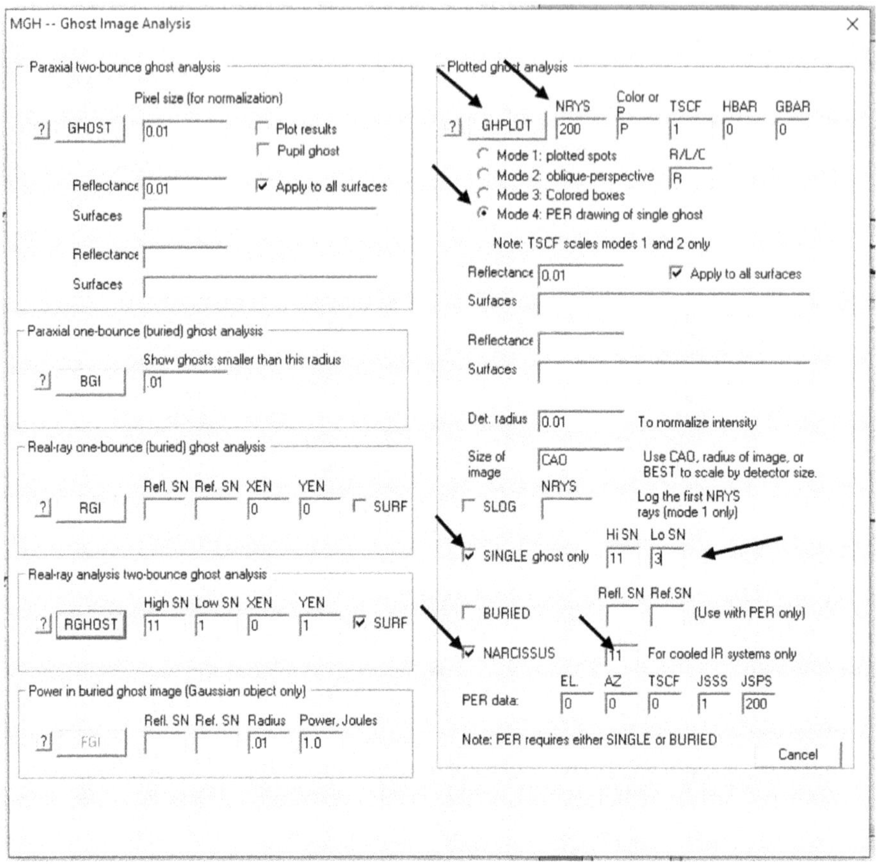

Figure 30.6. Data entered into the MGH dialog.

Select the area shown on the left and click inside the selection rectangle to zoom in to that portion, as shown in figure 30.8.

Light comes in from the left, drawn in red, until it reaches surface 11, the flat surface at the end of the lens, which is off the page to the right, and which we declared a NAR surface on the dialog. (That assigns a reflection coefficient of 1.0.) We requested a single ghost, from 11 back to 3, and from there returning to the last surface again. The signal flux there is collimated, and we assume that any light that returns there and is again collimated will show up on the detector sharply focused. After reflecting from 11, the rays are shown in blue, and following the second reflection, at surface 3, they are drawn in green. The retroreflected beam falls close to the incoming rays. This will be a noticeable narcissus unless we do something about it. How to control it?

To understand that question, you need to know the angular size of the Airy diffraction disk. Ask for a **PSPRD** analysis, on axis. Go to the dialog **MDI** and click the **PSPRD** button. You get the plot in figure 30.9.

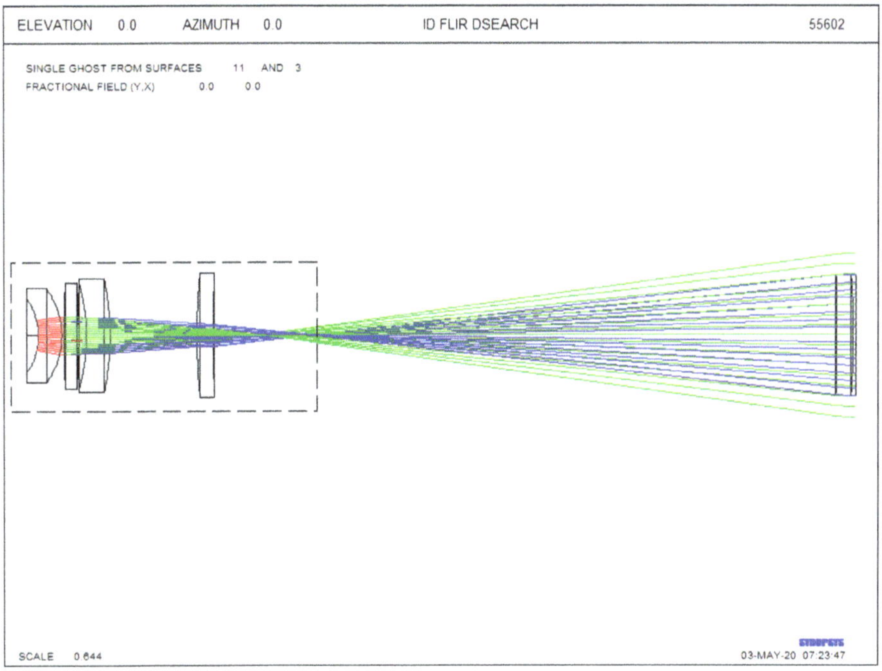

Figure 30.7. GHPLOT showing retroreflected beam at surface 1.

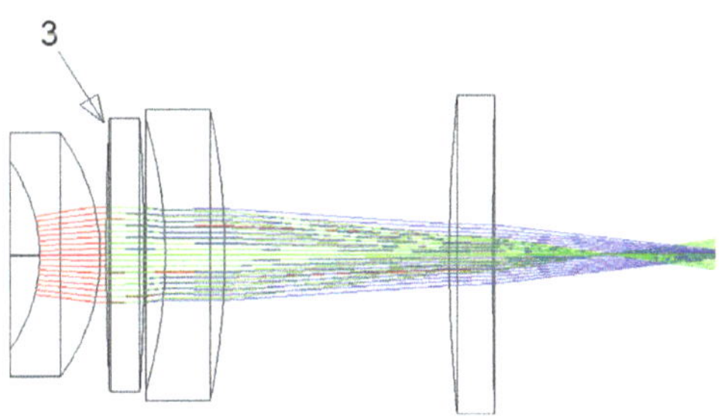

Figure 30.8. Portion of the lens responsible for a bad narcissus.

Note the value of the Airy disk radius: 0.000 24. That's an angular value, since the lens is declared AFOCAL. We know that if the narcissus beam returns with an angle equal or less than that value the beam will be in sharp focus and cause a very objectionable narcissus.

So what is the current value of the returning beam angle? Go back to the MGH dialog, and this time click the **RGHOST** button, with data shown in figure 30.10.

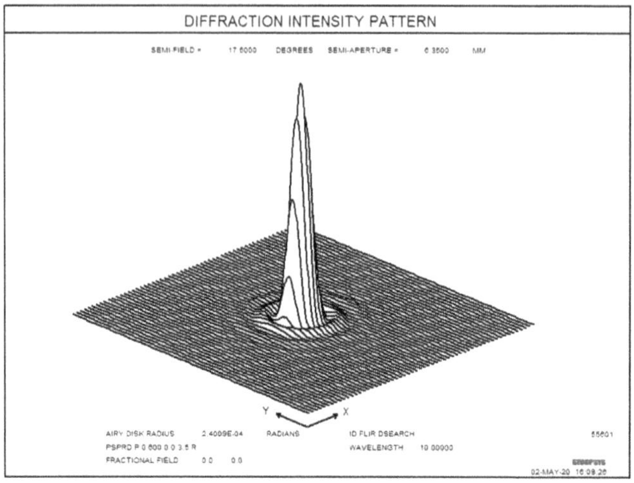

Figure 30.9. Diffraction point-spread image from the IR telescope.

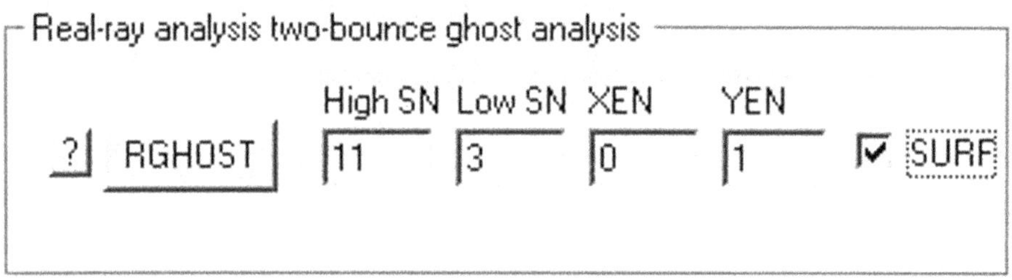

Figure 30.10. MGH selection to show a single ghost path.

```
SYNOPSYS AI> --- RGHOST 11 3 0 1 SURF
                          RAY VECTORS                (X DIR TAN)  (Y DIR TAN)  (INC. ANG.)
 SURF              X              Y              Z           ZZ            HH          UNI
------------------------------------------------------------------------------------------------
  OBJ       0.000000       0.000000       0.000000       0.000000       0.000000


   1       0.000000       6.350000      -0.771072       0.000000       0.143023     13.846911
   2       0.000000       7.760558      -0.921499       0.000000       0.007752      5.403949
   3       0.000000       7.775910       0.058978       0.000000      -0.009443      1.313247
   4       0.000000       7.718689      -0.092196       0.000000      -0.110335      1.909710
   5       0.000000       7.365350      -0.348873       0.000000       0.010154     11.720017
   6       0.000000       7.468152      -0.237194       0.000000      -0.065264      3.056502
   7       0.000000       5.002296       0.047351       0.000000      -0.030490      2.649402
   8       0.000000       4.781678       0.008387       0.000000      -0.112127      1.545407
   9       0.000000     -25.040452      -0.450767       0.000000      -0.054941      4.335096
  10       0.000000     -25.404563      -0.929646       0.000000   2.486630E-05      1.046706
--- RAY REVERSES AFTER NEXT SURFACE ---
  11       0.000000     -25.404515       0.000000       0.000000   2.486630E-05
  10       0.000000     -25.404467       0.929639       0.000000       0.054954
   9       0.000000     -25.040274       0.450760       0.000000       0.112178
   8       0.000000       4.795228      -0.008434       0.000000       0.030510
```

```
     7        0.000000      5.015983     -0.047610     0.000000      0.065188
     6        0.000000      7.479004      0.237884     0.000000     -0.010240
     5        0.000000      7.375330      0.349821     0.000000      0.110311
     4        0.000000      7.728514      0.092431     0.000000      0.009414
     3        0.000000      7.785558      0.059124     0.000000     -0.020965
     4        0.000000      7.658494     -0.090763     0.000000     -0.157242
     5        0.000000      7.152023     -0.328914     0.000000     -0.010216
     6        0.000000      7.048533     -0.211264     0.000000     -0.109986
     7        0.000000      2.899287      0.015905     0.000000     -0.035578
     8        0.000000      2.640941      0.002558     0.000000     -0.137767
     9        0.000000    -33.949175     -0.828789     0.000000     -0.070916
    10        0.000000    -34.390940     -1.705563     0.000000      0.014424
    11        0.000000    -34.351916      0.000000     0.000000      0.014424
  GHOST REFLECTED FROM SURFACES    3   11 AT SURFACE   12
        X                 Y                  ZZ                 HH
  -----------------------------------------------------------------
    0.00000        0.144227E-01      0.00000        0.144237E-01
   Type <ENTER> to return to dialog.
   SYNOPSYS AI>
```

The program creates and runs the RGHOST command for you, and you see the tangent of the ray (HH) when it returns to surface 11 equals 0.0144. The Airy disk radius is 0.000 24, and we want the narcissus blur to be much larger than that. Experience says that this lens will display a slight narcissus on the display. We have to do better.

Again, from experience, we learn that the minimum value of the YNI should be about 0.009 if the lens is in units of inches, and 0.229 for a lens in millimeters. (Even though the lens is AFOCAL and ray output is in angles—which are independent of lens units, the quantity YNI has units of length, and therefore scales with those units.)

Let us correct this lens, hoping to get a better narcissus value. Here are the PANT and AANT files (**C30M2**):

```
PANT
VY 0 YP1
VLIST RD ALL
VLIST TH ALL
END
AANT P
AEC
ACC
GSR      0.500000      5.000000       4   M      0.000000
GNR      0.500000      3.000000       4   M      0.750000
GNR      0.500000      1.000000       4   M      1.000000
M    0.350000E+03   0.100000E+00  A  TOTL
ACC 10 .1 1
AAC 26 .1 1
M 13.37 1 A P YA 1 0 0 0 2
M 0 1 A P YA 1 0 0 0 10
M -.076 10 A P HH 1
ADT 7 .1 10

LLL .23 1 .1 A NAR 3
END
SNAP    0/DAMP 1
SYNOPSYS  100
```

Run this job, and the lens changes very little. What happened to the narcissus?

SYNOPSYS AI>NAR

ID FLIR DSEARCH 55602 10-NOV-20 08:08:10

NARCISSUS ANALYSIS

SURF	YNI	Imarg/Ichief	CYNI	RATIO (M/C)
1	1.5688	2.5873	1.0082	1.5560
2	1.7327	1.4803	2.1432	0.8085
3	0.2295	0.0944	4.5598	0.0503
4	1.0226	-32.0901	0.0614	16.6548
5	1.5012	1.3930	2.2109	0.6790
6	1.0311	3.8484	0.5899	1.7480
7	0.2229	-0.2105	4.4805	0.0498
8	0.4903	3.3052	0.6506	0.7535
9	2.1366	1.2231	0.2087	10.2351
10	1.6251	-0.7430	0.2491	6.5249
11	0.0029	0.0015	0.2217	0.0131

SYNOPSYS AI>

The lens still works, and the narcussus on surface 3 has reached our target. Narcissus is usually quite easy to control.

Sometimes the narcissus from another surface will become worse when we improve the contribution from the target surface. That happens. In that case, just add a target for the new surface as well and reoptimize. In this way, all surfaces can usually be corrected to or above the limit. That's how it's done: indentify the problem and fix it. The final lens is in figure 30.11 (**C30L2**).

The above discussion models the case where the detector is at the end of a small cold finger. Now we will examine what happens if it is instead inside a dewar with a cold stop much larger than the detector itself. The narcissus will likely be more intense, simply because the reflected rays are more likely to wind up at something cold than before, when the cold area was small.

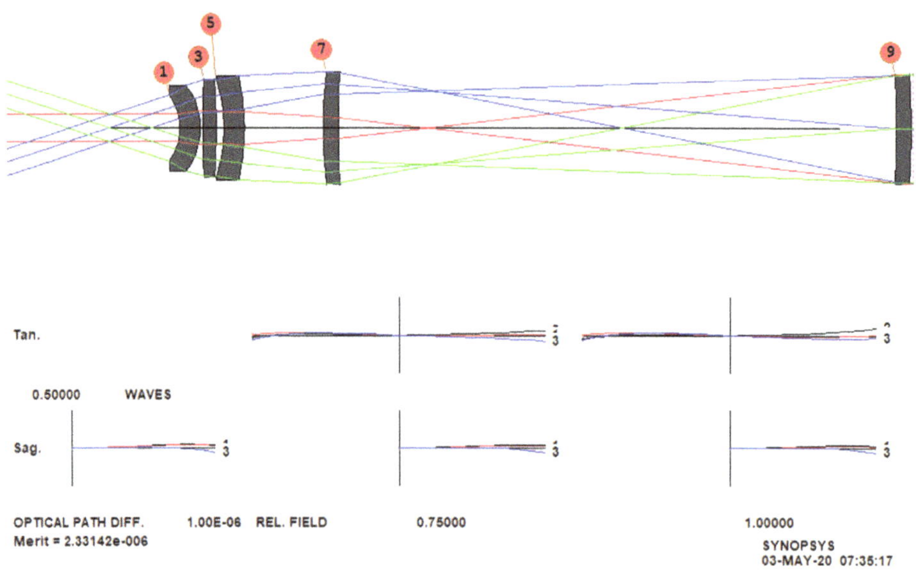

Figure 30.11. The final FLIR design.

In this case, we must examine how the narcissus effect varies over the field. In the discussion above, we were concerned with the value of YNI in the NAR listing, where a small value is to be avoided, since then the return rays are aimed near the cold finger. The last column on that listing, called RATIO (M/C) divides the YNI at the marginal ray by the same calculation at the chief ray. For the latter, a small value is to be *desired*, since it means that the narcissus will be relatively constant over the field and less noticeable to the eye. So if the ratio of the two is large, we have the best of both cases. To analyze this case, we use the FNAR feature, narcissus over field.

```
FNAR
NRYS 1000
ICOL P
JNAR 3 5 7
NSTEPS 50
REFL 0.01
CLOC 0 15.3
DISPLAY .8 .75 .85
END
```

(This feature requires the lens to be oriented with the detector on the left, as it is here. If the lens were designed the other way around, we would have to first REVERSE it.)

Here we will examine what the display would look like with the narcissus from three surfaces superimposed, the ones with the smallest ratio in the above NAR listing. The cold stop location is at 0.0, which means it is at surface 1, and the aperture radius is 15.3 mm. Here is the picture, in figure 30.12. The display is assumed to be adjusted so the background intensity level is 0.8, and the gain is set so a signal of 0.75 appears black and 0.85 white. We see a mild narcissus, spread over about half the screen.

The most serious contributions are from surfaces 3 and 7, which had the lowest ratio. If we wanted to further reduce the narcussus, we would add targets for the RNAR contribution of those surfaces in the AANT file, as you can read about in the User's Manual

But why did we select the *third* from the top of the DSEARCH results? This is a good place to point out a subtlty in the process. Look at the search results, part of which is shown in figure 30.13.

The top two, which are otherwise excellent designs, both positioned a lens surface rather close to the intermediate focal plane. When that happens, the narcissus from that surface is severe—and it cannot be corrected easily. To so so, the curvature would have to become so steep that the other aberrations would be excessive. So we chose the third, which places the focal plane nicely away from any lens surface.

And that's what the narcissus is all about. It is usually not difficult to control, but if you forget to look at the NAR listing and do not control the values, you might end up with a very poor display without expecting it.

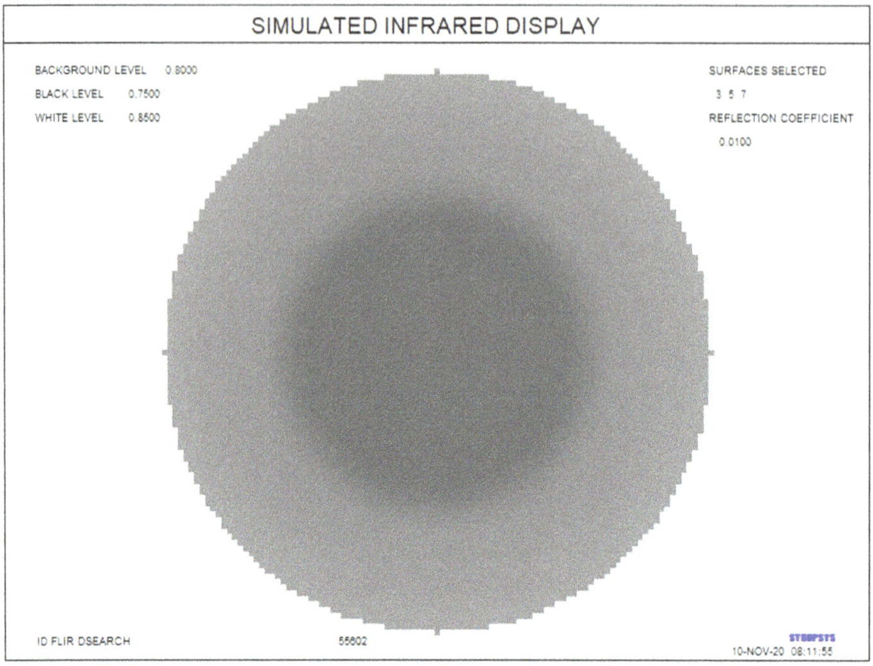

Figure 30.12. Display showing mild narcissus effect with simulated cold stop on surface 1.

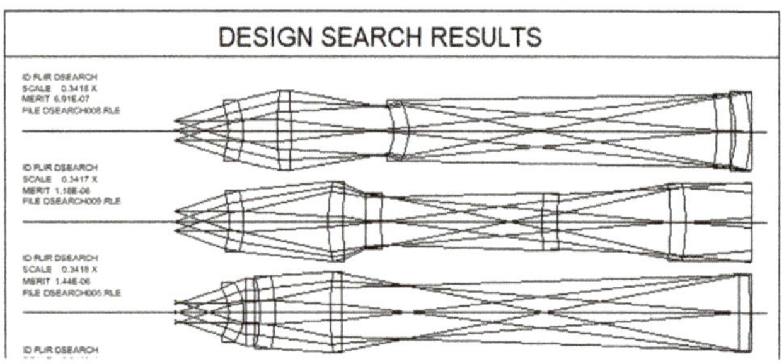

Figure 30.13. Portion of the returns from DSEARCH.

A final word about this system: Germanium is heavy and expensive, so you want the elements to be as thin as practical. Also there is some scatter, again calling for thin elements. And be careful the system does not get too hot; Germanium starts to absorb IR light when it gets hot, causing the heat to increase ever more, and so on. This is called ‘thermal runaway’, and you should be aware of it, especially if you design optics for a CO_2 laser, which can have high power. In this lens, element 3 is made of ZnSe, which is also expensive. So we would probably ask for a reduced thickness on that element.

IOP Publishing

Lens Design (Second Edition)
Automatic and quasi-autonomous computational methods and techniques
Donald C Dilworth

Chapter 31

Understanding artificial intelligence

Natural language input; changing lens parameters with AI; evaluation loops

In earlier chapters you saw a few of the `AI` features available in **SYNOPSYS**. This chapter will present a more complete picture of what this tool can do and how you use it. AI mode is turned on by typing the command **AI**, or by clicking on the `AI` button. It can be turned off by the command **INTERACTIVE**, or clicking the `AI` Off button.

Turn on AI, if it is not already on, and then **FETCH C31L1** and make a checkpoint. The lens is shown in figure 31.1.

What is the airspace distance after surface 3? Ask AI:

```
SYNOPSYS AI>3 TH?

The thickness or spacing of surface number  3  is      26.3666993
```

Change it to 27.0:

```
SYNOPSYS AI>TH 3 = 27

The thickness or spacing of surface number  3  is      27.00000000
```

What is the third-order spherical aberration? Well, you actually can ask just that question:

```
SYNOPSYS AI>What is the third-order spherical aberration?

The third-order spherical aberration sum (SA3) is      -0.02340108
```

Or you can type a much shorter sentence:

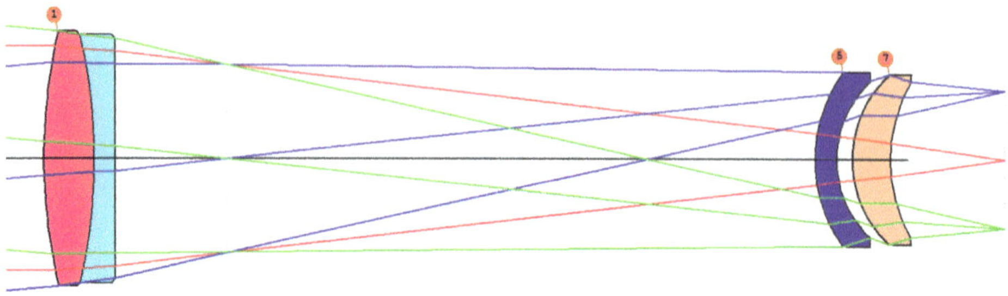

Figure 31.1. Example lens for AI exercises.

```
SYNOPSYS AI>SA3?

 The third-order spherical aberration sum (SA3) is       -0.02340108
```

The last question, "**SA3?**", is grammatically the same as the previous sentence, and of course you would prefer to do as little typing as possible, so that is what you would enter. But you see that the input is extremely flexible, and the way you type it is often not critical. The program parses the sentence, finds the subject and verb, satisfies any conditions, and then tries to answer the question.

Since many tasks in **SYNOPSYS** can be done in several ways, you naturally want to find the easiest. Suppose you want to know the global Z-coordinate of surface 7. You can type the **ASY GLOBAL** command and pick out the answer from the listing:

```
SYNOPSYS AI>ASY GLOB

THIS LENS HAS NO SPECIAL SURFACE TYPES
THIS LENS HAS NO TILTS OR DECENTERS
Global mode has been turned on.

GLOBAL COORDINATE DATA

GLOBAL COORDINATE SURFACE LOCATION IN COORDINATE SYSTEM OF SURFACE  1
```

SURF	X	Y	Z	NOTES	ALPHA	BETA	GAMMA
1	0.000000	0.000000	0.000000		0.00000	0.00000	0.00000
2	0.000000	0.000000	12.000000		0.00000	0.00000	0.00000
3	0.000000	0.000000	17.000000		0.00000	0.00000	0.00000
4	0.000000	0.000000	43.366699		0.00000	0.00000	0.00000
5	0.000000	0.000000	179.512319		0.00000	0.00000	0.00000
6	0.000000	0.000000	184.512319		0.00000	0.00000	0.00000
7	0.000000	0.000000	188.168005		0.00000	0.00000	0.00000
8	0.000000	0.000000	197.168005		0.00000	0.00000	0.00000
9	0.000000	0.000000	223.717528		0.00000	0.00000	0.00000

But it's easier to just ask AI:

```
SYNOPSYS AI>7 ZG?

Surface number  7  is not controlled by any tilt or decenter.
Surface number  7  has a global Z-coordinate of     188.16800509
```

Suppose you want to change that value. Surface 7 is not assigned global coordinates at present, so you could go to the SpreadSheet and enter data into sub-menus, or use the WorkSheet or a CHG file and enter the data in the correct format. In this case, however, AI is better:

```
7 ZG = 200
```

That simple sentence assigns the global coordinate.

Probably the most useful of the AI features is for making a plot of something against something else. Restore the lens to the checkpoint you made earlier, and then remove the paraxial solves.

```
CHG
NOP
END
```

Then type the sentence below to look at the color correction, shown in figure 31.2.

```
PLOT DELF FOR WAVL = .4 TO .8
```

Since the lens has no solves now, the paraxial defocus (DELF) varies with wavelength. If the lens were assigned a YMT solve, then DELF would have been zero at all

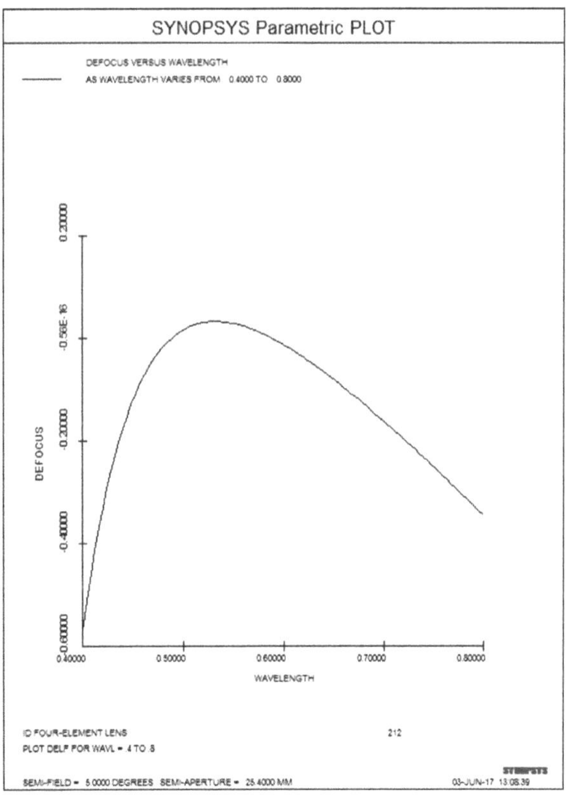

Figure 31.2. Color-correction curve plotted by the AI program.

wavelengths, and you would instead have plotted the back focus (BACK). (Note that we removed the *curvature* solve as well, by using the **NOP** entry, since we don't want the last radius to change with wavelength too!)

Suppose you are working on a lens where secondary color is an issue. You want to make this kind of plot more than once, but you don't want to type that long sentence every time. What can you do? Make a symbol, of course. Here's how to define a new command, "SC":

```
SC: PLOT BACK FOR WAVL = .4 TO .8
```

Now just type **SC**, and the program gives you the plot again. Put this definition in your CUSTOM.MAC MACro, and it will come back every time you start the program.

The principle is simple: ask AI for what you want with an English sentence. If that sounds too good to be true, you are right—probably all the sentences in Shakespeare would baffle the program. But applied to a limited range of issues, it can do a great deal. There are five categories of sentences that the program recognizes:

1. Questions about something
2. Changes to something
3. Looping, changing something and evaluating something else each time through, usually making a plot of the results.
4. Assigning symbols to equal a character string.
5. Evaluating an equation.

All of this is explained in chapter 15 of the User's Manual, and when you finish this chapter you should read the Primer that starts in section 15.2.

The AI vocabulary is a few hundred words, and you can see a list on the monitor if you ask AI to "**SHOW SUBJECTS**", "**SHOW VERBS**", or "**SHOW CONDITIONS**".

Changes are requested in an intuitive manner:

```
4 RAD = 123.456
Change radius 4 to 123.456
Increase 4 RD by 12.66
Increase 4 RAD to 33.5.
```

Note the difference between the last two examples. The last sentence actually will not work unless radius number 4 is currently *less* than 33.5. AI watches what you are doing and makes helpful suggestions if it sees what looks like an error.

31.1 Error correction

And speaking of errors, as a new user you will make lots of them. That's one reason the program has an extensive set of menus and dialogs. Those dialogs submit commands for you when you click a button, and the format is of course correct in that case. But some features can be run with a very simple command, and we usually get those things done faster by typing those commands manually. And, yes, we make mistakes sometimes. But don't worry; simple errors can be corrected in a jiffy—and you usually don't have to type the whole sentence again. Suppose by mistake you typed

```
4 RRD = 123.456.
```

The characters "**RRD**" are not in the vocabulary, and the program will immediately ask you to reenter four characters starting at "**RRD**". So you type "**RAD** " (notice the space after **RAD**: the program replaces *four* characters with whatever you type). It repairs the sentence and proceeds correctly. This error correction works both for AI sentences and for ordinary SYNOPSYS commands. Thus, typing

```
DDW 0 1 123 HBAR 0 1 -1
```

also produces the same error message, and if you then type "**DWG** ", the drawing command is properly executed. (Note the space after DWG; the program asks for *four* characters.)

And lastly, if your input is so garbled up that you just want to start over, simply hit the <Esc> key, and AI will discard the sentence.

31.2 MACro loops

The AI looping feature is powerful and general. Suppose you want a plot of the wavefront variance over the field. There are many kinds of diffraction image analysis, which you will see in the dialog **MDI**, but if the one you want is not already there, you can make it yourself. Here we will make a plot of the variance over field.

There actually *is* a command to do this analysis—but the lesson below is a good example of how to make your own feature using the tools of AI, for cases where there is no command.

Select the multicolor option on the VAR entry, and click the VAR button, as shown in figure 31.3.

The program prints the VAR value:

```
  VARIANCE      STD. DEV.       STREHL R.     XIP            YIP
 0.287577E-01  0.157605      0.428754     -0.492627E-20  0.874175E-21

VARIANCE IN EACH COLOR AT ABOVE IMAGE POINT:

WAVELENGTH, WEIGHT     0.587560     1.000000
```

```
  VARIANCE     STD. DEV.       STREHL R.
 0.591760E-01  0.243261       0.966967E-01
WAVELENGTH, WEIGHT      0.656270     1.000000
  VARIANCE     STD. DEV.       STREHL R.
 0.910363E-02  0.954130E-01  0.698097
WAVELENGTH, WEIGHT      0.486130     1.000000
  VARIANCE     STD. DEV.       STREHL R.
 0.179936E-01  0.134140       0.491468
 Type <ENTER> to return to dialog.
IMAGE>
```

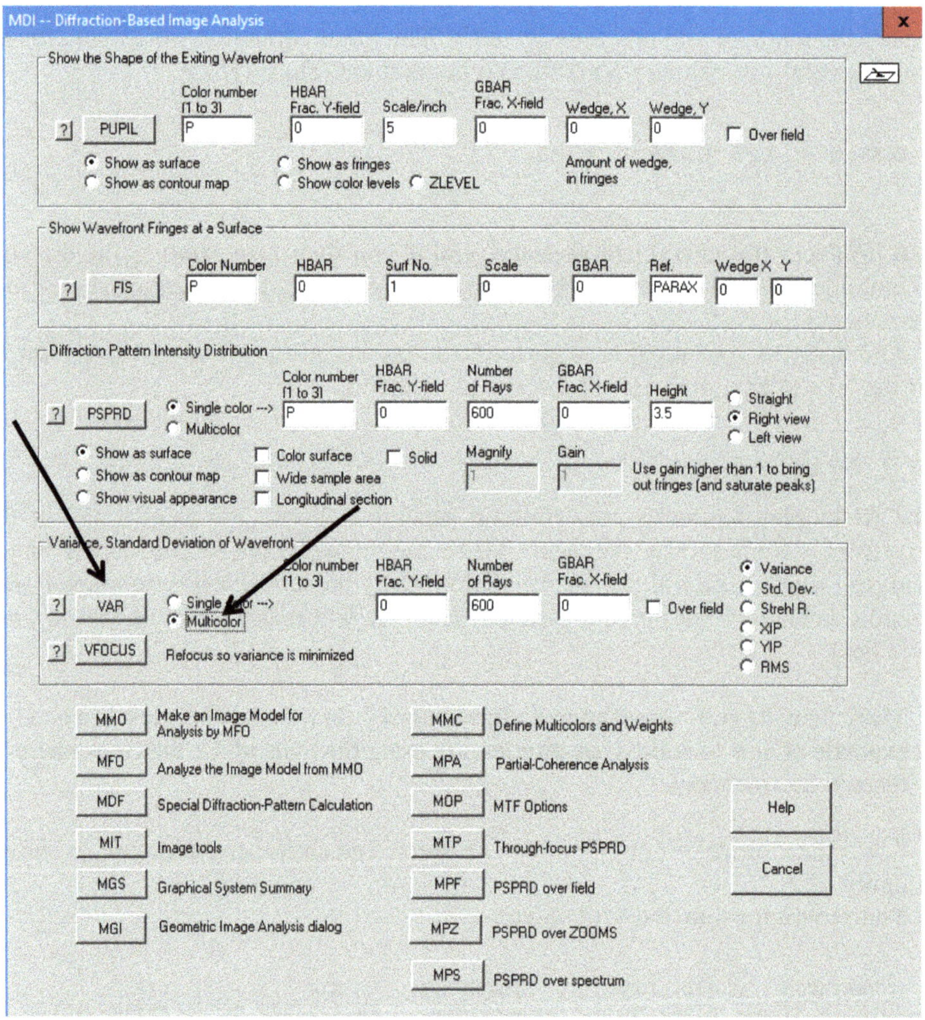

Figure 31.3. The MDI dialog, with multicolor variance selected.

Like many other features of SYNOPSYS, the VAR command puts a copy of its results into the AI buffer, and you can ask to see the contents with the question **BUFF?**.

```
IMAGE>BUFF?

The current FILE BUFFER contains
  1       0.02875775  VARIANCE
  2       0.15760478  STD. DEVIA.
  3       0.42875398  STREHL R.
  4      -4.92627452E-21  X IM. POINT
  5       8.74175408E-22  Y IM. POINT
  6       1.00000000  TRANS. FRAC.
  7       0.05917602  VARIANCE
  8       0.24326121  STD. DEVIA.
  9       0.09669675  STREHL R.
  10       0.58756000  WAVEL.
  11       0.00910363  VARIANCE
  12       0.09541296  STD. DEVIA.
  13       0.69809714  STREHL R.
  14       0.65627000  WAVEL.
  15       0.01799359  VARIANCE
  16       0.13414018  STD. DEVIA.
  17       0.49146805  STREHL R.
  18       0.48613000  WAVEL.
```

File location number 1 has the data you want. Now ask for a copy of the commands that were submitted by the VAR button. Type **LMM** (also found in the MACro menu dropdown list). The EE editor opens, with the **VARIANCE** command properly formatted, as shown in figure 31.4.

You need to tell AI to vary the relative field for every point on the plot. Which argument is that? *Select* the characters "VAR", and then look down at the tray, shown in figure 31.5.

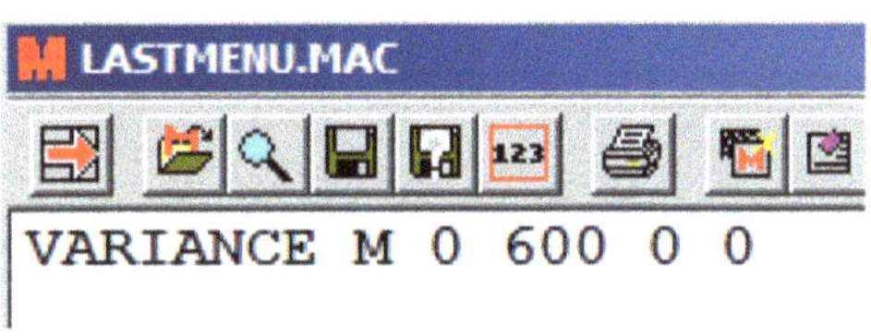

Figure 31.4. The VARIANCE command formatted automatically by LMM after running it from the **MDI** dialog.

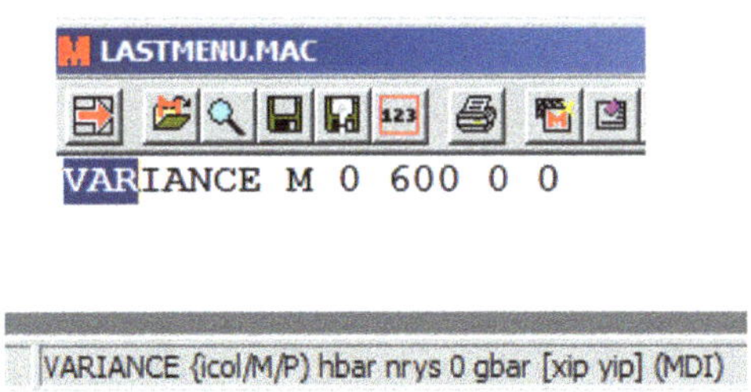

Figure 31.5. Selecting the characters VAR displays the command format in the Tray.

The program displays the format of the command, and you see that the relative field (hbar on the tray) is in word 3. Edit the command in the editor, replacing that word with the characters "AIP", which stands for "AI Parameter", as in figure 31.6. Then tell AI that the ordinate on the plot is to be taken from file location 1 in the AI output buffer:

Load this MACro into memory by clicking the **Load This** button. (Or you can click the **Run MACro** button, which also loads it, running the current value of AIP.) We're almost there. Now type the AI sentence

```
IMAGE>DO MACRO FOR AIP = 0 TO 1
```

The program loops for the default 100 cases and then displays the desired plot, shown in figure 31.7.

It's also easy to change the labels on the axes:

```
ALAB = "REL. FIELD"
AGAIN
```

(Using the built-in command to do this task, one would simply go to the MDI dialog, select the Over field checkbox, enter the appropriate data, and click the VAR button.)

You can loop over many kinds of things. For example, if you have designed a zoom lens, you could type

```
PLOT DISTORTION FOR ZOOM = 1 TO 9.
```

One more very useful feature of AI lets you do simple calculations involving output from other features. Fetch the lens saved as 4.RLE, in figure 31.8 (**FETCH 4**).

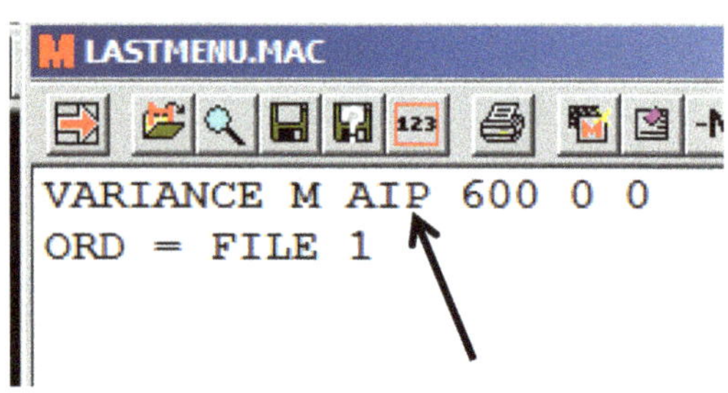

Figure 31.6. The relative field request is replaced by the characters AIP, and the result will come from the first file location in the AI buffer.

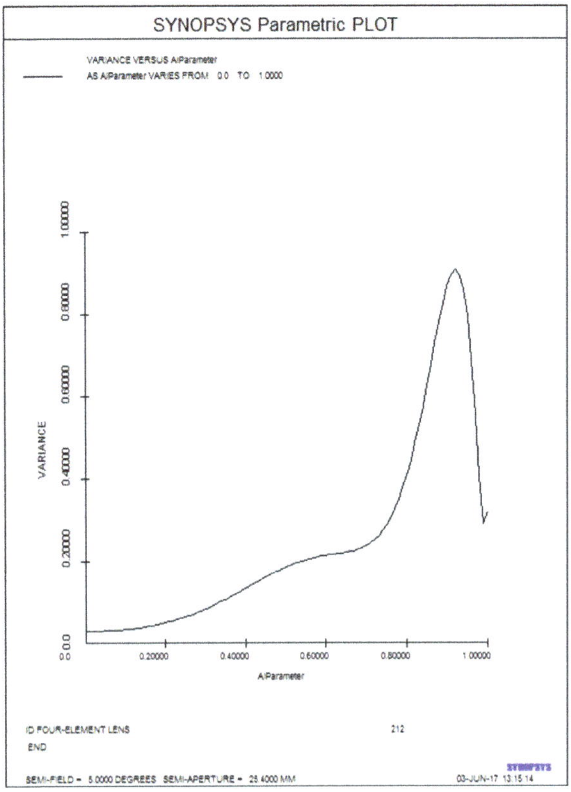

Figure 31.7. Plot of wavefront variance over field, produced by `AI`.

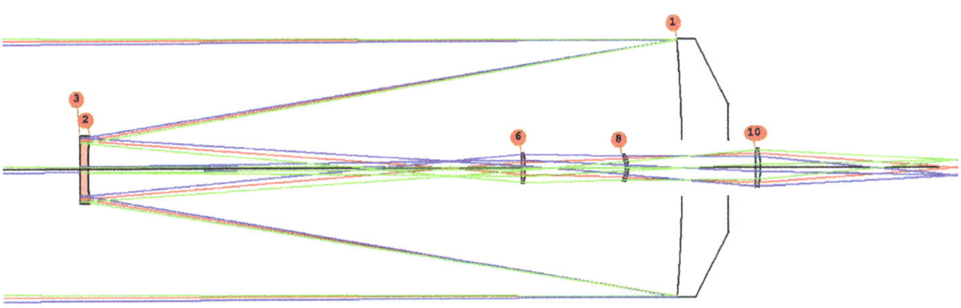

Figure 31.8. A telescope with an obscuration.

Look at the current clear apertures with the **CAP** command.

```
SYNOPSYS AI>CAP

ID RELAY FLAT                                  141

CLEAR APERTURE DATA
```

SURF	X OR R-APER.	Y-APER.	REMARK	X-OFFSET	Y-OFFSET	EFILE?
1	8.0014		Soft CAO			*
1	1.7500		*User CAI			*
2	2.0456		Soft CAO			*
3	2.0070		Soft CAO			*
4	1.9644		Soft CAO			*
5	0.4628		Soft CAO			
6	0.8797		Soft CAO			*
7	0.8849		Soft CAO			*
8	0.7373		Soft CAO			*
9	0.7781		Soft CAO			*
10	1.1225		Soft CAO			*
11	1.1340		Soft CAO			*
12	0.4898		Soft CAO			

Right now, the mirror has an assigned clear aperture inside (CAI) of 1.75. Suppose you want that to equal the outside aperture of surface 2. Here's how AI can do it:

```
SYNOPSYS AI>Z1 = CAO OF 2

The semi-aperture on surface number  2  is        2.04561850

SYNOPSYS AI>CAI OF 1 = Z1

Surface number  1  has an inside semi-clear aperture        2.04561850
```

Here we use one of the 20 Z-parameters to transfer the value from one place to another. Now the CAI on 1 equals the CAO of 2, as desired.

Lastly, AI can do simple calculations. Just enter a sentence that *starts* with an equals sign "=", and involves only constants, the Z-parameters, and any currently defined symbols that equate to a number. For example, if the variable Z1 currently equals 2.055619 and AIP has the value 3.66, one could calculate as follows:

```
SYNOPSYS AI>AA: 4.147


SYMBOL  43 DEFINED: AA*
4.147
SYNOPSYS AI>= Z1 + AIP + AA

= Z1 +     3.66000000000     + AA
= Z1 +     3.66000000000     + 4.147

The composite value is       10.76261900
```

You are encouraged to read chapter 15 of the User's Manual, where you will find other examples of how to use AI.

IOP Publishing

Lens Design (Second Edition)

Automatic and quasi-autonomous computational methods and techniques

Donald C Dilworth

Chapter 32

The annotation editor

Annotating graphics drawings; adding tolerances to element drawings

In this chapter, you will learn how to use the Annotation Editor of SYNOPSYS, a tool that can add many kinds of symbols and text to a graphics drawing. Get out the lens saved as 1.RLE, and make a drawing, as in figure 32.1.

```
FETCH 1
DWG
```

Let's add a warning message on the drawing. Click the Annotate button Ab on the Graphics Window toolbar. The Annotation editor toolbar opens up. Click the leftmost button, as in figure 32.2.

Then click in the drawing above element 3. Enter the text shown in figure 32.3, select size 14, and click OK.

Now the text shows up on your drawing, as in figure 32.4.

We're not quite done. Click on the arrow button, as in figure 32.5.

And then click below the text line and drag down to element 3. You have added an arrow, shown in figure 32.6.

Now click the red box on the toolbar and draw another arrow pointing at the last element, and add some more text there, as seen in figure 32.7.

Now click the red box again (to turn it off), and click the leftmost hashmark button, shown in figure 32.8.

Click several times in the area showing element 2. The lens element now shows hashmarks, as in figure 32.9.

Try the right hashmark button and add marks to element 3. If you hold down the <Ctrl> button when you click the element, the hash marks are smaller, which would be appropriate for smaller elements, as shown in figure 32.10.

doi:10.1088/978-0-7503-3695-6ch32

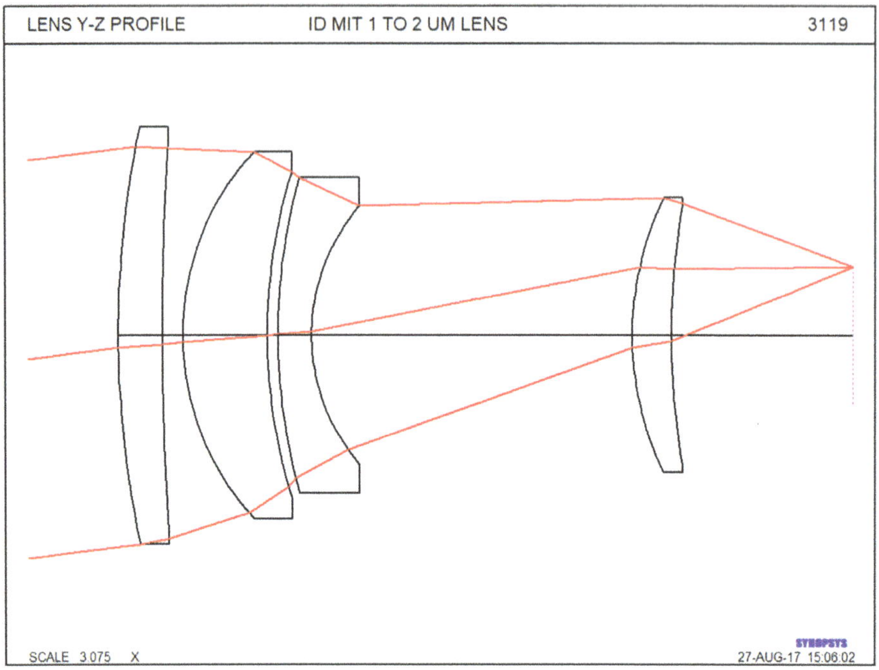

Figure 32.1. Sample DWG drawing.

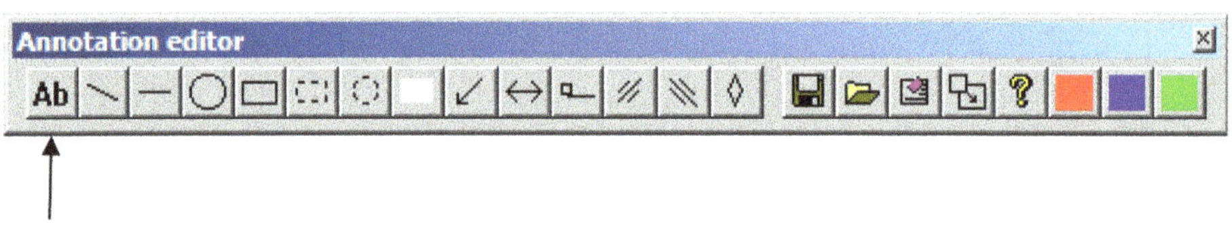

Figure 32.2. Opening the annotation editor.

You can probably figure out how to use the line, circle, and rectangle buttons. Give them a try. Those all let you drag in the drawing to define the size and position of the annotation. What if you change your mind? Simple: click on the Edit button, shown in figure 32.11.

All of the annotations you have added sprout an edit handle. Click on one and it turns black. You can then press the <Delete> key to remove it—or drag it to a different location. All of this is pretty simple. Now we'll get more advanced. Make a MACro as follows, run it, and then open the **MPL** dialog and enter the data as shown in figure 32.12.

```
CCW
FETCH 1
BTOL 2
```

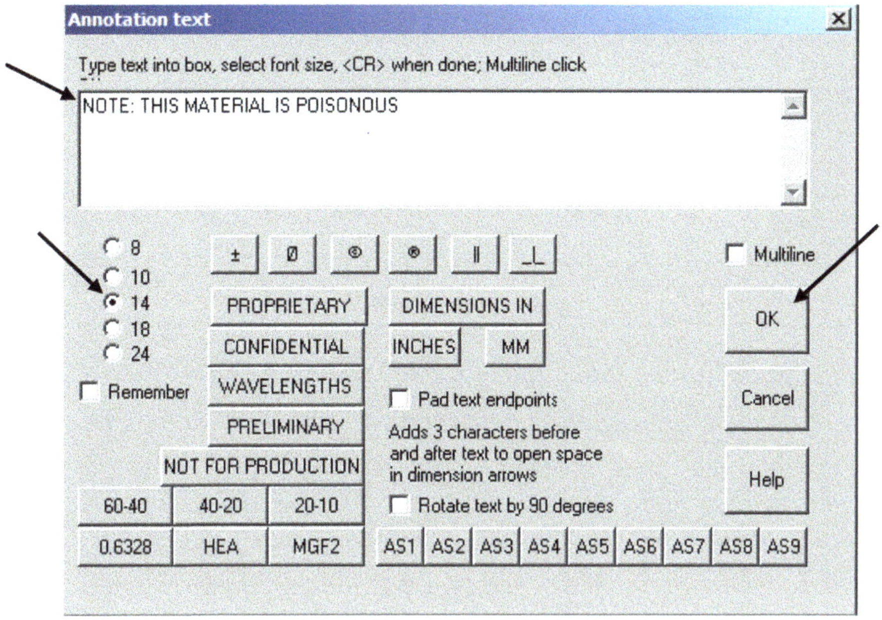

Figure 32.3. Entering text in the annotation editor.

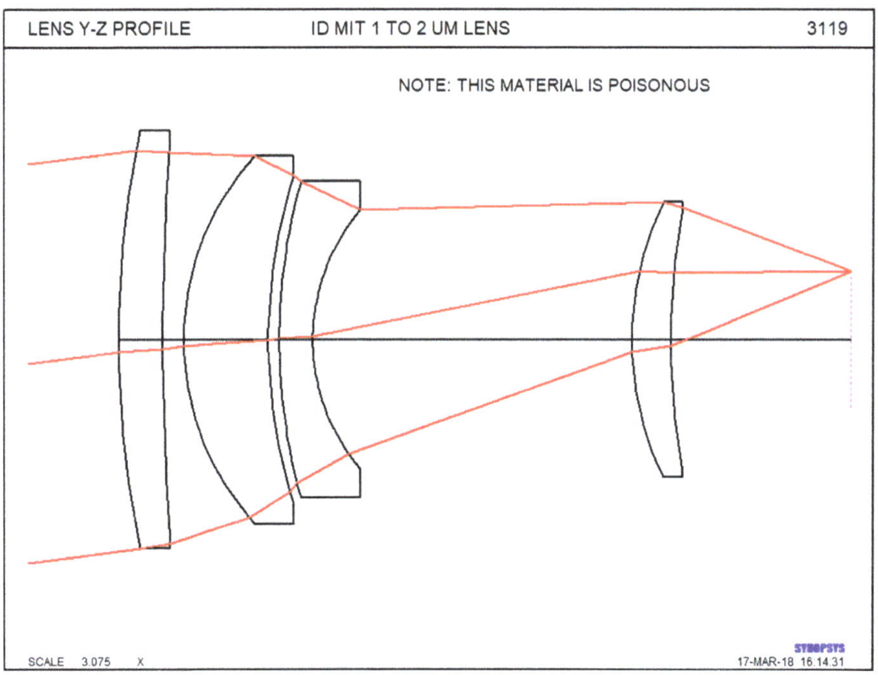

Figure 32.4. DWG drawing with added annotation.

Figure 32.5. Selecting the Arrow option.

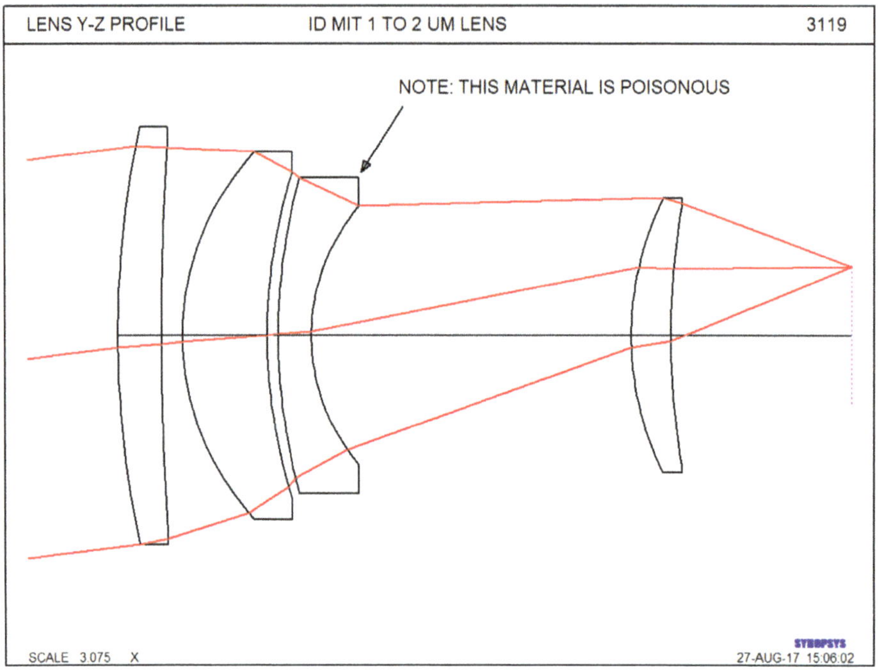

Figure 32.6. DWG drawing with added arrow and line of text.

```
TPR ALL
DEGRADE WAVE 0.2
GO
```

The program has made a tolerance budget for the lens, and the ELD command makes a drawing for element 2.

Open the dialog MPL and select the USE BTOL check box. Click the ELD button. The drawing shows up, complete with the tolerances for that element produced by BTOL, as shown in figure 32.13. That is the function of the **USE BTOL** command.

All of the tolerances are added by the program as *annotation* rather than graphics text, so if you want to change or customize anything, you can do it with the Edit button. The drawing at the moment does not specify the surface finish or

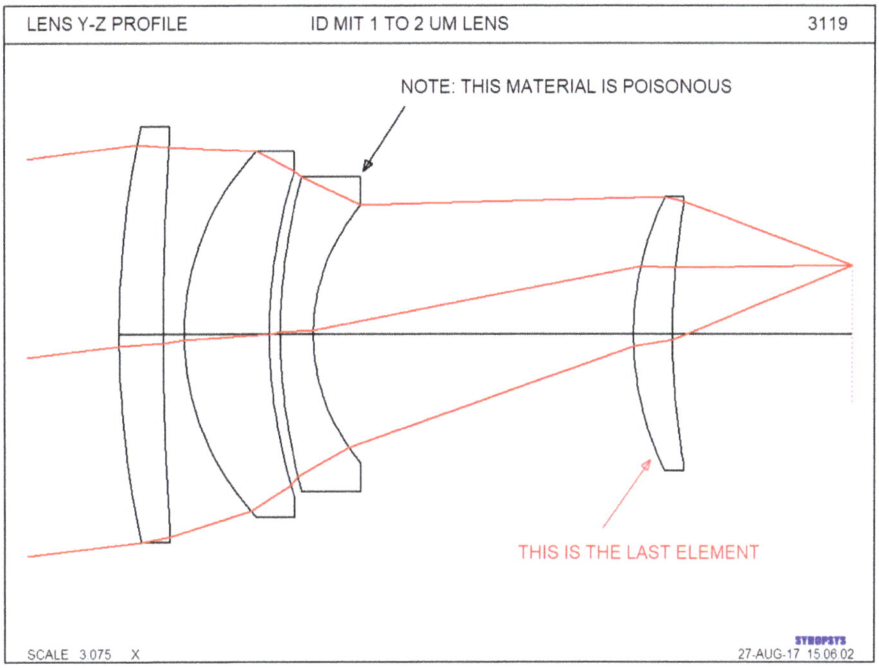

Figure 32.7. DWG drawing with added text in red.

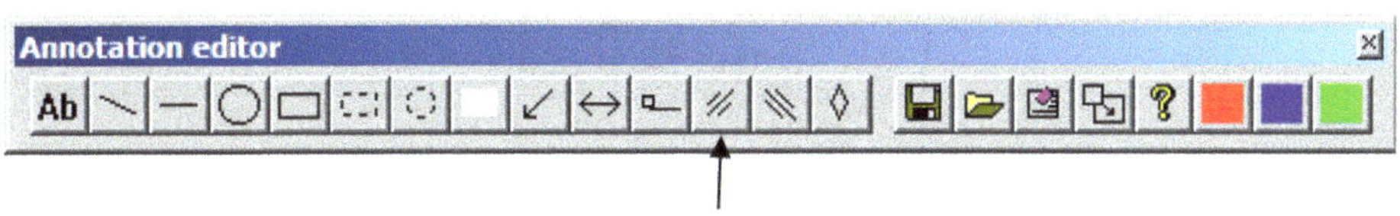

Figure 32.8. Selecting the hashmark option.

coating, but you can add those data with the annotation editor as well. Finish is usually specified by a scratch-dig specification, such as 60-40, which is a standard quality good for most lenses, or 20-10, which is a very high quality used mainly for reticles, where any defect would be sharply in focus. An uncoated glass surface reflects about 4% of the incident light, so all surfaces except those to be cemented to another element are routinely coated to minimize unwanted reflections. For a multi-element lens, the light loss would quickly become unacceptable otherwise, and the reflected light has to go somewhere, often ending up as veiling glare at the final image. The cheapest coating is a ¼-wave layer of MgF_2, but today one usually specifies a high-efficiency antireflection coating (HEA) and gives the wavelength range over which it should be designed. Such coatings can achieve less than 0.1% loss per surface.

You can ask ELD to annotate the coating and finish boxes automatically too. It will accept input of the form

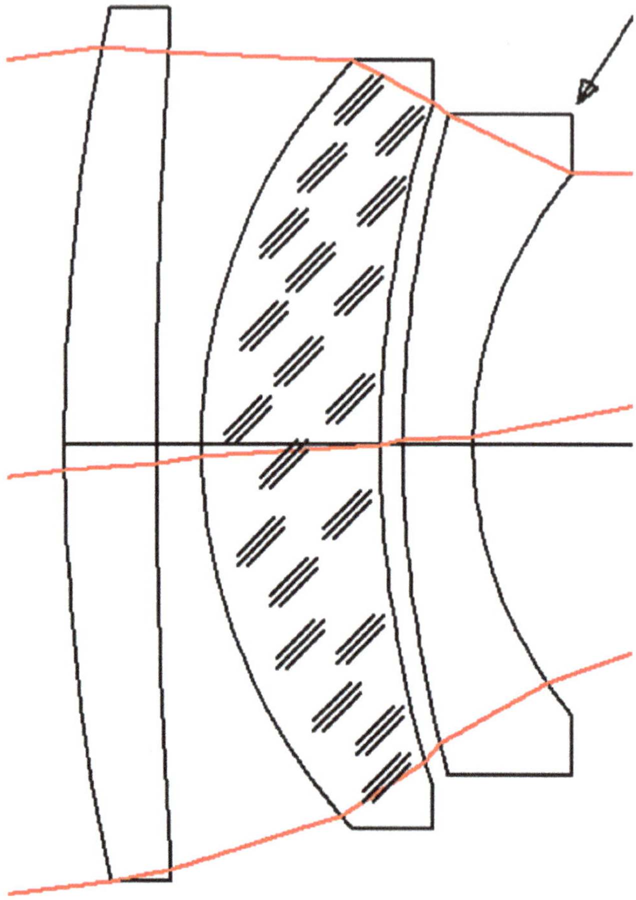

Figure 32.9. Lens element with added hashmarks.

```
ELD …
COAT "MGF2"
FINISH "60-40"
DESIGNER "JOE ENGINEER"
CHECKER "JOHN CHECKER"
TWAV "0.6328"
PLOT
```

and then add those annotation strings to the appropriate box on the drawing.

Here's a nice trick: We often want to add a comment or note to every element drawing. Here's how you can type it only once. There are nine annotation strings that you can define with a command. We will define the first one (note the quotation marks here and above; the program needs to know that those characters are not commands):

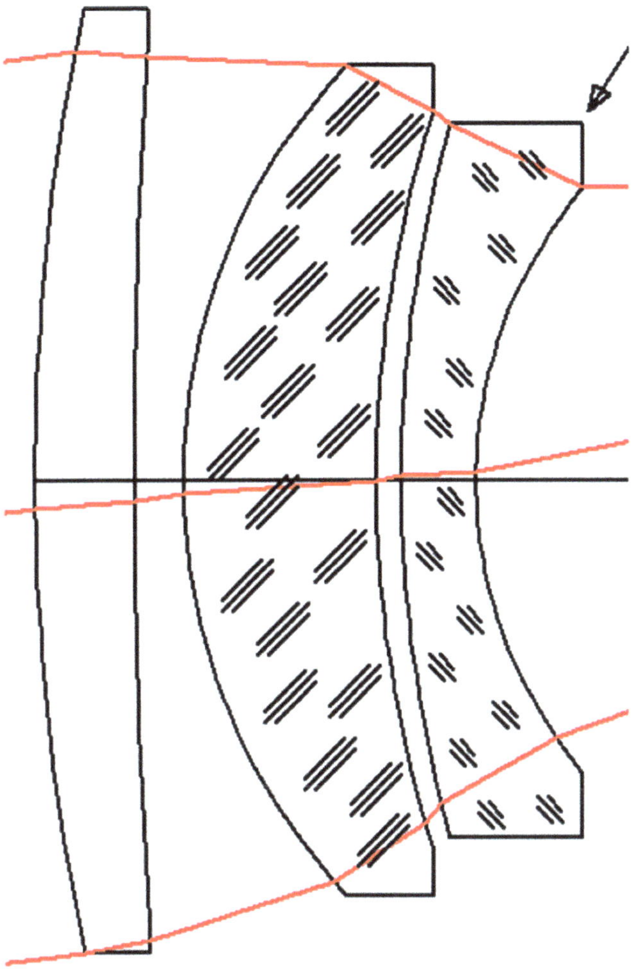

Figure 32.10. Lens elements with two kinds of hashmarks.

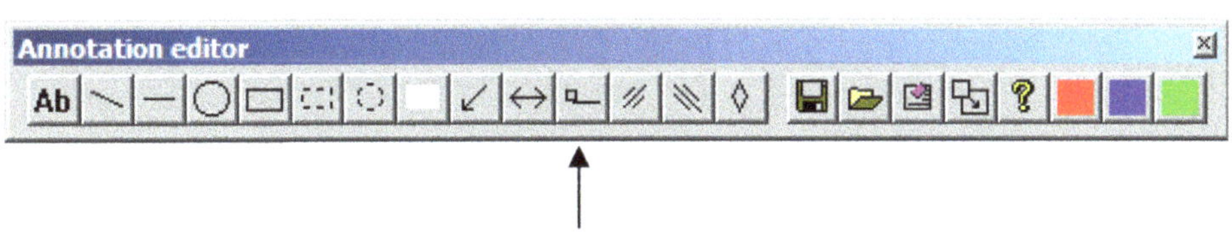

Figure 32.11. Selecting the edit handle.

```
AS1 "GET MELT DATA FOR ALL ELEMENTS"
```

Now, open the Annotation text editor, click in the drawing, and then click the `AS1` button AS1. Your string pops up in the text window. Click `OK` and it's on the drawing, as shown in figure 32.14.

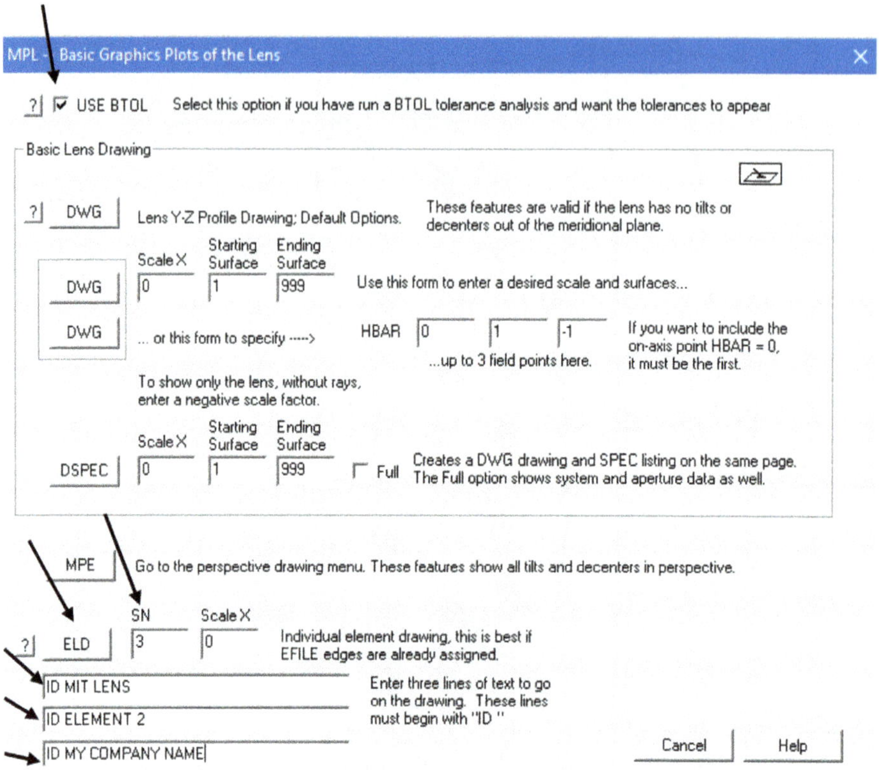

Figure 32.12. Selections in the MPL dialog to make an element drawing with tolerances.

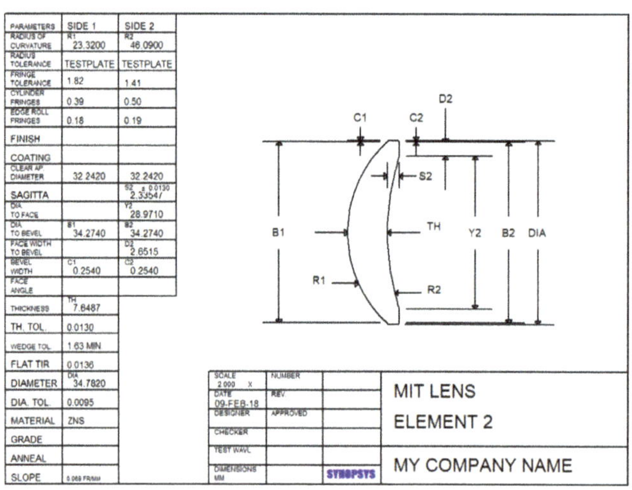

Figure 32.13. Element drawing with tolerances added automatically.

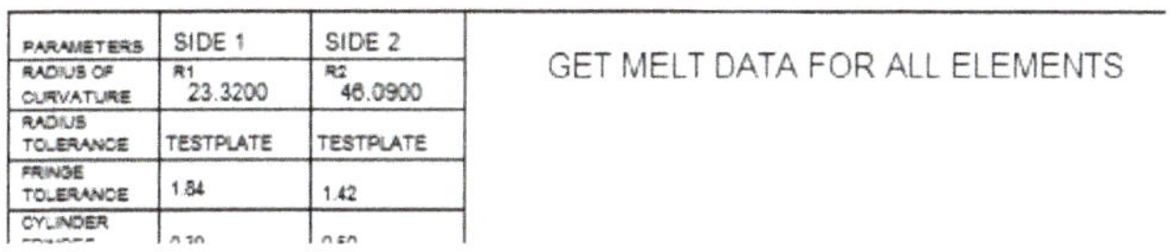

Figure 32.14. Annotation string added automatically by the annotation editor.

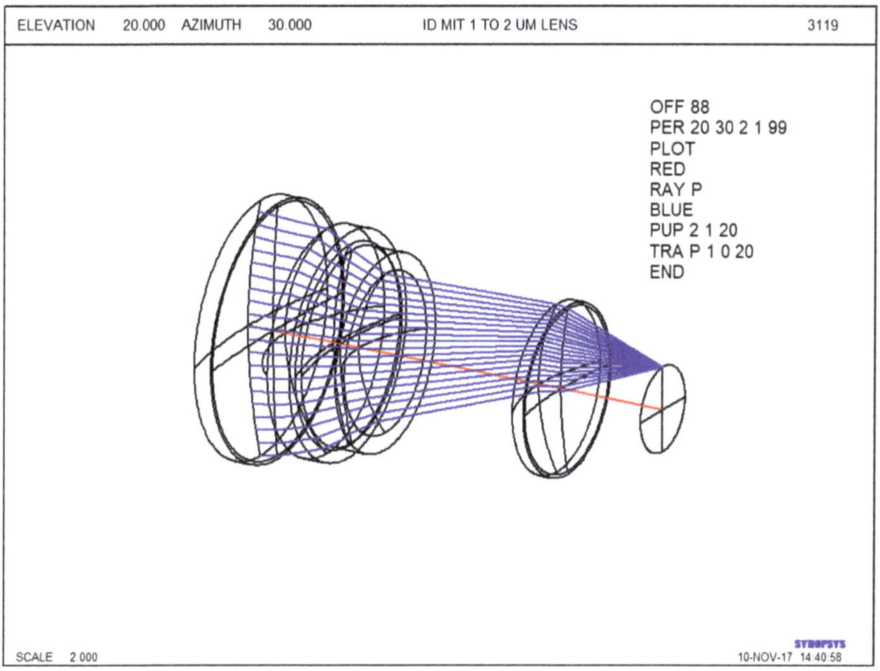

Figure 32.15. Drawing with command strings annotated.

Here's another nice trick. Sometimes you would like to list the MACro that produced a particular drawing. Make a MACro as follows:

```
OFF 88
PER 20 30 2 1 99
PLOT
RED
RAY P
BLUE
PUP 2 1 20
TRA P 1 0 20
END
```

Select the text and press <Ctrl>+C to put a copy on the clipboard. Now run the MACro, and when the picture shows up, open the Annotation Editor, select Text, and click in an empty space. Now paste the clipboard into the edit pane, with <Ctrl>+V, select size 14 and click OK. The annotation shows up, as seen in figure 32.15.

If you save your picture (with the Save button), the annotation is saved too. Once you have saved a picture, you can easily copy the annotation and put it on a new picture. Just make the new picture, and then, in the Annotation Editor, click the Fetch button and select the file you wish to copy. All the annotation comes back. That button erases any annotation that was previously there, but if you want to *add* to what was there before, use the Copy button instead.

Chapter 33

Understanding Gaussian beams

Propagation of Gaussian beams; modeling with real rays

Lasers, which are frequently used as light sources for all kinds of optical systems, produce a beam that is usually very small in diameter. The intensity of such beams is nonuniform, following a Gaussian profile in the ideal case, hence the name Gaussian beam, and deviating from that profile in a characteristic manner in many real cases. There are two issues that must be accounted for when designing and analyzing systems with this kind of illumination: the shape of the profile and the fact that beams of very small diameter exhibit strong diffraction effects as they propagate.

33.1 Gaussian beams in SYNOPSYS

As with most sophisticated features, the program aims at getting accurate results with as few complications as possible. For this reason, it addresses the peculiar properties of such beams in a novel fashion.

The main issue is the fact that, if the diameter is small, diffraction plays a role all along the beam. Rays going through normal lenses, on the other hand—where the beam diameter is much larger than the wavelength of the light—follow straight lines to a very good approximation, and we can then deal with 'rays' of light. That is not the case with Gaussian beams wherever the beam is small. Then the path of a light ray is *curved* and requires special attention in ray tracing.

Consider the following system (**C33L1**), shown in figure 33.1.

```
RLE
ID OBG DEMO
OBG .15 2
UNI MM
WA1 .6328
1 TH 50
2 RD -2.55 TH 2 GTB S
```



```
BK7
2 CAO 2
3 CAO 2
3 RD -55 TH 100
4 RD 100 TH 2 PIN 2
5 TH 50 UMC
4 CAO 10
5 CAO 10
7
AFOC
END
```

Following the rules for this kind of beam, the object is declared type **OBG**, with a waist on surface 1 and a radius of 0.15 mm. We are concerned with rays out to a point that is twice the $1/e^2$ point, according to the third word on the **OBG** line. The marginal rays shown in the picture above originate at that point in the beam. The example also includes two simple lenses, to expand and then recollimate the beam.

If we took the beam at surface 1 to be exactly collimated, the ray intercept on surface 2 would be the same as on surface 1. But that would be incorrect, since diffraction enlarges the beam by the time it gets there. To account for this effect, the program considers the beam at the waist to be slightly curved, just enough so that a *real ray*, traced from surface 1, hits surface 2 at about the same place as the diffracted Gaussian beam, and with the same divergence angle. From that point, we can treat a diffracting beam with the usual raytracing methods, provided diffraction thereafter is minimal. How accurate is this trick?

Ask for a beam trace, which evaluates the beam everywhere according to paraxial Gaussian beam theory.

```
SYNOPSYS AI>BEAM

ID OBG DEMO                                33262            13-MAY-13   14:16:08

GAUSSIAN BEAM ANALYSIS

SURF     BEAM RADIUS  WAIST LOCATION     WAIST RADIUS        DIVERGENCE
------------------------------------------------------------------------
  1         0.150000 -7.5157030E-15         0.150000          0.001343
  2         0.164341       -7.368983        0.005965          0.022287
  3         0.208892       -6.563589        0.006332          0.031811
  4         3.389933     -357.899054        0.014036          0.009472
  5         3.408876    -2087.561971        3.406641    5.9127598E-05
  6         3.408985    -2137.561971        3.406641    5.9127598E-05
  7         3.408985    -2137.561971        3.406641    5.9127598E-05
```

Notice that the beam radius on surface 2 is larger than on surface 1 because of diffraction. Now trace a real ray at pupil point (0, .5), which is at the $1/e^2$ point.

Figure 33.1. Laser system using a Gaussian beam.

```
SYNOPSYS AI>RAY P 0 0 .5 SURF

INDIVIDUAL RAYTRACE ANALYSIS

FRACT. OBJECT HEIGHT               HBAR      0.000000   GBAR      0.000000
FRACT. ENTRANCE PUPIL COORD.        YEN      0.500000    XEN      0.000000
COLOR NUMBER                          1

                         RAY VECTORS        (X DIR TAN)  (Y DIR TAN)  (INC. ANG.)
SURF          X            Y           Z          ZZ           HH          UNI
-------------------------------------------------------------------------------------
 OBJ      0.000000     0.000000     0.000000    0.000000      0.000549
   1      0.000000     0.136910     0.000000    0.000000      0.000549     0.031434
   2      0.000000     0.164338    -0.005301    0.000000      0.022307     3.663636
   3      0.000000     0.209062    -0.000397    0.000000      0.031846     1.060103
   4      0.000000     3.395560     0.057666    0.000000      0.009449     3.769940
   5      0.000000     3.413463    -0.047616    0.000000 -5.576629E-05     1.057009
   6      0.000000     3.410672     0.000000    0.000000 -5.576629E-05     0.003195

         REDUCED RAY ANGLES IN RADIANS AT IMAGE SURFACE
          PSI (X)         PHI (Y)              Z
          0.000000 -5.576629E-05       0.000000
```

The path of this real ray closely approximates the **BEAM** trace. We now have a tool that lets you analyze and optimize such a system using real rays. As long as the beam is expanded early on in the system, so diffraction has little effect thereafter, this real-ray approximation is useful and simple to set up.

33.2 Complications

But there are sometimes complications. Suppose, for example that there is an element right at the waist. If thickness number 1 is zero, or if that surface is not a dummy, the program cannot make the adjustments described above. Instead, it then adjusts the geometry so that it can trace an OBA object (finite object distance) where

```
THO = 1.0E14
YPO = THO * DIV
YMP1 = WAIST * RBS
YP1 = 0.0
```

Thus, the object is then at infinity and the entrance pupil radius is a function of the input OBG waist. In this situation, the program can still run the BEAM analysis, but diffraction is not taken into account with real rays, as it was before. If the first element expands the beam, however, then diffraction plays little role anyway and this is still a useful approach.

But what if there are one or more surfaces or elements in the beam where it is still very small? Suppose an expander is located, say one meter from the waist, and there are several fold mirrors along the way. The trick first described only operates between surfaces 1 and 2, and diffraction between other surfaces would be neglected in that case but should not in this one. Fortunately, there is yet another trick, and it's very simple.

What you do is assign a thickness of one meter to surface 1 (or whatever the distance is to the expander), put a dummy surface 2 at that distance, and then assign a thickness of *minus* a meter (or whatever is needed to get back to the first element or mirror) to surface 2. Now the program can adjust the beam properties at the waist so that diffraction is taken into account at the dummy surface 2. If you trace a real ray, it will hit surface 2 at the same place as the Gaussian beam does, and the path, once the beam actually gets to the expander, will be correct thereafter.

33.3 Beam profile

Let us look at that Gaussian profile. Type the following `AI` sentence to get the Gaussian profile shown in figure 33.2.

```
STEPS = 100
PLOT TRANS FOR YEN = -1 TO 1
```

This shows a beautiful Gaussian shape. There are other ways to see the shape too. Chapter 15 showed how to make a MACro using the `COMPOSITE` aberration format to plot the profile, and how to design a simple system to expand the beam and yield

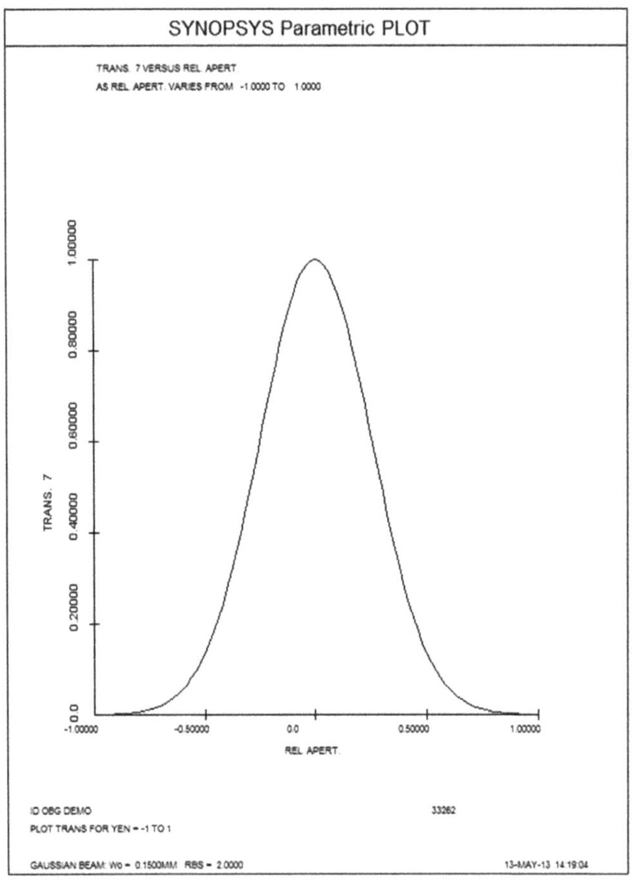

Figure 33.2. Intensity profile of a Gaussian beam.

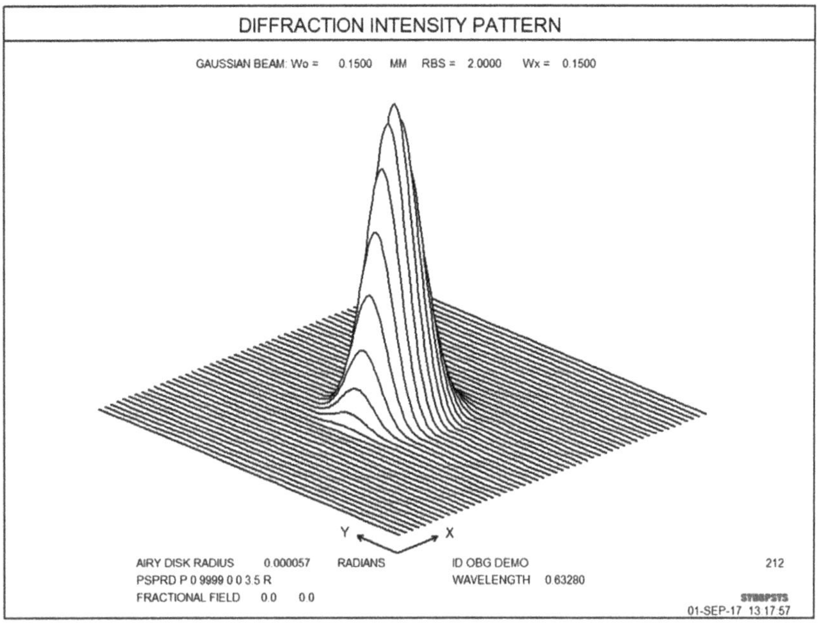

Figure 33.3. Diffraction pattern of Gaussian beam.

uniform intensity at the same time. It also showed how the diffraction-propagation program DPROP can analyze the improved energy distribution, giving yet another way to analyze this kind of beam.

33.4 Effect on image

To complete this chapter, let us make a diffraction pattern of the output. Since the beam is Gaussian, the far-field image is also Gaussian in shape. Go to the **MDI** dialog, ask for a PSPRD plot, shown in figure 33.3, and specify 9999 rays. (The energy is all concentrated near the center of the beam, and analyzing the image with fewer rays will be less accurate.)

Indeed, we see there are no diffraction rings at all! That's a property of Gaussian beams. Diffraction takes place mainly around the edge of a beam, and if that edge is very fuzzy, having fallen off to a much lower value than at the center, then diffraction at the edge plays no role.

To read about other subtleties regarding Gaussian beams, including noncircular beams and the effects of beam quality, type **HELP OBG** in the Command Window.

IOP Publishing

Lens Design (Second Edition)
Automatic and quasi-autonomous computational methods and techniques
Donald C Dilworth

Chapter 34

The superachromat

Correcting chromatic aberration with Herzberger's theory

This chapter will explore a unique feature of SYNOPSYS that can be helpful when you need exceptional color correction, better even than an apochromat. Chapter 12 showed how to select three glass types that make it possible to correct axial color at three wavelengths, and for many tasks that is as good as you will need.

But not always. Suppose you are designing a lens to be used over the range 0.4–1.0 μm. Can you do it with an apochromat? Let's find out. Here is the RLE file for a starting system (**C34L1**), where all surfaces are flat except for the last, which will give you an F/8 telescope objective of six-inch aperture.

```
RLE
ID WIDE SPECTRAL RANGE EXAMPLE
OBB 0 .25 3
UNITS INCH
1 GLM 1.6 50
3 GLM 1.6 50
5 GLM 1.6 50
6 UMC -0.0625 YMT
7
1 TH .6
2 TH .1
3 TH .6
4 TH .1
5 TH .6
END
```

This file did not specify the wavelengths yet, so you get the default CdF lines. We have to change this. Open the Spectrum Wizard (), and change the data indicated in figure 34.1.

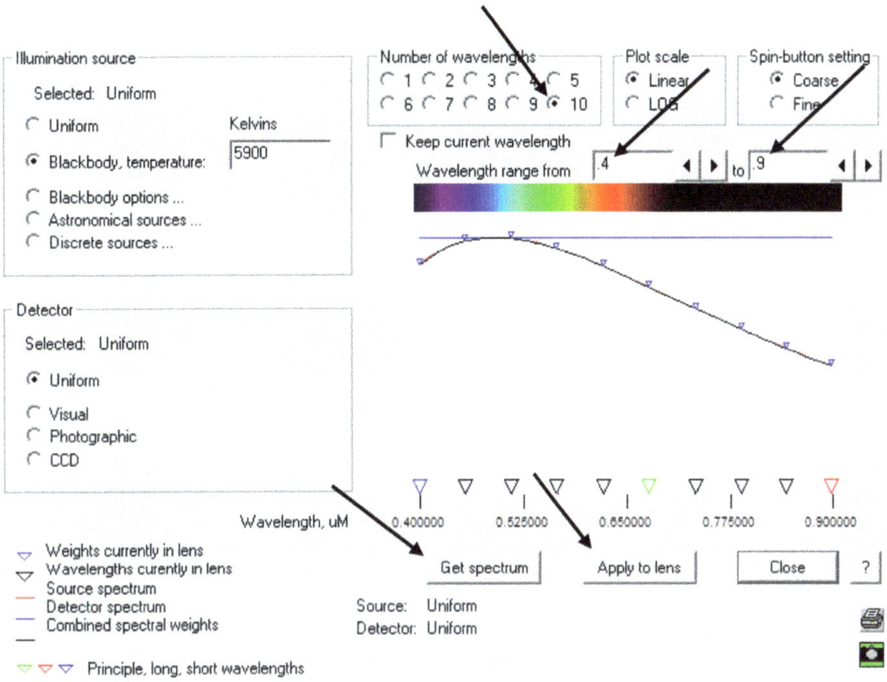

Figure 34.1. Spectrum Wizard with 10 wavelengths selected over the range 0.4–0.9 μm.

After clicking the **Get spectrum** button, click the **Apply to lens** button. The lens now has a wider spectrum. Here is the starting lens, shown in figure 34.2.

Uggh! Yes, it's really awful. (Of course: all curves are flat except for the last one.) Let's optimize it, varying the glass models, radii, and airspaces. Make a MACro:

```
LOG
STO 9
OFF 1
PANT
VLIST RAD 1 2 3 4 5
VLIST TH ALL
VLIST GLM ALL
END

AANT
ACM .5 1 1
LUL 5 1 1 A TOTL

END

SNAP
SYNOPSYS 50
```

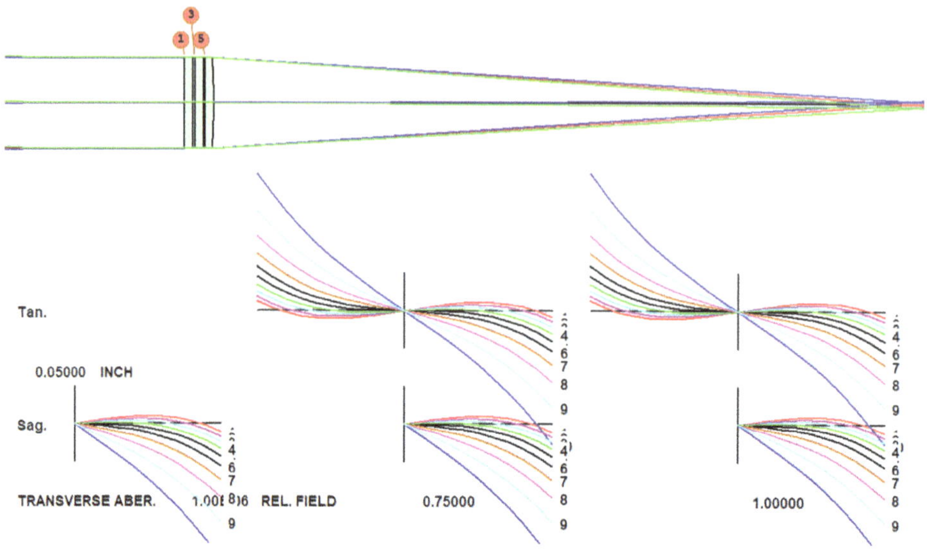

Figure 34.2. Starting lens with 10 wavelengths defined.

Now put the cursor on the blank line in the AANT section and click the button. Merit function number 6 is selected by default, so just click the Back to MACro editor button. This gives you a simple merit function, and the requirement on TOTL is there because the airspaces will vary and you want the lens to remain a compact group. Modify the AEC monitor so lens edges stay above 0.2 inches in thickness.

```
...
AANT
ACM .5 1 1
LUL 5 1 1 A TOTL
AEC .2 .01 1
ACC
GSR .5 10 5 M 0
GNR .5 2 3 M .7
GNR .5 1 3 M 1

END
...
```

Here, you correct all 10 colors via the '**M**' in the AANT file. Okay, it's time to optimize. Run the MACro and then open the annealing dialog. In this case, select the **Free GLM** option in the dialog, because the glass models are likely to become pinned to either the crown or flint boundary of the glass map almost immediately, and we want them to be free to leave the boundary later on as the design form changes. This option is generally not advisable if the lens already has a reasonable construction at the start, but that is not the case here. Select a temperature of 50, cooling 2, 50 passes. The lens is vastly better, as shown in figure 34.3.

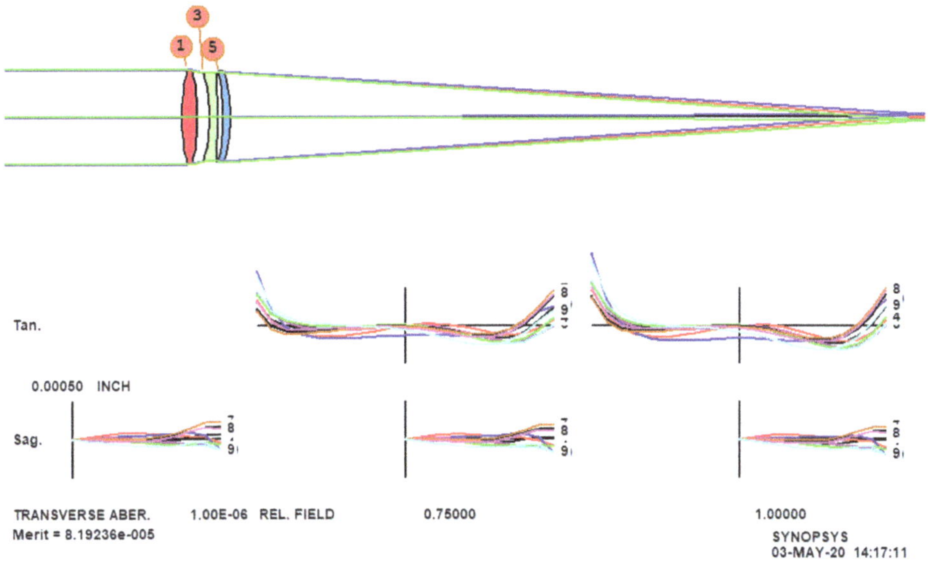

Figure 34.3. Lens optimized at 10 wavelengths with glass model variables.

How good is this state of correction? We can ask AI to show us the defocus over wavelength—but that would be unwise at the moment. This lens has a curvature solve, and at each wavelength the program would recalculate it. So instead, we make a second MACro, as follows:

```
STORE 9
STEPS = 50
CHG
NOP
END
PLOT DELF FOR WAVL = .365 TO 0.9
GET 9
```

This file removes all of the solves (and pickups, if there are any) via the **NOP** entry, and then plots the defocus. Afterwards, it gets back the lens the way it was. Here is the color correction curve, in figure 34.4.

This lens already looks like a 'superachromat', a term coined by Max Herzberger in 1963. His theory[1] says that, if you make a graph of the glass catalog where the axes are the values of P* and P**, and then select three glasses that lie on a straight line, it is possible to correct at *four* wavelengths at the same time. The term P* refers to the partial dispersion (NF–N*)/(NF–NC), where F and C are the Fraunhofer lines at 0.4861 and 0.6563 μm, and N* is the IR line at 1.014 μm. N** is the UV line at

[1] Herzberger M 1963 *Appl. Optics* **2** 553

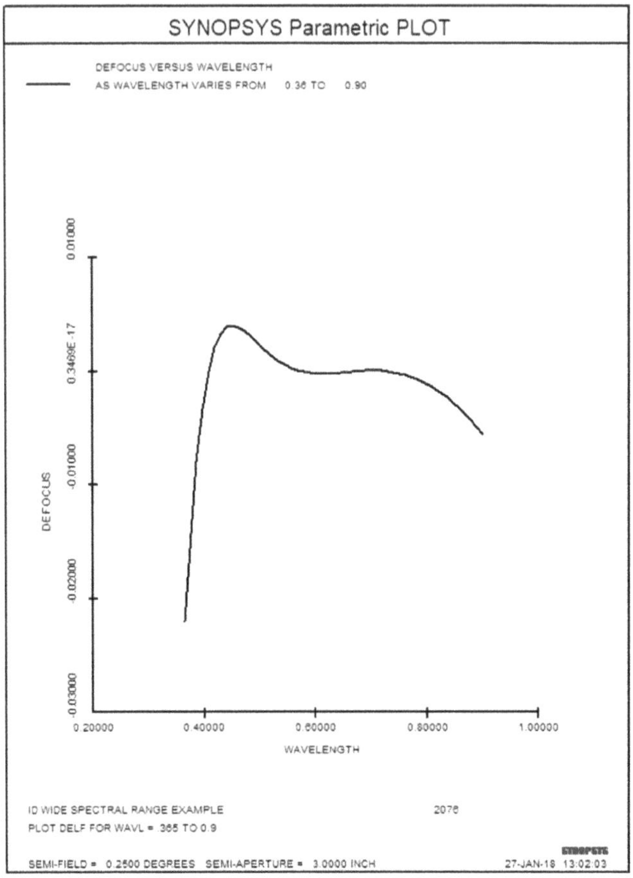

Figure 34.4. Color correction curve for the reoptimized lens.

0.365 μm, giving you a similar equation for P**. In this case, the program found a great combination of glass models automatically.

But now it's time to learn about making a superachromat yourself, for cases when the program does not find one automatically.

First, we show how to find suitable glass combinations by hand, using the glassmap feature of SYNOPSYS. Then we will show another way program can do the task automatically, which is a real time-saver.

The on-screen glassmap of SYNOPSYS can show just the kind of plot you need. Type **MGT** to open the Glass Table Selection dialog and select the O (Ohara) catalog, and when the map is displayed, click the Graph button and select the bottom option as shown in figure 34.5 to see the display shown in in figure 34.6.

On this graph, you see the current location of the model for each element (the red circles). They are lined up pretty well, but the line between them is quite short. What you have to do is to adjust the line so it connects three glass types, preferably with a line as long as practical. Select a glass near the bottom, where the flints tend to be, and <Ctrl>click one of them. That puts the bottom of the black line on that glass

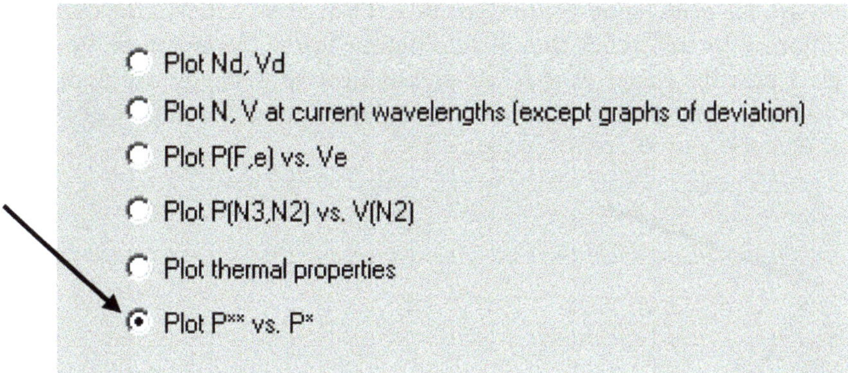

Figure 34.5. Selecting the graph of P** versus P* on the glass map

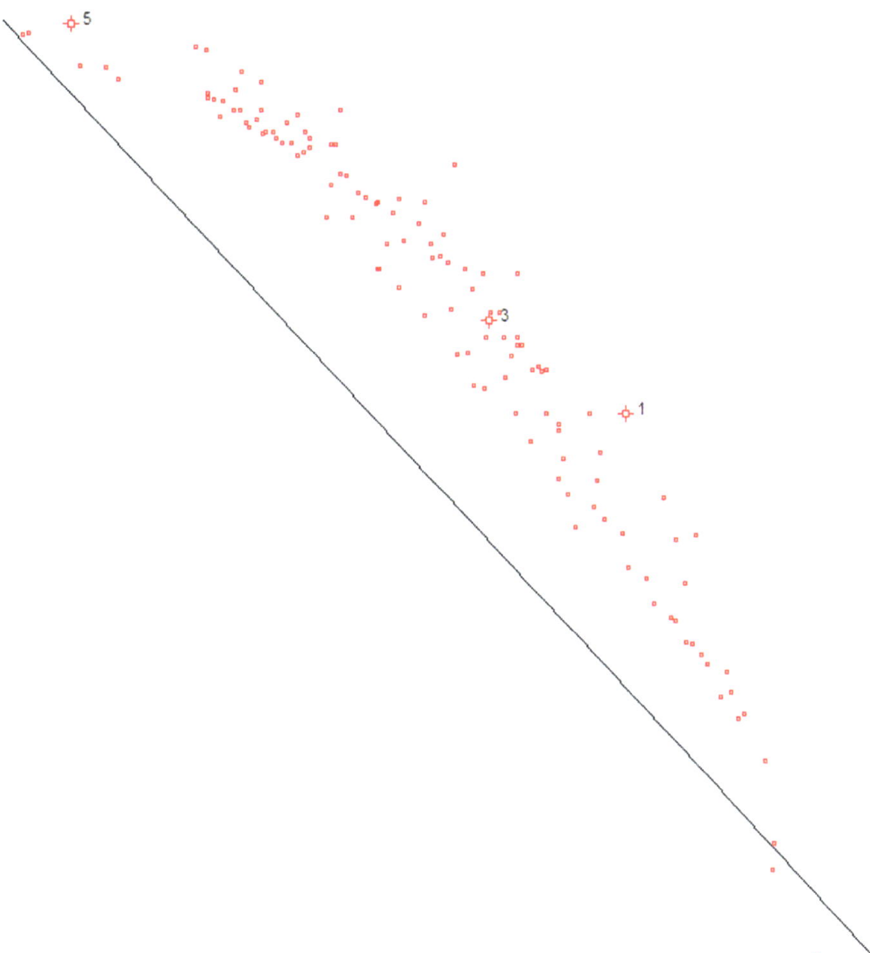

Figure 34.6. Glass map display showing plot of P** versus P* for the Ohara glass catalog.

and displays the glass name in the Glass box. Then select a glass near the top of the distribution and <shift>click that one, to put the top of the line there. Now select a third glass near the center of that line and as close to it as you can find. Click the symbol so you can see that glass name too. Write down the names of those three glasses. Our selection is shown in figure 34.7.

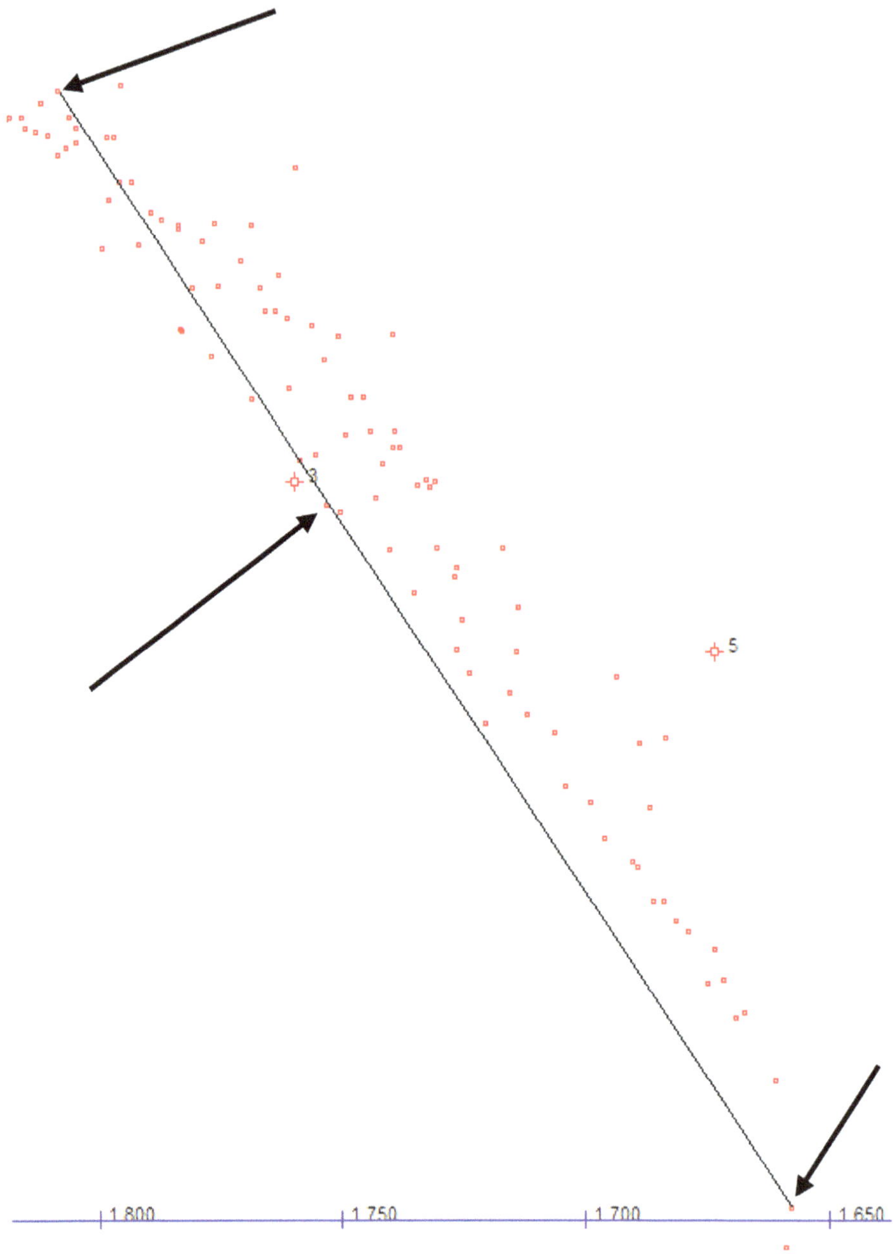

Figure 34.7. Glass map display with three glasses selected.

We have three potential glasses for our superachromat. They are types S-PHM52, S-NPH5, and S-TIL27. You can also display the relative cost and other properties, to help you select three acceptable glasses. Then you insert those three glasses into the lens and optimize. If that does not yield a satisfactory lens, you select a different three according to the same procedure. This process is rather tedious but works.

The other procedure is to let the program select glass combinations for you. Make and run a new MACro containing the following:

```
FST
PREF
CAT O
CAT S
GO
```

FST means Find Superachromat Triplets. This input will examine all combinations of glass types from the Ohara and Schott catalogs and rate the 10 most suitable for a superachromat, rejecting all but the preferred glass types. The program finds the following:

```
--- FST
 --- PREF
 --- CAT O
 --- CAT S
 --- GO
SUPERACHROMAT GLASS SEARCH RESULTS (LOWER SCORES ARE BETTER)

      SCORE       UPPER              MIDDLE             LOWER                    OFFSET
 1    0.02171385 O S-FPL55           O S-LAM54          S SF57                0.00000914
 2    0.02008505 O S-FPL53           O S-LAL8           O S-NPH1W             0.00000923
 3    0.02082027 O S-FPL55           S N-KF9            S SF10                0.00000567
 4    0.02008505 O S-FPL53           O S-LAL8           O S-NPH1              0.00000923
 5    0.01881642 O S-FPL55           O S-TIL27          O S-TIH23             0.00000071
 6    0.02026308 S N-FK58            S N-SSK8           S SF4                 0.00000296
 7    0.02120605 O S-FPL53           O S-LAL13          O S-TIM28             0.00000424
 8    0.02139100 O S-FPL53           S N-SK4            S SF56A               0.00000909
 9    0.02171385 O S-FPL55           O S-LAM54          S SF57HTultra         0.00000914
10    0.02147608 O S-FPL55           S N-SSK8           S SF1                 0.00000460
 SYNOPSYS AI>
```

This method is superior to doing it by hand since it can combine glasses from different manufacturers. Combination number 10, for example, is made from one Ohara glass and two from Schott. Let us try that combination. Go back to the original lens, with 10 wavelengths selected. Edit the optimization MACro as shown below (**C34M1**). (Here, we use ready-made merit function number 8, which corrects a combination of transverse and OPD aberrations, and then adjusted the weights.)

```
LOG
STO 9

CHG
1 GTB O 'S-FPL55'
3 GTB S 'N-SSK8'
```

```
5 GTB S 'SF1'
END

PANT
VLIST RAD 1 2 3 4 5
VLIST TH ALL
END

AANT
ACM .5 1 .1
LUL 5 1 1 A TOTL

AEC .1 1 1
ACC
GSR .5 10 5 M 0
GNR .5 5 3 M .7
GNR .5 4 3 M 1
GSO 0 0.003916 5 M 0
GNO 0 0.003 3 M .7
GNO 0 0.002 3 M 1

END

SNAP
SYNOPSYS 90
```

After running this and then annealing (**50, 2, 50**), you get a lens corrected to about 1/10 wave on axis and ½ wave at full field, although color 10 (at 0.4 μm) is not as well corrected as the others, as shown in figure 34.8.

But we only guessed the order of our three glasses. There are six possible combinations, and next we will try the order 5, 1, 3. The result is shown in figure 34.9, after annealing. This lens is superb.

Now the lens (**C34L2**) is corrected to about a quarter wave over the entire (very wide) spectral region. What does the second MACro show now? The curve is in figure 34.10.

Well, it's corrected at three wavelengths for sure—but we were aiming for four. How come the curve doesn't go up again at the right end—as a true superachromat would? Simple. As usual, the program is balancing everything in the merit function, not just axial color, and the other aberrations make it depart slightly. Still, this is a great lens! And if you look at the `OPD` fans of the first lens, where the program found

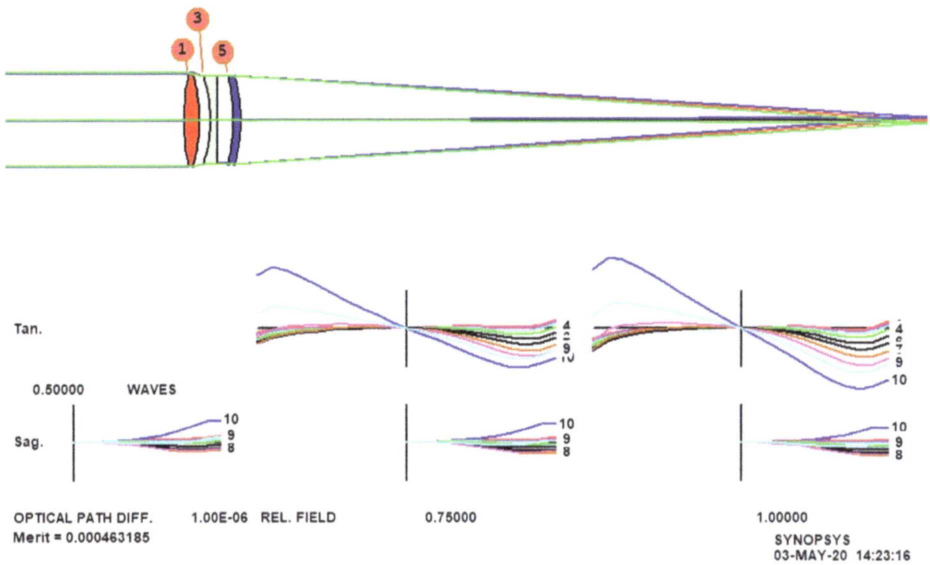

Figure 34.8. Lens optimized with three glasses found by `FST`.

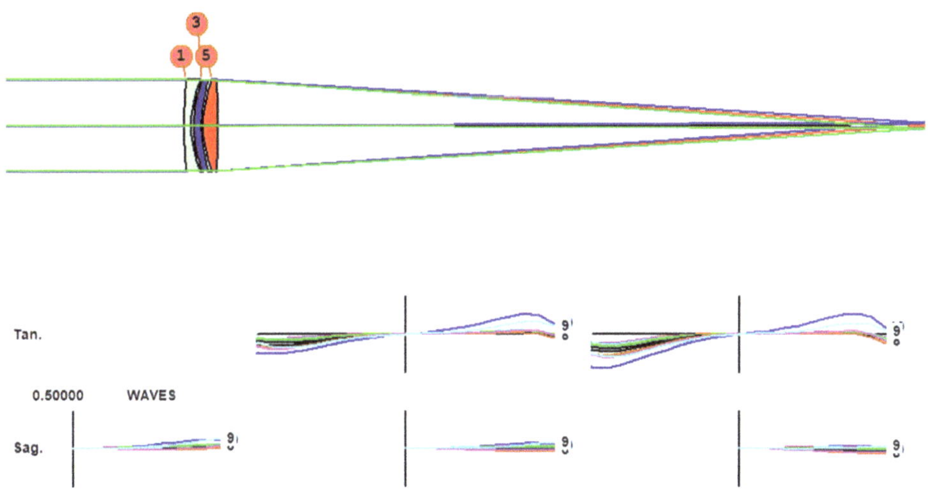

Figure 34.9. Lens optimized with a different order of glasses from `FST`.

a superachromat by itself (with model glasses), you see that the `OPD` errors in that design are slightly *larger*. The beautiful plot for that lens showed the *paraxial* focus, and when you consider real rays over the whole pupil it's not necessarily the best solution. Here's another example of when classic theory may not tell you all you need to know.

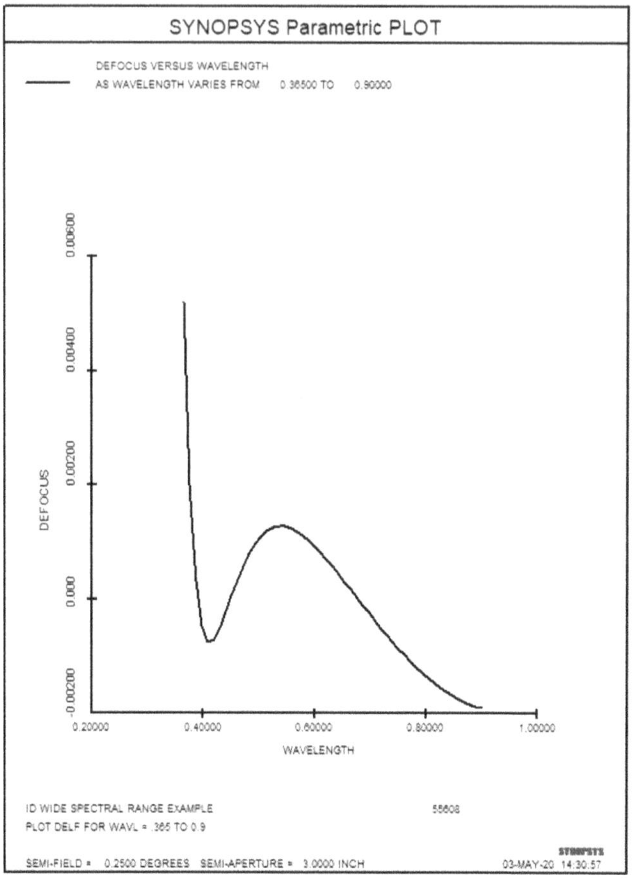

Figure 34.10. Color correction of the final lens.

The next chapter will design an even more demanding superachromat, and chapter 47 shows how `DSEARCH` and `GSEARCH` together can come up with great glass combinations that correct secondary color.

IOP Publishing

Lens Design (Second Edition)

Automatic and quasi-autonomous computational methods and techniques

Donald C Dilworth

Chapter 35

Wide-band superachromat microscope objective

Challenges correcting a very wide spectral range; polarization; vector diffraction

In this chapter we will undertake a rather difficult lens design task that will utilize some of the many powerful tools that you have learned about in previous chapters (You will need a license to run this example, since it requires more than 12 surfaces and requires saving lens files). As you read each of the paragraphs below, be sure to look up any topics you are not yet familiar with in the help file so you understand what the arguments mean and what other possibilities you have available.

This lens must work over the wavelength range of 0.38–0.9 μm—which is a challenge right off the bat. In addition, we want the lens to work at a speed of F/0.714. That's also not too easy. Here are the requirements:

Object at infinity, 0.8 degree semi-field, 1.26 mm semi-aperture.
Spectral range 0.38–0.9 μm.
F/number 0.714
Total track length not more than 25 mm.
Good distortion correction.
Telecentric at image.
No feathered edges, center thicknesses not over 8 mm.

We guess that this job will require perhaps 10 elements, but want to get there gradually. Set up the input for `DSEARCH`, asking for eight elements, as shown below. That will give you some potential configurations, and you can increase the complexity as needed once you see how things are going. Since the spectral range is so wide, specify five wavelengths instead of the usual three in order to avoid large focus errors at in-between wavelengths. Here is the `MACro` (**`C35M1`**).

doi:10.1088/978-0-7503-3695-6ch35

```
CORE 64
OFF 1
OFF 99
TIME
DSEARCH 3 QUIET

SYSTEM
ID EXAMPLE WIDE-SPECTRUM FAST LENS
UNI MM
OBB 0 0.8 1.26
WA1 0.9 0.77 0.64 0.51 0.38
CORDER 3 1 5
END

GOALS
ELEMENTS 8
FNUM 0.7143 100
BACK 0 0
TOTL 0 0
STOP FREE


COLORS M
RSTART 20
THSTART 1
ASTART 1
RT 0.0
OPD
PASSES 50
QUICK 50 50
ANNEAL 200 20 Q 100
END

SPECIAL PANT
SLIMIT 100 0.1          ! SMALL ELEMENTS; CAN BE CLOSE TOGETHER
END

SPECIAL AANT
AEC .1 1 .05            ! edge monitor
ACM .1 1 .05            ! minimum element TH
ACC 8 1 0.5             ! maximum TH
ACA 70 1 1              ! avoid critical-angle refraction

LUL 25 1 1 A TOTL       ! limit track length
A BACK
M 0.5 1 A BACK          ! want image clearance of 0.5 mm
M 0 1 A P YA 1          ! control distortion
S GIHT
M 0 1 A P HH 1          ! and make telecentric
END

GO
TIME
```

Notice the use of a weighting factor on the FNUM request in word 3. This has subtle consequences: If it were left out, the program would satisfy the request exactly, with a UMC solve on the last radius. But with a lens with a very low F/number like this one, that is likely to yield a very short radius of curvature on that surface and produce ray failures when real rays are traced. So in cases like this, it is better to enter a weighting factor. Then the radius becomes an ordinary variable and the F/number is controlled via an entry in the MF.

Run this file, and in two minutes DSEARCH returns a set of potential starting points. It also creates an optimization MACro, and after running it and then annealing **(50, 2, 50)**, you get the design shown in figure 35.1 (If you turned on the **Free GLM** option in the previous chapter, be sure to turn it off for this one; the example below was prepared with that option off).

Since color correction is going to be a challenge, the next step is to find some glasses that have the potential to make a superachromat. We will do this in two ways: first using the superachromat theory, and then by letting GSEARCH find glass combinations automatically. Save this version so you can go back to it later.

```
STORE 1
```

Chapter 34 explained the theory of superachromats; for now, open the glass map with the command **MGT**, select the Schott catalog, click the Graph button, and select the bottom option, **plot P* vs. P****. We need three glasses that lie on a long line. <Ctrl> click the glass P-SF68, which defines the bottom of the line, and then <Shift> click the glass N-PK52A, defining the top, as shown in figure 35.2.

See the glass N-F2? It's near the center of the line. That gives us three types, but we don't know which glass to assign to which element yet. Never mind; GSEARCH can tell us.

Next, create two files. The first is a normal optimization file. Using the MACro that DSEARCH has created, just edit it a little: request that the optimization program run the automatic ray-failure fixing routine if any of the combinations will not trace initially (**C35M2**) (And well they might; large changes to the index of refraction send rays in a different direction, which can cause failures).

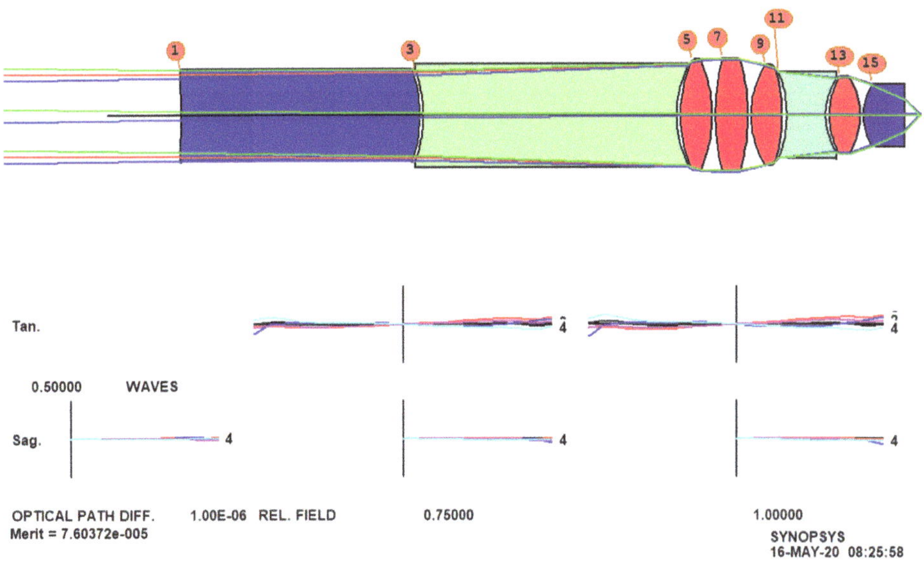

Figure 35.1. Lens returned by DSEARCH, optimized and annealed, with model glasses.

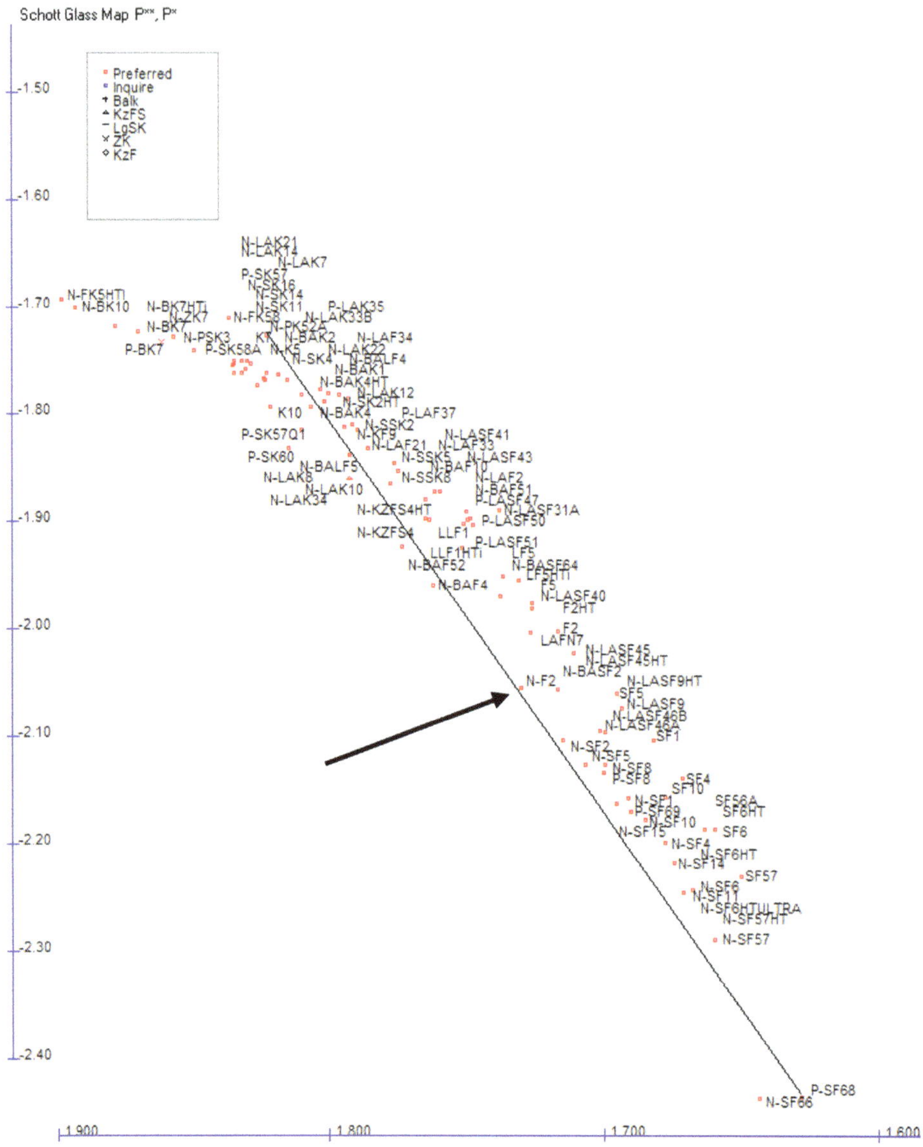

Figure 35.2. Glass map display showing p* versus P**.

```
PANT
SLIMIT 100 0.1   ! SMALL ELEMENTS; CAN BE CLOSE TOGETHER
VY 0 YP1
VLIST RD ALL
VLIST TH ALL
VLIST GLM ALL
END
AANT P
```

```
AEC
ACC
M   0.139997E+01  0.100000E+03 A CONST 1.0 / DIV FNUM
GSO      0.000000      0.039182      4  M      0.000000
GNO      0.000000      0.023509      4  M      0.750000
GNO      0.000000      0.007836      4  M      1.000000
AEC .1 1 .05   ! EDGE MONITOR
ACM .1 1 .05   ! MINIMUM ELEMENT TH
ACC 8 1 0.5    ! MAXIMUM TH
ACA 70 1 1    ! AVOID CRITICAL-ANGLE REFRACTION
LUL 25 1 1 A TOTL  ! LIMIT TRACK LENGTH
A BACK
M 0.5 1 A BACK  ! WANT IMAGE CLEARANCE OF 0.5 MM
M 0 1 A P YA 1  ! CONTROL DISTORTION
S GIHT
M 0 1 A P HH 1  ! AND MAKE TELECENTRIC
END
SNAP/DAMP 1
SYNOPSYS  50 FIX 30
```

Save this file with the name GSOPT.MAC. You don't run this file yourself; it will be run by GSEARCH on each combination it tests. Then create a second MACro (**C35M3**) to tell GSEARCH what you want to do.

```
GSEARCH 3 QUIET LOG
SURF
1 3 5 7 9 11 13 15
END
NAMES
S N-PK52A
S N-F2
S P-SF68
END
USE 3        ! only allow cases that use all three glass types
GO
```

Then run this file.

With 64 cores activated, this runs for about two minutes, producing the design in figure 35.3, after optimization and annealing (**C35L1**).

This is a rather good design, which is somewhat surprising since the super-achromat theory applies only to thin lenses, and these elements are clearly not thin. Let's see what happens if we let GSEARCH find its own glasses. Go back to the version you saved, and then edit your MACro so GSEARCH searches combinations of the three nearest glasses from the Guangming catalog instead of the three that we picked out above (Note the **SKIP** directive, which ignores input up to the **EOS** line; the USE directive will not apply when the **NEAREST** option is used).

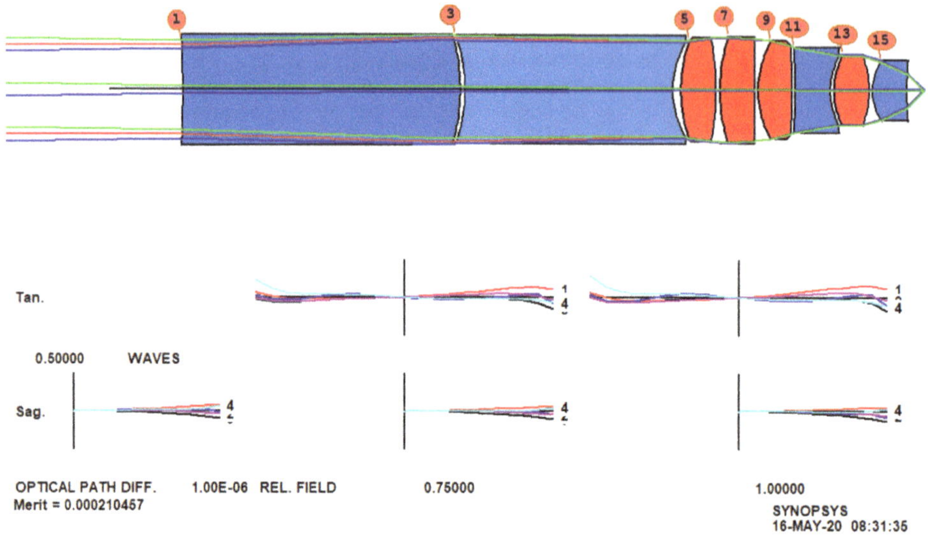

Figure 35.3. Lens returned by GSEARCH when three glass types were specified according to superachromat theory.

```
GSEARCH 3 QUIET LOG
SURF
1 3 5 7 9 11 13 15
END
SKIP
NAMES
S N-PK52A
S N-F2
S P-SF68
END
EOS
NEAREST 3 P
G
END
USE 3        ! only allow cases that use all three glass
types
GO
```

This version produces a lens of very similar quality—and you do not have to find suitable glasses yourself when you do it this way. Optimize and anneal again, and you get the result shown in figure 35.4. GSEARCH does not make any thin-lens assumptions, using numerical methods that do not depend on superachromat theory. This is another example of when autonomous methods can be as good as or better than classic techniques.

This lens is essentially perfect. But we instinctively ask, can we do it with fewer elements? It's easy to find out with the Automatic Element Deletion feature. Go back to the version you saved, with model glasses, and add a new line at the top of the optimization MACro:

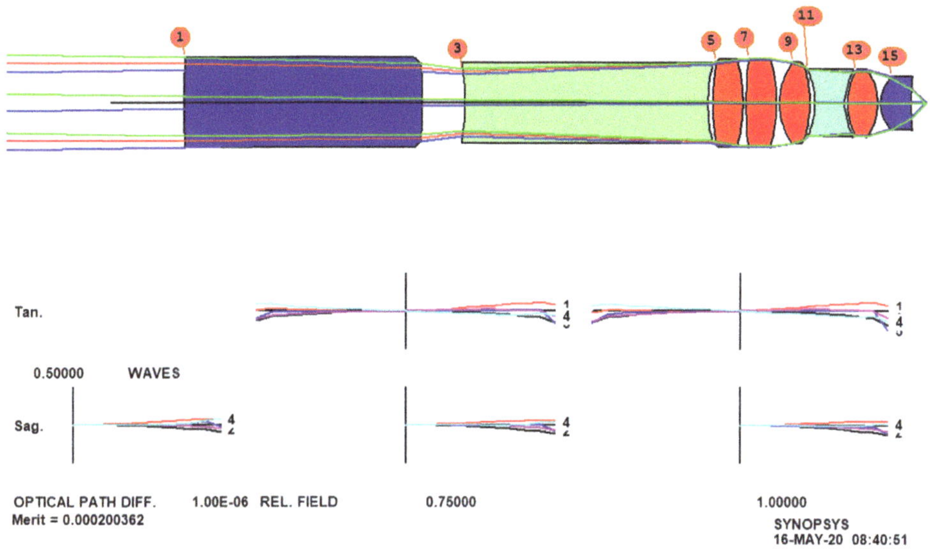

Figure 35.4. Results when GSEARCH matched the three nearest glasses.

```
AED 3 QUIET 1 16.
```

Then run it again. The program detects that you can remove element 4. Accept the suggestion (which deletes that element), remove the AED line from the MACro, and reoptimize and anneal again. Then match this version to the G glass catalog as you did before (delete surface number 15 from the list, since it no longer exists), optimize and anneal, and you get the lens in figure 35.5 (**C35L2**).

Again, nearly perfect—and requiring only seven elements! Let's see what the MTF looks like over the field. Type **MMF**, select Multicolor, and click Execute.

The result is shown in figure 35.6.

This is as close to perfect as most people will ever need.

Are we done? Let's see how stable the back focus position is as a function of wavelength. Enter the AI sentences

```
STEPS = 100
PLOT BACK FOR WAVL = .38 TO .9
```

Since the lens has a YMT solve on the final airspace, this will show us the color-correction curve. The plot is shown in figure 35.7.

Indeed! The paraxial focus position varies by just over one micrometer over this wide range. Yes, this is an excellent lens! Before you actually make the lens, it would be a good idea to move the stop to surface 8—but that is enough for now. Even a difficult challenge like this responds well to these new tools.

We also tried asking DSEARCH for a seven-element lens, rather than finding an 8-element lens and then deleting an element with AED as we did above. It is hard to

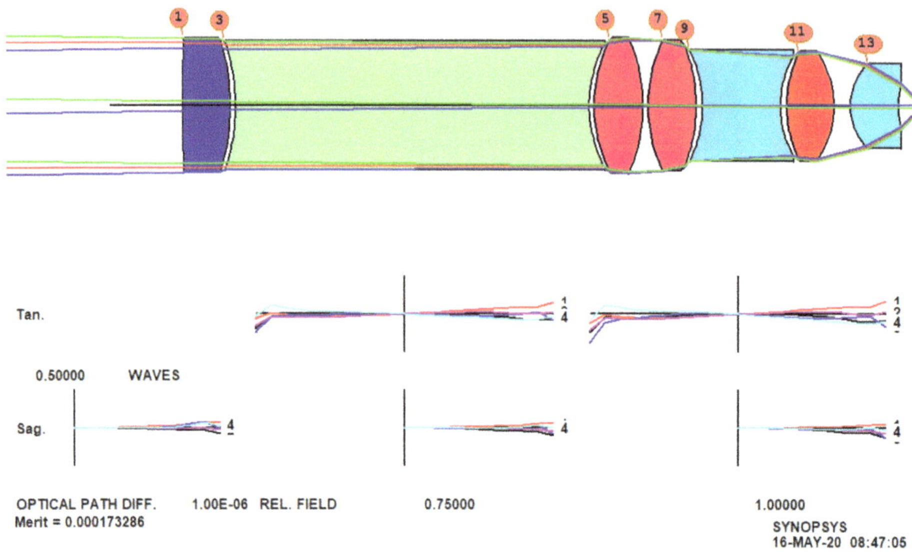

Figure 35.5. Lens reoptimized after an element is removed by AED and the design matched to real glass.

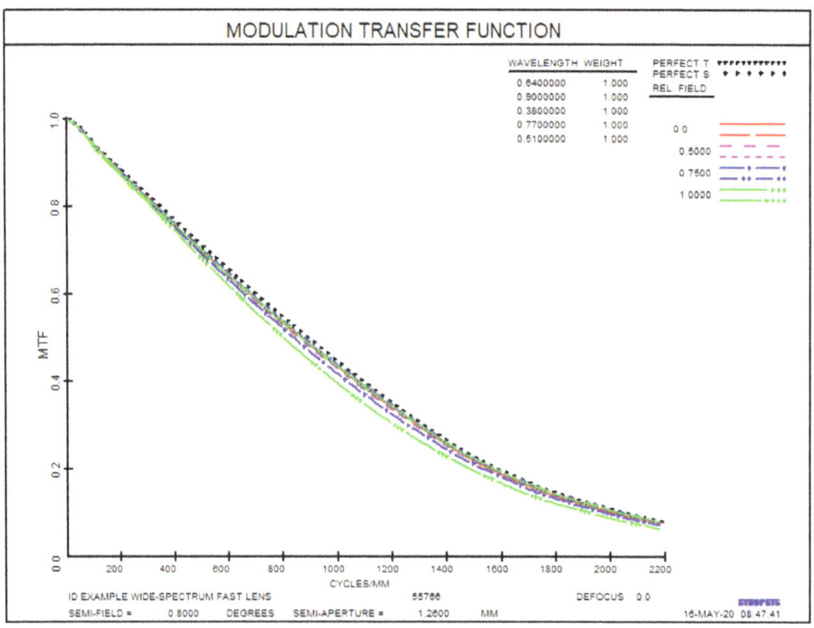

Figure 35.6. MTF of the lens in figure 35.5.

predict which path through the chaotic design tree will turn out best, and all one can do is to try. The results in this case were almost as good as the previous, as shown in figure 35.8 (**C35L3**), when we tried several of the lenses and matched them to the Ohara glass table.

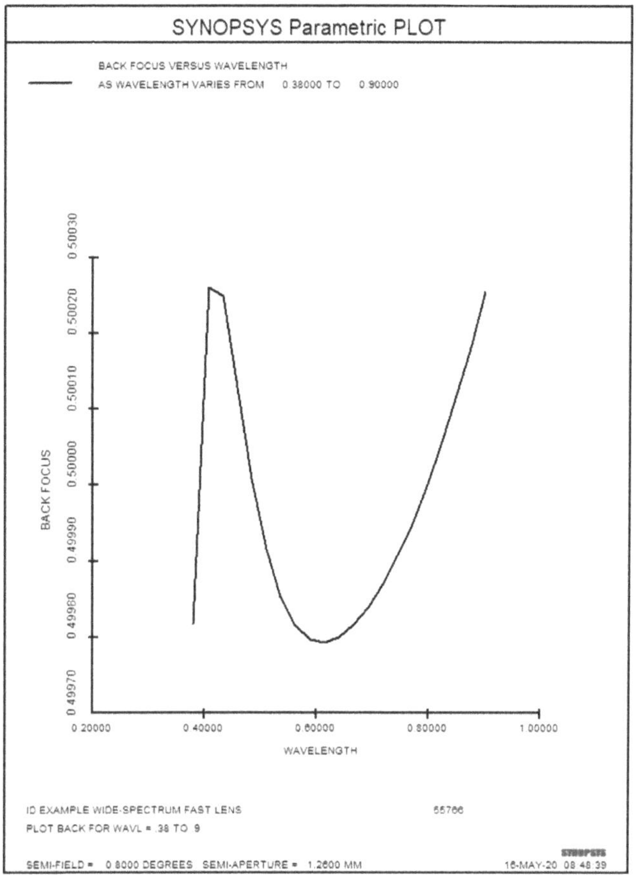

Figure 35.7. Color-correction curve for the lens in figure 35.5.

There are other things we could have tried. If the results were not good enough with seven elements, we could try the Automatic Element Insertion feature, adding the line

```
AEI 3 1 14 CONLY 100 1 10 50
```

to the top of the MACro. That will add a cemented element on each side of all the current lenses in sequence and then come back with the combination that worked best. With these tools, you can go either way. If you also want to try airspaced elements, change **CONLY** to **CEMENT**. Then they will be tested as well (the entry '0', which we have used before, tries only airspaced elements).

This is a very good lens, and we can illustrate that by going to the dialog **MPS** and showing the diffraction pattern plotted over wavelength, visual appearance, full field, magnify 10 times, and Execute. The result is shown in figure 35.9.

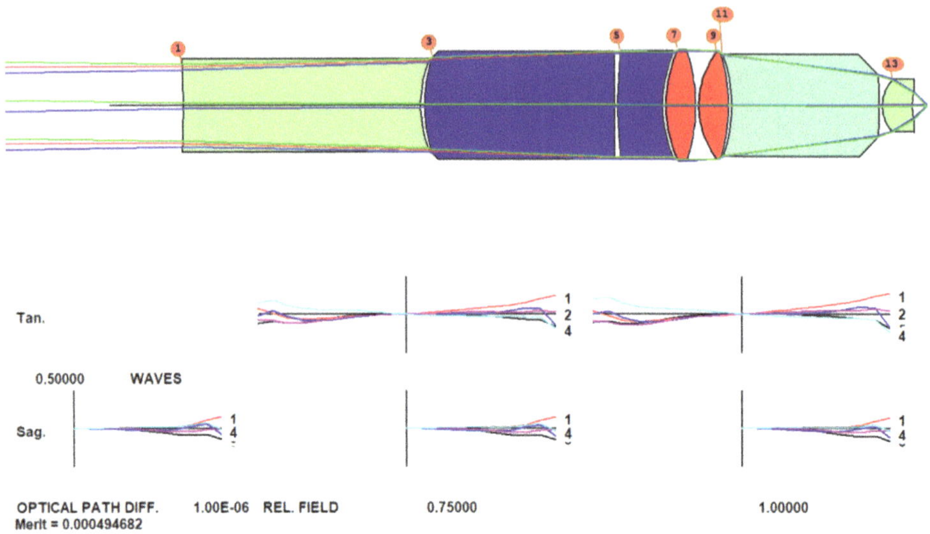

Figure 35.8. Lens returned by `DSEARCH` with seven elements, optimized, annealed, and matched to Ohara glasses.

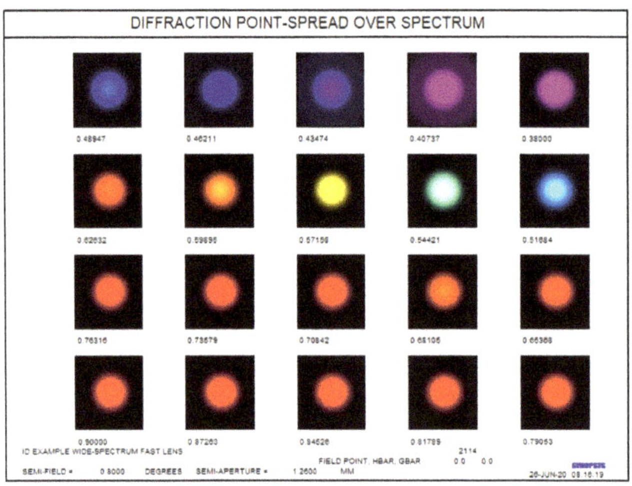

Figure 35.9. Diffraction pattern plotted at 20 wavelengths.

35.1 Vector diffraction, polarization

This is a good place to discuss some of the finer points of image analysis. Look closely at the rays converging to the image in the above designs, shown in figure 35.10. This lens works at a very 'fast' F/number, the term originating in the early days of photography, when it was discovered that a lens with a low F/number did not require the long exposure needed by other lenses and was therefore 'fast'.

Ordinary diffraction-based image analysis employs what is called 'scalar diffraction' theory, which says that if two wavefronts are in phase they add, and if out of phase they cancel. But the electric vectors E1 and E2 in figure 35.10 are almost at right angles to each other, and they cannot add or cancel completely. To accurately analyze the image in such a system one must use *vector diffraction* theory, which involves resolving the electric vector into three (x,y,z) components and performing the diffraction calculation three times, adding the results scalarly. But now we have to take into account the polarization of the light. If this beam is formed with light polarized in the y direction (in the plane of the picture), the y-components of the E vectors add if they are in phase, while the z-components, which are almost in opposite directions, will cancel. Light polarized in the x-direction, on the other hand, follows scalar rules in this case since the vectors on those two rays point in and out of the page and are parallel to each other.

To model this case, first put the lens in polarization mode, with the input

```
CHG
POL LIN Y
END
```

Now one can do a vector-diffraction analysis, using the Fourier-transform version of the modulation transfer function (MTF) calculation.

```
DMTF M 0 6000 1 0 P
```

The result, shown in figure 35.11, tells us that the MTF in the y-direction is lower than in X (The MTF rises past the diffraction cutoff frequency because the short

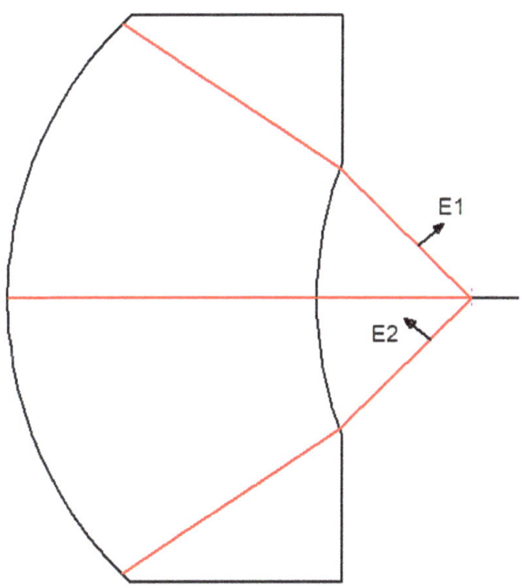

Figure 35.10. Rays converging at the image of a very fast lens.

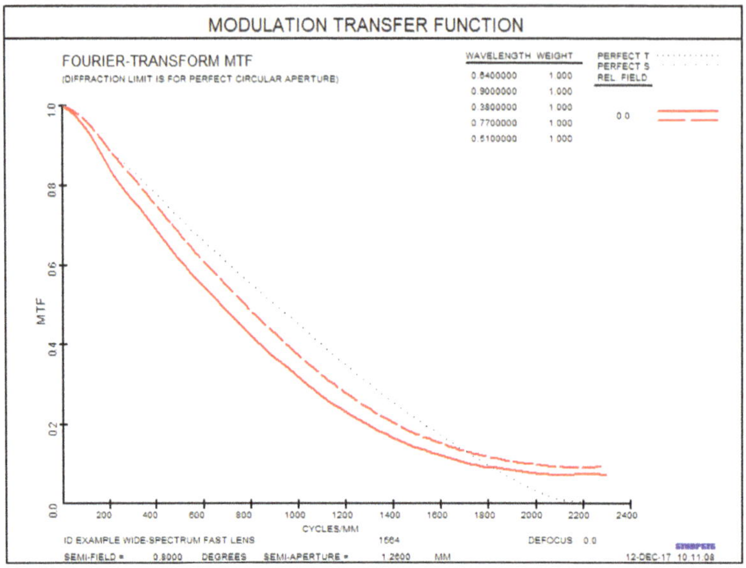

Figure 35.11. Vector-diffraction calculation of the on-axis MTF of the fast lens, in y-polarization.

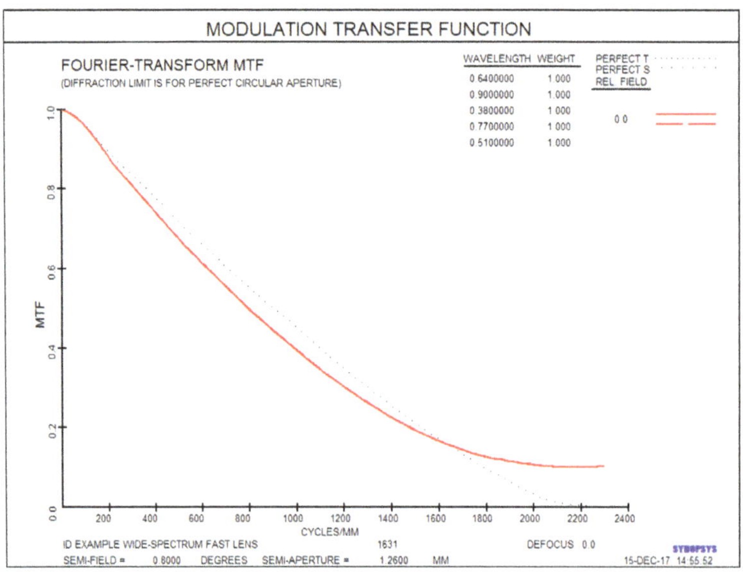

Figure 35.12. Vector-diffraction MTF calculation with unpolarized light.

wavelengths have their own cutoff at a higher frequency). This form of MTF can also be accessed from the menu **MFM**.

If the light is *unpolarized*, the results are a little different, as shown in figure 35.12. The calculation is again based on vector-diffraction theory, but is performed in both x and y polarizations and the results added scalarly.

```
CHG
POL UNPOLAR
END
DMTF M 0 6000 1 0 P
```

While we're on the subject of polarization, there is another subtle effect worth knowing about. Fetch the file **AMICI.RLE**, shown in figure 35.13, where an SFAN of rays is traced. This one of several kinds of prisms containing a roof surface, where rays are reflected from one side and then immediately from the other.

Special tools are required to analyze this kind of system. First, the system is in NONSEQUENTIAL mode, because the order in which the surfaces are encountered differs depending on which side of the roof is hit first. That issue can be ignored in optimization, since the MF is the same either way if the roof angle is perfect, but not when the final image is analyzed. Then one must account for the effect of the roof on the polarization of the light. If you send in light linearly polarized in the *y* direction and make a map of the polarization coming out, you get the picture in figure 35.14. Here again, the polarization vectors on each side of the pupil are at a steep angle with respect to those on the other side, and again the MTF suffers, as shown in figure 35.15. Someone unaware of this effect might wonder why the design did not yield the MTF that was expected (and might blame the prism maker!)

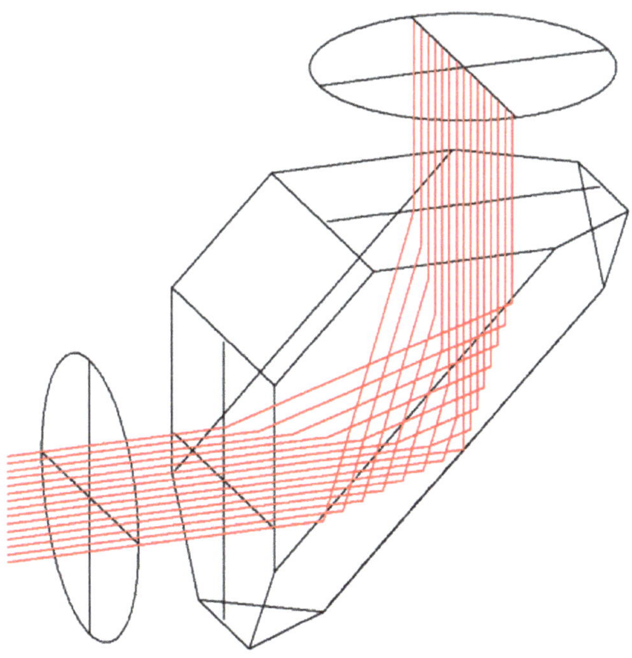

Figure 35.13. An Amici prism with an SFAN of rays.

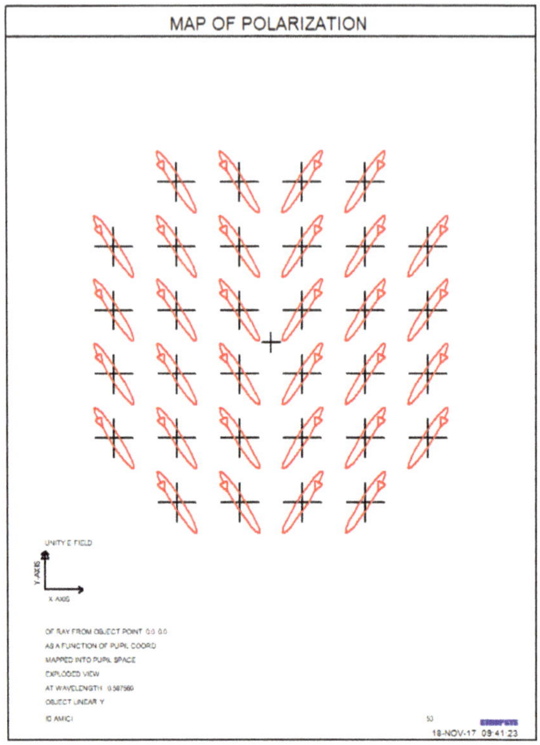

Figure 35.14. MAP of polarization emerging from the Amici prism when the roof surfaces are uncoated.

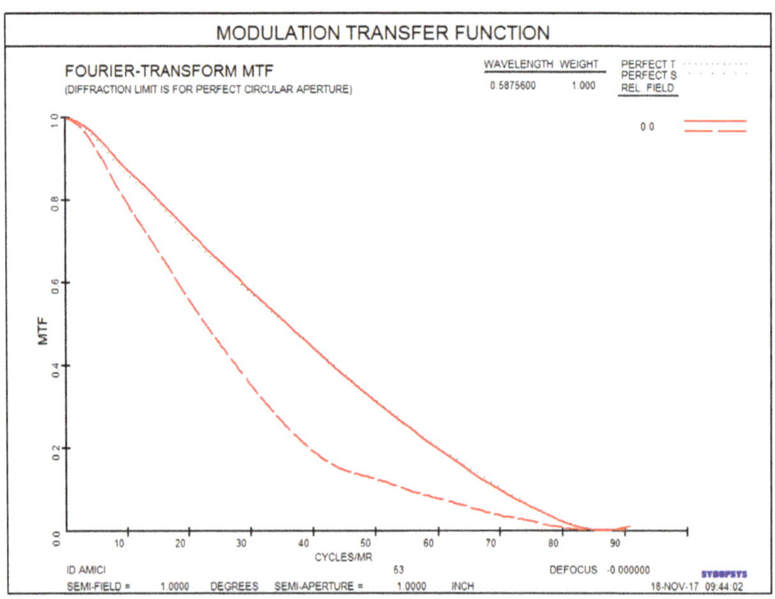

Figure 35.15. Effect of polarization rotation on the MTF of the Amici prism when the surfaces are uncoated.

The command **PCOAT** shows that the roof surfaces are uncoated at the moment.

```
SYNOPSYS AI>PCOAT

SURF. NO.   COATING
-------------------
   1 Dummy surface
   2 UNCOATED
   3 UNCOATED REFLECTOR
   4 UNCOATED REFLECTOR
   5 UNCOATED
   6 Dummy surface
```

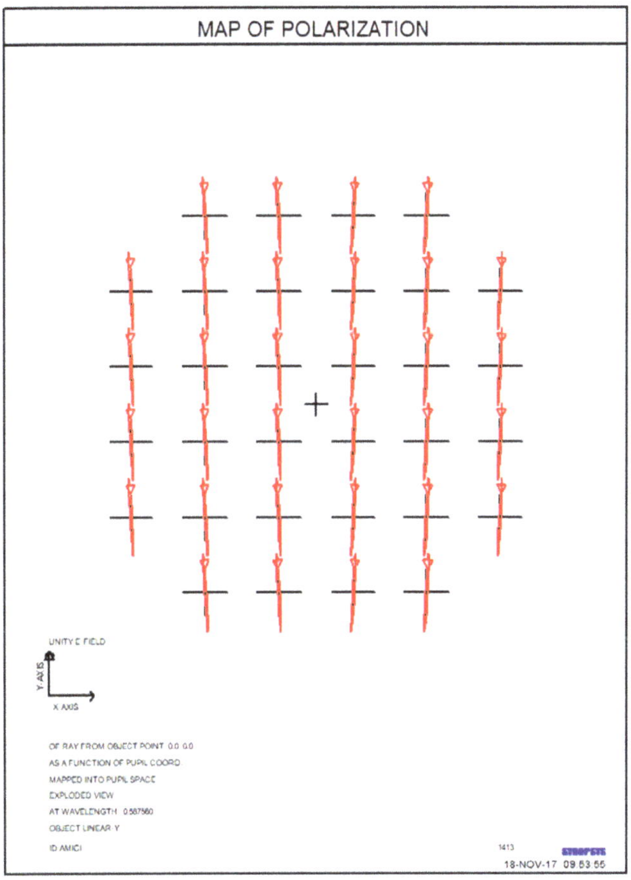

Figure 35.16. Polarization from the Amici prism when the roof surfaces are coated with aluminum. This yields a nearly perfect MTF curve.

But if you assign a reflective coating to the roof, the polarization is restored almost perfectly, as seen in figure 35.16, and the MTF is nearly perfect as well. This input assigns an aluminum coating to the roof surfaces:

```
SYNOPSYS AI>CHG
RLE>3 COAT AL
RLE>4 COAT AL
RLE>END
```

IOP Publishing

Lens Design (Second Edition)

Automatic and quasi-autonomous computational methods and techniques

Donald C Dilworth

Chapter 36

Ghost hunting

Ghost images; analyzing; correcting in the merit function

Your lens looks great on paper and the shop did a fine job of building it. But when you test it, you see a horrible ghost image whenever a bright source enters the field. Not a good scenario, but one that happens all too often. The customer will not be pleased.

To prevent this sort of surprise, use the tools in the **MGH** dialog (Menu, GHost image), with which you can find problems such as this early in the design process and correct them as you go.

Put briefly, a ghost image is a concentration of light at the image arising from two unwanted reflections within the lens system. If you have 3 lenses, there are 15 possible ghosts. With 6 elements you have 66, and so on. That's a lot to keep track of. To see what some of these tools can do, **FETCH** the lens `1.RLE`. Then look at the `PAD` display, shown in figure 36.1.

It's not obvious that you'll be bothered with ghost images, but then it seldom is. Let's see about that. Open the dialog **MGH**, shown in figure 36.2.

At the upper left is the **GHOST** button. That feature is used to find ghosts using only paraxial raytracing, and the ghosts it finds of course differ somewhat from those formed by real rays. Nonetheless, the results are usually close enough for you to see where problems will show up. You can assign reflectance coefficients to any or all surfaces in the lens, and the program will take the values into account when it estimates the intensity of each of the ghosts it finds. The dialog opens with a default reflectance of 1%, applied to all lens surfaces. This is about what you can expect with some antireflection coatings.

Click the **GHOST** button. You get two tables of numbers. The first analyzes all combinations of surfaces; a portion of the output is shown below.

doi:10.1088/978-0-7503-3695-6ch36

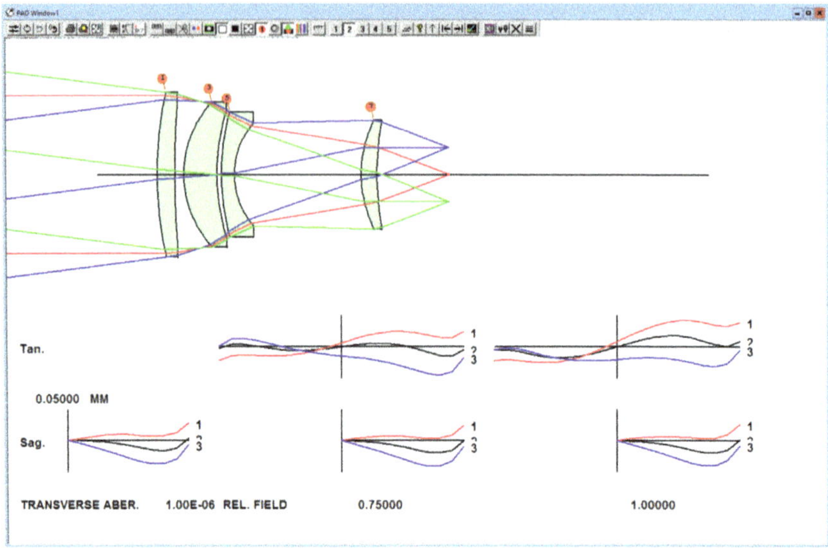

Figure 36.1. PAD display of a lens with a bad ghost image.

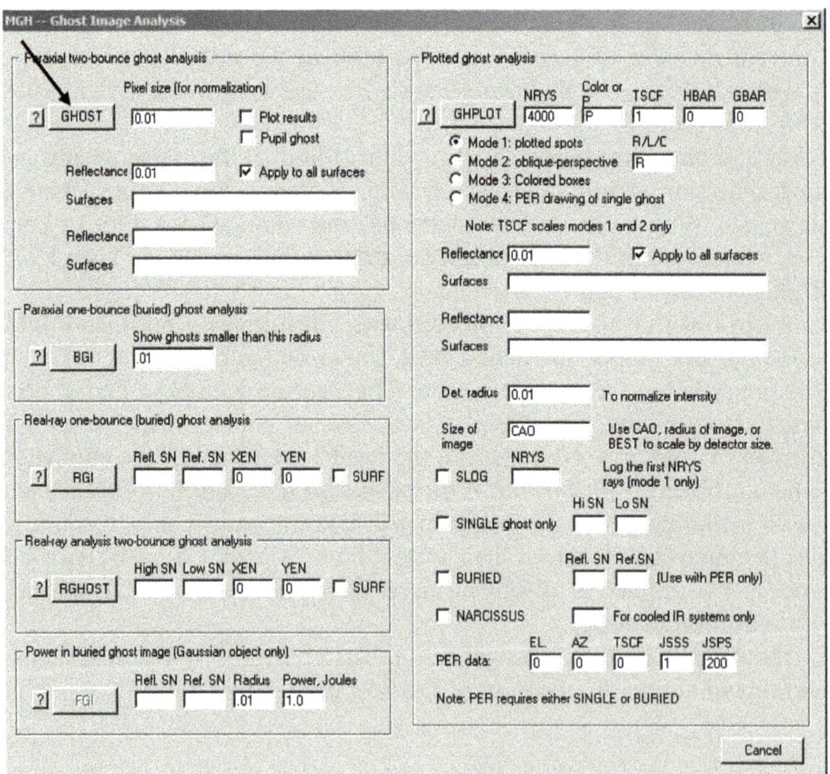

Figure 36.2. MGH dialog for analyzing ghost images.

```
--- GHOST R 0.01
 ID MIT 1 TO 2 UM LENS

 GHOST IMAGE ANALYSIS

 --- R 0.01 ALL
 --- END
```

NO.	GHOST SURF	Ymarg	U'marg	Ychief	INTENSITY
1	2 - 1	-30.0244	-0.3778	7.3095	1.10930E-11
2	3 - 1	27.2543	-0.3984	4.6907	1.34627E-11
3	3 - 2	63.5945	-0.2721	4.1819	2.47264E-12
4	4 - 1	-46.5167	-0.2920	4.9249	4.62149E-12
5	4 - 2	-25.1674	-0.3088	6.6272	1.57879E-11
6	4 - 3	-75.6813	-0.4937	4.4156	1.74591E-12
7	5 - 1	-45.8712	-0.2619	4.0455	4.75247E-12
8	5 - 2	-27.0481	-0.2839	6.4039	1.36686E-11
9	5 - 3	-72.6068	-0.4577	2.7558	1.89691E-12
10	5 - 4	-3.4550	-0.3402	6.5304	8.37735E-10
11	6 - 1	0.5515	-0.5074	4.2824	3.28758E-08
12	6 - 2	44.1259	-0.4028	8.4078	5.13585E-12
13	6 - 3	-38.8865	-0.5677	1.4386	6.61305E-12
14	6 - 4	67.9018	-0.0930	9.4364	2.16889E-12
15	6 - 5	66.9185	-0.0241	8.9874	2.23310E-12
16	7 - 1	-26.2150	-0.5336	-37.0598	1.45513E-11
17	7 - 2	17.4775	-0.4677	31.7392	3.27372E-11
18	7 - 3	-71.3906	-0.6925	-107.3067	1.96208E-12

. . .

Under the heading Ymarg, notice the smallest value, 0.5515, for the combination of surfaces 6 and 1. This tells you that light reflected from surface 6, and then from surface 1, will arrive at the image plane forming a (paraxial) blur of about ½ mm in radius. That may be a problem.

If your lens is lengthy, it is easier to pick out the problem ghosts by examining the second listing:

```
CUMULATIVE GHOST DISTRIBUTION
NORMALIZED FOR DETECTOR SEMI-APERTURE          0.0100
```

NO.	GHOST INTENS.	ACCUM. INTENS.	SURFACES	
6	1.74591E-12	1.74591E-12	4	3
9	1.89691E-12	3.64282E-12	5	3
18	1.96208E-12	5.60491E-12	7	3
14	2.16889E-12	7.77380E-12	6	4
15	2.23310E-12	1.00069E-11	6	5
3	2.47264E-12	1.24795E-11	3	2
20	3.94955E-12	1.64291E-11	7	5
19	4.15611E-12	2.05852E-11	7	4
4	4.62149E-12	2.52067E-11	4	1
7	4.75247E-12	2.99592E-11	5	1
12	5.13585E-12	3.50950E-11	6	2
13	6.61305E-12	4.17081E-11	6	3
21	6.62606E-12	4.83341E-11	7	6

```
 1   1.10930E-11   5.94271E-11   2   1
 2   1.34627E-11   7.28898E-11   3   1
 8   1.36686E-11   8.65584E-11   5   2
16   1.45513E-11   1.01110E-10   7   1
 5   1.57879E-11   1.16898E-10   4   2
25   2.20749E-11   1.38973E-10   8   4
26   2.25477E-11   1.61520E-10   8   5
17   3.27372E-11   1.94257E-10   7   2
24   5.08281E-11   2.45085E-10   8   3
23   5.72162E-11   3.02302E-10   8   2
28   8.28536E-11   3.85155E-10   8   7
10   8.37735E-10   1.22289E-09   5   4
27   3.67946E-09   4.90235E-09   8   6
22   1.09304E-08   1.58327E-08   8   1
11   3.28758E-08   4.87085E-08   6   1
```

Here the ghosts are sorted, with the most severe ones at the bottom, and the accumulated intensity is calculated, printed, and summed. Indeed, the cumulative ghost intensity, 4.87E-08 is mostly due to that single ghost, which has the intensity 3.29E-8. Now you know where the ghost is coming from. Let's see the effect.

To do this, we will use a MACro that exercises several ghost analysis features. Here is that MACro (**C36M1**):

```
; GHPLOT.MAC
; THIS EXAMPLE EXAMINES THE GHOST IMAGE IN A LENS
; IT RUNS GHPLOT IN ALL FOUR MODES.

CCW            ! CLEAN UP FIRST; CLEAR COMMAND WINDOW
KAG            ! AND CLOSE GRAPHICS WINDOWS
FET 1
CHG
CFIX           ; FIX CLEAR APERTURES TO DELETE VIGNETTED GHOSTS
VIG            ; AND TURN ON VIG MODE IF OFF
END

OFF 27         ! SPOTS SHOWN AS SYMBOL
SSS .01        ; SMALL SPOT SIZE HERE

GAW            ; NEED NEW WINDOW FOR EACH PICTURE (GRAPHICS ADD
               ! WINDOWS)
GHPLOT 4000 P 10 .5 0 1      ; SELECT MODE 1, INDIVIDUAL RAYS
R .01 ALL            ; THIS IS ALSO THE DEFAULT REFLECTANCE
PLOT

GHPLOT 20000 P 1 .5 0 2 L ! NOW GET AN OBLIQUE !PERSPECTIVE VIEW
DRAD .0004
PLOT

GHPLOT 20000 P 1 .5 0 3 L    ; THIS MAKES COLORED BINS
DRAD .0004
PLOT

GHPLOT 400 P 1 .5 0 4 L             ! AND THIS DRAWS A SINGLE GHOST
                                    ! WITH PERSPECTIVE
SINGLE 6 1
PER 0 0 0 1 99
PLOT

GRW    ; RESTORE GRAPHICS OPTION (GRAPHICS REUSE WINDOW)
```

GHPLOT has four modes, and you are encouraged to read about them before you go any further. Load this MACro in the editor, *select* the characters GHPLOT, and then look at the TrayPrompt, shown in figure 36.3.

Since this is a multiline command, the prompt cannot show the entire format, but if you press the **F2** key when the prompt is displayed, the help file opens to that section in the index.

You will use all four modes in this exercise. The first call to GHPLOT uses mode 1, producing a picture of all the ghost images, superimposed for an object point at HBAR = 0.5, at the image plane, as shown in figure 36.4. There is indeed a dark blob

Figure 36.3. The TrayPrompt display when the characters GHPLOT are selected.

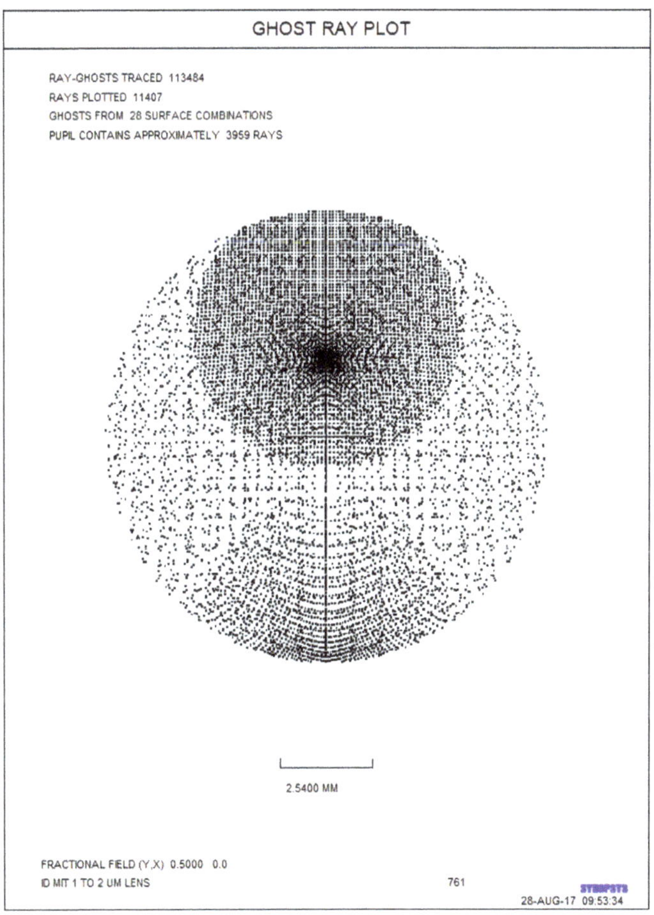

Figure 36.4. Mode 1 GHPLOT plot, showing all ghosts superimposed.

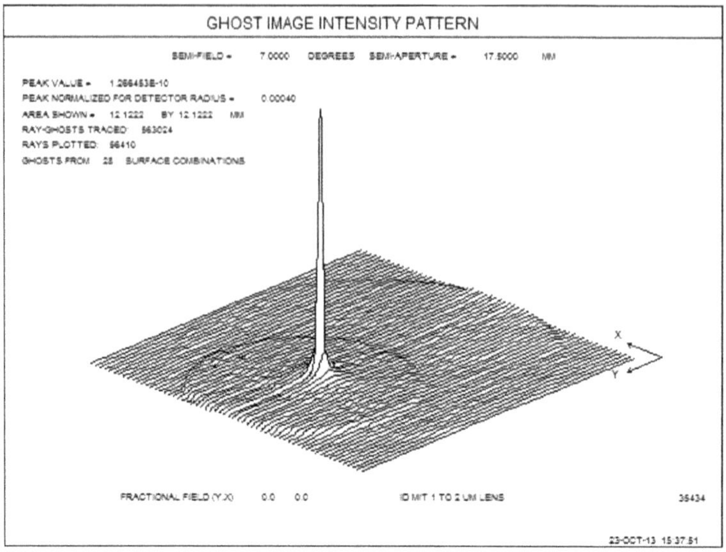

Figure 36.5. Superimposed ghosts plotted in oblique perspective with Mode 2.

about halfway up in the field. That is certainly the ghost we flagged earlier. The mode-2 analysis shows the same energy distribution as an oblique-perspective plot, as seen in figure 36.5.

That sharp peak is indeed the ghost. Yet another way to view it is the mode-3 plot, shown in figure 36.6.

And lastly, the mode-4 plot singles out that particular set of reflections (which is what we asked for) and draws a tangential fan of rays. This is shown in figure 36.7.

Here, light comes in from the left drawn in red, turns blue after reflecting from surface 6, and then turns green after the second reflection, at surface 1. It is indeed almost in focus at the image, but there is a lot of spherical aberration and coma so the ghost is not very sharp.

Looking again at the **MGH** dialog, you see four more features we have not used yet. Let's trace the path of a real ghost ray at zone 0.5. Fill in the boxes as shown in figure 36.8, and click the **RGHOST** button.

This produces the output shown below.

```
--- RGHOST 6 1 0 .5 SURF
                     RAY VECTORS         (X DIR TAN)  (Y DIR TAN) (INC. ANG.)
 SURF          X             Y             Z            ZZ           HH          UNI
------------------------------------------------------------------------------------------
  OBJ      0.000000     0.000000     0.000000     0.000000    8.750000E-12
   1       0.000000     8.750000     0.442564     0.000000      -0.056702     5.790967
```

```
   2      0.000000    8.540210    0.142402    0.000000    -0.086503    1.334779
   3      0.000000    8.256701    1.510615    0.000000    -0.246653   15.791928
   4      0.000000    6.624679    0.478579    0.000000    -0.384916    5.591649
   5      0.000000    6.274262    0.388953    0.000000    -0.226690   13.957837
 --- RAY REVERSES AFTER NEXT SURFACE ---
   6      0.000000    5.481374   -0.886627    0.000000     0.444814
   5      0.000000    6.994957   -0.483893    0.000000     1.206445
   4      0.000000    7.951450   -0.691073    0.000000     0.498903
   3      0.000000   10.791512   -2.647176    0.000000     0.470102
   2      0.000000   12.783436   -0.319172    0.000000     0.221387
   1      0.000000   13.505397    1.05809     0.000000    -0.095157
   2      0.000000   13.193103    0.339970    0.000000    -0.151262
   3      0.000000   12.414364    3.579028    0.000000    -0.404621
   4      0.000000   10.296378    1.164810    0.000000    -0.674892
   5      0.000000    9.767857    0.947929    0.000000    -0.369535
   6      0.000000    8.241826    2.077523    0.000000    -0.148082
   7      0.000000    4.247415    0.334421    0.000000    -0.153884
   8      0.000000    3.743737    0.107527    0.000000    -0.282387
GHOST REFLECTED FROM SURFACES     1    6 AT SURFACE    9
    X              Y               ZZ              HH
----------------------------------------------------------------
   0.00000      -0.828738       0.00000       -0.282387
```

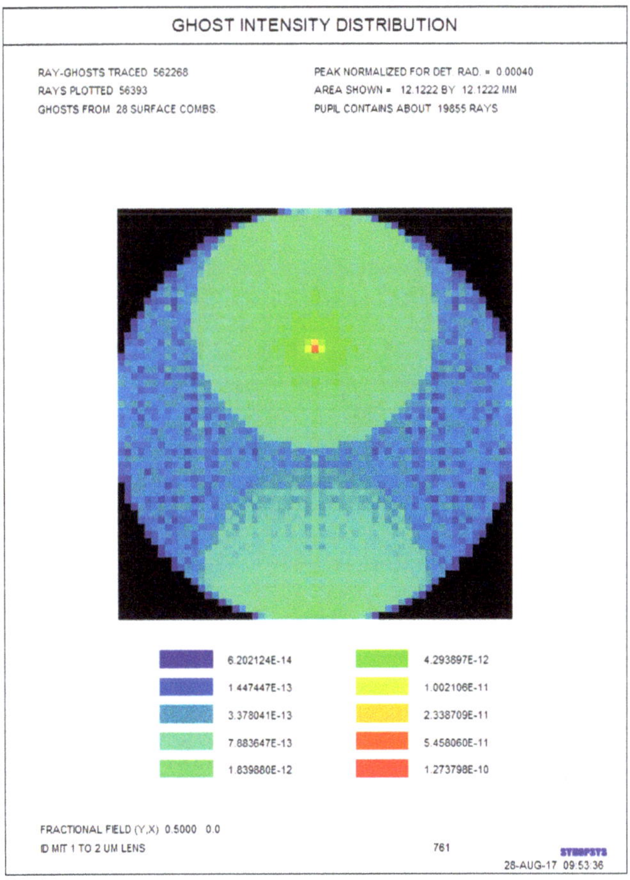

Figure 36.6. Superimposed ghosts plotted with color scale, with mode 3.

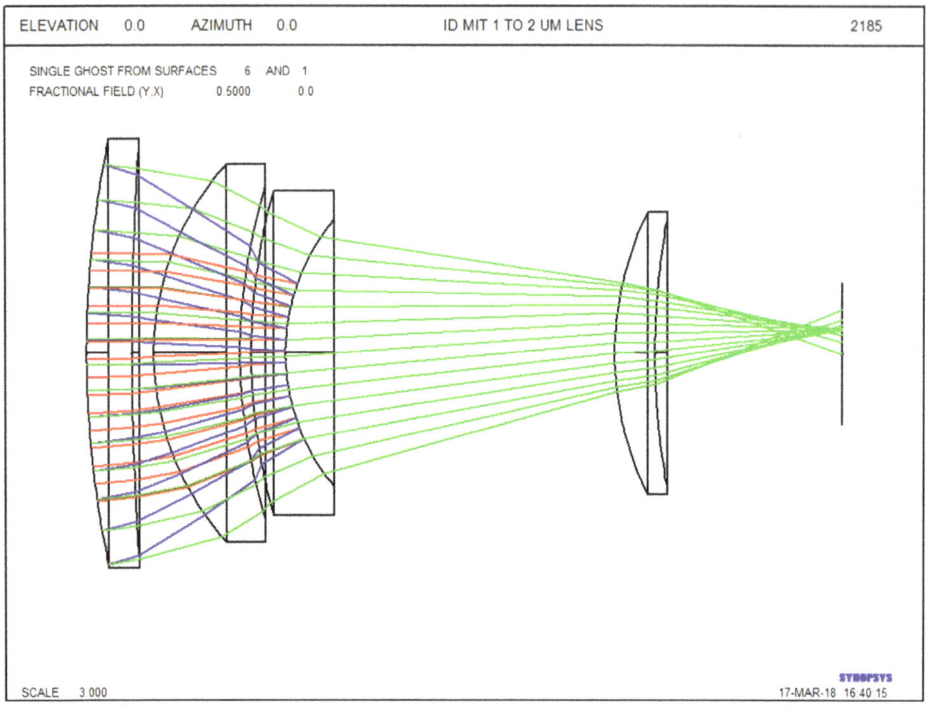

Figure 36.7. Path of a single ghost plotted by GHPLOT in mode 4.

Figure 36.8. Data for the RGHOST command.

The ray reflects from surface 6, then again at 1, and proceeds to the image, where its y-coordinate is –0.829 mm. That may indeed be a serious ghost.

If you catch this problem early in the design process, it is usually easy to control. Type **HELP GHOST**, and select the link that describes controlling the ghost, as shown in figure 36.9.

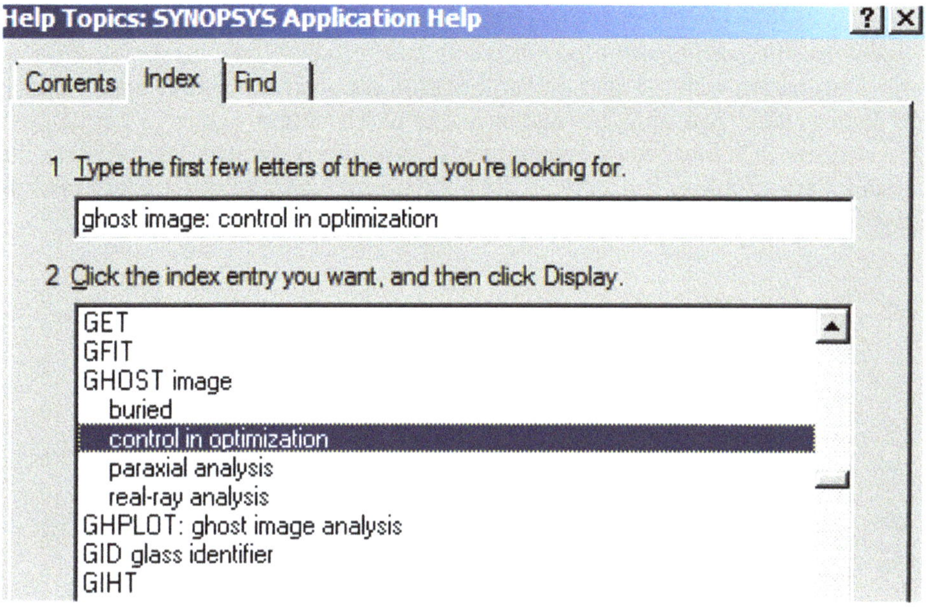

Figure 36.9. Selecting a ghost option in the help file.

This opens a page that describes how to control ghosts.

10.3.1.5 Ghost-image control Next Previous TOC

A ghost image is caused by a reflection from one or more refracting surfaces. SYNOPSYS can evaluate and control two types: The GHOST program can show which combinations of surfaces are responsible for ghost images at the image surface, and BGI can evaluate the properties of a ghost image that is formed at another place within a lens system.

To control the size of the blur at the image from a selected paraxial ghost, the input is

M TAR WT A PGHOST JREFH JREFL

...

Here you see the simple input required to control the ghost. A suitable request in your AANT file might be

```
M 5 0.1 A PGHOST 6 1
```

with the weighting adjusted to balance nicely with your other aberrations. If you achieve this goal, the ghost will be about 1% as intense as before. The target of five is somewhat arbitrary; a larger ghost is a weaker ghost, and this is a good guess to start with.

This procedure usually yields a great improvement in the specified ghost. Often, however, another combination of reflections then produces its own ghost, requiring another evaluation with `GHOST` and another `PGHOST` aberration in the merit function. Add them as they show up until you reach a point where many ghosts are roughly the same intensity. We have never encountered a situation where this intensity was high enough to be a problem. If it were, it would be time to invest in better coatings on the problem surfaces.

IOP Publishing

Lens Design (Second Edition)

Automatic and quasi-autonomous computational methods and techniques

Donald C Dilworth

Chapter 37

Importing a Zemax[1] file into SYNOPSYS

Converting a file from one program to another requires user input

SYNOPSYS can open most lens files created by the programs Zemax and Code-V[2]. As is true with most conversions from one protocol to another, however, the results are often incomplete, and the user must then edit the lens file to restructure certain parameters according to the rules of the target program.

But some things cannot be converted. The two programs utilize very different descriptions of the entrance pupil—although both in the end achieve much the same result. And not all surface shapes that can be defined in Zemax can be defined in SYNOPSYS (and vice-versa). Nonetheless, all of the most popular shapes work well in either program, so most users will not encounter difficulties for that reason. And while a Zemax file contains a whole lot more information than does a SYNOPSYS file —such as definitions of variables, the merit function, and tolerances—the conversion will capture only the basic lens data, since the RLE file in SYNOPSYS is just a lens description and the other data are stored as separate files. Anyone converting from one program to the other will naturally want to exploit the advantages of SYNOPSYS and will create their own data files, so makes no sense to attempt to import those other items.

A more common question is properly identifying the names of commercial glass types. The two programs have extensive glass tables, but the names often differ. So the most common user task, after importing a .zmx file, is to edit the RLE file and insert the correct glass names. An example will illustrate some of these issues.

(You are encouraged to read section 5.42 of the User's Manual before importing a file, where you will find even more information.)

[1] Zemax™ is a trademark of Focus Software.

[2] Code-V is a trademark of synopsys, Inc. (not affiliated with Optical Systems Design LLC)

doi:10.1088/978-0-7503-3695-6ch37

To illustrate this feature, we will convert a file describing a diffractive optical element that is stored in the `Dbook-ii` directory with the name `doe.zmx`. This file contains the following lines:

```
VERS 91012 185 25430
MODE SEQ
NAME Achromatic singlet
NOTE 0 Notes...
NOTE 4
NOTE 0
NOTE 4
NOTE 0
UNIT MM X W X CM MR CPMM
ENPD 5.0E+1
ENVD 2.0E+1 1 0
GFAC 0 0
GCAT SCHOTT
RAIM 0 0 1 1 0 0 0 0 0
PUSH 0 0 0 0 0 0
SDMA 0 1 0
FTYP 1 0 3 3 0 0 0
ROPD 2
PICB 1
XFLD 0 0 0
XFLN 0 0 0 0 0 0 0 0 0 0 0 0
YFLD 0 3.5 5.0
YFLN 0 3.5 5.0 0 0 0 0 0 0 0 0 0
FWGT 1 1 1
FWGN 1 1 1 1 1 1 1 1 1 1 1 1
ZVDX 0 0 0
VDXN 0 0 0 0 0 0 0 0 0 0 0 0
ZVDY 0 0 0
VDYN 0 0 0 0 0 0 0 0 0 0 0 0
ZVCX 0 0 0
VCXN 0 0 0 0 0 0 0 0 0 0 0 0
ZVCY 0 0 0
VCYN 0 0 0 0 0 0 0 0 0 0 0 0
ZVAN 0 0 0
VANN 0 0 0 0 0 0 0 0 0 0 0 0
WAVL 4.861E-1 5.876E-1 6.563E-1
WAVN 4.861E-1 5.876E-1 6.563E-1 5.5E-1 5.5E-1 5.5E-1 5.5E-1 5.5E-1 5.5E-1 5.5E-1
5.5E-1 5.5E-1
WWGT 1 1 1
WWGN 1 1 1 1 1 1 1 1 1 1 1 1
WAVM 1 4.861E-1 1
WAVM 2 5.876E-1 1
WAVM 3 6.563E-1 1
WAVM 4 5.5E-1 1
WAVM 5 5.5E-1 1
WAVM 6 5.5E-1 1
WAVM 7 5.5E-1 1
WAVM 8 5.5E-1 1
WAVM 9 5.5E-1 1
WAVM 10 5.5E-1 1
WAVM 11 5.5E-1 1
WAVM 12 5.5E-1 1
WAVM 13 5.5E-1 1
WAVM 14 5.5E-1 1
WAVM 15 5.5E-1 1
WAVM 16 5.5E-1 1
WAVM 17 5.5E-1 1
WAVM 18 5.5E-1 1
WAVM 19 5.5E-1 1
```

```
WAVM 20 5.5E-1 1
WAVM 21 5.5E-1 1
WAVM 22 5.5E-1 1
WAVM 23 5.5E-1 1
WAVM 24 5.5E-1 1
PWAV 2
POLS 1 0 1 0 0 1 0
GLRS 1 0
GSTD 0 100.000 100.000 100.000 100.000 100.000 100.000 0 1 1 0 0 1 1 1 1 1 1
NSCD 100 500 0 1.0E-6 5 1.0E-6 0 0 0 0 0 1 1000000 0
COFN COATING.DAT SCATTER_PROFILE.DAT ABG_DATA.DAT PROFILE.GRD
SURF 0
  TYPE STANDARD
  CURV 0.0 0 0 0 0 ""
  HIDE 0 0 0 0 0 0 0 0 0 0
  MIRR 2 0
  SLAB 1
  DISZ 2.5E+2
  DIAM 5.0 0 0 0 1 ""
  POPS 0 0 0 0 0 0 0 0 1 1 1 1 0 0 0
SURF 1
  STOP
  TYPE STANDARD
  CURV 7.5762934618539999000E-003 0 0 0 0 ""
  HIDE 0 0 0 0 0 0 0 0 0 0
  MIRR 2 0
  SLAB 2
  DISZ 2.5E+1
  GLAS BK7 0 0 1.69673 5.6419998E+1 -7.4E-3 1 1 1 0 0
  DIAM 3.0E+1 1 0 0 1 ""
  POPS 0 0 0 0 0 0 0 0 1 1 1 1 0 0 0
  FLAP 0 3.0E+1 0
SURF 2
  TYPE BINARY_2
  CURV -6.6766952605729997000E-003 0 0 0 0 ""
  HIDE 0 0 0 0 0 0 0 0 0 0
  MIRR 2 0
  SLAB 3
  PARM 0 1
  PARM 1 0
  PARM 2 0
  PARM 3 0
  PARM 4 0
  PARM 5 0
  PARM 6 0
  PARM 7 0
  PARM 8 0
  XDAT 1 3.000000000000E+000 0 0 0.000000000000E+000 0.000000000000E+000 0 ""
  XDAT 2 3.000000000000E+001 0 0 0.000000000000E+000 0.000000000000E+000 0 ""
  XDAT 3 -2.993832387049E+003 0 0 0.000000000000E+000 0.000000000000E+000 0 ""
  XDAT 4 1.135544608547E+003 0 0 0.000000000000E+000 0.000000000000E+000 0 ""
  XDAT 5 -5.932105454300E+001 0 0 0.000000000000E+000 0.000000000000E+000 0 ""
  DISZ 2.5073834507E+2
  DIAM 3.0E+1 1 0 0 1 ""
  POPS 0 0 0 0 0 0 0 0 1 1 1 1 0 0 0
  FLAP 0 3.0E+1 0
SURF 3
  TYPE STANDARD
  CURV 0.0 0 0 0 0 ""
  HIDE 0 0 0 0 0 0 0 0 0 0
  MIRR 2 0
  SLAB 4
  DISZ 0
  DIAM 5.175465768436 0 0 0 1 ""
  POPS 0 0 0 0 0 0 0 0 1 1 1 1 0 0 0
BLNK
TOL TOFF   0   0              0              0   0 0 0
MNUM 1 1
MOFF   0   1 "" 0 0 0 1 1 0 0.0 ""
```

Figure 37.1. Warning message displayed by ZMCONVERT.

Begin by typing the command **ZMC** (ZeMax Convert). A warning message is displayed, shown in figure 37.1.

(This is to prevent naïve users from blindly selecting a file and expecting the lens to open every time exactly as it does in Zemax. Yes, some users are that naïve. But life is not simple.) If you click the No button, you go immediately to the help file describing ZMC.

When you click the Yes button, you are shown a list of the .zmx files in the current directory, and you can select the one mentioned above. (The file to be imported must reside in the current directory.) The above lines scroll past on the Command Window, and you see a warning message:

```
**************************************************
****************** WARNING *****************
***** THE DOE EMULSION INDEX IS NOT GIVEN ****
*** IN THE ZEMAX FILE AND MUST BE ENTERED BY **
*** HAND IN THE RLE FILE AFTER CONVERSION ****
****** IN WORD THREE OF THE DOE ENTRY ********
**************************************************
```

This is an example of the difference in protocols. In a SYNOPSYS RLE file, the exact index values of a material are given along with the glass-catalog name, if any, and then the properties of the DOE are listed. Because of this protocol, anyone reading an RLE file created by SYNOPSYS knows the index of the material, even if, years from now, that glass type has become obsolete and is no longer in the catalog. Zemax lists the glass name but not the index values. So when the DOE input is read (and converted by ZMC) the index data are not yet known. The glass name shows up later, but the conversion is then way past that stage already. Since the DOE specification in SYNOPSYS requires the index of the emulsion when the input is first read in, the program inserts a dummy index of 1.517 just to avoid input errors. It turns out that this DOE is in fact made from BK7, so the index, just by chance, is correct. Otherwise you would have to edit the file and change that number to the correct index for the correct glass. (In SYNOPSYS, the index values are retrieved from the glass tables *after* the RLE file is completely processed, and are not available while ZMC is running.)

At the end of the conversion, the program displays an informative message:

```
NOTE: OBJECT AND PUPIL DEFINITIONS MAY DIFFER. THE PROGRAM PUTS THE
WAP 3 PUPIL IN EFFECT TO BE SAFE. BUT THIS LENS MAY OR MAY NOT REQUIRE
THAT OPTION. YOU SHOULD DELETE IT IF IT IS NOT NECESSARY.
IF ANY GLASS-TABLE GLASSES WERE NOT FOUND, IT MAY BE DUE TO DIFFERENT
SPELLINGS. CHECK THE LISTING ABOVE TO SEE WHAT THE NAME WAS, AND CHANGE
TO THE APPROPRIATE SPELLING IF THAT GLASS IS IN ONE OF THE GLASS TABLES.
```

Here you learn that the program has (by default) implemented the WAP 3 option, which is sometimes a safe bet but is almost never actually required. By all means try to understand the geometry that underlies the pupil definition used in Zemax, and if it does not really require WAP 3, try the simpler WAP 0 instead.

The next job is to look at the RLE file it has created and loaded into a MACro editor.

```
RLE
ID ACHROMATIC SINGLET
ID1 NOTES...
ID2
ID3
 UNITS MM
TEMPERATURE 20.000
PRESSURE 100.000
 GTZ
WT1 1.00000    1.00000    1.00000
WA1 0.486100   0.587600   0.656300
CORDER 2 3 1
 POLAR OFF
  0 CV 0.0
OBC 250.000    5.00000    25.0000    0.00000    0.00000    0.00000    0.00000
APS -1
 WAP 3
  1 RD 131.991
  1 TH 25.0000
  1 GTB S
BK7
  1 CAO 30.0000
```

```
  3 RD  -149.775
  3 TH   250.738
  3 CAO   30.0000
  2 PIN   1
  2 DOE  0.587600     1.51700
RNORM   30.0000
A11   476.483     -180.728      9.44124      0.00000      0.00000
  3 DC1 0.0000000E+00 0.0000000E+00 0.0000000E+00 0.0000000E+00 0.0000000E+00
  4 CV 0.0
  4 TH   0.00000
END
```

To see this lens, change `WAP 3` to `WAP 0`, run the `RLE` file, and then open **`PAD`**. The lens in figure 37.2 appears.

Converting this file was rather simple (as it should be). If you now ask `SYNOPSYS` to create a proper `RLE` file (with the command **`LEO`**), you get this:

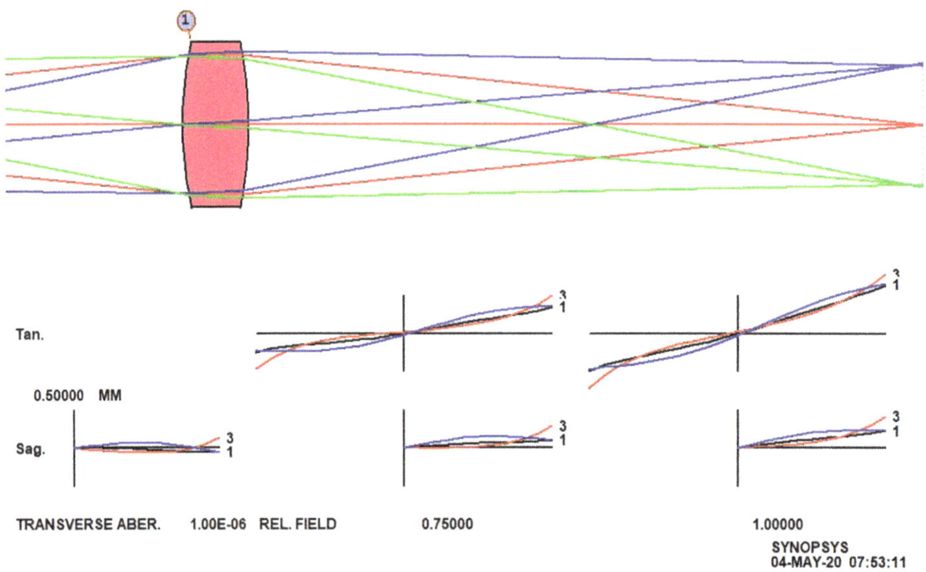

Figure 37.2. Lens converted from a Zemax file to SYNOPSYS format.

```
RLE
ID ACHROMATIC SINGLET                         378
ID1 NOTES...
ID2
ID3
 LOG      378
 WAVL .4861000 .5876000 .6563000
 CORDER   2   3   1
 APS              -1
 GTZ
 UNITS MM
 OBC  250.  5.  25. 0 0 0  25.

   0 AIR
   1 CAO      30.00000000       0.00000000       0.00000000
   1 RAD     131.9906634000000    TH      25.00000000
   1 N1 1.52237223 N2 1.51679274 N3 1.51431609
   1 CTE    0.710000E-05
   1 GTB S     'BK7                    '
   2 N1 1.52237223 N2 1.51679274 N3 1.51431609
   2 CTE    0.710000E-05
   2 GID 'BK7                    '
   2 DOE        0.587600       1.517000      55.000000
   RNORM     30.0000
    A11  4.7648E+02 -1.8073E+02  9.4412E+00  0.0000E+00  0.0000E+00
    A12  0.0000E+00  0.0000E+00  0.0000E+00  0.0000E+00  0.0000E+00  0.0000E+00
    A13  0.0000E+00  0.0000E+00  0.0000E+00  0.0000E+00  0.0000E+00  0.0000E+00
   2 PIN     1
   3 CAO      30.00000000       0.00000000       0.00000000
   3 RAD    -149.7746955999999    TH     250.73834510 AIR
   3 DC1  0.00000000E+00 0.0000000E+00 0.0000000E+00 0.0000000E+00 0.00000000E+00
  3 DC2  0.00000E+00 0.00000E+00 0.00000E+00 0.00000E+00 0.00000E+00 0.00000E+00
  3 DC3  0.00000E+00 0.00000E+00 0.00000E+00 0.00000E+00 0.00000E+00 0.00000E+00
  3 DC4  0.00000E+00 0.00000E+00 0.00000E+00 0.00000E+00 0.00000E+00
   4 CV       0.0000000000000    TH       0.00000000 AIR
 END
```

Notice that the OPD coefficients have been altered. Zemax expresses the coefficients in units of *radians*, while all OPD expressions in SYNOPSYS are in units of *cycles*, or waves. So the values must be different, as you can see.

This example gave us little trouble, so let us show a more difficult one. We open a file describing an IR lens (**IR_EXAMPLE.ZMX**), and see an error message when we run the conversion, shown in figure 37.3.

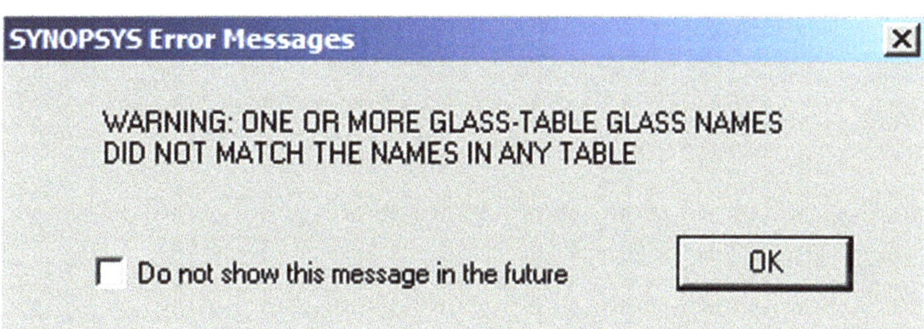

Figure 37.3. Warning message displayed by ZMCONVERT when a glass name does not match.

Reading the listing that scrolled by, you see some details:

```
SURF 12
  COMM OBJ EL1
  TYPE STANDARD
  CURV -4.127115146513000200E-001 0 0.000000000000E+000 0.000000000000E+000 0
  HIDE 0 0 0 0 0 0 0 0
  MIRR 2 0.000000000E+000
  SLAB 4
  DISZ -2.362204724409E-001


 GLAS CLEARTRAN_WANDA 0 0 3.46217496 0.000000 0.000000 0 0 0 0.000000 0.000000
************************************************
***********   GLASS TYPE NOT FOUND *************
CLEARTRAN_WANDA

*****  A GLASS MODEL (GLM) IS USED INSTEAD  ****
***  SOME GLASS TABLES USE DIFFERENT SPELLING **
***  CHECK THE NAME CAREFULLY.  GLM DATA MAY  **
*************  NOT BE APPROPRIATE  *************
************************************************
```

This surface needs a material whose name is not found in the SYNOPSYS glass tables. The program assigns a glass model since it has no other information at this point, but when you edit the resulting RLE file, you have to change to the right material. Here is what is assigned at this point:

```
 12 SID 'OBJ EL1                    '
 12 RD    -2.42300
 12 TH  -0.236220
 12 GLM     1.50000           55.0000
 12 CAO    0.745000
 13 RD     7.82870
```

If you do not know the name of the material to use, you might look at the Unusual glass catalog. Type **HELP UNUSUAL** and follow the link. In the list that opens you find a likely candidate:

```
...
NACL        Sodium chloride                      0.2          22.3
NAFL        Sodium fluoride                      0.186        17.3
PBFL        Lead fluoride                        0.2909       11.9
SAPPHIRE    Aluminum oxide                       0.193        5.263
SILICON     Silicon; see SILICON-NIR, below      1.4          16.0
ZNS         Zinc sulfide                         0.42         18.2
CLEARTRAN   Zinc sulfide, higher grade           0.4047       13.0
ZNSE        Zinc selenide                        0.54         18.2
CRQUARTZ    Crystal quartz, ordinary ray  0.198        2.053
```

Now you can edit the RLE file:

```
12 SID 'OBJ EL1                    '
12 RD    -2.42300
12 TH   -0.236220
12 GTB U
CLEARTRAN
12 CAO    0.745000
13 RD     7.82870
13 TH   -0.100000E-01
13 CAO    0.745000
```

The same error showed up on several other surfaces, and you can correct them as well, perhaps with a PIN 12 directive. Yet other surfaces wanted a material with the name SILICON_FIT. Of course you change this to SILICON. Proceeding in this manner, you identify all materials whose names you must update, and then run the MACro with the corrected RLE file.

Be especially wary of glasses in Zemax from the Chinese company Guangming (CDGM). That company uses many of the same glass names as the Schott company, even though the index and dispersion are very different. Clearly, this is a disaster waiting to happen, and you must verify carefully which glass from which catalog you want to use. (SYNOPSYS identifies all of the Chinese glasses that have conflicting names with a prefix. Thus F2 becomes G-F2 and so on.)

Chapter 38

Improving a Petzval lens

Designing a diffraction-limited spy camera

Joseph Max Petzval (1807–1891) was a German mathematician whose name is attached to a type of camera lens consisting of two separated positive groups. (Legend says he used artillery gunners to perform *six months* of ray tracing for the first examples.) This lens form was widely used during the Cold War for spy cameras that flew over the Soviet Union. A typical example is shown in figure 38.1.

This example has two lenses added near the focal plane to correct field aberrations. Can `DSEARCH` design this kind of lens? The original required eight elements, but since we like a challenge, we will attempt it with only seven.

Here are the requirements,

13-inch focal length
F/3.5
Field +/− 6 degrees
Spectral range 0.7–0.52 μm
Total length 17.06 inches
Back focus 0.7 inches

and here is a suitable input MACro (**C38M1**):

```
CORE 64
OFF 1
OFF 99
TIME
DSEARCH 3 QUIET
 SYSTEM
 ID DSEARCH SAMPLE
 OBB 0 6 1.857
 WAVL 0.7 .6 .52
```

doi:10.1088/978-0-7503-3695-6ch38

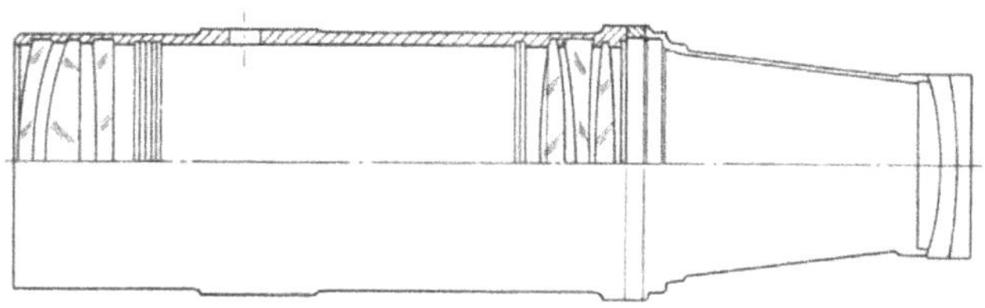

Figure 38.1. A Petzval spy camera design. Reproduced with permission from Berge Tatian.

```
 UNITS INCH
 END
 GOALS
 ELEMENTS 7
 FNUM 3.5 10
 BACK 0.7 SET
 TOTL 17 .1
 STOP FIRST
 STOP FREE
TSTART .5
ASTART 1.0
 RT 0.5
 FOV 0.0 0.75 1.0 0.0 0.0
 FWT 5.0 3.0 3.0 1.0 1.0
 NPASS 55
 ANNEAL 200 20 Q 100
 OPD
 SNAPSHOT 10
 QUICK 40 40
 END
SPECIAL PANT
END
SPECIAL AANT
ADT 7 .01 10
M 0 10 A GIHT
S P YA 1
END
GO
TIME
```

Run this file, and in about two minutes the program returns a lens corrected to about 0.2 waves, which, after optimization and annealing **(50, 2, 50)** is even better, as shown in figure 38.2.

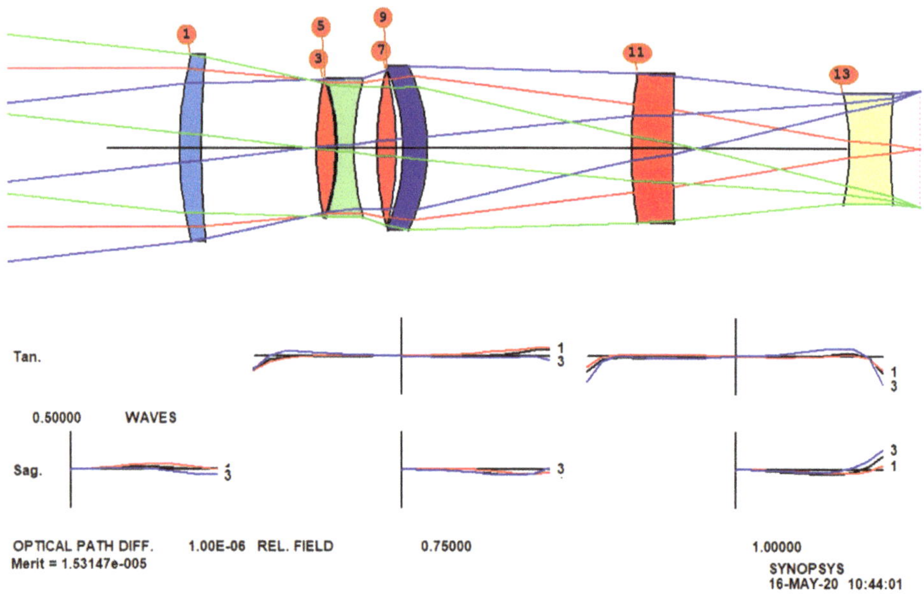

Figure 38.2. Lens returned by DSEARCH, optimized and annealed.

This lens is already at the diffraction limit, and it does not resemble the Petzval configuration at all. It is always pleasant to discover a lens form the old masters missed.

The lens is superb, with almost no secondary color. Note the positive flint element at surface 1. The program often finds this a good way to reduce secondary color. Next, we will insert real glasses with ARGLASS—but first, save this version so you can go back if necessary.

Open the **MRG** dialog, select Schott and Sort and … the lens comes back with excessive secondary color. That program assigns glasses by finding the closest real glass to the model and optimizing the result. It often works very well, but in this case we need more horsepower.

Let us see what GSEARCH can do. Get back the saved version, save the optimization MACro with the name GSOPT.MAC, and prepare another MACro:

```
GSEARCH 5 QUIET LOG
SURF
1 3 5 7 9 11 13
END
NEAREST 4 P
G
END
CLIMIT 20000
GO
```

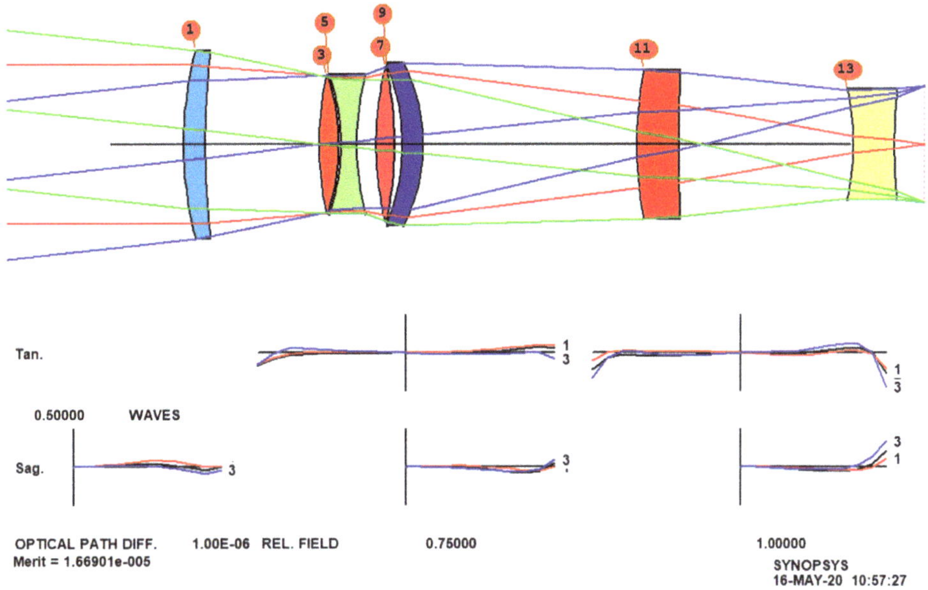

Figure 38.3. Lens with real glasses found by GSEARCH.

Run this, and GSEARCH finds an excellent selection of glass types from the Guangming catalog. The program will examine, in this case, 4^7 combinations, or 16 384 cases altogether. (And we need to override the CLIMIT to increase the default of 9999 cases.) The result (**C38L1**), after more optimization and annealing, is shown in figure 38.3.

We also tried matching this design to the Schott and Ohara catalogs, but those, in this case, were not quite as good. This is no criticism of those vendors, by the way; sometimes the reverse is true. When running GSEARCH, it pays to try several catalogs, since the results depend on which glasses are close to which GLMs in the lens, and they all differ.

Open the dialog **MMF**, select Muilticolor, and click Execute. The MTF curves for this lens are shown in figure 38.4. The performance is considerably better than that of the original Petzval design, and with only seven elements. Enough said.

We should mention why the DSEARCH file contained the lines

```
FNUM 3.5 10
BACK 0.7 SET
```

Those options create a lens which does *not* use a curvature or thickness solve on the last surface. While the program can find an excellent design using those solves, we ran into trouble when matching real glass when they were active. Since the image is close to the last element, a small change in the index up front can change the paraxial ray path–and the curvature solve can produce a very short radius since the beam diameter can become very small. And this can lead to ray trace failure. By defeating

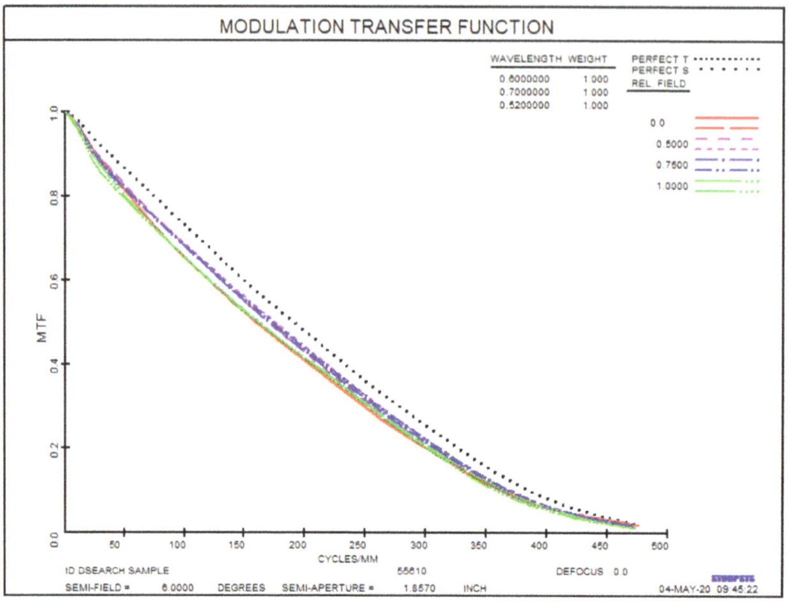

Figure 38.4. MTF of the seven-element lens with real glass.

the solves, the F/number will be a target in the MF instead of the result of a UMC solve on the last element, and the back focus distance will be set at 0.7 inches and will not be controlled by a YMT solve. With these changes, when real glasses were inserted with GSEARCH, no ray failures were encountered.

Another thing to try is giving a value for RSTART different from the default 100 inches. That will explore different branches of the design tree.

In preparing this chapter, we tried several combinations of the paramters RT, ASTART, and RSTART. Each produced different results, and while most were in the ¼ wave region, some were not. This is the nature of the task we are addressing. As Kingslake observed, 'I tried everything …'. It would be great if our algorithm always found the best of all possible results, but we are not there yet. Still, the tools we have are already impressive, even considering their limitations. We are searching a very large and very bushy tree, and those tools are able to find excellent designs most of the time.

We are curious what would happen if we ask DSEARCH for a six-element lens. Can it do as well? Here is the input file (**C38M2**):

```
CORE 64
TIME
DSEARCH 3 QUIET
 SYSTEM
 ID DSEARCH SAMPLE
 OBB 0 6 1.857
```

```
 WAVL 0.7 .6 .52
 UNITS INCH
 END
 GOALS
 ELEMENTS 6
 FNUM 3.5 10
 BACK 0.7 SET
 TOTL 17 .1
 STOP FIRST
 STOP FREE
TSTART 1.0
ASTART 1.0
 RT 0.5
 FOV 0.0 0.75 1.0 0.0 0.0
 FWT 5.0 3.0 3.0 1.0 1.0
 NPASS 90
 ANNEAL 200 10 Q 90
 TOPD
 SNAPSHOT 10
 QUICK 40 40
 END
 SPECIAL PANT
 END
SPECIAL AANT
ADT 7 .01 10
M 0 10 A GIHT
S P YA 1
END
GO
TIME
```

Run this, and when DSEARCH finishes, save the file DSEARCH_OPT as GSOPT as you did before. Then make a new MACro to run GSEARCH and find the best glass combination from the Schott catalog:

```
GSEARCH 5 QUIET LOG
SURF
1 3 5 7 9 11
END
NEAREST 4 P
S
END
CLIMIT 20000
GO
```

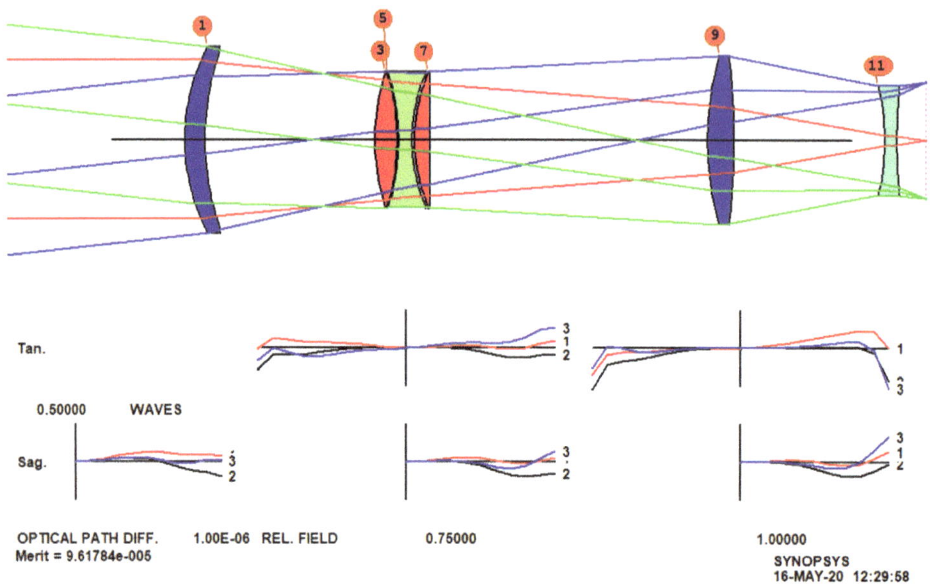

Figure 38.5. Six-element lens found by DSEARCH and then GSEARCH.

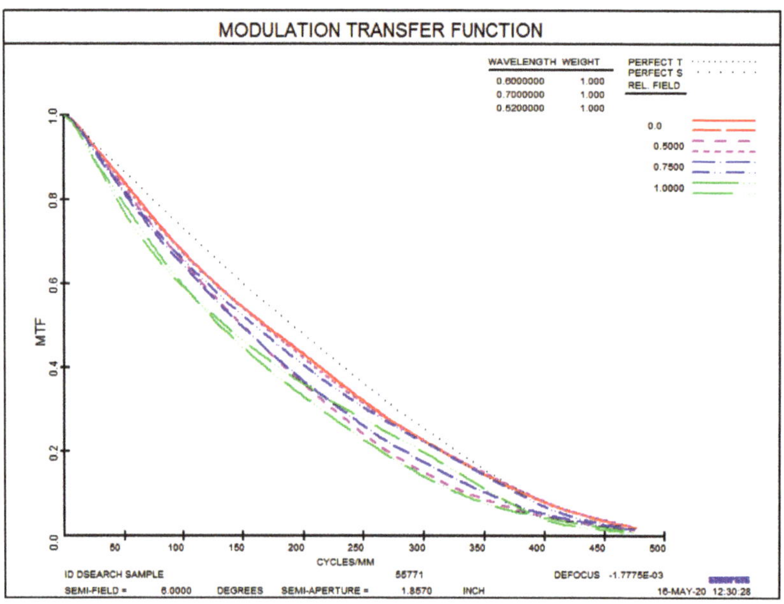

Figure 38.6. MTF of the final six-element lens, with real glass.

Run this `GSEARCH MAC`ro, and then do the same with the Ohara and Guangming catalogs. The best one came back with Ohara glasses when optimized and annealed (**50, 2, 50**). The lens is shown in figure 38.5, and the modulation transfer function (MTF) is in figure 38.6. This is a superb design (**C38L2**). Only six elements, and better than the original eight-element Petzval!

Different glasses have different partial dispersions, so it makes sense to try more than a single catalog. And if you get a great design with one catalog, you can sometimes then change to equivalent but less expensive glasses from a different vendor with little loss of quality.

In this chapter we obtained an excellent six-element lens with better performance than the original eight-element lens shown in figure 38.1. This is not a criticism of the designers of that lens, however. These new designs were developed with better software than was available in the 1960s, and the results depend strongly on the properties of many glass types that were also not then available. (I should know: I was in the group that worked on those designs.)

IOP Publishing

Lens Design (Second Edition)
Automatic and quasi-autonomous computational methods and techniques
Donald C Dilworth

Chapter 39

Athermalizing an infrared lens

Calculating spacings and materials so thermal effects are minimized

In this chapter, we will examine what happens to the image of a mid-infrared lens as the temperature changes. Start with lens in figure 39.1 (**C39L1**).

Here is the RLE file for this example:

```
RLE
ID FOUR ELEMENT INFRARED OBJECTIVE
 WAVL 4.000000 3.250000 2.500000
 APS              1
 UNITS MM
 OBB  0.000000   3.0000  30.0000   0.0000   0.0000   0.0000   30.0000
 MARGIN       1.270000
 BEVEL       0.254001
   0 AIR
   1 RAD    163.0500000000000   TH      4.50000000
   1 N1 3.42403414 N2 3.42836910 N3 3.43782376
   1 DNDT  1.336E-04 1.336E-04 1.336E-04 1.4000E+00 7.5000E+00 1.6000E+01
   1 CTE   0.255000E-05
   1 GTB U    'SILICON          '
   1 EFILE EX1    31.417334    31.417334    31.671335     0.000000
   1 EFILE EX2    31.014427    31.417334     0.000000
   2 RAD    255.4500000000000   TH      5.55000000 AIR
   2 AIR
   2 EFILE EX1    31.014427    31.417334    31.671335
   3 RAD   -721.5000000000000   TH      3.60000000
   3 N1 4.02415626 N2 4.03741119 N3 4.06419029
   3 DNDT  4.100E-04 4.100E-04 4.100E-04 2.0500E+00 1.1000E+01 2.200E+01
   3 CTE   0.550000E-05
   3 GTB U    'GE               '
   3 EFILE EX1    30.633643    30.633643    30.887644     0.000000
   3 EFILE EX2    30.633643    30.633643     0.000000
```



```
  4 RAD  -1590.0000000000000   TH     65.70000000 AIR
  4 AIR
  4 EFILE EX1    30.633643    30.633643    30.887644
  5 RAD    145.5000000000000   TH      3.15000000
  5 N1 4.02415626 N2 4.03741119 N3 4.06419029
  5 DNDT  4.100E-04 4.100E-04 4.100E-04 2.0500E+00 1.1000E+01 2.200E+01
  5 CTE   0.550000E-05
  5 GTB U     'GE                  '
  5 EFILE EX1    27.236976    27.236976    27.490977     0.000000
  5 EFILE EX2    26.712556    27.236976     0.000000
  6 RAD    120.4500000000000   TH     13.20000000 AIR
  6 AIR
  6 EFILE EX1    26.712556    27.236976    27.490977
  7 RAD    255.0000000000000   TH      4.50000000
  7 N1 3.42403414 N2 3.42836910 N3 3.43782376
  7 DNDT  1.336E-04 1.336E-04 1.336E-04 1.4000E+00 7.5000E+00 1.600E+01
  7 CTE   0.255000E-05
  7 GTB U     'SILICON             '
  7 EFILE EX1    27.355510    27.355510    27.609511     0.000000
  7 EFILE EX2    27.165926    27.355510     0.000000
  8 RAD   2025.0000000000000   TH    107.272545 AIR
  8 AIR
  8 EFILE EX1    27.165926    27.355510    27.609511
  9 RAD   -405.0000000000000   TH      0.00000000 AIR
  9 AIR
END
```

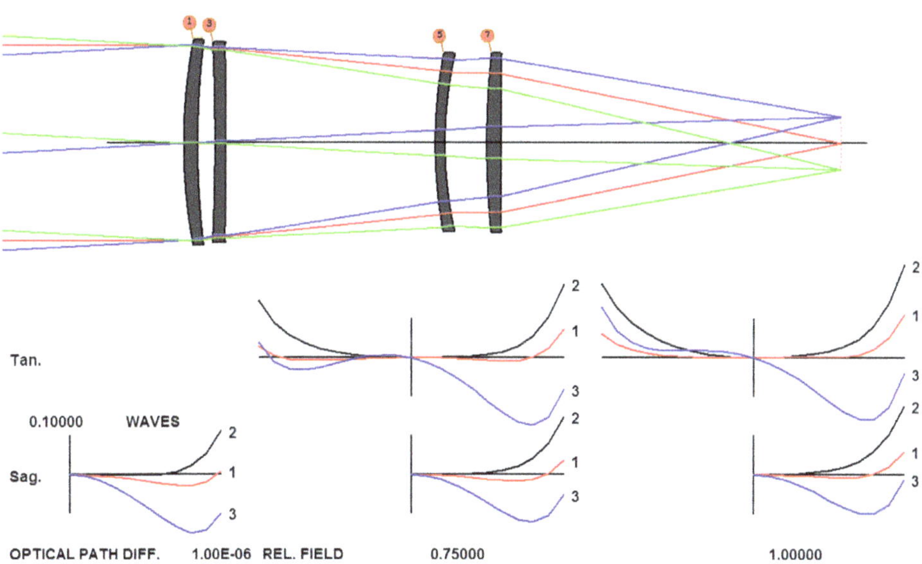

Figure 39.1. A NIR telescope to be athermalized.

Let us assume this lens must stay in focus over the temperature range of 20 to 100 C. What does is do now? To find out, run the THERM program, first testing to see if all required coefficients are present.

```
SYNOPSYS AI>THERM TEST

WARNING - NO DEFAULT CTE HAS BEEN ASSIGNED TO AIRSPACES
ALL GLASSES IN THIS LENS HAVE BEEN ASSIGNED THERMAL-
INDEX COEFFICIENTS
```

Indeed, this lens was never assigned a coefficient for the airspaces. Fix that with a CHG file, assigning the coefficient of aluminum type 6061:

```
CHG
ALPHA A6061
END
```

Now you can activate thermal shadowing. Create and run a new MACro:

```
THERM
ATS 100 2
END
```

This puts a copy of the lens in configuration 2, with all parameters altered as required by the change in temperature from the default 20 to 100 degrees. Figure 39.2 shows what ACON 2 looks like at that temperature.

Ouch! The lens is out of focus. We have to correct that.

Here's an easy way to tell where an axial shift of an element might do some good. First, make a checkpoint in ACON 2, by clicking the button. Now open the WorkSheet (click on the button), and then click on surface 4 in the PAD display. We suspect that a change of that airspace might alter the focus position. To be practical, the required motion must be quite small, so slide the speed slider closer to the bottom, and then slide the Spacing slider to the left, as shown in figure 39.3.

Indeed, the image comes back nearly in focus, and the motion was quite small, from 65.827 to 65.53. We're getting close.

Now we have to figure out a way to make element 3 move in that way with temperature. One trick that sometimes works is to design the cell with an outer sleeve that extends from surface 4 to the right, past the next elements, and then with an inner sleeve that comes back partway and holds those elements. If the outer sleeve is made of aluminum and the inner one of plastic, the net motion of element 3 will be less than it would be with an all-aluminum cell.

Go back to ACON 1 again, with the WorkSheet still open, make a checkpoint, and click the **Add Surface** button. Now click on the axis in the lens drawing in between surfaces 4 and 5. A dummy surface is inserted, shown in figure 39.4.

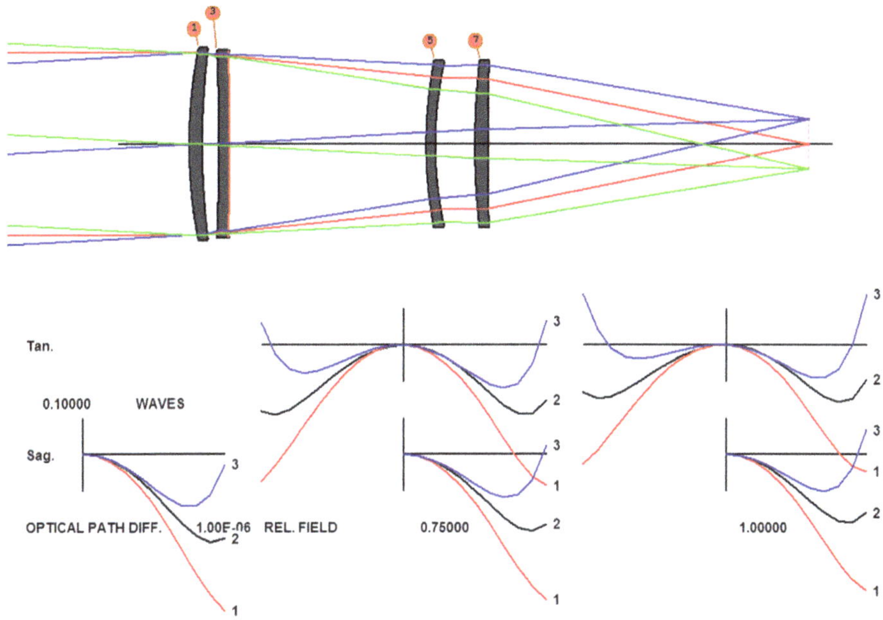

Figure 39.2. NIR telescope at 100 degrees, before correcting.

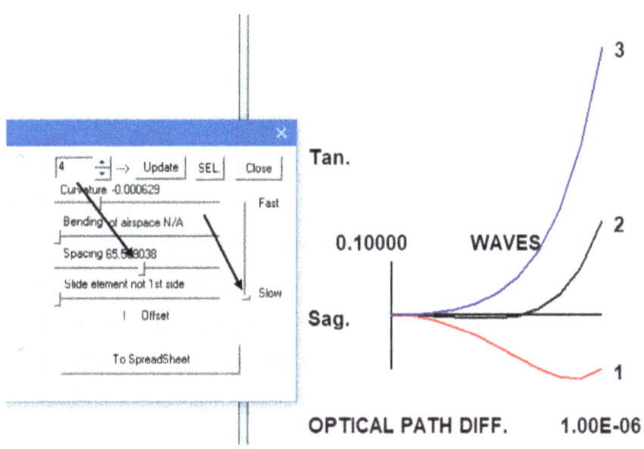

Figure 39.3. WorkSheet sliders for adjusting an airspace in the lens.

Now you must tell the program that the expansion coefficient from 5 to 6 is different from the default aluminum, which we assigned above to all airspaces. Close `WS` and make a new `THERM` file:

```
THERM
COE 1 STYRENE
TCHANGE 1
5
ATS 100 2
END
```

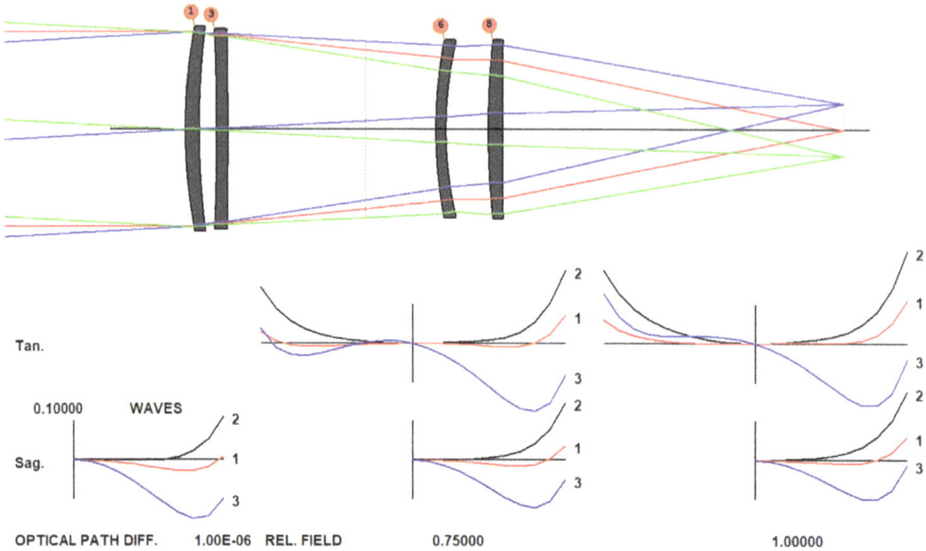

Figure 39.4. NIR lens with dummy surface inserted in preparation for athermalization.

This file says to define coefficient number 1 as that of styrene, and then assign thickness coefficient number 1 to surface 5. Then alter the lens to 100 degrees C and put the result in ACON 2.

Run this, and ACON 2 has indeed changed. The trick now is to find the length of the outer and inner sleeves that will best compensate for this thermal change. For this task, we use the optimization program. Here is the MACro (**C39M1**):

```
ACON 1
PANT
VY 4 TH 1000 -1000
VY 5 TH 1000 -1000
END

AANT
ACON 1
M 0 1 A DELF
M 8.103249 1 A P YA 1

GSO 0.5 5.332000 3 M 0
GNO 0.5 1 3 M 0.5
GNO 0.5 1 3 M 1.0

ACON 2
M 0 1 A DELF
```

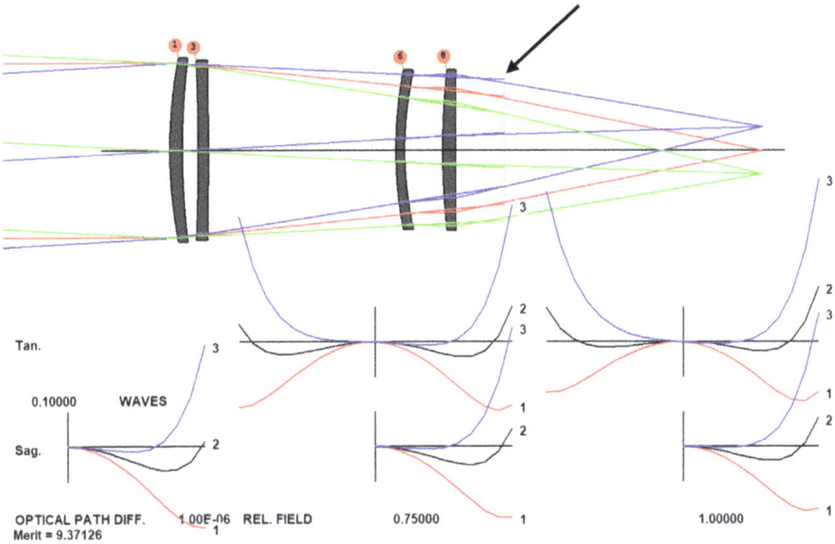

Figure 39.5. Lens at 100 degrees with athermalization.

```
GSO 0.5 5.332000 3 M 0
GNO 0.5 1 3 M 0.5
GNO 0.5 1 3 M 1.0
END

SNAP
SYNO 20 MULTI
```

This will attempt to keep the system in focus at both temperatures and try to hold image quality at the same time, while varying thicknesses four and five. Note the `MULTI` declaration on the optimization command. That allows the program to optimize more than a single configuration. Run this, and now the lens in `ACON 2` is better than before, as shown in figure 39.5.

There is some image degradation, but within reason, and the focus remains where it should even with this change in temperature. Note the position of surface 5 now. That tells you where the two sleeves must extend to and where they should connect. Athermalization does not have to be difficult.

Some comments are in order. We entered explicit limits for the TH variables since the program will not let a positive TH become negative otherwise. To keep the magnification constant, we added a target for the YA of the chief ray. We did not implement the options to account for whether the cells hold the lenses on the right or left side of an element, because for this example the expansion is applied in the right place by default. Yes, sometimes athermalization is more complicated, and you are referred to the User's Manual for a complete description of the options you have available for more demanding tasks.

IOP Publishing

Lens Design (Second Edition)

Automatic and quasi-autonomous computational methods and techniques

Donald C Dilworth

Chapter 40

Edges

Defining lens edges and bevels

Before you send element drawings to the shop, you have to carefully define the shape and dimensions of the edges. This is done with the Edge Wizard[1] (**MEW**).

Get out lens shown in figure 40.1 (**C40L1**).

This lens has already been assigned reasonable edges with the Edge Wizard. To show how it works, we will first delete all the edge definitions—and then show how to put them back. Type in the Command Window

```
EFILE
ERASE
END
```

Now you see the default edges in figure 40.2, assigned by the program so they clear the upper and lower marginal rays at full field. Those are reasonable edges to use during lens optimization, so you can see what is going on, but when you manufacture the elements they have to be larger and defined more carefully to allow for mounting in a cell. Open the Edge Wizard, either by typing **MEW** or by clicking the button on the PAD toolbar. At the moment, the lens has no edge definitions anymore. Click the **Create all** button, and you get a set of reasonable edges. Click **Yes** to the prompt, and the picture has changed. Enter the number 1 in the surface-number box and you can see the dimensions that have been applied to the first element, shown in figure 40.3.

The program creates five reference points at the edge of each element, labeled A through E in the diagram on the dialog. You usually have to edit those default dimensions, and the data for element 1 show one reason why. The first surface is convex, and you may not want a bevel on that surface. The program defined the

[1] Edge Wizard™ is a trademark of Optical Systems Design LLC, an Arizona, USA corporation.

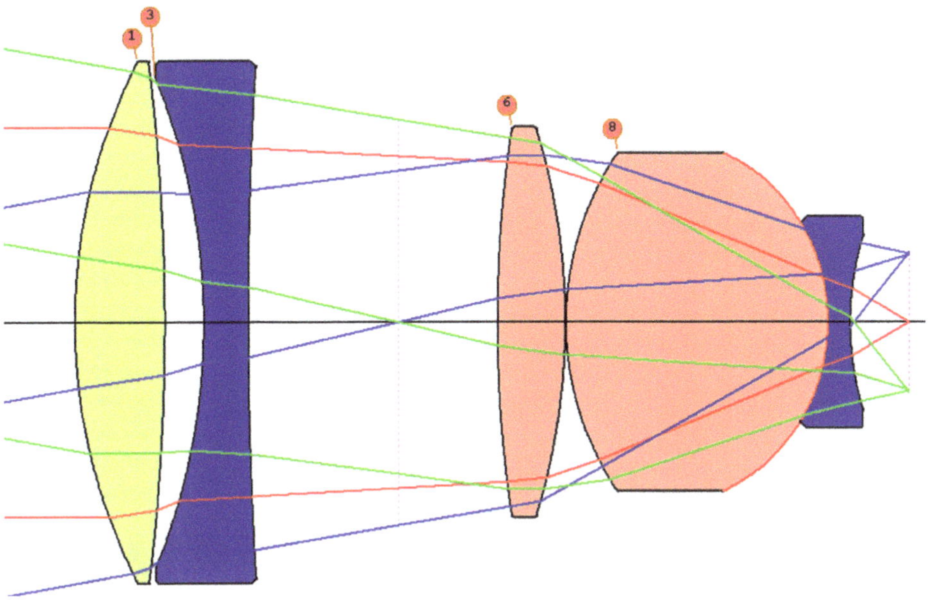

Figure 40.1. Lens with EFILE edges assigned.

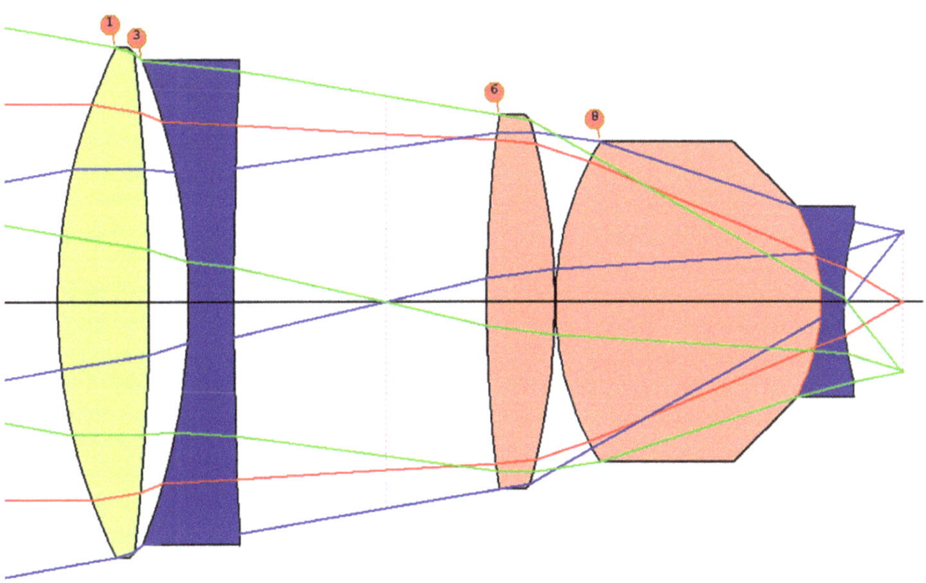

Figure 40.2. Lens with default edges.

default edges and put into effect the Explicit rules, which work for most lenses, with data you can edit with the edit boxes and spin buttons on the dialog if you want to.

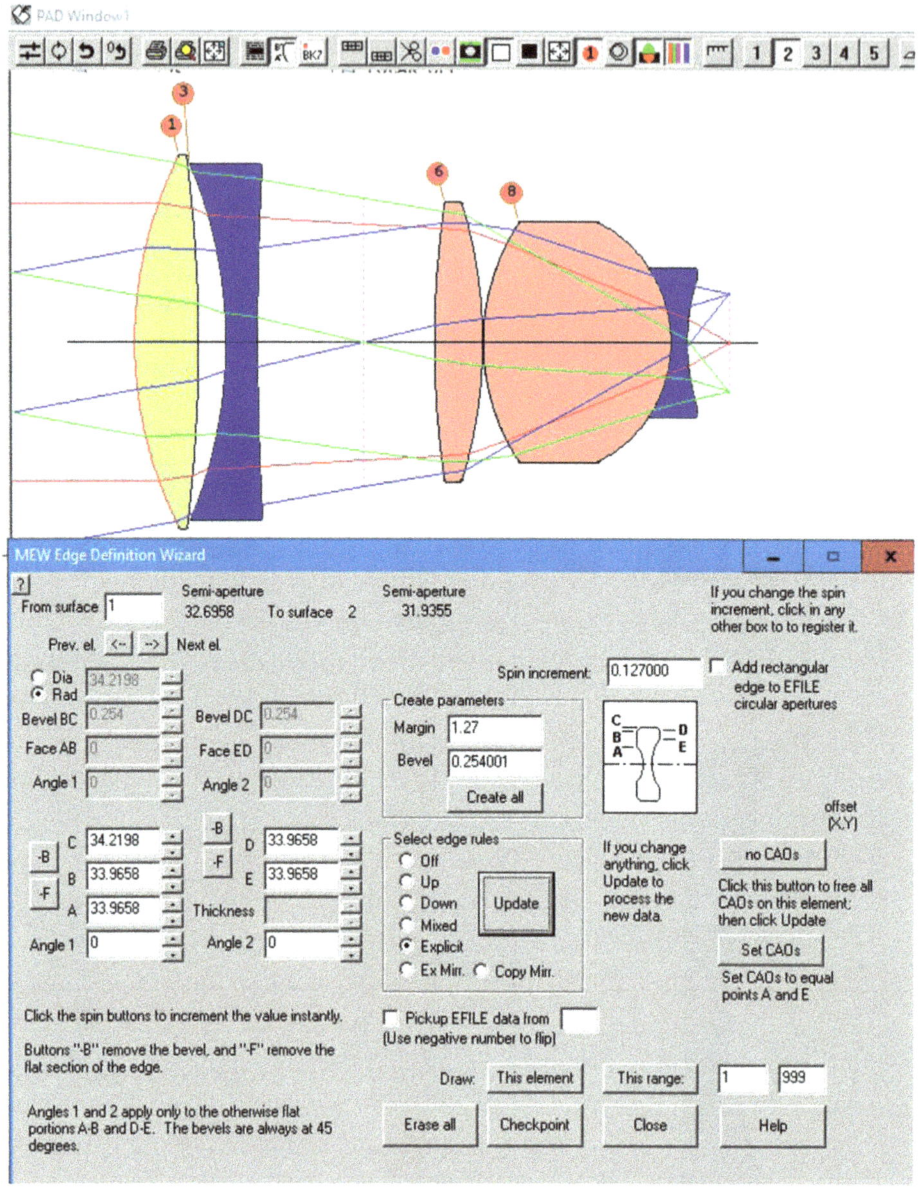

Figure 40.3. Edge Wizard, showing data for element 1.

The default point C is currently at 34.2198 mm from the axis, while the clear aperture on surface 1 is 31.9355. This element has a rather thin edge, so let us reduce the diameter slightly. Enter the number 34 in the box for dimension C and click Update. We will also remove the bevel on that surface. Click the box '-B' to the left of the C dimension. '-B' means remove the bevel on that side. Then click the '-F' button too. When you remove the bevel, you leave point A where it was, which may be appropriate for some plastic elements that have a mounting flange molded in place, but is not welcome here. That button removes the flat portion from A to B. Side 1 is now reasonable, and you may want to remove the bevel and flat on side 2. (For thicker positive elements with shallow curves, we usually leave the bevels in place.)

Element 2 is negative, and here we want a flat portion on side 1 and a bevel—but no flat—on side 2. Click the **Next el.** button to see the data on that element, shown in figure 40.4 on the left.

Let us assume you want the outer diameter of element 2 the same as on element 1. Just enter the same dimension, 34 in the C box and click **Update**. The edge changes, as shown in the center.

This increases the element diameter, but again has left dimensions A and B where they were. Let us reduce the size of the bevel on surface 3. To the right of the edit box for dimension B there are two spin buttons (see figure 40.3). Click the upper of the two a number of times, watching the bevel get smaller and the flat portion getting larger, shown on the right in figure 40.4. By clicking on the two spin buttons while watching the picture, you can define the edge exactly as you want.

If you also want a smaller bevel on side 2 of that element, adjust it with the spin buttons for points D, and then click the **–F** button for that side to remove the flat portion. The edges are now as shown in figure 40.5.

Now the edges of the first two elements look about right. It would be a good idea at this point to click the **Checkpoint** button on the MEW dialog. As you work

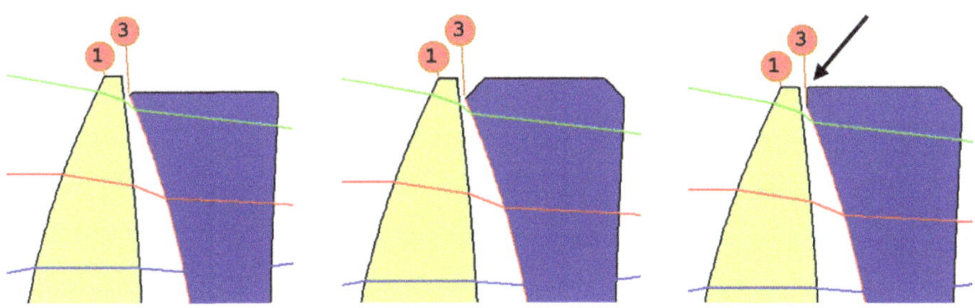

Figure 40.4. Edges to be modified by the Edge Wizard.

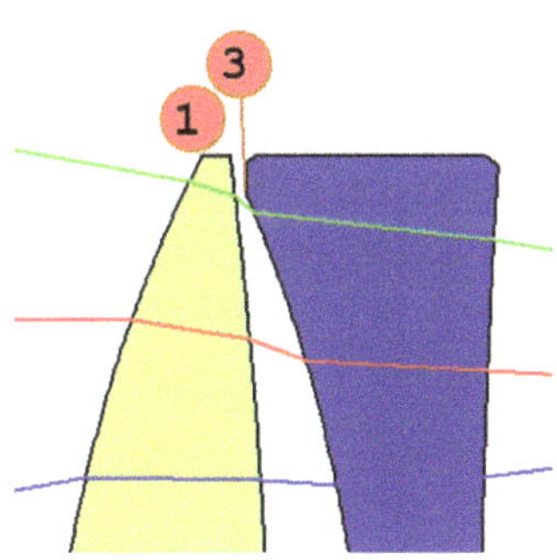

Figure 40.5. Finished edges of first two elements.

on the other elements, you might make a mistake and want to go back to a previous version.

Proceeding in this fashion you can define all the edges just as you want them. When you have finished, close the Wizard and type **ELIST** in the Command Window. This listing shows the currently defined points A and C on each element, as well as the distances from A to B, B to C, and so on.

```
SYNOPSYS AI>ELIST

CURRENT EFILE DATA:

Surf.   A            AB            BC              C             ANG           CAO      TYPE
        E            ED            DC              C             ANG2          CAO

--------------------------------------------------------------------------------------------
  1    34.000       0.0000        0.0000         34.000        0.0000        32.696    EXPL
       33.966       0.0000      0.34200E-01      34.000        0.0000        31.936

  3    32.345       1.5240       0.13071         34.000        0.0000        31.075    EXPL
       33.615     0.12192E-05    0.38471         34.000        0.0000        29.602

  6    25.242       0.0000       0.25400         25.496        0.0000        23.972    EXPL
       25.242       0.0000       0.25400         25.496        0.0000        23.349

  8    21.809       0.0000       0.25400         22.063        0.0000        20.539    EXPL
       21.809       0.0000       0.25400         22.063        0.0000        12.174

  9    13.444       0.0000       0.25400         13.698        0.0000        12.174    EXPL
       11.590       1.8532       0.25400         13.698        0.0000        10.320

CURRENT BEVEL IS   0.2540010
CURRENT MARGIN IS    1.270000
SYNOPSYS AI>
```

These edges become part of the lens file and show up in the RLE data as EFILE parameters. For element 2, those data look like this:

```
...
    3 RAD    -81.3505230000000    TH      6.00000000
    3 N1 1.83648474 N2 1.84664080 N3 1.87201161
    3 CTE   0.830000E-05
    3 GTB S    'SF57                '
    3 EFILE EX1     32.345300     33.869287     34.000000      0.000000
    3 EFILE EX2     33.615288     33.615289      0.000000
    4 RAD    553.8617899999995    TH     19.92504900 AIR
    4 AIR
    4 EFILE EX1     33.615288     33.615289     34.000000
...
```

While it is *possible* to edit the edge dimensions in the WorkSheet, it is not recommended. Some of them are coupled to others and the result is not always intuitive. Use the Wizard to edit the data if you need to. Everything is shown on the dialog and it is very simple to use.

40.1 A mirror example

This has been a useful example, but now we'll look at a system with fold mirrors. Those can be assigned edges—and thicknesses—too. The lens is in **C40L2**.

Fetch this lens and type **CAP** to look at the current apertures.

```
SYNOPSYS AI>CAP

 ID EXAMPLE FOLDED SYSTEM                        28301           01-SEP-17   14:25:22

 CLEAR APERTURE DATA
 (Y-coordinate only)

 SURF    X OR R-APER.      Y-APER.    REMARK      X-OFFSET     Y-OFFSET   EFILE?
 ------------------------------------------------------------------------------
    1          0.2621              *User CAO
    2          0.6611               Soft CAO                                 *
    3          0.6870               Soft CAO                                 *
    4          0.7014               Soft CAO                                 *
    5          0.7064               Soft CAO                                 *
    6          0.7071               Soft CAO                                 *
    7          0.6302               Soft CAO                                 *
    8          0.5566               Soft CAO
    9          1.2000       1.6000 *User RAO                                 *
   10          0.6779               Soft CAO
   11          0.9358               Soft CAO                                 *
   12          0.9595               Soft CAO                                 *
   13          1.5000       2.2000 *User RAO                                 *
   14          0.9705               Soft CAO
   15          2.0000       2.4000 *User RAO                                 *
   16          0.9793               Soft CAO
   17          1.0306               Soft CAO                                 *
   18          1.0406               Soft CAO                                 *
   19          1.0424               Soft CAO
   20          1.0424               Soft CAO

 NOTE: CAO, CAI, EAO, and EAI input is semi-aperture.
       RAO and RAI input is full aperture.
 SYNOPSYS AI>
```

This system currently has EFILE edges, as indicated by the '*' on the listing above. Now open the Edge Wizard again and click the **Erase all** button on the Wizard to revert to default edges. A portion of the system is shown in figure 40.6.

Surface 9 is a fold mirror that has been assigned a rectangular outside aperture of dimensions 1.2×1.6 inches. (Those are the full dimensions of the rectangle: circular apertures are given by the radius, rectangular by the side lengths.) But without assigned EFILE data, it shows up as just a straight line on the PAD display. The Edge Wizard can create suitable dimensions on this mirror, with the aspect ratio taken from the RAO data. In the Wizard, navigate to surface 9, and you see that nothing is assigned yet. Select the **Ex Mirr** option and click **Update**. A default edge is created, and now the mirror has a thickness, as shown in figure 40.7.

Let's assume you want it to be thicker. You can enter a larger number in the Thickness edit box or click the upper spin button on that box. The thickness increases, as shown in figure 40.8. The amount by which the spin buttons change the dimension is given in the Spin increment box.

Change the increment to 0.02, click Update, and use the lower spin button in box D to add a bevel to the back side of the mirror. The bevel is shown in figure 40.9.

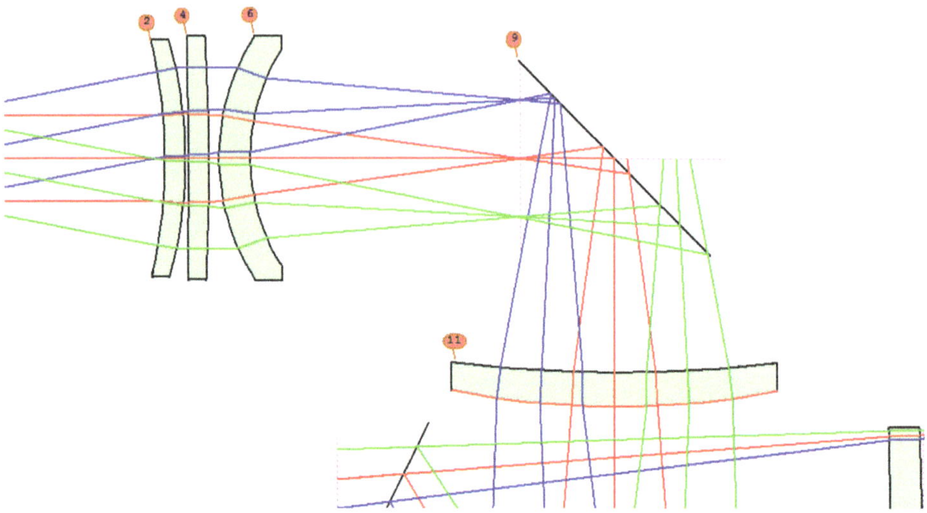

Figure 40.6. System with fold mirrors and default edge definitions.

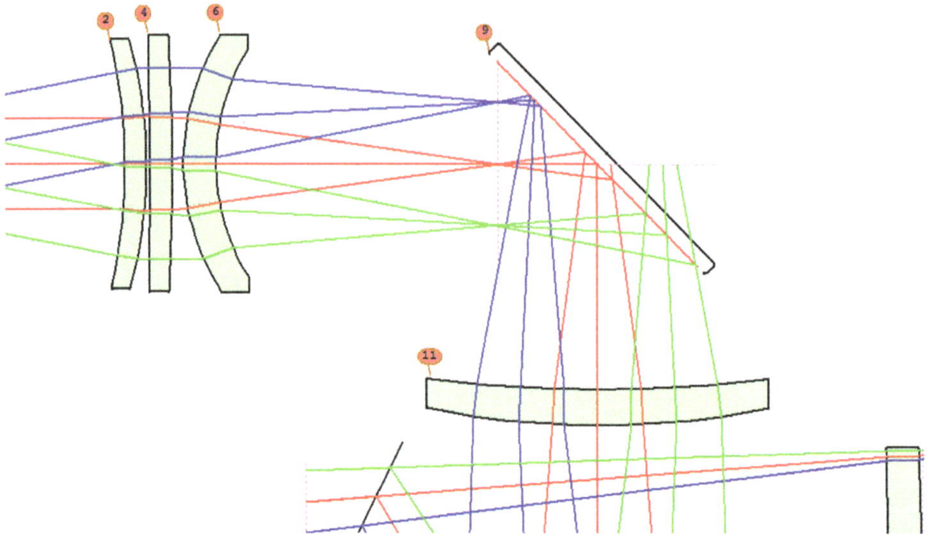

Figure 40.7. A thickness has been assigned to the fold mirror at surface 9.

Add edges to the other fold mirrors, at surfaces 13 and 15, in the same way, and close the Wizard. Now make an RSOLID picture with the dialog **MPE**, shown in figure 40.10 (**C40L3**).

Your mirrors are shown, beveled exactly as you wanted.

Go back to the Wizard and define the edge for element 3, at surface 6, as shown in figure 40.11.

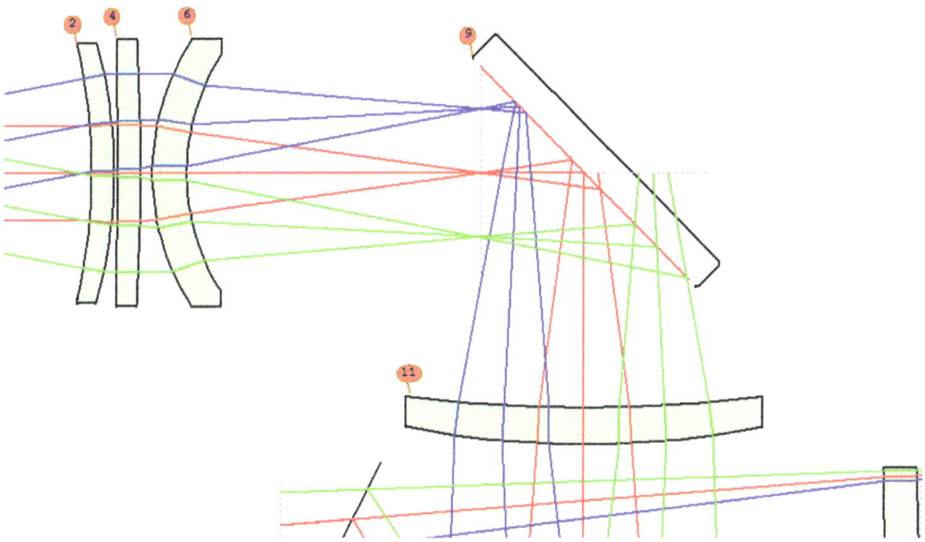

Figure 40.8. Mirror thickness increased with the Edge Wizard.

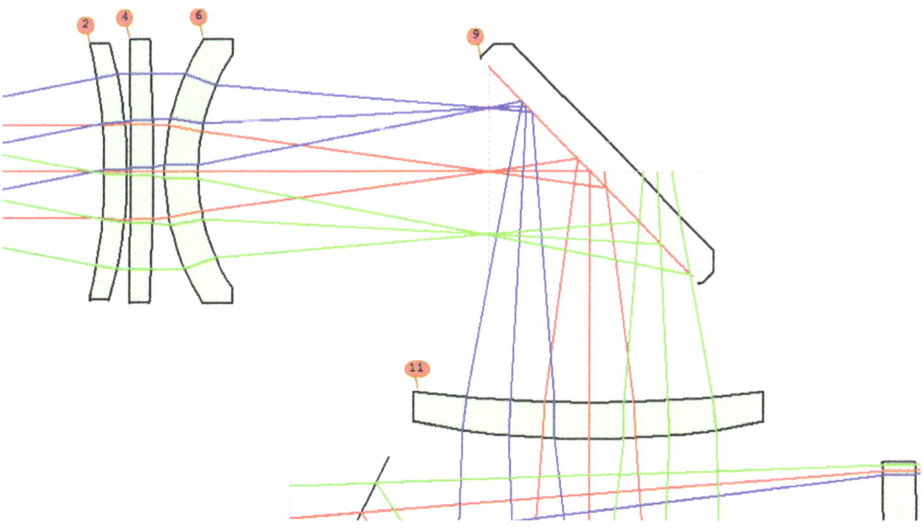

Figure 40.9. A bevel has been added to the back side of the fold mirror.

Now open the **MPL** dialog and enter data for an ELD drawing at surface 6, as shown in figure 40.12.

Click the **ELD** button, and your drawing shows up—with all of the edge dimensions shown and documented, as in figure 40.13.

This has been a brief introduction to the Edge Wizard. There are many more options, including some that will automatically change dimension C if the lens CAO changes. You may never need that much horsepower, but by all means read

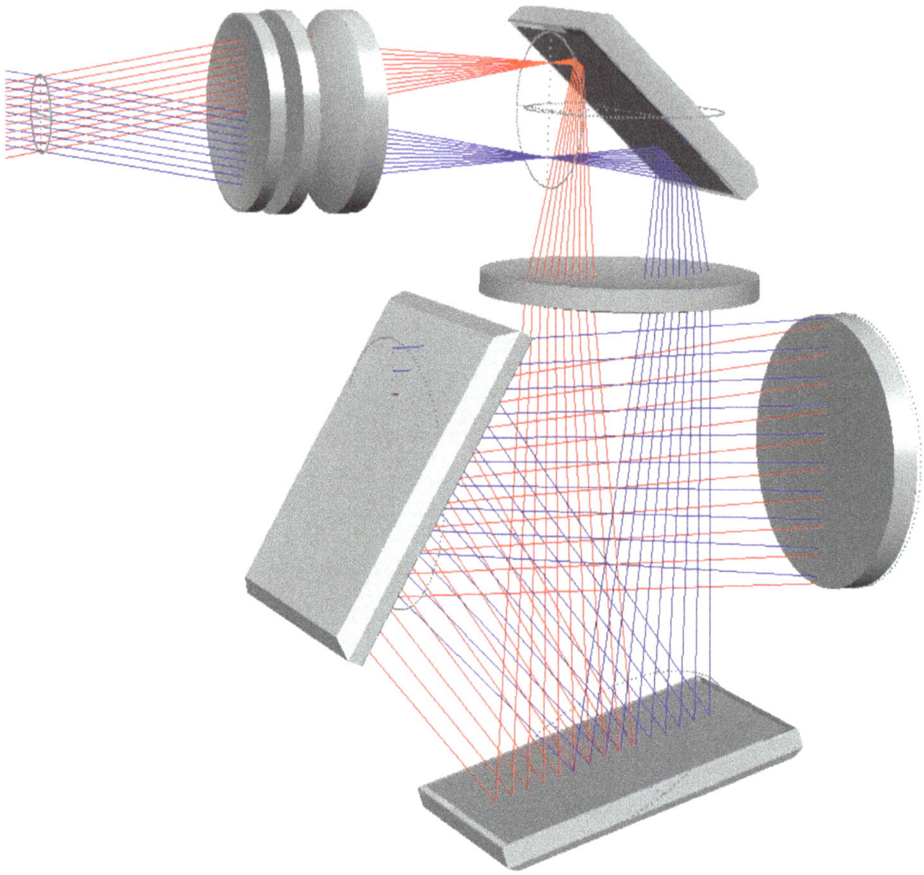

Figure 40.10. RSOLID drawing of the system.

Section 7.8 of the User's Manual. There you will learn how to create edges like this (**C40L4**):

```
RSOLID 22 33 0 0 0
PLOT
PUPIL 1
RED
TRACE P 0 0 200
END
```

The RSOLID drawing is shown in figure 40.14. This lens has a combination of CAO, RAO, and EFILE data, some decentered, as you will learn if you study the **LEO** listing.

Figure 40.15 shows a system with an off-axis aspheric mirror (**C40L5**). Study the RLE file to see how the edges were defined. This system has a real stop on surface 5,

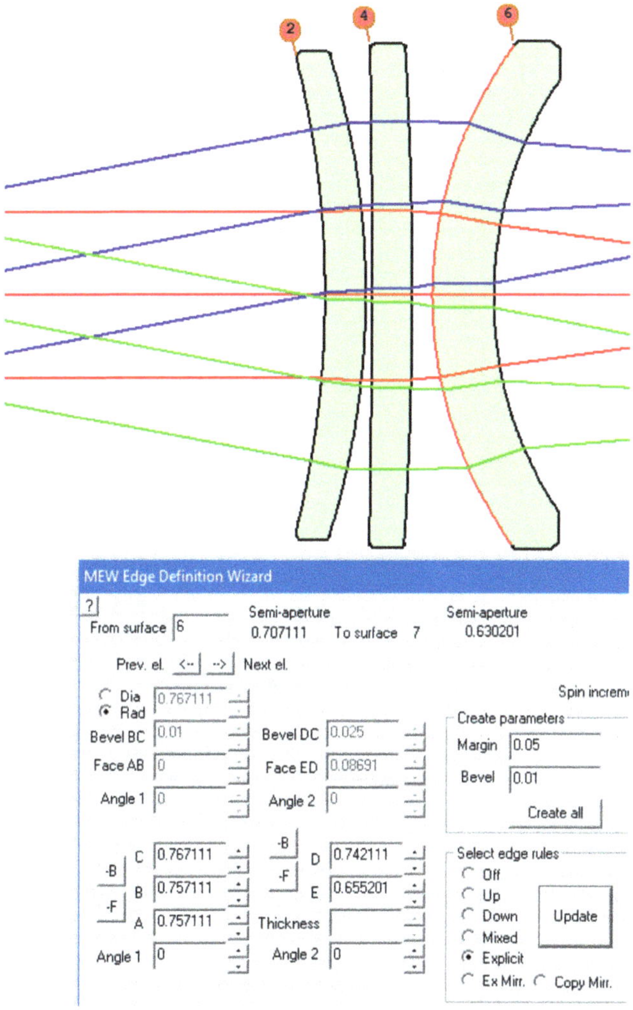

Figure 40.11. Edge Wizard with data for adjusting edges on surfaces 6 and 7.

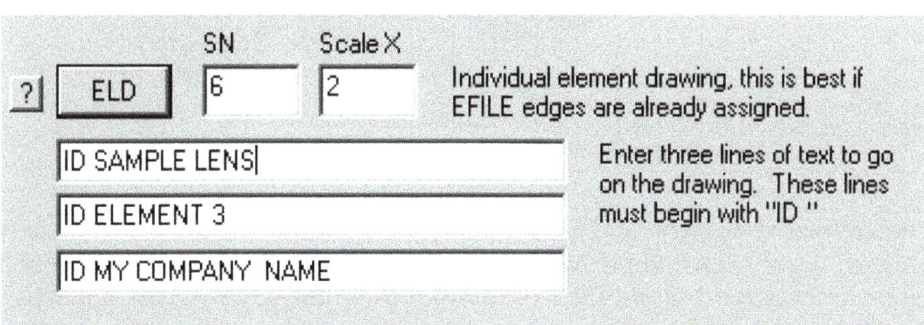

Figure 40.12. MPL data for drawing a lens element.

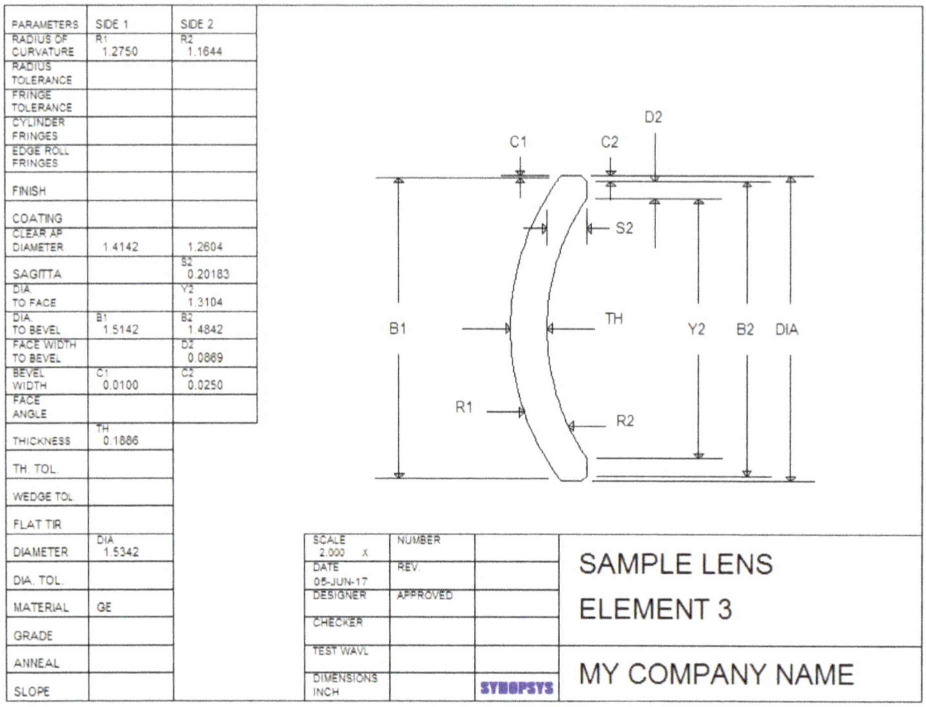

Figure 40.13. Sample element drawing with user-defined edge geometry.

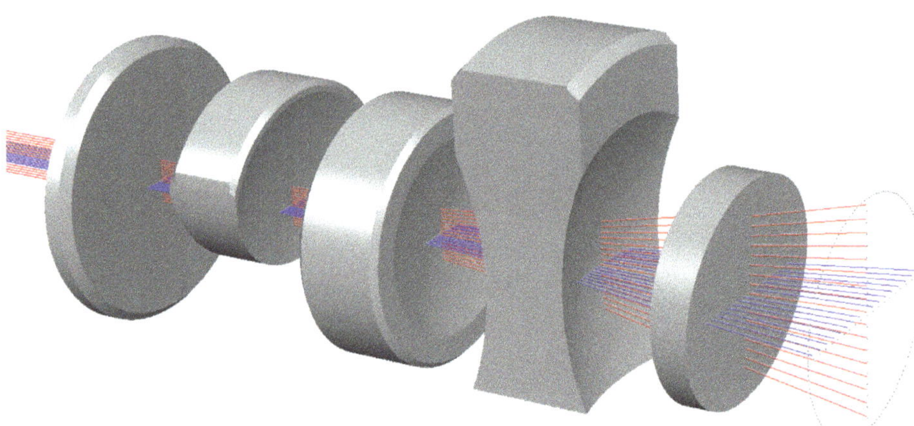

Figure 40.14. RSOLID drawing showing a variety of edge shapes.

the first refractive element, and a DCCR directive on mirrors at 2 and 3. Surface 4 is decentered downward, and the two mirrors share a common axis. The chief ray has to come into surface 2 above the axis in order to hit the center of the stop when it gets to 5. The real pupil takes care of that. The clear aperture on two is decentered so as to fit the marginal rays at the upper and lower field points, and the same applies to

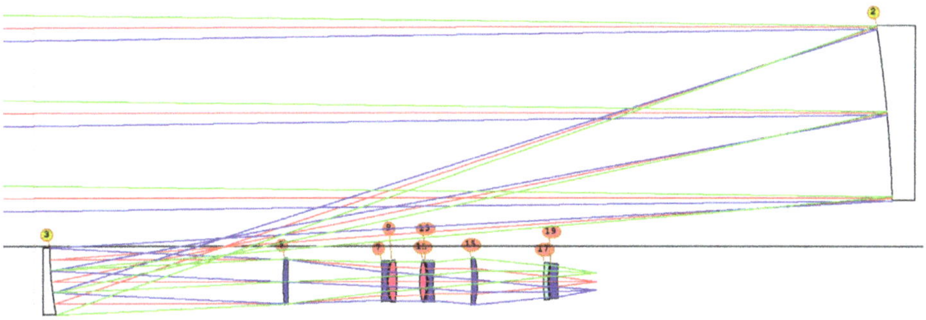

Figure 40.15. A system with a decentered aperture on an off-axis mirror.

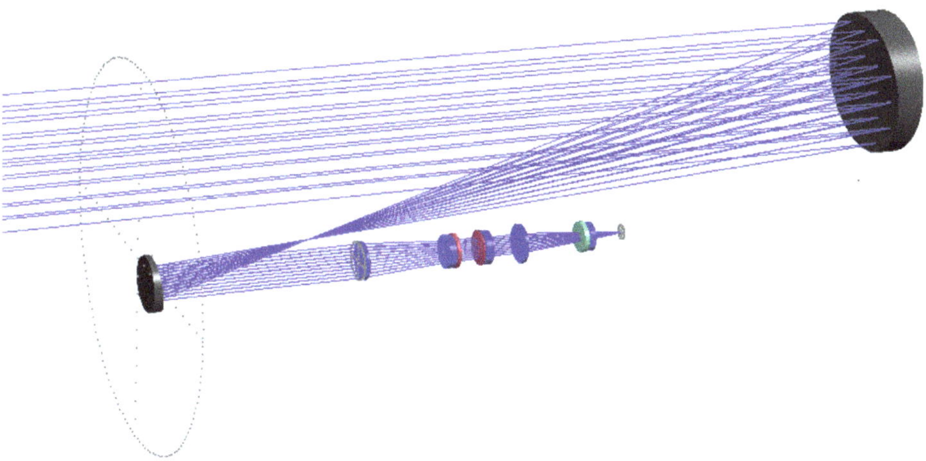

Figure 40.16. RSOLID drawing of the system.

the second mirror. This common geometry is set up with rather simple input, and the DCCR directives take care of the decentered apertures on the mirrors.

The RSOLID view in figure 40.16 was created with the input

```
RSOLID 22 -15 .1 0 0
ACOLOR
PLOT
PUPIL 1
BLUE
TRACE P 0 0 20
END
```

Chapter 41

A 90-degree eyepiece with field stop correction

Correcting an intermediate image; controlling exit pupil aberrations

If you have read the previous chapters, you are already familiar with the tools we will use in this exercise—but you may not know how excellent the results can be when you combine them. Here we present a challenging problem and then show how those tools can find an excellent design in a fraction of the time needed by those using classical design methods. Time is money, after all, and this is worth knowing about. In this lesson, you will use DSEARCH to derive a starting configuration, and then use other features to modify the lens construction, always improving its performance. We want the eyepiece to be diffraction-limited, and also have to make sure that the image of the field stop is sharp to the eye. That is more complicated and makes a good exercise.

The problem is to design a wide-angle eyepiece according to the goals summarized below.

Field of view: 90-degrees total at the eye.
Eye relief: 15 mm or greater
F/number of beam from telescope objective: F/7.
Visible spectrum: C, d, and F Fraunhofer lines.
Correction to ¼ wave or better in d light at 0.58756 um.
Correction to ½ wave or better in C (0.6563 um) and F (0.4876 um) light.
Pupil aberration at eye point no greater than 1/2 mm.
An internal field stop, where tangential image errors must be no greater than twice the Airy disk at the local F/number of the beam.
The telescope objective to be 2000 mm away.
The eyepiece must have no more than 10 elements.
Total length of the eyepiece no more than 200 mm.

This is not an easy problem—nor should it be. If the problem could be solved entirely by the computer, the results would appear in a manner of minutes and the lesson would not be very interesting. But these requirements cannot all be input to any lens design code with a single set of commands, so here we will use the computer

doi:10.1088/978-0-7503-3695-6ch41

to do the job that it does best, and the skill of the designer to fill in where needed and to direct the program as necessary.

We will start with absolutely nothing and require the computer to come up with a starting design all by itself, with **DSEARCH**, which you have seen in earlier chapters. The input is listed below. Most of this input can created in the dialog **MDS**, which will create a MACro (**C41M1**) for you. Then you can edit it as you wish; here is a useful example.

```
TIME
CORE 64
OFF 1 99
DSEARCH 5 QUIET
SYSTEM
ID EYEPIECE EXAMPLE
OBD 1.0E9 45 1.27
UNI MM
WAVL CDF
WAP 1
END

GOALS
ELEMENTS 9
TOTL 200 .01
BACK 0 0
FNUM 7.0 10
ASTART 5
THSTART 5
RSTART 100
RT 0.25
NPASS 80
ANNEAL 100 10 Q 100
SNAP 10
TOPD
STOP FIRST                    ! keep the stop at the eye point
STOP FREE
QUICK 50 100
FOV 0 .3 .6 .75 .9 1.               ! correct over five field points
FWT 3 1 1 1 1 1
END

SPECIAL AANT
ACA 50 1 1
ADT 10 .1 10
M 15 1 A P YA 1 0 0 0 1       ! control eye relief
M -.008 10 A P HH 1           ! aim light at objective to right
M -.004 10 A P HH .5! and control pupil aberrations
M -.0064 10 A P HH .8
M 0 1 A P YA 1                ! and distortion too
S GIHT
END
GO
TIME
```

Run this, and in about five minutes the program displays a picture of the 10 best configurations it found. Running the optimization MACro prepared by DSEARCH on the top lens and annealing (**50, 2, 50**) produces a rather good lens, as shown in figure 41.1.

The OPD errors are all less than ¼ wave, straight out of DSEARCH, and the intermediate image is not located too close to a lens surface (where dust would be visible). So far, so good—but one must also watch and correct pupil aberrations in these wide-angle eyepieces. If those aberrations are too large, the eyepiece will suffer from the notorious 'kidney bean' effect, which causes portions of the field to go black as the user moves his eye about. We have to check.

Prepare a new MACro as follows:

```
STO 9
CHG
NOP
18 TH 2000
19 YMT
20
END
STEPS = 100
PLOT YA ON 19 FOR HBAR = 0 TO 1
GET 9
```

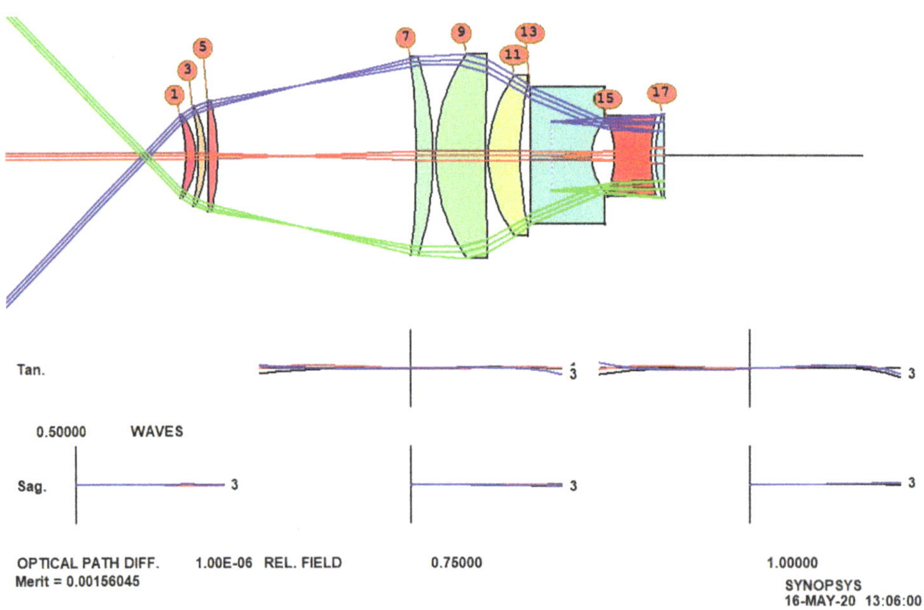

Figure 41.1. Eyepiece design returned by DSEARCH, optimized and annealed.

and run it. This will do the following:

1. Remove the YMT solve on surface 18 (via NOP, which removes all solves).
2. Put surface 19 at a distance of 2000 mm. This will model the telescope objective, assumed at that distance.
3. Assign a YMT solve to surface 19, which then focuses on 20.
4. Declare surface 20, so it exists.
5. Make a plot of the chief-ray intercept on 19 over the field. If the rays all hit near the center of surface 19, the aberrations will be under control.

Run this MACro, and you see the pupil aberration at the objective, shown in figure 41.2. At F/7, at a distance of 2000 mm, the diameter of the objective will be 285.7 mm. A chief-ray error of 4 mm is therefore less than 2% of the size of the objective, and we are allowed 1/2 mm on an entrance pupil of 2.54 mm, or about 20%, so we judge this degree of correction satisfactory. That did not come for free, of course; the HH targets in the SPECIAL AANT section penalized any solution that showed large pupil aberrations. You should feel free to adjust the weights on those targets to balance the errors as you prefer.

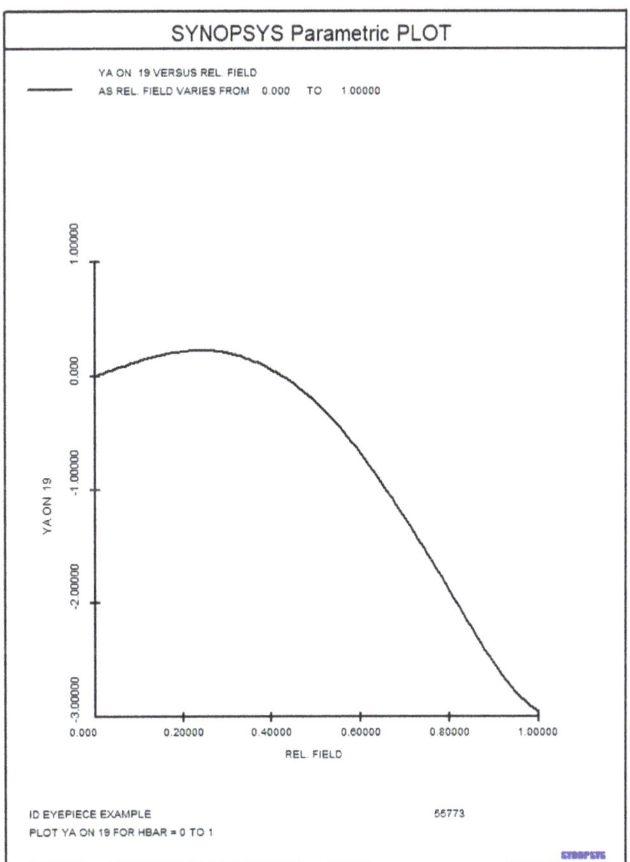

Figure 41.2. Pupil aberration at the objective lens, calculated for the eyepiece.

The eyepiece is already at the diffraction limit but is not yet finished since we have not controlled image quality at the field stop.

We have to do that—but the lens does not even have a field stop yet. In the WorkSheet, click the Add Surface button, shown in figure 41.3, and then click on the axis between surfaces 6 and 7 (or wherever the intermediate image is in your lens). A surface is added, as shown in figure 41.4.

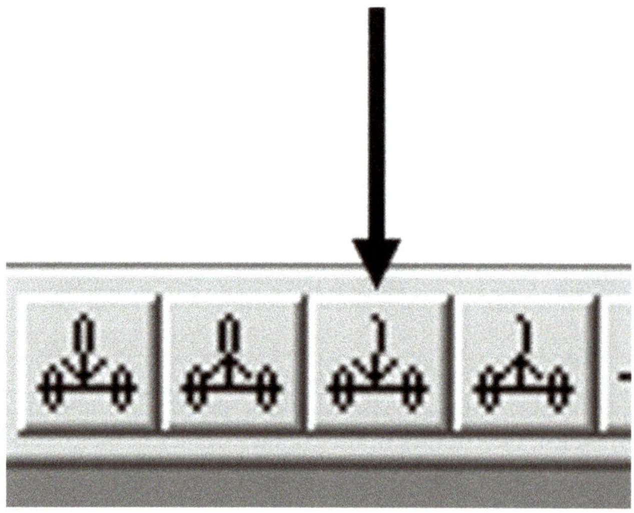

Figure 41.3. Selecting the Add Surface button.

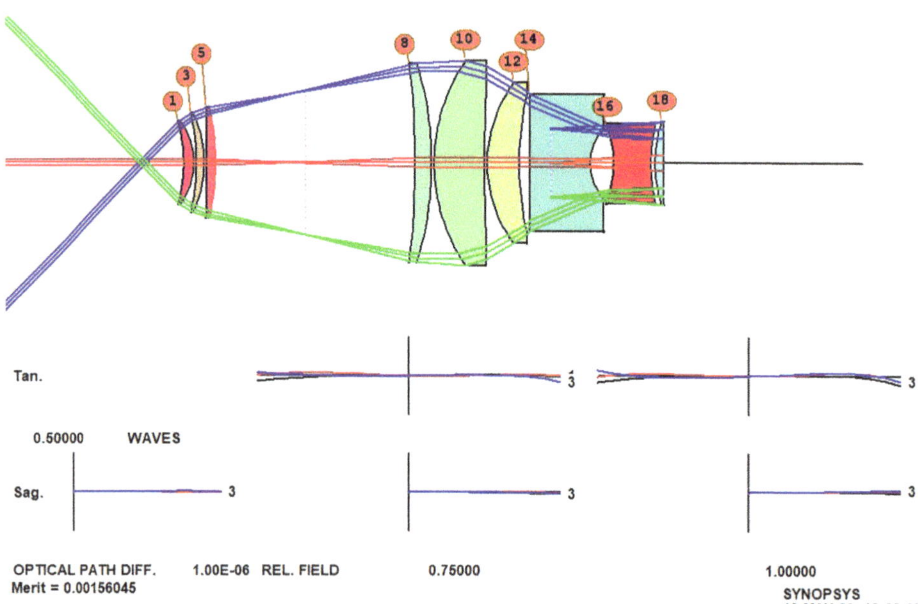

Figure 41.4. Lens with added surface for the field stop.

Now type, in the WS edit pane,

```
7 FLAG
```

and click Update. Now you can refer to that surface in the AANT file with that name.

Edit the MACro that DSEARCH prepared for you. Here, you add some GTR ray sets to control the tangential blur at the flag surface. We don't care about errors in the X-direction since they don't affect the sharpness of the field stop as seen by the eye. Also, correct the difference between the full-field chief rays in colors 1 and 3 at the stop, so the image of the stop will not show objectionable lateral color.

```
PANT
VY 0 YP1
VLIST RD ALL
VLIST TH ALL
VLIST GLM ALL
END

AANT P
AEC 3 1 1
ACM 3 1 1
ACC
GTR 0 2 4 P 1 0 FLAG
GTR 0 2 4 1 1 0 FLAG
GTR 0 2 4 3 1 0 FLAG
M 0 10 A 1 YA 1 0 0 0 FLAG
S 3 YA 1 0 0 0 FLAG

M    0.142857E+00  0.100000E+02 A CONST 1.0 / DIV FNUM
GSR      0.500000      3.000000       4  M      0.000000
GNR      0.500000      1.000000       4  M      0.300000
GNR      0.500000      1.000000       4  M      0.600000
GNR      0.500000      1.000000       4  M      0.750000
GNR      0.500000      1.000000       4  M      0.900000
GNR      0.500000      1.000000       4  M      1.000000
GSO      0.500000      0.246460       4  M      0.000000
GNO      0.500000      0.082153       4  M      0.300000
GNO      0.500000      0.082153       4  M      0.600000
GNO      0.500000      0.082153       4  M      0.750000
GNO      0.500000      0.082153       4  M      0.900000
GNO      0.500000      0.082153       4  M      1.000000
M    0.200000E+03  0.100000E-01 A TOTL
ACA 50 1 1
ADT 10 .1 10
M 15 1 A P YA 1 0 0 0 1 ! control eye relief
M -.008 10 A P HH 1  ! aim light at objective to right
M -.004 10 A P HH .5 ! and control pupil aberrations
M -.0064 10 A P HH .8
M 0 1 A P YA 1  ! and distortion too
S GIHT
END
SNAP  10/DAMP    1.00000
SYNOPSYS   80
```

Run this MACro, and the image gets worse. The program has made some tradeoffs. DSEARCH does not know about field stops, so the lenses it returned were not selected for that property. The lens is shown in figure 41.5. Image quality is not good.

Can we improve this lens?

It's time to run the **Automatic Element Insertion** feature. An expert would observe that you cannot correct lateral color at the field stop without a flint element to the left. Let's see if AEI can figure that out. Add the line

```
AEI 6 1 123 0 0 0 10 2
```

before the PANT command, and run the MACro again. The program adds a flint element at surface 5, and the image is much better. Comment out the AEI line, optimize again, and then anneal. The merit function comes down, as shown in figure 41.6.

One also has to monitor intermediate field points in wide-angle designs like this. Run the PAD scan |↑|, and you see the correction remains below ¼ wave everywhere.

Make a checkpoint and type **MRG** to open the Real Glass Menu. Select the Ohara catalog, Library 6, QUIET, SORT, PREFFERED, and then OK. The lens is assigned real glass everywhere.

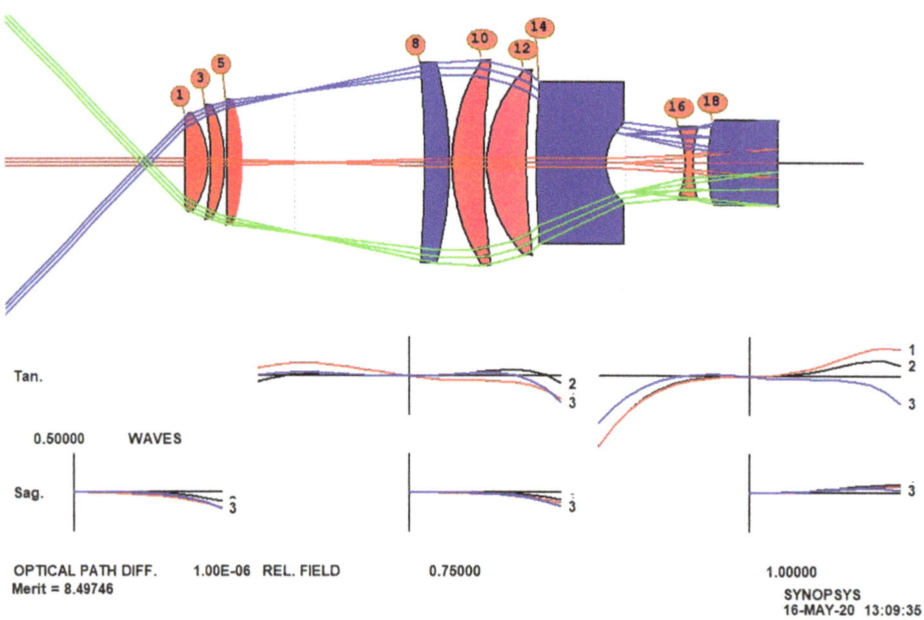

Figure 41.5. Lens optimized with field stop, before adding a new element.

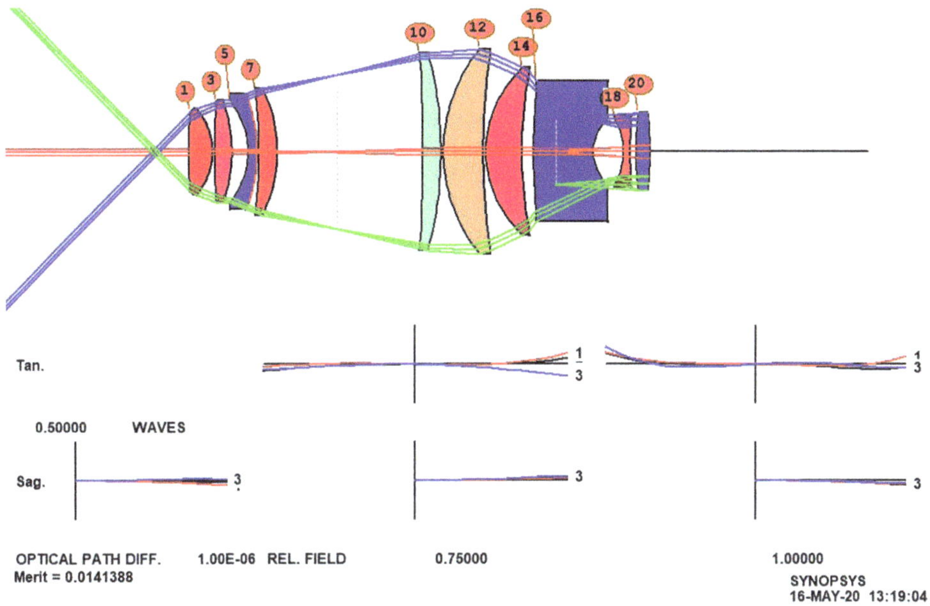

Figure 41.6. Lens reoptimized with added element.

```
--- ARGLASS 6 QUIET
Lens number      6 ID EYEPIECE EXAMPLE
 GLASS S-FPL51          HAS BEEN ASSIGNED TO SURFACE   1; MERIT =   0.137816E-01
 GLASS S-FPL51          HAS BEEN ASSIGNED TO SURFACE   7; MERIT =   0.124168E-01
 GLASS S-FPM2           HAS BEEN ASSIGNED TO SURFACE  18; MERIT =   0.120108E-01
 GLASS S-PHM52          HAS BEEN ASSIGNED TO SURFACE  14; MERIT =   0.122779E-01
 GLASS S-BSM15          HAS BEEN ASSIGNED TO SURFACE   3; MERIT =   0.129175E-01
 GLASS S-LAH64          HAS BEEN ASSIGNED TO SURFACE  12; MERIT =   0.134646E-01
 GLASS S-LAH95          HAS BEEN ASSIGNED TO SURFACE  10; MERIT =   0.129732E-01
 GLASS S-NPH4           HAS BEEN ASSIGNED TO SURFACE   5; MERIT =   0.100322E-01
 GLASS S-NPH4           HAS BEEN ASSIGNED TO SURFACE  16; MERIT =   0.259975E-01
 GLASS S-TIH11          HAS BEEN ASSIGNED TO SURFACE  20; MERIT =   0.171688E-01
 Type <ENTER> to return to dialog.
 SYNOPSYS AI>
```

The lens is now almost perfect. Let's look at the distortion. Type **GDIS 21 G**. The eye will not notice any distortion at all, as shown by figure 41.7.

Now we have to check how well the image is corrected at the field stop. Make a checkpoint and type

```
CHG
9 MXSF
END
```

This truncates the lens at surface 9 (temporarily, so we can evaluate the image at the field stop). Only the TFAN affects the sharpness of the field stop as seen by the eye. We corrected those rays above, but now want to see how well it worked.

Use the Spectrum Wizard to model 10 wavelengths, visible spectrum, bright light, Get spectrum, and Apply to lens.

Then open the Image Tools Menu (**`MIT`**), select a reference dimension of 0.2 mm, Coherent effect, Multicolor, Point source at HBAR = 1, and click Process, as shown in figure 41.8.

Indeed, the blur at the field stop is close to the diffraction limit in the *y*-direction. Restore your checkpoint so you can evaluate the final image.

This lens seems to meet every one of our goals. To verify, run the **Spectrum Wizard (MSW)** again to define 10 wavelengths spaced across the visible spectrum, and then run the **`OFPSPRD`** feature to show the diffraction pattern over the field. (This is best done with the **`MPF`** dialog; select Show visual appearance, Magnify 4.) The result is in figure 41.9. This is as close to a perfect eyepiece as you will likely ever see.

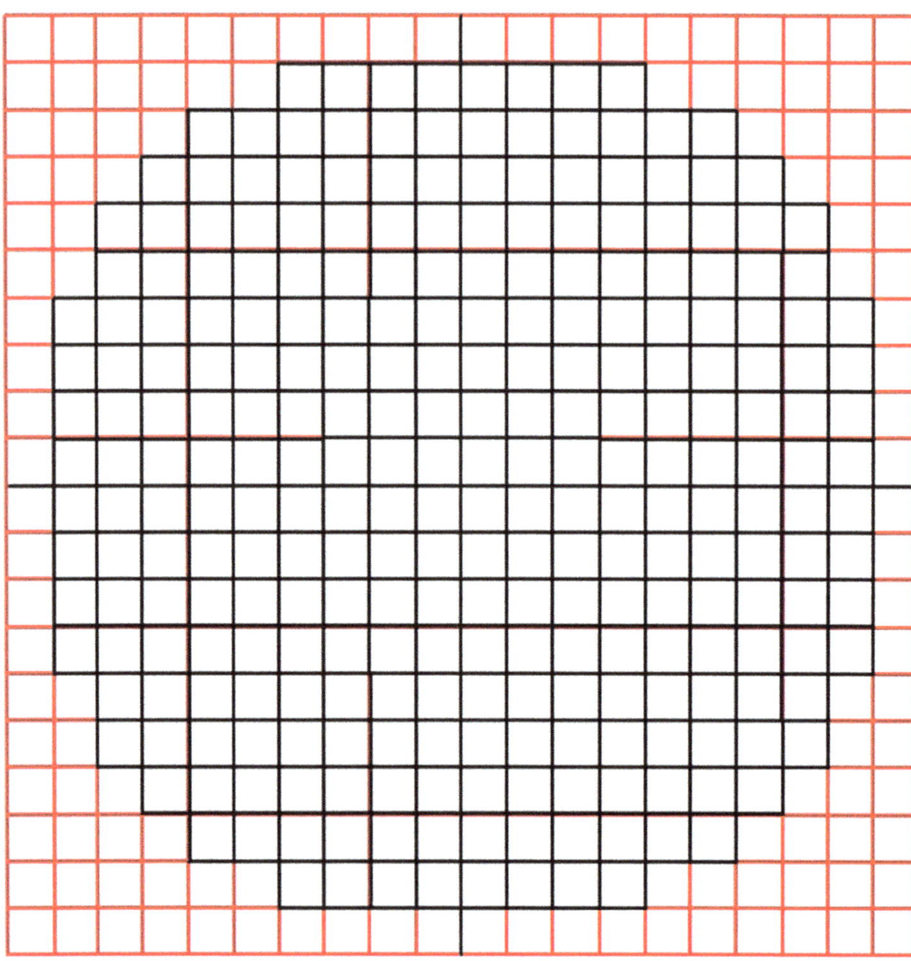

Figure 41.7. Distortion plot for the final design.

To the eye, this eyepiece will produce an image that is essentially perfect and undistorted. A check of the pupil aberrations shows that pupil wander is less than the allowed ½ mm over the field. It is unlikely you will ever see a telescope as well corrected as this, and if you do, this will be a superb combination!

Since this is a paper exercise, I will stop here. A design for a real market would need more attention, perhaps adjusting the thickness of some of the elements. But the point has been made. The final lens is in **C41L1** and is shown in figure 41.10. We can show a graphical system summary by opening the dialog **MGS**, selecting Plot fringes, Wedge X of 3, and clicking OK. The result is shown in figure 41.11.

This result comes back when switch 98 is turned on, but as noted before, for real design work you want that switch turned off. Then you may get a different set of design forms every time.

We ran it several times with the switch off. Sometimes the results were not quite as good as this, and in one case we got a lens of only *nine* elements that was very nearly as good as this 10-element lens. DSEARCH can explore hundreds of the branches of the design tree in a matter of seconds, and with slightly different input will explore yet others. For investigating the design space, this is the tool to use.

What happens if you run AED on this lens? We got a nine-element lens that way with almost the same performance. And considering that we are adding a

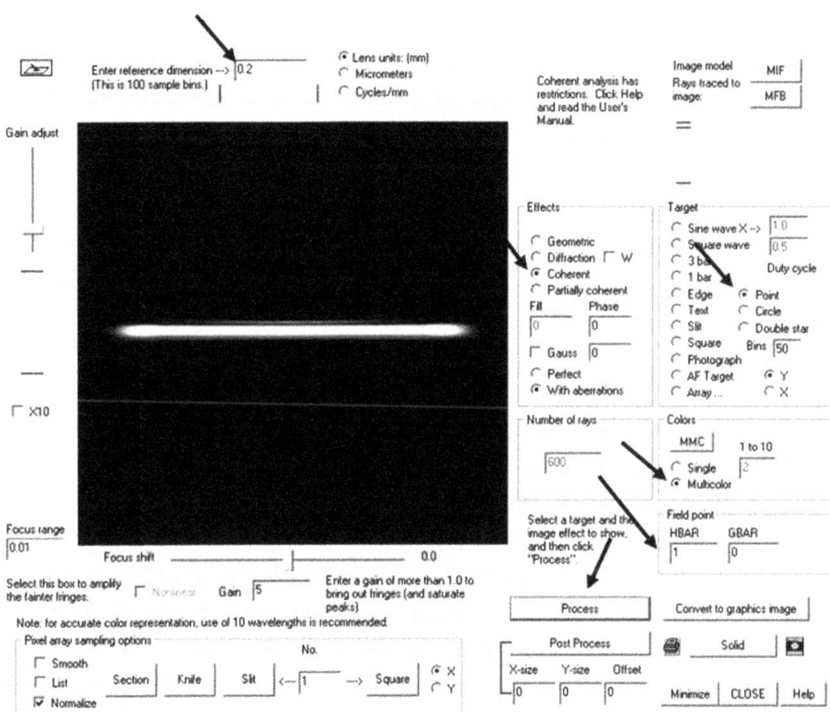

Figure 41.8. MIT dialog with image of a point on the edge of the field stop. This will look sharp to the eye in the *y*-direction.

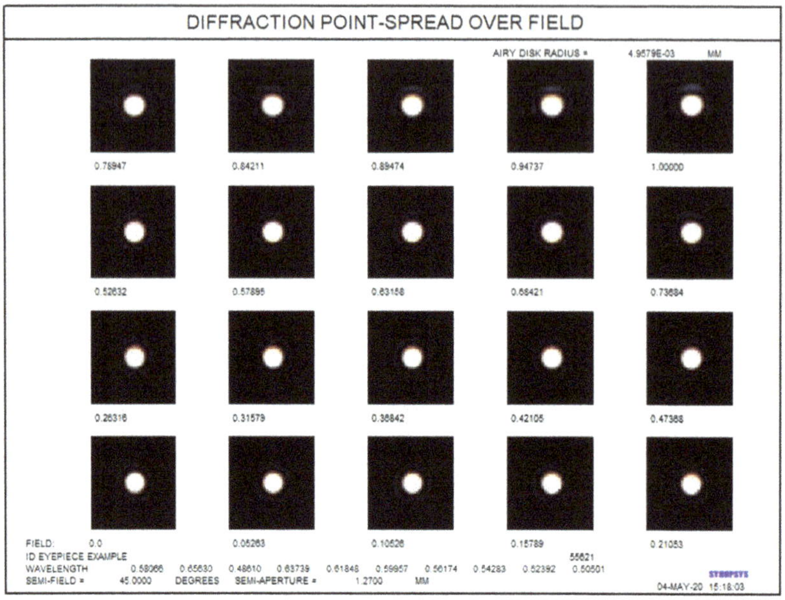

Figure 41.9. Diffraction pattern across the field of the eyepiece.

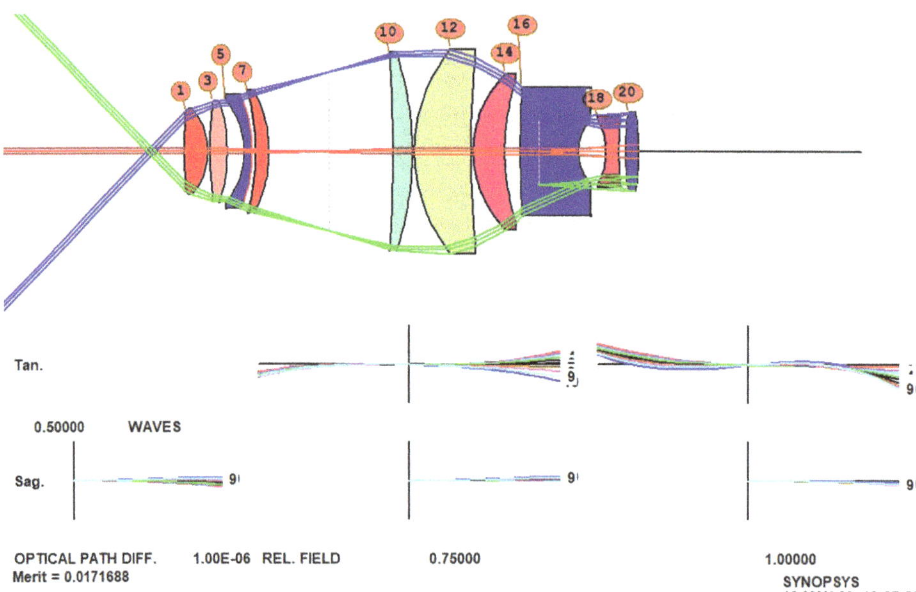

Figure 41.10. The final eyepiece design.

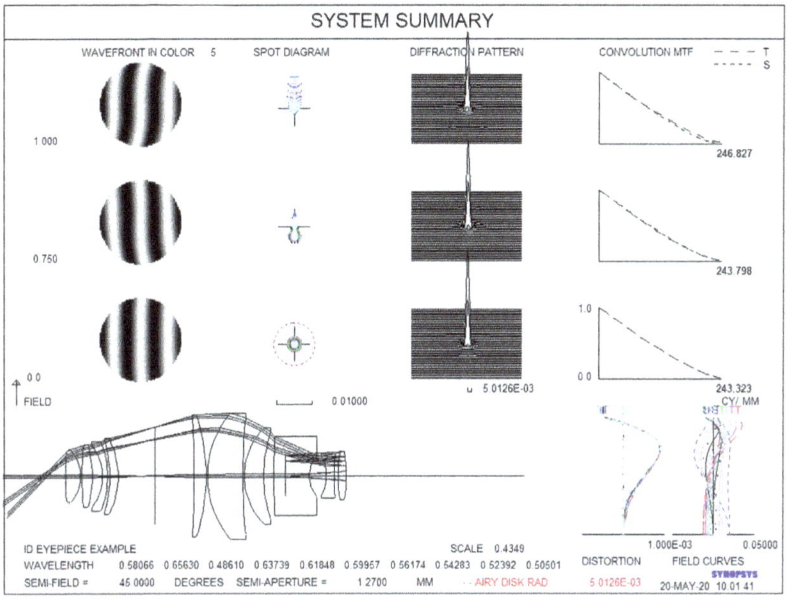

Figure 41.11. Graphical system summary of the final eyepiece design.

requirement (correction at the field stop) that is not expressed in the DSEARCH input, it would make sense to go back and try some of the other configurations the program returned. It's possible that a design other than the one we chose will turn out to be superior when you are done.

New users may wonder why this lesson called for object type OBD and activated the WAP 1 option. There is some optics here, and it's a good idea to understand it. When designing an eyepiece like this, you are after what's called an 'F-theta' lens. In a normal camera lens, one wants the image height to be proportional to the object height; then there is no distortion. But that would not work in an eyepiece, for which one wants the *angles* of the object and image to be proportional, not the heights. Object OBD specifies an object angle (here 45 degrees, tracing from the eye point) and then the field parameter *HBAR* refers to a fractional *angle*, instead of height, as well. When distortion is corrected, the angles are proportional, and the apparent angular separation between a pair of a double stars is constant, no matter where they appear in the field of view, as one would expect. Since the angular magnification is constant over the field, by Lagrange's law the diameter of the entering beam (at the eye) should be constant as well. The WAP 1 option takes care of that.

What did we learn from this exercise? It is clear that the numerical approach works. Designers of the classic school will work for many days on a design like this and will rightly be proud of the result if they succeed. They will have developed some insight into which elements are correcting which aberrations and the like. The numerical tools used in this lesson, on the other hand, will produce an excellent design in a fraction of the time. If your goal is to get a product out the door at minimal cost, never mind how it works, then the numerical approach is clearly superior. And if you do want to know how it works, check out the **THIRD CPLOT** feature. It's all there, in living color.

IOP Publishing

Lens Design (Second Edition)

Automatic and quasi-autonomous computational methods and techniques

Donald C Dilworth

Chapter 42

A zoom lens from scratch

Designing a zoom lens with no starting design; changing zoom numbers

It's noon on Friday. Your boss runs in. 'The customer wants a proposal for an 8x zoom lens by 8:00 Monday morning'. You've never designed a zoom lens in your life. Your job hangs in the balance. He gives you a list of requirements and heads out. What do you do now?

You might go to a patent database and try to find a lens that resembles what you need. That would probably take all weekend. But you have a better plan.

1. Start SYNOPSYS.
2. Type **HELP ZSEARCH** in the Command Window. Open that chapter, which is 10.7.3.
3. Read the whole chapter. Yes, read it. To be a professional lens designer you have to know what you are doing. But if you already know how to do other tasks on SYNOPSYS you are almost there. So it's an easy read. And now you know something about zoom lenses—so it's time well spent.
4. Set up your input to ZSEARCH. The lens is to be F/3.5 with a semi field angle of 14 degrees in the wide setting and a GIHT of 5 mm. This implies a focal length of 20.05 mm, and the semi-aperture is therefore 2.85 mm. The lens must be able to focus over a range of object distances from four meters to infinity.

(This is a good place to underscore what you learned in the manual: SYNOPSYS does not use multiconfigurations to do zoom lenses. A single configuration can model up to 20 zoom settings.)

doi:10.1088/978-0-7503-3695-6ch42

Here is what will be in your MACro (**C42M1**):

```
LOG                 ! to keep track of things later
OFF 1
ON 99
TIME                ! to see how long this run took
CORE 64             ! to run faster

ZSEARCH 3 QUIET     ! save results in library location 3

SYSTEM
ID ZSEARCH TEST
OBB 0 14 2.85         ! infinite object, 14 degrees semi field, 2.85 mm
                      ! semi aperture.
                      ! This defines the wide-field object
UNI MM
WAVL CDF
NOVIG
END

GOALS
ZOOMS 5
GROUPS 2 3 3 3      ! lens has four groups with 11 elements altogether
ZGROUP 0 Z Z 0      ! groups 2 and 3 will zoom

FINAL                 ! declare the desired object at the last zoom position
                      ! which is the narrow field zoom
OBB 0 1.7545 22.8     ! object is 1.7545 degrees semi field and 22.8 mm


                       ! semi aperture.
                       ! This implies an 8X zoom.

ZSPACE APERT           ! other zoom apertures will be evenly spaced between
                       ! the first and last
ZFOCUS 4000 4 15       ! also correct for object at 4 meters
APS 17                 ! put the stop on the first side of the last group
RT 0.25
GIHT 5 5 10            ! the image height is 5 mm for all zooms, with a
                       ! weight of 10.
BACK 5 .01      ! the back focus is 5 mm and will vary.  A target
                ! will be added to the merit function with a low weight.
FOV 0 .4 .6 .85 1      ! correct five field points
FWT 5 4 3 3 3
COLOR M                ! correct all defined colors
ANNEAL 50 10  Q 40     ! anneal the lens as it is optimized in quick and
                       ! real-ray modes
QUICK 50 100           ! 50 passes in quick mode, 100 in real-ray mode
END

SPECIAL AANT
AAC 30 1 5      ! request a maximum semi aperture on all elements of 30 mm
ACA 50 1 1      ! monitor rays to keep away from the critical angle.
M 212 .01 A SECTION FOCL 1 4
END
GO              ! start ZSEARCH
TIME
```

This zoom lens will consist of four groups, the first with two elements and the others with three. The first group will be used for range focus, and the last is fixed to provide a constant F/number over the zoom range. It is likely you will need more than 11 elements—and you can ask ZSEARCH for more if you want to—but it will run faster if you start here and add elements later where needed.

Run this MACro and watch the progress on a set of windows that monitor each of the cores that have been authorized, a portion of which is shown in figure 42.1.

When quick mode is finished, the program optimizes each of the best 10. In about 21 minutes you see the results shown in figure 42.2.

A glance at the merit function values shows that most of these are promising candidates. The program displays the best one in PAD.

Run the MACro **ZSS**, which ZSEARCH has created, to look at all 10 matches. We like the top one. Not perfect yet, to be sure—but not bad considering we gave the program nothing but a list of goals and constraints. This will be our starting lens, shown in figure 42.3.

The program has created an optimization MACro already loaded with a starting merit function definition and a set of variables. But you have to adjust some things. Many of the elements are too thin, so add two monitors:

```
AEC
ACC
AZA
ACA
ACM 3 1 1
ADT 7 .1 10
```

The **ADT** monitor will penalize any variable thickness that is less than 1/7 of the aperture diameter. The paraxial stop on 17 also needs to be changed to a real stop, with **APS -17** in **WS** or a **CHG** file.

Then run the MACro and anneal (**20, 2, 50**). The MF is now at 0.831, as shown in figure 42.4.

Often one wants to know why the MF won't go any lower. Type **FINAL 5,** (or the AI symbol **FF**, which will substitute that string of characters), to show the five largest items in the MF.

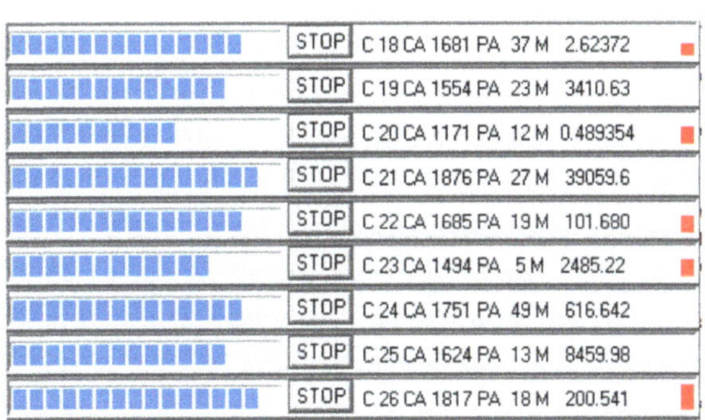

Figure 42.1. Progress bars showing multicore operations. The red blocks show the progress of the annealing stage.

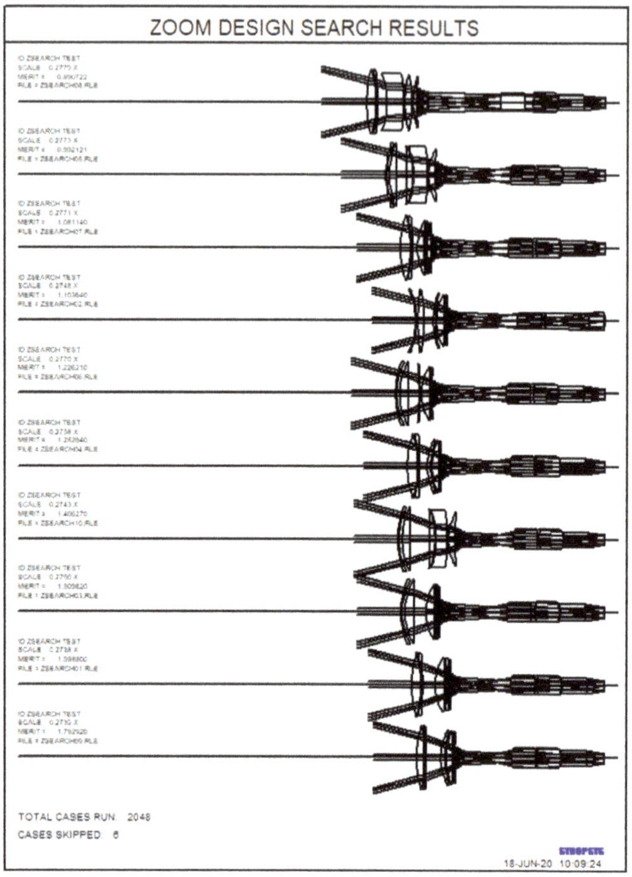

Figure 42.2. 10 best zoom lenses found by ZSEARCH.

```
SYNOPSYS AI>FF

 FINAL 5
 ABERRATION LIST
        NAME              TARGET          WEIGHT              RAW VAL.  FINAL ERROR   R. EFFECT

      1 AEC                 1.0000000       1.0000000            ------     0.886927E-01   0.009470

      3 AZA                 1.0000000       1.0000000            ------     0.621713E-01   0.004653

      7                     5.0000000       0.0100000           21.4424   0.164424         0.032546
         A    BACK    ACON 2

      8 AAC                30.0000000       1.0000000            ------     0.261205         0.082135

   2588                     0.0000000       4.0721936 SR         -0.0151  -0.614933E-01   0.004552
     A    1  YC     0.40000    0.37500    0.87500   0.00000        ACON 2      ZOOM 5
 *** ZFOCUS CHANGED TO      4000.00       4   15.0000        5.00000       ***
```

Here you notice that the **AAC** monitor wants an aperture larger than 30 mm. That aberration is larger than any of the others. Let us assume we can tolerate a larger size.

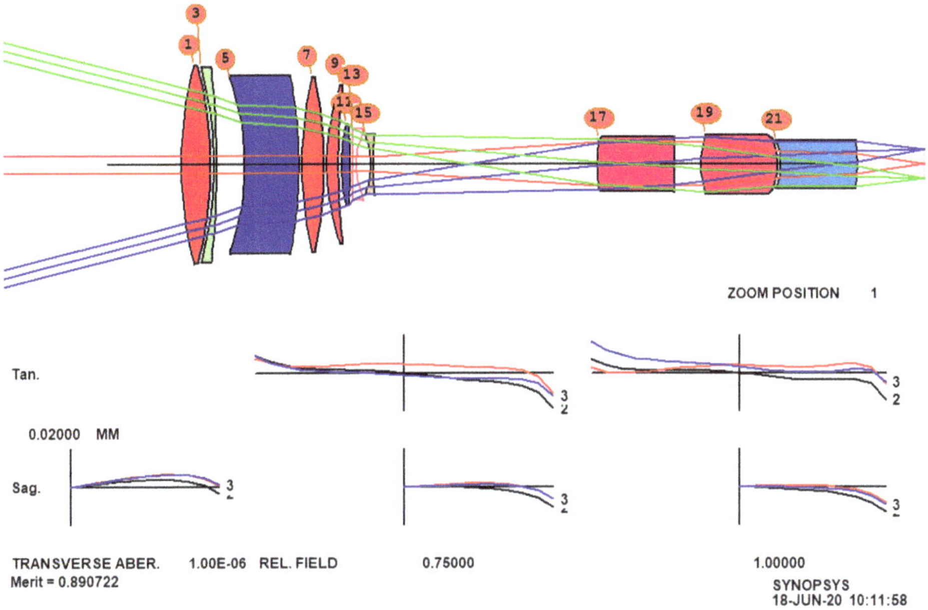

Figure 42.3. A starting zoom lens found by ZSEARCH.

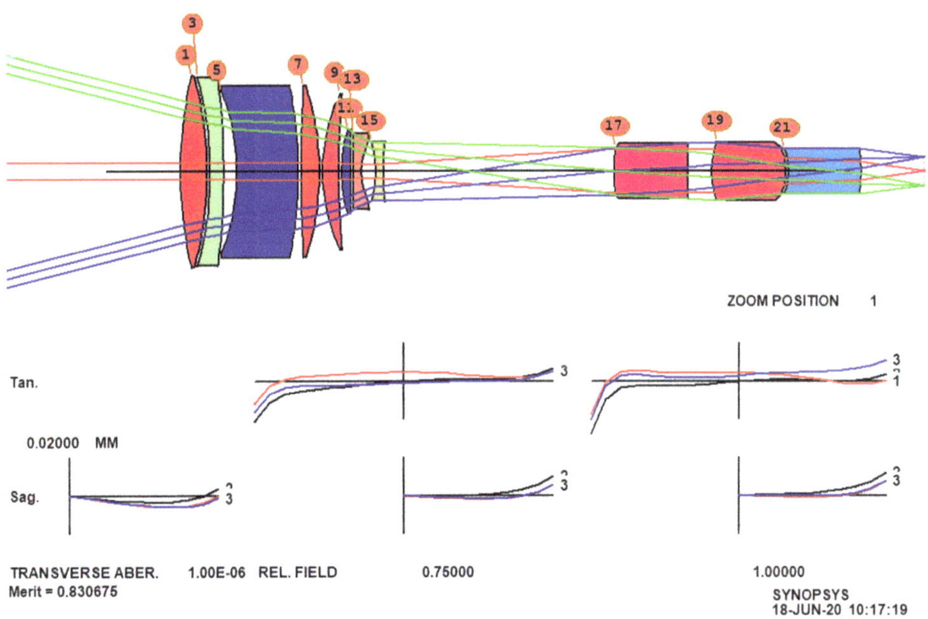

Figure 42.4. Zoom lens reoptimized and annealed.

In the AANT file, change the line

AAC 30 1 1

to

AAC 35 1 1.

Here's a useful trick: When you get to a stage that you might want to come back to later (if some ideas do not work out as well as you hope) click the **ACON copy** button in the top toolbar. If the current lens is in configuration 1, as it is by default, this puts a copy in alternate configuration 2, or ACON 2. (This is called *bumping* the lens.) Then make a checkpoint in that ACON and develop the design further—and you can instantly go back to ACON 1 with the '1' button should you want to. I use all six configurations this way quite often.

Make a new checkpoint, then run the MACro and anneal again. The MF comes down to 0.624, as shown in figure 42.5.

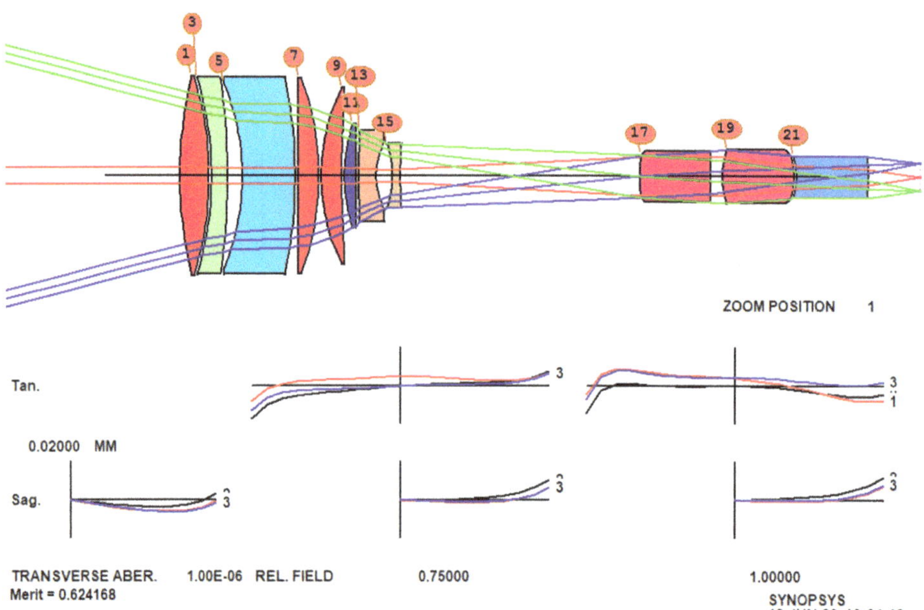

Figure 42.5. Zoom lens reoptimized with larger target on the aperture monitor.

Figure 42.6. The ZoomBar[1] buttons.

It's time to take stock. Click each of the zoom buttons, shown in figure 42.6, to see how the correction holds up over the zoom range.

The image is not bad; not as sharp at the wide-aperture setting, which is normal, but okay for now. We also should check how well the range focus is going to work.

Open a second EE editor (type **AEE**) and enter the following

```
CHG
VIG
END
ACON BUMP
ZFOCUS 4000 4 15 5
PAD/U
```

Save this as a MACro named BUMP, and run it. This will turn on VIG mode (so rays outside feathered edges get vignetted), put a copy of the lens in the next-higher ACON, move the object point from infinity to 4 meters, and then move the first two elements to the left by 15 mm, which we specified as the focus adjustment at that range, and adjust the object height so the GIHT remains at 5 mm.

Now look at all five zooms in this configuration. All are quite well corrected, with the largest residual aberrations showing up at the end of the zoom range, as before. figure 42.7 shows the lens in zoom 5 at the near conjugate.

Let us see if adding an element improves things. Go back to the previous ACON (at infinity conjugate), make a checkpoint, add the line

```
AEI 9 1 123 0 0 0 20 2
```

before the PANT file, and repeat the optimization. The program adds an element to group 4 and the MF comes down to 0.221. Aberrations are nicely reduced. figure 42.8 shows the lens at near conjugate in zoom 5. We could probably get a small improvement with yet another element, but this is already a pretty good lens.

[1] ZoomBar™ is a trademark of Optical Systems Design LLC, an Arizona, USA corporation.

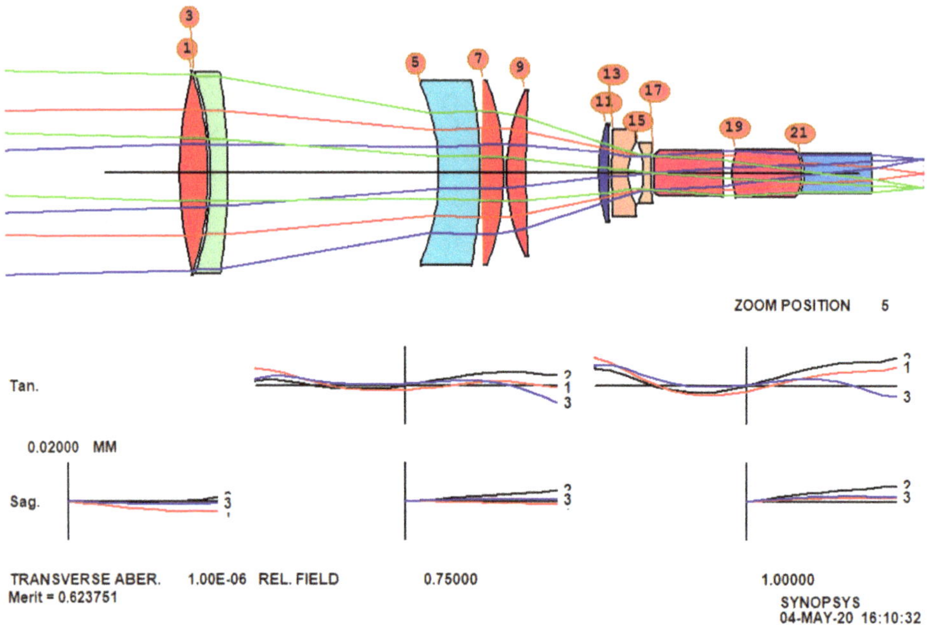

Figure 42.7. Lens in zoom 5 at 4000 mm conjugate.

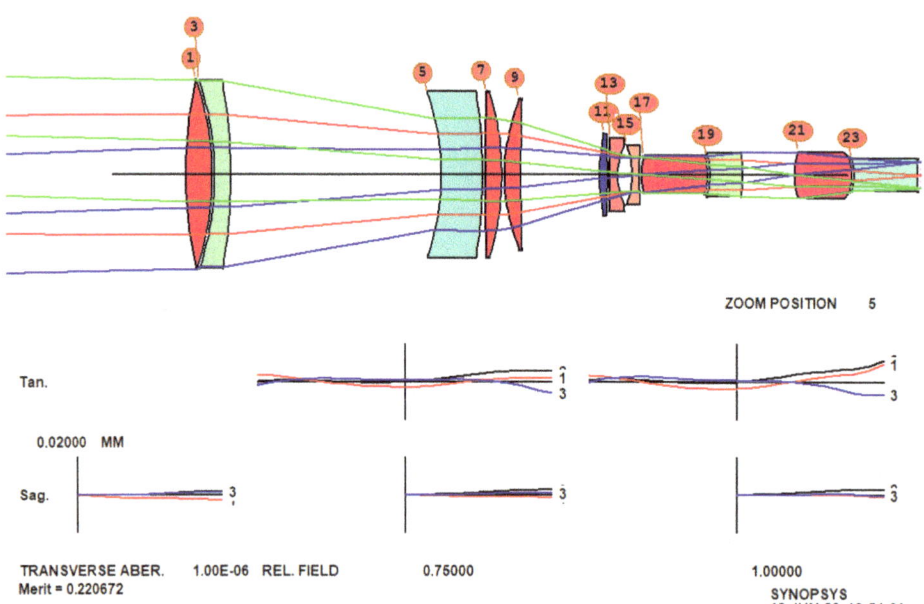

Figure 42.8. Zoom lens with element added to group 4, by AEI, at near conjugate, in zoom 5.

The lens works at five zoom positions, but what happens in between those zooms? We have to find out. Maybe the lens will be poorly corrected somewhere and we will have to add more zooms. The program uses a cam curve to interpolate the defined zooms at in-between points, and there are two types: power-series and piecewise cubic. The latter is often better, so open WS and click on the **ZFILE** button. **Z** This loads the zoom definitions into the edit pane. Type **CUBIC** where shown in figure 42.9, add the word **DFOCUS** after the group designation 5 to 10, and click the Update button and Close.

This says that the first zoom group is to be used for focusing at all zoom positions. (This is what zoom lens makers do in practice: move one group and refocus with another, then mark the locations and cut the cam curves accordingly.) We have requested that group 1 (from 5 to 10) adjust itself at each cam position so that the image remains at the same focus relative to the paraxial focus as in the designed lens.

Run the ZoomSlider and examine the image in both conjugates. The image is great everywhere–but, horrors! The lenses overlap in between zooms 1 and 2, as shown in figure 42.10. Group 2 has zoomed to the left and overlaps with group 1.

How can that happen? Doesn't ZSEARCH know how to watch out for this kind of error? Well, ZSEARCH only avoids interference at the vertex at the zooms it knows about, honoring the AZA monitor—and those zooms clear as they should. But in this case, settings *in between* zooms 1 and 2 have a problem. No problem. Here is how to proceed. Increase the number of zooms to 10 with **CAM 10 SET.**

```
WS -- WorkSheet Lens-Edit Window
 CAM EXPONENT    0.800000
  5  10  DEFOCUS
 11  16
 RSOLVES
CUBIC
 ZFOCUS    0.400000E+04    4   0.150000E+02    0.500000E+01
 ZOOM     2
 OBB  0.350825E-09      5.18054      7.83750    0.00000    0.00000    0.00000     7.83750
 ZDATA   -2.1944574E+00   2.8492506E+01
 ZOOM     3
 OBB  0.413335E-09      3.17130     12.82500    0.00000    0.00000    0.00000    12.82500
 ZDATA    1.9077426E+01   5.2315483E+01
 ZOOM     4
 OBB  0.350825E-09      2.28446     17.81250    0.00000    0.00000    0.00000    17.81250
 ZDATA    4.0467604E+01   7.2337683E+01
 ZOOM     5
 OBB  0.163293E-09      1.78510     22.80000    0.00000    0.00000    0.00000    22.80000
 ZDATA    6.0210323E+01   8.9317356E+01
```

Figure 42.9. The WorkSheet edit pane for changing zoom settings. Here, we have added a CUBIC declaration and specified that group 1 is to be adjusted to maintain the nominal paraxial defocus.

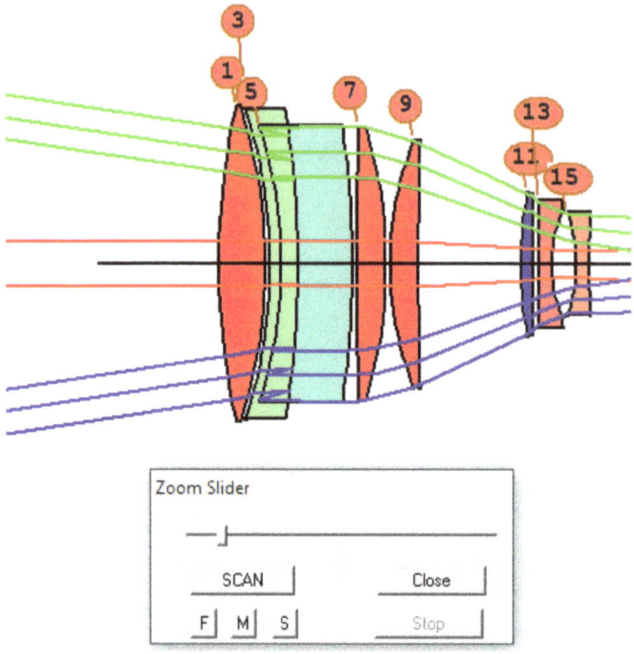

Figure 42.10. Interference between zoom groups when set in between defined zooms.

```
CAM CURVE
 ZOOM OFFSET, BY GROUP
 INDEX     1          2          3          4          5
 ______________________________________________________________
    1-6.661E-16  7.105E-15
    2    -9.6402    13.2954
 *** OVERLAP DETECTED *** GROUP   1
    3    -4.2174    25.5732
    4     4.5386    36.9094
    5    14.1902    47.3805
    6    23.9432    57.0666
    7    33.4963    66.0391
    8    42.7349    74.3607
    9    51.6256    82.0979
   10    60.1797    89.3174
 SYNOPSYS AI>
```

Now edit the optimization MACro. At the moment, it calls for correcting ray sets at each of the 5 zooms at both conjugates. Change the AANT file so it corrects *all* defined zooms instead of each one individually, with the ZGROUP ALL option. Note that the ZROUP sections require their own END line, in addition to the END that ends the AANT file.

```
PANT
VLIST RD ALL
VLIST TH ALL
VLIST GLM ALL
VLIST ZDATA ALL
END
AANT
AEC
ACC
AZA 3 10 5
ACA
ACM 3 1 1
ADT 7 .1 10
M     5.00000       0.100000E-01 A BACK
AAC 35 1 5             ! REQUEST A MAXIMUM SEMI APERTURE ON ALL ELEMENTS OF 30
ACA 50 1 1             ! MONITOR RAYS TO KEEP AW
ZGROUP ALL
M    0.500000E+01  0.100000E+02 A GIHT
GSR       0.500000      5.000000       4  M      0.000000
GNR       0.500000      4.000000       4  M      0.400000
GNR       0.500000      3.000000       4  M      0.600000
GNR       0.500000      3.000000       4  M      0.850000
GNR       0.500000      3.000000       4  M      1.000000
END
 ZFOCUS     0.400000E+04    4    0.150000E+02    5
ZGROUP ALL
GSR       0.500000      5.000000       4  M      0.000000
GNR       0.500000      4.000000       4  M      0.400000
GNR       0.500000      3.000000       4  M      0.600000
GNR       0.500000      3.000000       4  M      0.850000
GNR       0.500000      3.000000       4  M      1.000000
END
END
SNAP/DAMP 1
SYNOPSYS    50
```

Here we changed the AZA monitor to require 3 mm clearance at the vertex at all defined zooms, with a higher weight:

```
AZA 3 10 5
```

Now delete the AEI line, optimize and anneal again. The lens is corrected at 10 zooms, and there is no overlap anymore. This version is shown in figure 42.11. **(C42L1)**

We now have a pretty good lens, but elements 4 and 5 look somewhat strange and might be redundant. We might be able to remove one of them. The search routines don't care if something looks strange; they just crunch the numbers and call home. We have to try. Add a line before the PANT file

```
AED 5 QUIET 1 123
```

and reoptimize. The program reports that the best place to remove an element is at surface 23, but the lens is then not quite as good, so we reject this solution, shown in figure 42.12 in zoom 3.

Execute your BUMP MACro, and you see that the near conjugate is also quite well corrected. It's time for some finishing touches.

Since we have come a long way, it's time to save this lens so you can easily get back to it if necessary. Click the button in the top toolbar, and the lens is saved as

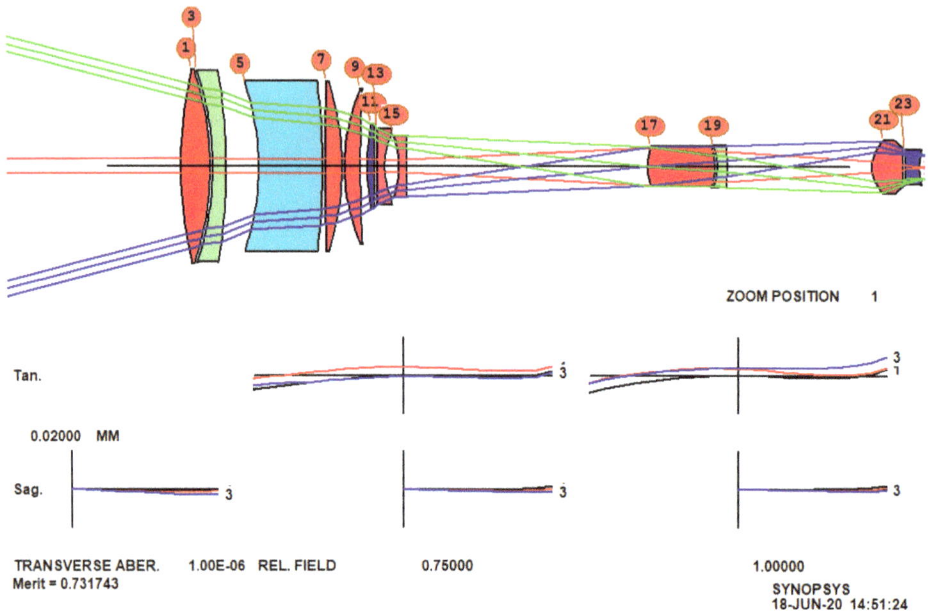

Figure 42.11. Lens corrected at 10 zooms, with no interference.

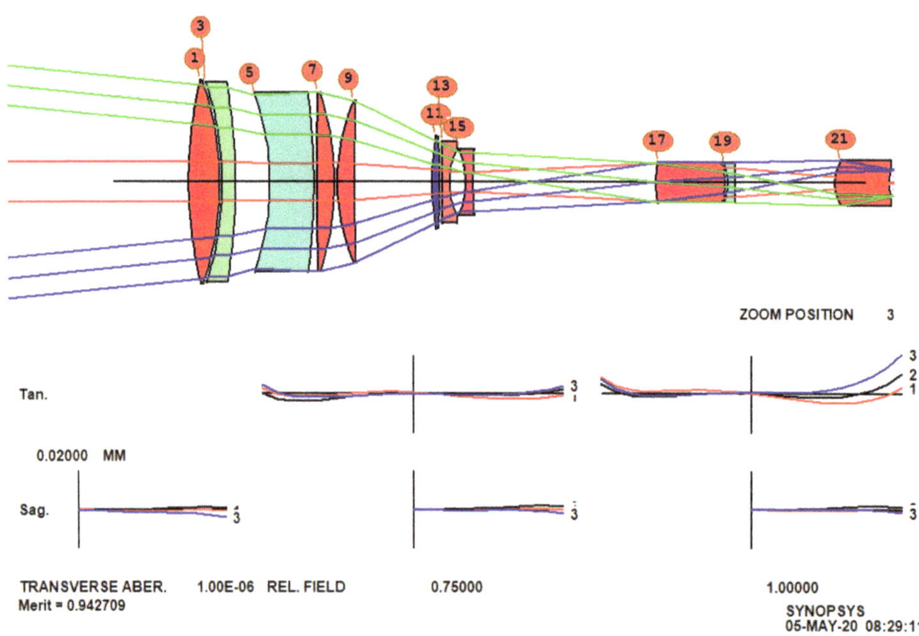

Figure 42.12. Zoom lens with element removed by AED, in zoom 3. The quality is degraded, so we revert to the previous design.

an RLE with a name taken from the current log number. ZSEARCH has already added a LOG command to your MACro, so each run increases this number. That way, whenever you get a version that you might want to go back to, just click that button and it's saved. And if you also want to save the optimization MACro, so you can document the file that created the lens, just click the same button on the editor toolbar. The MACro will be saved, with the same number, as a .MAC file.

We have to replace the model glasses with real glass, using ARGLASS (Automatic Real GLASS). It's a good idea to run the optimization once more before running ARGLASS, since it reuses the same merit function and variable list, and you want those to be current. (If you changed ACONs, you *must* run the optimization again in the current one, since ARGLASS uses the variables and merit function that apply to *that* ACON.)

Type **MRG** and select the Guangming catalog, QUIET, and SORT. Since geometric aberrations are the main issue here, not secondary color, we expect that the simpler ARGLASS will suffice and we don't need the performance of GSEARCH for this lens. The **MRG** dialog will prepare the input for, and run, ARGLASS.

Click OK, and you get the lens shown in figure 42.13. (**C42L2**)

Okay, we know the last group needs four elements, so let's run ZSEARCH again, ask for groups of 2, 3, 3, 4, and allow an aperture of 35 mm radius instead of 30, instead of working our way there in stages as we did above. Will the result be as good? Better?

Doing just that, then assigning a real stop to surface 17, adding monitors for lens thicknesses, optimizing and annealing once again, we got the lens in figure 42.14. In

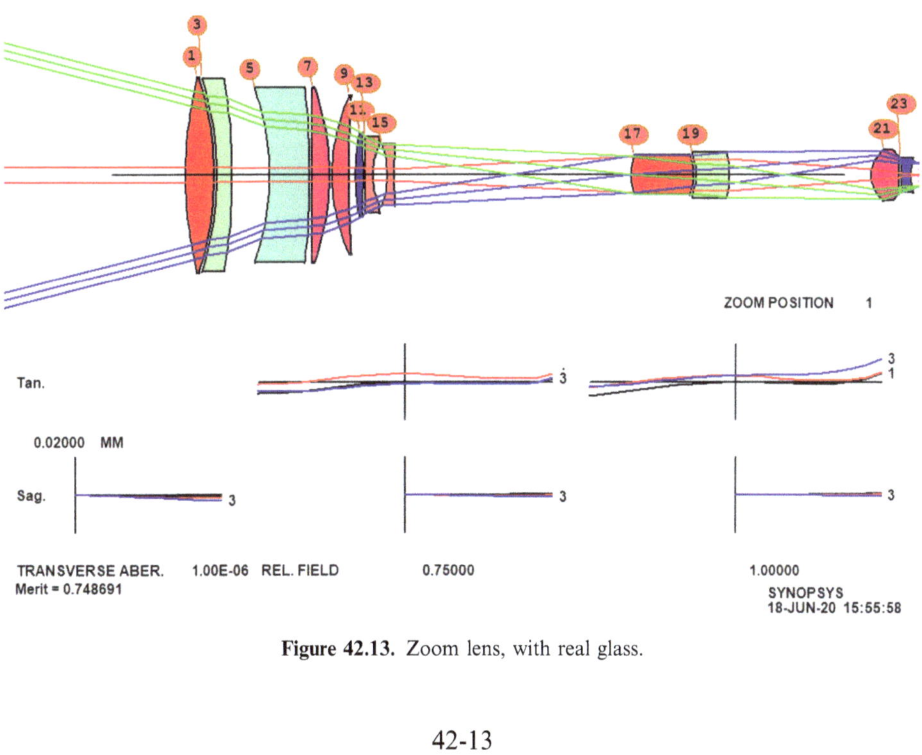

Figure 42.13. Zoom lens, with real glass.

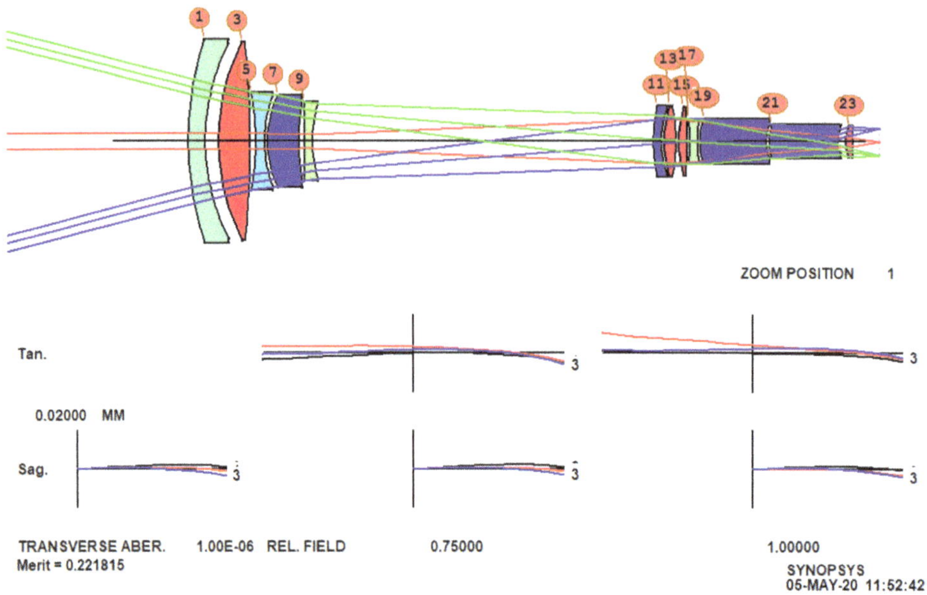

Figure 42.14. Zoom lens found when input parameters for ZSEARCH were altered per previous results.

this case, there was no overlap of zoom groups, and the ray fans were somewhat better. But the BUMP MACro shows that zoom 5 at the near conjugate is vignetted on the right side, as shown in figure 42.15. Element 2 is feathered at that zoom, and the vignetted rays are not well controlled anyway, so we elect to assign a hard aperture on that element at the infinite conjugate setting so those rays will not get through.

You will notice that, when ARGLASS matched real glass to the lens in figure 42.13, it selected glass type S-LAH58 for some elements, including the second. This is a very expensive glass and that is a large element, so this time, specify a price limit no more than 12 times that of BK7 when you run **MRG**.

Do just that, and when it finishes, tweak up the aperture on 11 as you did above. You get the lens in figure 42.16. **(C42L3)**

Now check the spot sizes. First, type

```
OFF 27
SSS .003
```

To show spot symbols at the entered size. Then type **MSF** to open the dialog shown in figure 42.17.

Change the scale to 50 μm as shown, and click OK. Do this for all zooms at both conjugates. The spot sizes are fairly constant over all cases. figure 42.18 shows the spots at near conjugate in zoom 5, which gives the largest spots, and figure 42.19 shows the spots at infinity conjugate in zoom 1, which is more representative of the other cases.

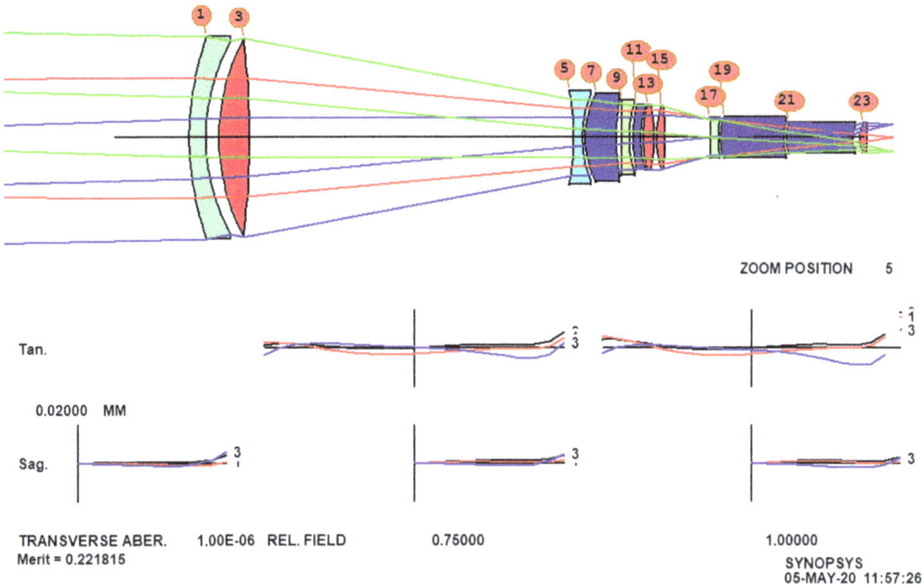

Figure 42.15. Lens at near conjugate in zoom 5. Rays are vignetted by a feathered edge at surface 3.

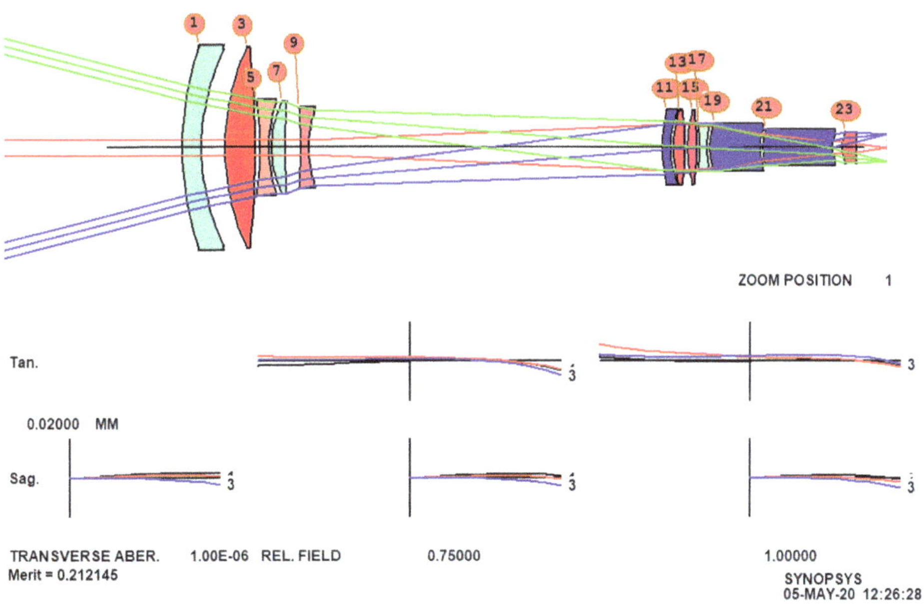

Figure 42.16. Final zoom lens, with real glass.

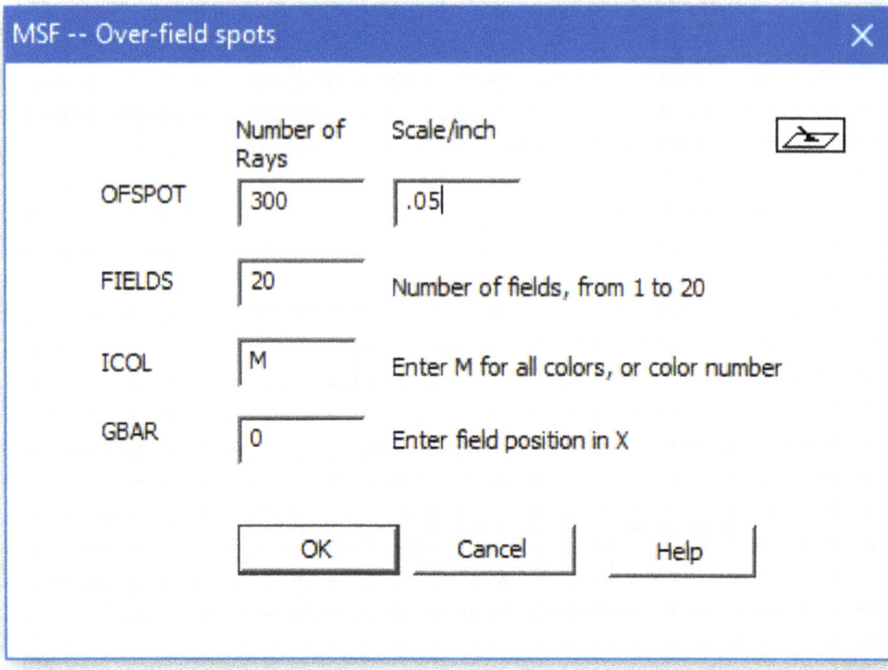

Figure 42.17. The **MSF** dialog for plotting spot diagrams over the field.

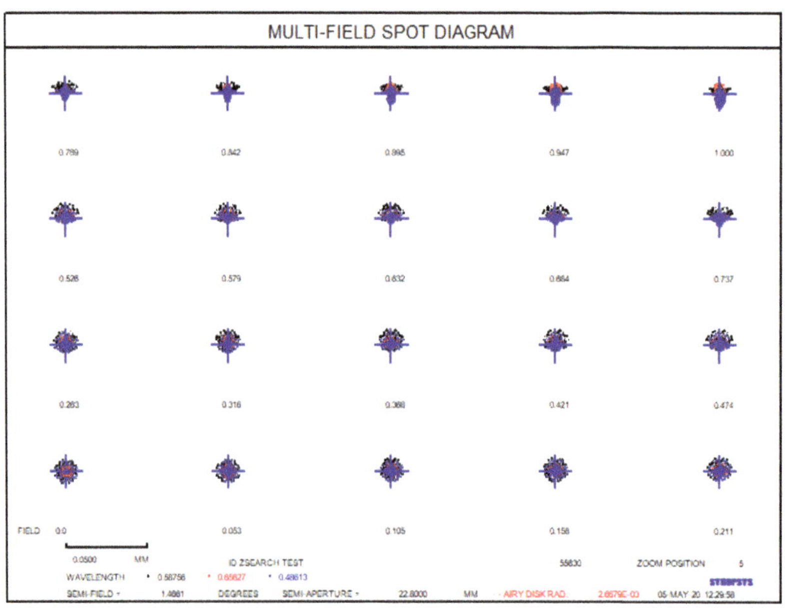

Figure 42.18. Spots plotted over field for zoom 5 at near conjugate.

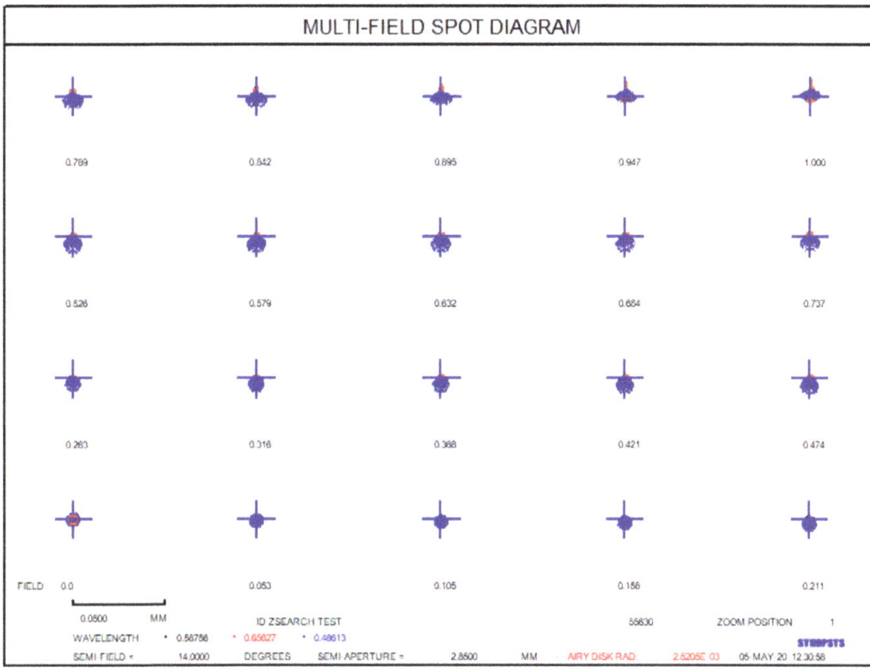

Figure 42.19. Spots in zoom 1 at infinity conjugate.

You quickly learn that the benefit of adding yet another element goes down as the number of elements gets larger. That makes sense: adding an element to a doublet increases the variable count by 50%. Adding one to a 10-element lens increases the count by 10%, and so on.

Type

```
CAM 10 SET
OFF 65
ZDWG .25
```

to see a picture of the lens at 10 zooms, as shown in figure 42.20.

We can also show the spot diagrams plotted as a function of defocus. Type in the CW:

```
SSS .005
OFF 27
```

Activate zoom 5, open the dialog **MTS**, enter a scale of 0.02, and click Execute. The plot is shown in figure 42.21. Switch 27 governs how the individual spots are drawn; here we elect to show them with symbols designating wavelength instead of single dots, and the command **SSS** sets the size of the symbol. Click in the picture on the monitor and drag a box around the center diagram. Then click inside the box. You see an enlarged image of that section, shown in figure 42.22.

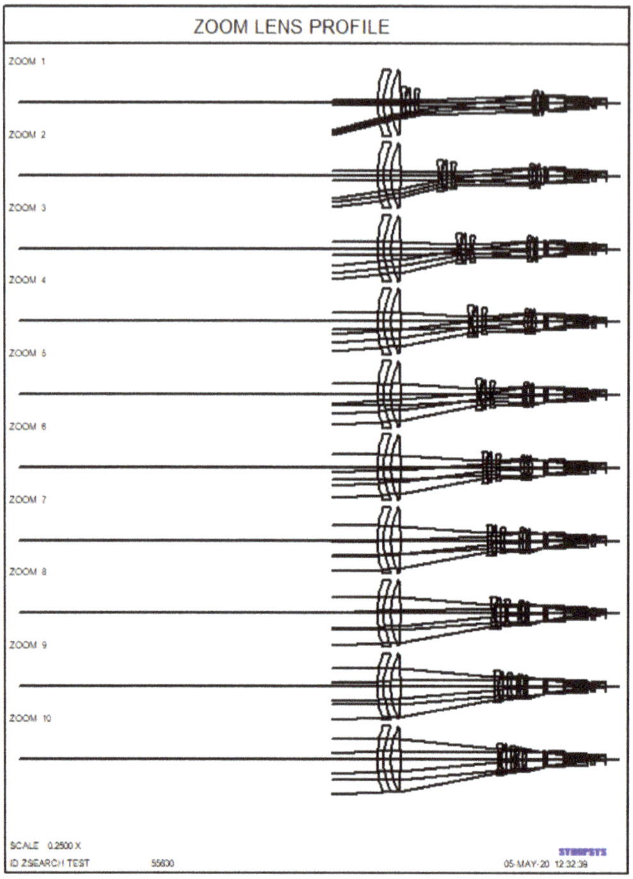

Figure 42.20. The zoom lens at 10 zoom positions.

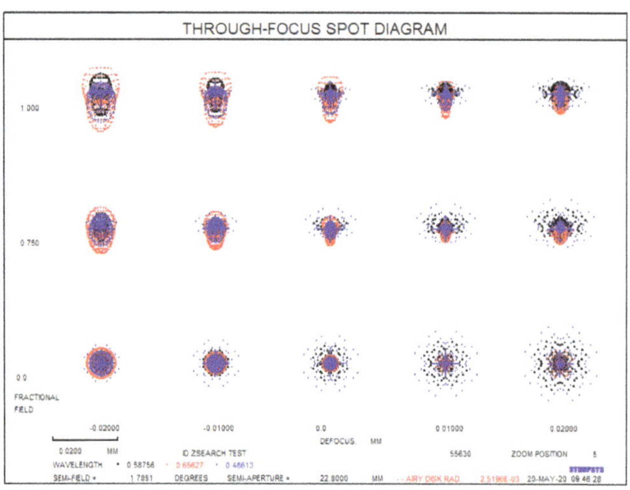

Figure 42.21. Spot diagrams at zoom 5, through focus.

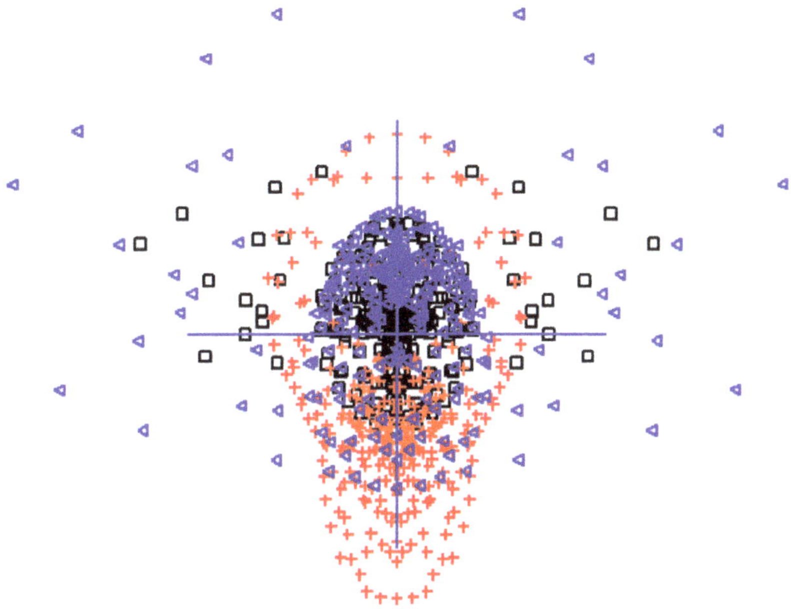

Figure 42.22. Center image enlarged, showing symbols designating wavelengths.

42.1 Zoom spacing

So far, we have said nothing about how the zoom settings were spaced over the zoom range.

In this example, they were evenly spaced over the aperture, as requested in the ZSEARCH file, and as shown in Figure 42.23. This plot is produced by the command **CAM 100 APERTURE**. The fiducial marks on the abscissa are the currently defined zooms

If we instead ask for **CAM 100 ANGLE**, the plot indicates that the zooms are spaced further apart at the wide-angle end and closer together at the other, as shown in figure 42.24. But this may not be the most advantageous arrangement, since we noted that the aberrations were changing most quickly at the wide-angle end. It would seem we need the zooms to be spaced closer at that end and do not need them so close at the other.

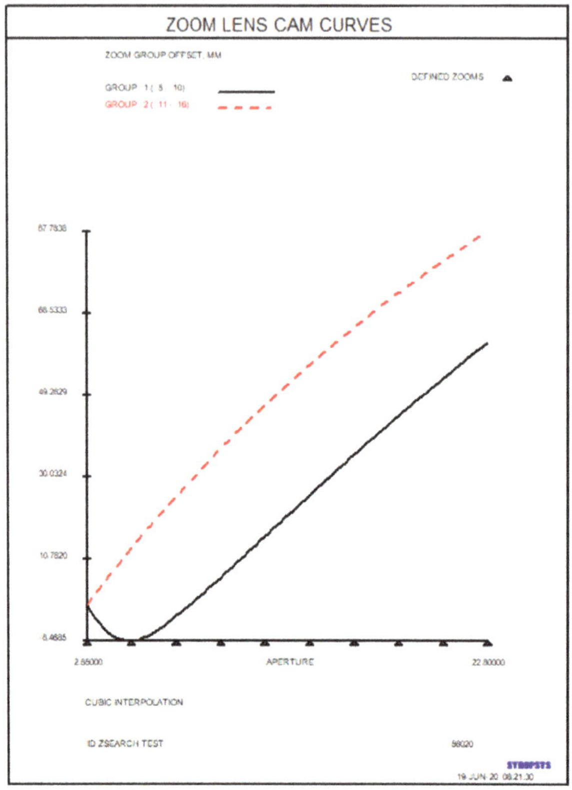

Figure 42.23. Cam curves showing zoom spacings from ZSPACE APERTURE, plotted over aperture.

Having learned that, we ran this job again, this time specifying **ZSPACE ANGLE** instead of **ZSPACE APERTURE**, and got a design with more zooms at the wide end —and found the narrow end was not as well corrected as before. It seems we went too far. The program has other settings, and in this case, when we asked for **ZSPACE ENDS**, they were spaced just about right, as shown in figure 42.25. This version is shown in figure 42.26. That setting spaces the zooms closer at both ends of the range and not as close near the center. (**C42L4**)

Now the lens does not show any overlap, even though only five zooms are defined. Still, we would increase the number to 10 zooms and reoptimize, just to get the best balance.

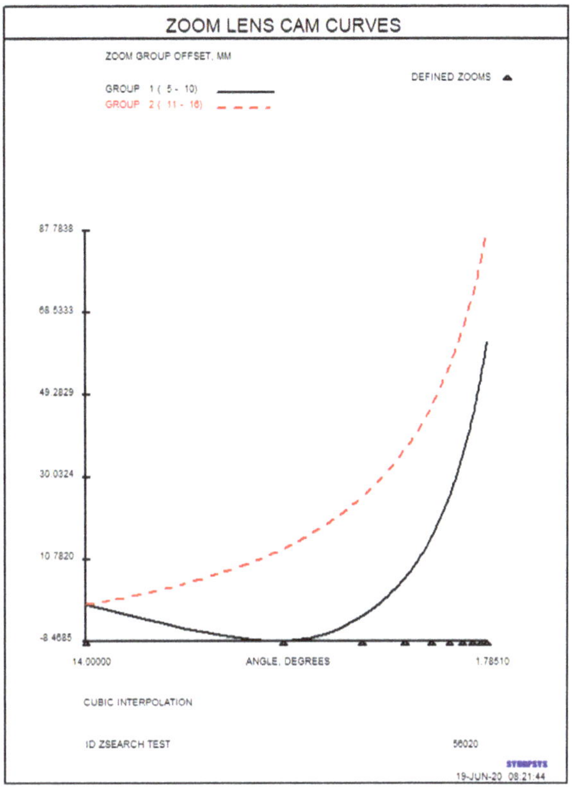

Figure 42.24. Zoom spacings plotted over angle. The wide-angle end has too few zooms.

We should point out that by selecting the best zoom spacing, one can get a surprisingly wide range of focal lengths with rather few zooms. One customer begged us to provide 45 zoom positions, unaware of the flexibility of this setting. When it is used properly, one does not need so many. We have gone as far as a 50:1 zoom with only 15 zooms defined. A number as large as 45 is quite impractical when you count how many rays one would have to correct over all colors, fields, and zooms. (And conjugates, if that is a requirement.) So it makes sense to understand how to use these options and request only the minimum number of zooms that will do the job.

It is also possible to respace the zooms after a design is well along, with the **CAM ZMAG** … command, which you should read about in the UM.

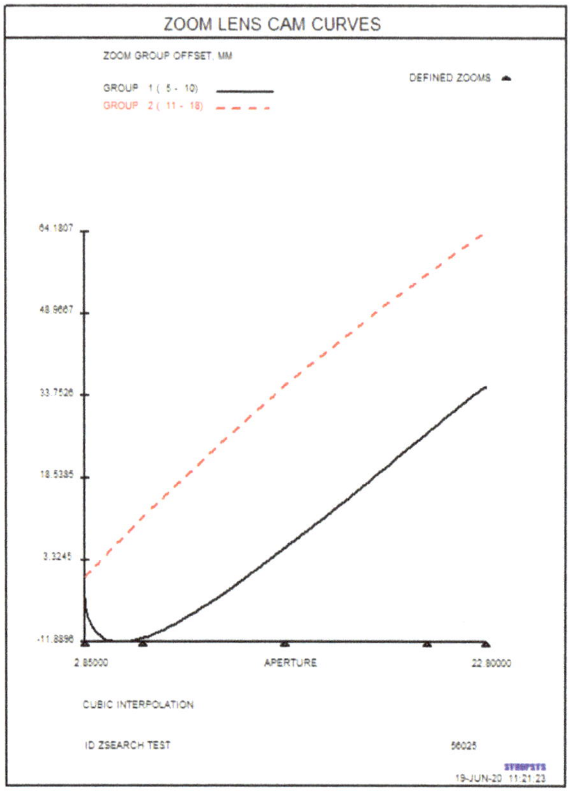

Figure 42.25. Zoom spacing when ZSPACE ENDS was requested.

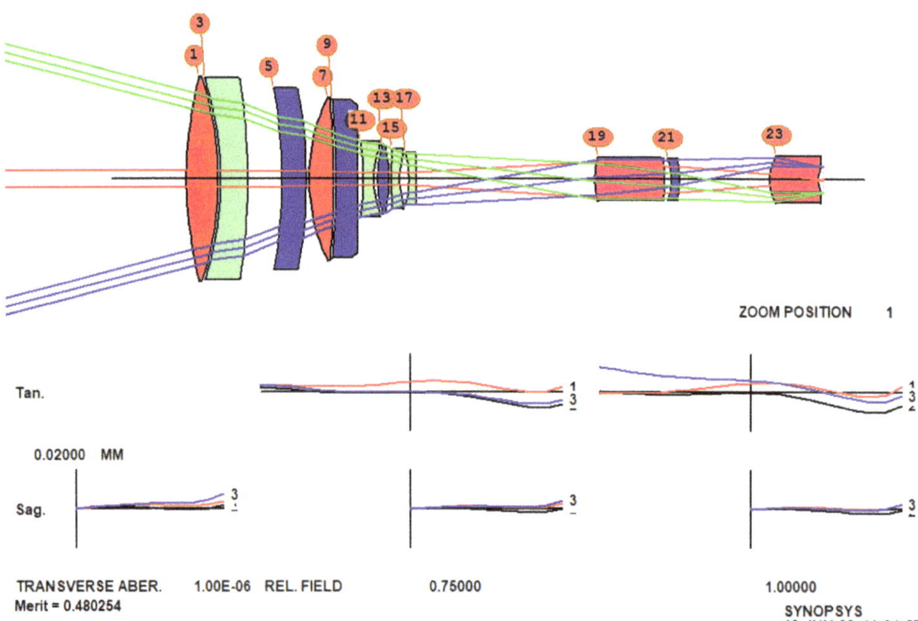

Figure 42.26. Lens returned with ZSPACE ENDS requested.

That's enough for this lesson. You get the idea. If the lens is still not good enough, another run with AEI should help. Your boss will be delighted to learn that you have a good design—and it's still only 3:00 on Friday afternoon. You can head out to the golf course.

IOP Publishing

Lens Design (Second Edition)

Automatic and quasi-autonomous computational methods and techniques

Donald C Dilworth

Chapter 43

Designing a free-form mirror system

Correcting the image; avoiding beam interference

Optics fabrication is becoming more and more sophisticated as better methods of shaping and measuring optical surfaces become available. One result is the emergence of 'free-form' optics, which consist of lenses or mirrors whose shape is not axially symmetric about the center of the part. A simple example is the off-axis paraboloid, where the parent is polished to the required aspheric shape and then the desired part is cut from that parent. More elaborate shapes may involve high-order aspheric terms described with either power series, Zernike, Forbes, or other polynomials. As interest shifts to such systems, it becomes important to be able to design them.

SYNOPSYS provides features that can simplify that process, and you should read about **FFBUILD** in the help file before you go any further. This lesson assumes you have read that chapter and shows how to design a system with free-form mirrors.

The first step is to lay out the rough geometry you are after. Here is an example with three mirrors, shown in figure 43.1.

Light will come in from the left at surface 1, hit mirrors located at 2, 3, and 4, and then go to the image plane at 5. Here is the input for FFBUILD (**C43M1**):

```
FFBUILD
SYSTEM
ID EXAMPLE FFBUILD
OBB 0 2 12 2 2
WAVL CDF
UNI MM
CFOV
END
GEOM
2 MIRROR 0 0 140
```


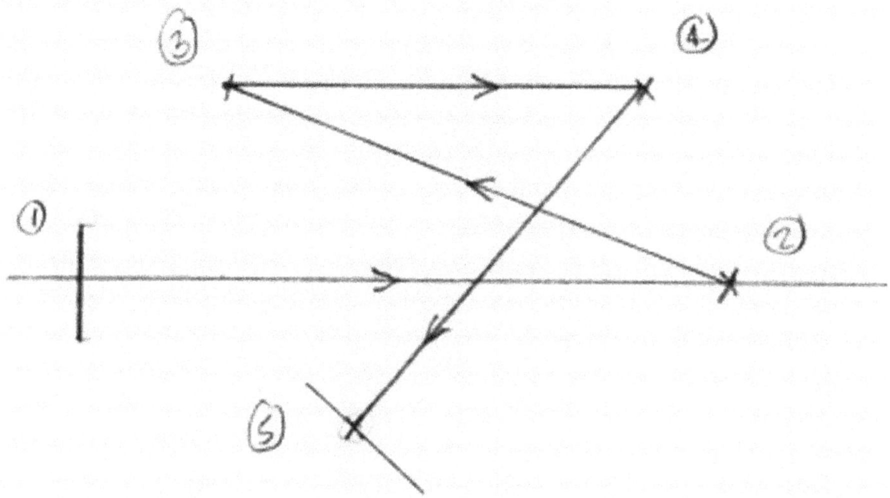

Figure 43.1. Proposed location of free-form mirrors.

```
3 MIRROR 0 40 30
4 MIRROR 0 40 120
5 IMAGE 0 -30 60 -7 7
 END

SHAPES
2 ZERN
3 ZERN
4 ZERN
END
```

In this example, the mirrors will be assigned Zernike polynomials, which accept up to 36 coefficients that are functions of polar coordinates on the surface. Since FFBUILD only supports designs with bilateral symmetry, terms asymmetric in X will not be used.

The above input contains part of the requirements: the semi-field angle is two degrees, the field is circular, and the semi-aperture is 25 mm; other requirements will be added as we go along. But first run the above input file, which produces two results: a mirror system (with flat surfaces at the moment) and an optimization MACro that contains most of the input needed to refine this design. Here is that system, shown in figure 43.2.

The optimization MACro is quite long and contains variables for the angles and global positions of the mirrors and image plane in Y and Z, as well as for the Zernike coefficients on the mirrors. Most of the variables are commented out, however, since

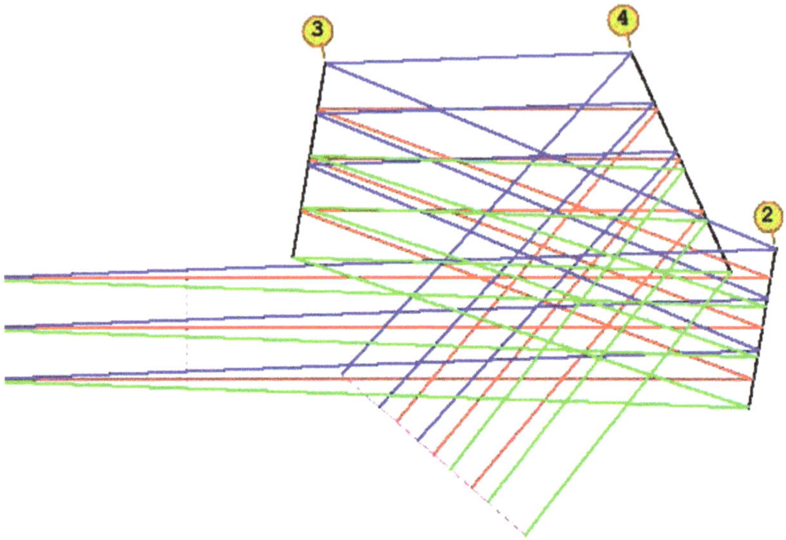

Figure 43.2. System returned by FFBUILD, before optimization.

the process works best if you first rough out the design with only the radii and angles varying, and then add the other variables gradually as they are needed. Here is part of that MACro; the lines in green are commented out.

```
PANT
 SKIP
VY    2 YG
VY    2 ZG
VY    3 YG
VY    3 ZG
VY    4 YG
VY    4 ZG
VY    5 YG
VY    5 ZG
 EOS
VY    2 RAD
! VY     2 CC 10 -10
! VY     2 G   3
! VY     2 G   4
! VY     2 G   7
! VY     2 G   8
! VY     2 G  10
! VY     2 G  11
! VY     2 G  14
! VY     2 G  15
! VY     2 G  16
! VY     2 G  19
! VY     2 G  20
! VY     2 G  23
! VY     2 G  24
! VY     2 G  26
! VY     2 G  27
! VY     2 G  30
! VY     2 G  31
! VY     2 G  34
! VY     2 G  35
! VY     2 G  36
...
```

Most of the file consists of operands that will control the clearance of the beam as it bounces between mirrors. Here is an example portion:

```
LLL 1.0000 1 1.0000
A P CCLEAR 1 0 1 0   1  3
S CAO    3
LLL 1.0000 1 1.0000
A P CCLEAR 1 0 -1 0   1  3
S CAO    3
LLL 1.0000 1 1.0000
A P CCLEAR -1 0 1 0   1  3
S CAO    3
LLL 1.0000 1 1.0000
A P CCLEAR -1 0 -1 0   1  3
S CAO    3
LLL 1.0000 1 1.0000
A P CCLEAR 0 0 1 0   1  3
S CAO    3
LLL 1.0000 1 1.0000
A P CCLEAR 0 0 -1 0   1  3
S CAO    3
LLL 1.0000 1 1.0000
A P CCLEAR 1 0 1 0   1  4
S CAO    4
LLL 1.0000 1 1.0000
A P CCLEAR 1 0 -1 0   1  4
S CAO    4
...
```

The first three lines of this excerpt tell the program to trace the full-field upper rim ray and then look at the segment of the ray path between surfaces 1 and 2. Calculate where that segment intersects surface 3, find the absolute distance from the vertex, and subtract the clear aperture radius of surface 3. If the difference is more than 1 mm, the intersection is outside the clear aperture and the aberration is zero. But if it is less than 1 mm, the merit function gets a penalty. The program has also assigned the DCCR surface property to the mirrors, so the default clear aperture is centered between the extreme points on the surface required by the rays in the meridional plane, instead of at the vertex, which is the default.

The remaining **CCLEAR** entries control the clearance between upper and lower rim rays at the top and bottom of the field between each mirror pair and the other mirrors. There are many combinations, and they all must be controlled.

The merit function contains GNR requests for seven points in the *y-z* plane and three in the skew field direction (because of the **CFOV** directive in the **SYSTEM** file, declaring a circular field), and controls distortion in both *x* and *y* with the **GDR** request (according to the desired image size in words 6 and 7 of the **IMAGE** line in the **GEOM** section).

```
GDR 0 10 4 P 0.700000E+01 -0.700000E+01
```

It is a good idea to run this MACro just as it is, the first time around. That will get the design roughed out well enough so you can gradually improve things. After running it, the system starts to look more reasonable, as shown in figure 43.3.

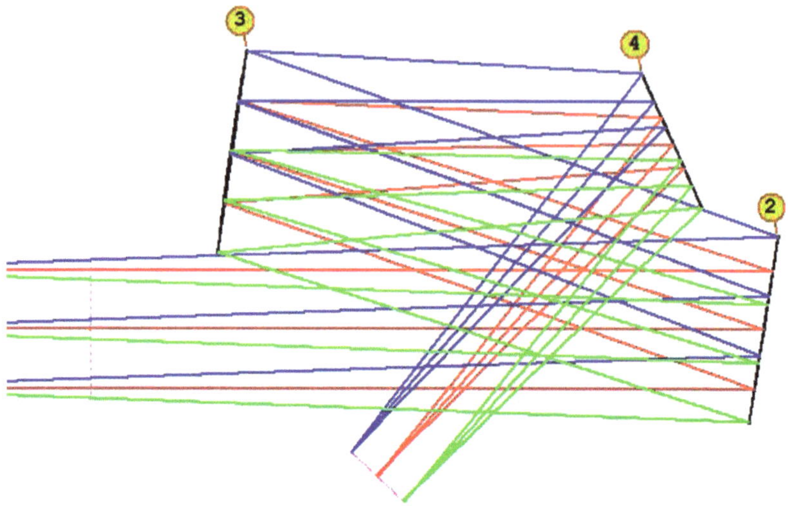

Figure 43.3. Free-form system after initial optimization.

The image is formed in the right place and the beams clear all of the mirrors rather well.

We have specified the focal length we are after indirectly in this example. It would not be a good idea to control FOCL itself, since that is a paraxial property that does not have much meaning with a folded system such as this one. What we want is for the distance between the upper and lower field points at the image to be 14 mm. The program controls that with the **GDR** request, as noted above.

The system is roughed out, but the image quality is not at all good; now you have to free up some of the other variables. It is wise to sneak up on the system little by little, so go slowly. If you go too fast, the system will sometimes jump to a bizarre configuration that is nowhere near the nice region you are in now. Control low-order aberrations with low-order terms, and save the higher orders for when you need them. Remove the comment character ('!') before the variables for G 3 through G 10 on each surface by deleting the '!' character on those lines.

```
VY   2 RAD
! VY   2 CC 10 -10
 VY   2 G   4
 VY   2 G   7
 VY   2 G   8
 VY   2 G  10
! VY   2 G  11
! VY   2 G  14
```

Do the same for surfaces 3 and 4. (Select those lines and then click to uncomment them quickly.)

Also, comment out the SKIP directive that bypassed the global *y* and *z* location variables, so they will become active,

```
PANT
! SKIP
VY     2 YG
VY     2 ZG
VY     3 YG
VY     3 ZG
VY     4 YG
VY     4 ZG
VY     5 YG
VY     5 ZG
```

and then run the MACro and anneal (**50, 2, 50**). The MF comes down. Now free up the remaining G variables and optimize again. The MF goes down to 0.00016. Anneal, and the MF comes down to 0.00012. Pretty good.

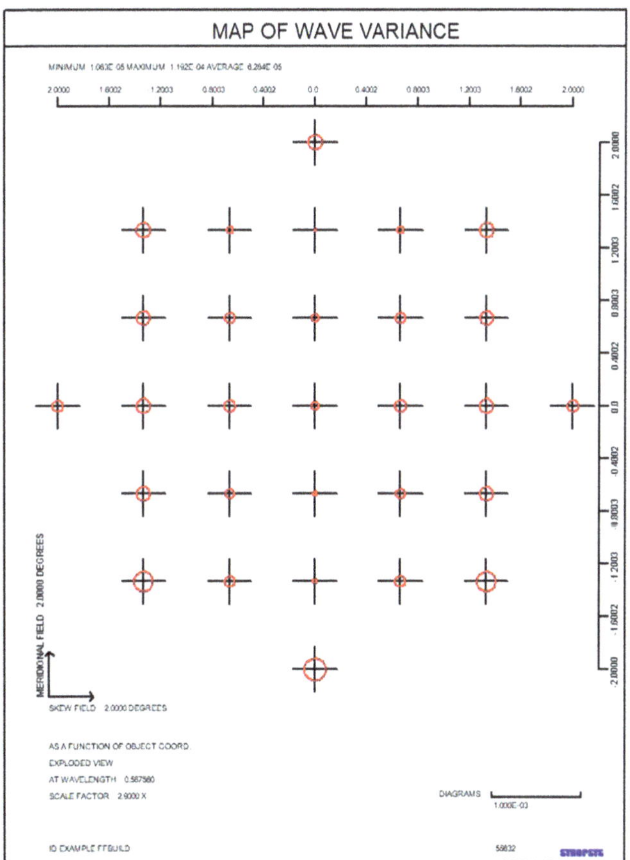

Figure 43.4. Wavefront variance over field.

You are probably wondering why we did not let the CC (conic constant) vary. You can experiment with that too, if you want—but we find that the existing Zernike terms can generate much the same shape as the conic constant, and it is not wise to use duplicate variables.

It's time to assess where we are. Go to the MAP dialog (**MMA**) and ask for a map of the wavefront variance over a grid of object points,

object points CREC,
ray pattern CREC 9,
Show circles,
EANALOG scale 0.001,
and Execute.

The result is in figure 43.4.

Let's examine the results. The worst field point is now at HBAR -1, as indicated by the MAP analysis. Here is that image, created by the **MDI** dialog, shown in figure 43.5.

All other points are better. This is a superb design. Let's assume for this application we will use a CCD array sensor with pixels of 10 micrometers on a side, so this looks good.

You can get a better view with RSOLID, which shows only the portion of the surfaces within the decentered CAO. But first you should go to the Edge Wizard

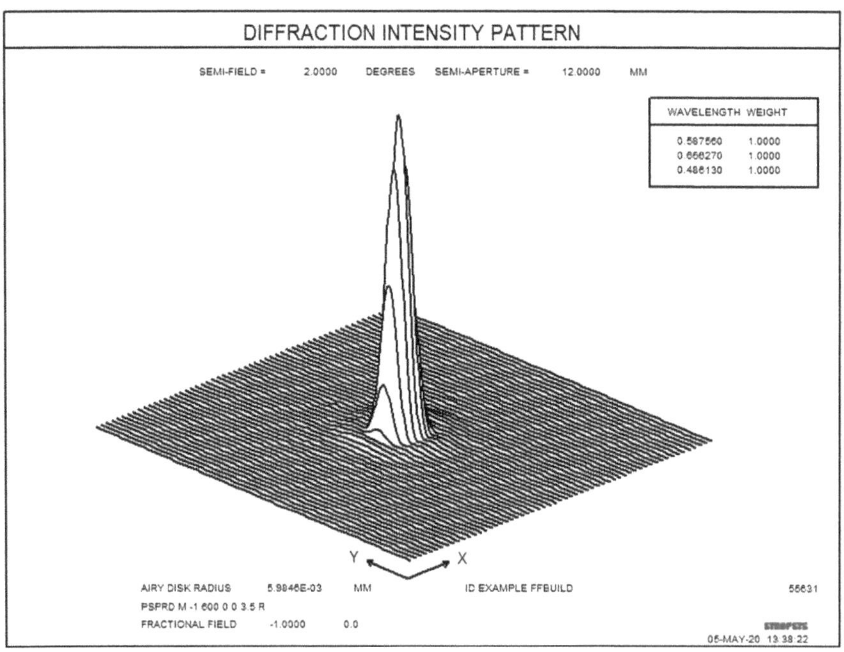

Figure 43.5. Diffraction pattern at worst field point of free-form mirror design.

(**MEW**), select Create All, and adjust the mirror thicknesses as you want, as you did in Chapter 40. Now the mirrors are assigned realistic edges and thicknesses. Then create an RSOLID picture, shown in figure 43.6 (Type **MPE** and select that option, or click the button |⚙|).

Your free-form system is designed (**C43L1**).

Now you can look at the shapes that resulted, with the FreeForm Analysis tool (**FFA**). The command

```
FFA 2 0 RSAG SURF
```

produces the picture in figure 43.7, which shows the shape when all rotationally-symmetric terms are excluded. This tells you how much the surface varies from a symmetric curve.

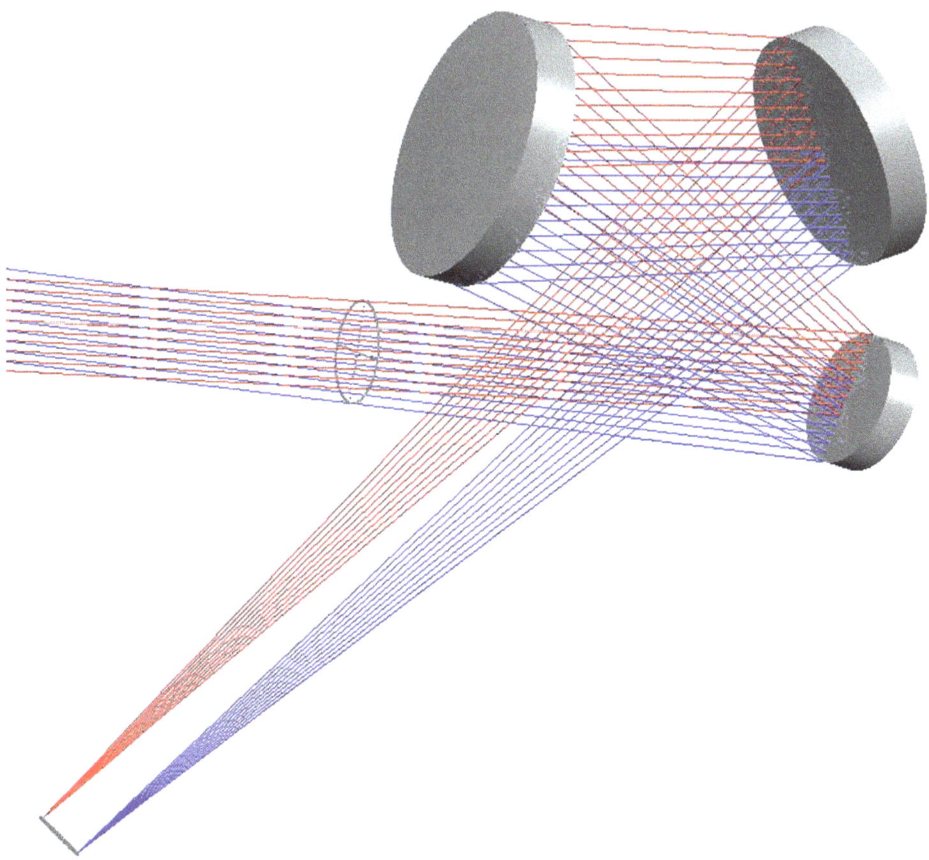

Figure 43.6. RSOLID view of the final design.

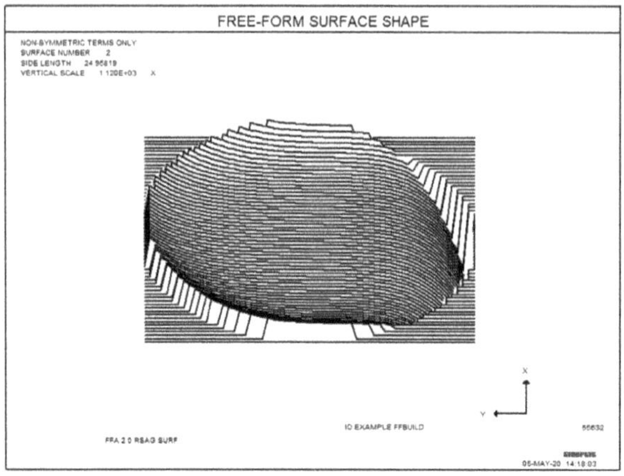

Figure 43.7. Shape of free-form mirror on surface 2, with symmetric terms subtracted.

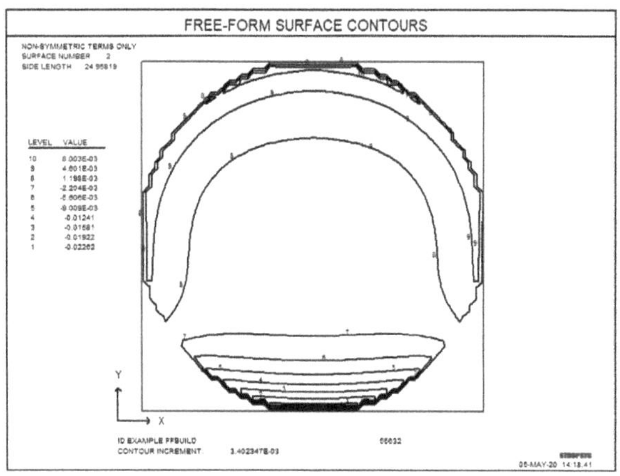

Figure 43.8. Contour plot of the non-symmetric terms on surface 2.

To see contours instead, type **FFA 2 0 RSAG CONTOUR**, and you get the picture in figure 43.8.

The shape of the actual surface is given by **FFA 2 0 SAG CONTOUR**, shown in figure 43.9. It is rather close to a spherical surface. Proceeding in this manner, you can look at the shapes of all of the mirrors.

What about distortion? The GDR request has nicely handled that too. Here is the picture produced by the command **GDIS 31**, in figure 43.10. Not bad at all.

One more question remains: How does one test these mirrors? The easiest way is to look at the fringes when tested in double-pass in an interferometer against a

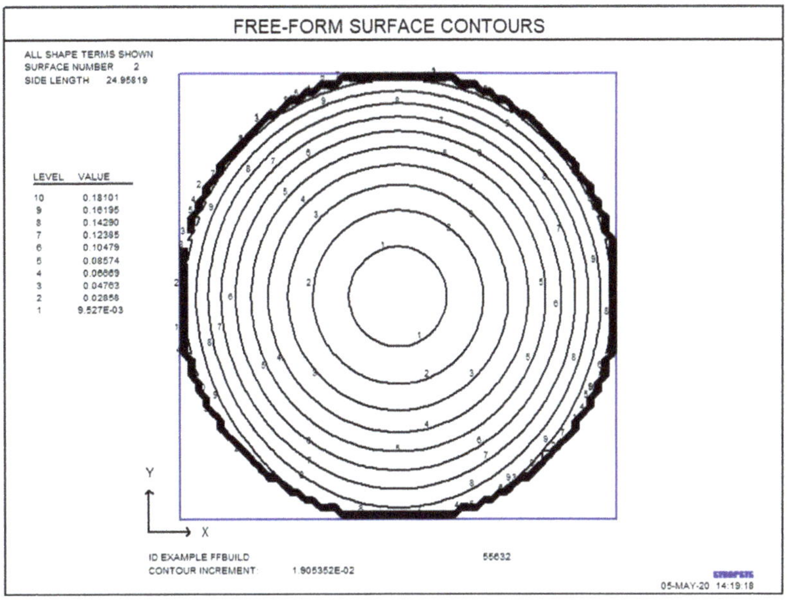

Figure 43.9. Contours of complete shape of surface 2.

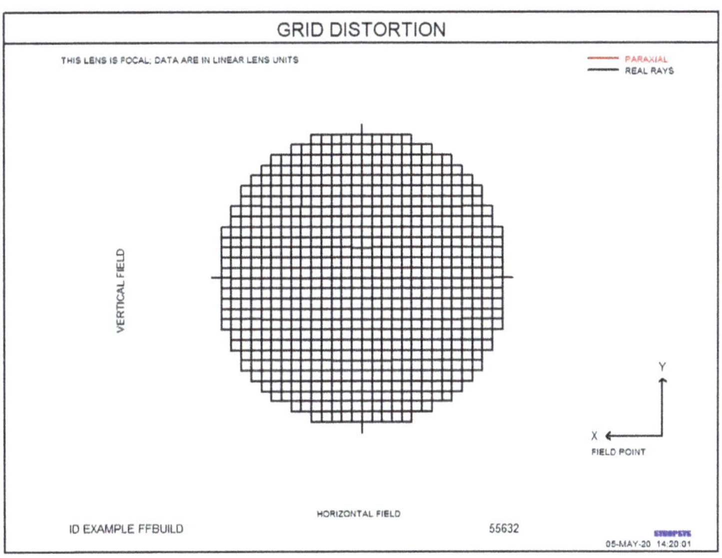

Figure 43.10. Grid distortion of the free-form mirror design.

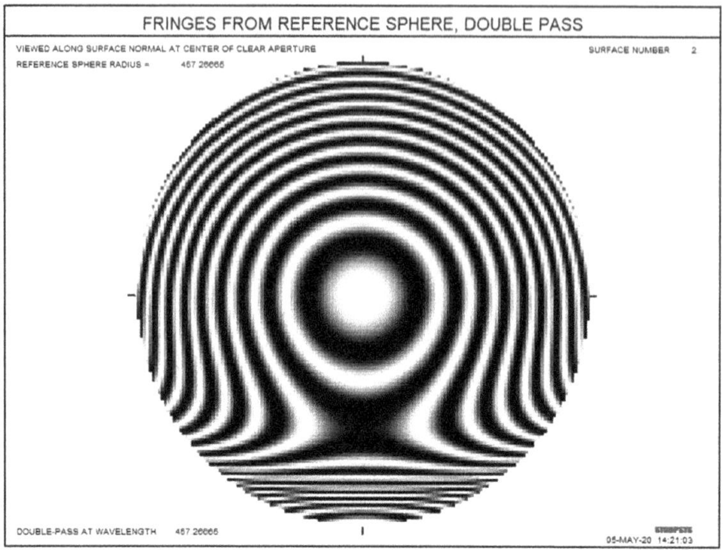

Figure 43.11. Fringe pattern of free-form mirror on surface 2 relative to a reference sphere.

reference surface of a known radius. **FFA** can show that too. Here is the output from the command **FFA 2 0 RFRINGES**, in figure 43.11.

If you see that fringe pattern, the mirror is perfect.

One more task: The mechanical engineers will want to model the system, and likely will need to know the location of many points on each mirror in global coordinates. Here is the input that will generate a table of coordinates over the surface of surface 4:

```
MAP GSAG OVER SURFACE ON SURFACE 4
FGRID POINT 0 0
RGRID CREC 7 7
SCALE AUTO
DIGITAL
ACTUAL
PRINT FULL
```

And here is the output:

```
MAPPING PROGRAM OUTPUT
     X-COORD.        Y-COORD.          DATA

 -0.34180425E-06  0.38423206E+02  0.13948858E+03
 -0.22938095E+02  0.48609554E+02  0.13420832E+03
 -0.11469048E+02  0.48698662E+02  0.13439271E+03
 -0.34180425E-06  0.48728447E+02  0.13445435E+03
  0.11469047E+02  0.48698662E+02  0.13439271E+03
  0.22938093E+02  0.48609554E+02  0.13420832E+03
 -0.22938095E+02  0.58866894E+02  0.12907496E+03
 -0.11469048E+02  0.58955116E+02  0.12925753E+03
```

```
MAPPING PROGRAM OUTPUT
     X-COORD.        Y-COORD.          DATA

  -0.34180425E-06   0.38423206E+02   0.13948858E+03
  -0.22938095E+02   0.48609554E+02   0.13420832E+03
  -0.11469048E+02   0.48698662E+02   0.13439271E+03
  -0.34180425E-06   0.48728447E+02   0.13445435E+03
   0.11469047E+02   0.48698662E+02   0.13439271E+03
   0.22938093E+02   0.48609554E+02   0.13420832E+03
  -0.22938095E+02   0.58866894E+02   0.12907496E+03
  -0.11469048E+02   0.58955116E+02   0.12925753E+03
```

So that's how you can design a free-form mirror system with these advanced tools. The computer does most of the work for you.

Now it is up to you and the shop to communicate well enough so they understand the results and can make the parts correctly. Here are some pointers:

1. In this example, the surfaces are defined by Zernike terms, as requested. Variable G 39 can vary the center point of the expansion, but we did not use that variable here. Although it is sometimes helpful, the center of the expansion would then not be at the vertex, which is a complication we like to avoid. And the vertex of the surface is not at the center of the clear aperture either, which is something we cannot avoid. So be careful here. There are two center points to consider. Also, variable G 51 can vary the Y-scale of the expansion, which distorts the Zernike zones and can sometimes be helpful. But this complication is also to be avoided unless it really makes a difference.
2. When presenting these data to the shop, be sure they understand the coordinate systems and locations of the relevant parameters.

Check out the other features of the **FFA** program. You can create a table of sags across the surface parallel to the surface normal at the center of the CAO that will be of great interest to the technician running the precision milling equipment.

IOP Publishing

Lens Design (Second Edition)
Automatic and quasi-autonomous computational methods and techniques
Donald C Dilworth

Chapter 44

An aspheric camera lens from scratch

Spy camera; aspheric plastic elements; restricting glass model

When developing a modern cell-phone camera lens or a pinhole spy camera, designers are resorting more and more to using multiple aspheric surfaces. These are typically embodied as small plastic elements, and even though the molds are expensive to make, the lenses can be produced in quantity at very low cost. It is even possible to mold mounting flanges directly onto the elements, simplifying assembly and enabling some dimensions to be held to very tight tolerances.

To help in designing such systems, DSEARCH can do a global search for systems with aspherics. Here is an example, a five-element lens with plastic elements and a glass cover plate before the image:

This will be the input to DSEARCH (**C44M1**):

```
TIME
CCW              ! clear command window
CORE 32          ! use 32 cores for speed
OFF 1
ON 99

DSEARCH 1 QUIET        ! start DSEARCH; put best lens in library location 1
SYSTEM                 ! define the system specs
ID DSEARCH ASPHERIC CAMERA LENS      ! identification
OBB 0 41.3 .285        ! infinite object, semi field 41.3 degrees, semi ap. 0.285
UNI MM                 ! lens will be in millimeters
WAVL CDF               ! use visual wavelengths at C, d, and F lines
END                    ! end of system section

GOALS                  ! define the goals here
ELEMENTS 5             ! we want a four-element lens with a cover glass
BACK 0.4 SET           ! ask for 0.4 mm back focus distance
FNUM 2.7 10            ! ask for F/2.7, weight of 10
THSTART .5             ! global search use thicknesses .5 mm
ASTART .1
RSTART 30              ! and starting radius of 30 mm
```



```
ASPH R          ! use all terms in real mode; quick mode spherical
ASPH Q          ! vary CC in quick mode too
ASPH 3          ! allow three aspheric terms: CC, 4th, 6th power of aperture

ANNEAL 40 2 Q 10       ! anneal each case, temp 40 degrees, cool 2, including
quick
SNAP 1                 ! redraw PAD screen every pass
STOP FIRST             ! put the stop in front
STOP FIXED             ! and keep it there

RT 0.5                 ! ray weight higher near center of aperture

QUICK 100 20           ! run quick mode 100 passes, then real mode 20
NGRID 6                ! 6x6 grid of rays in pupil
NPASS 50               ! 50 passes in the MACro when finished
TOPD                   ! correct both transverse aberrations and OPDs
FOV 0 .2 .45 .7 .85 1  ! correct six field points
FWT 5 4 3 3 3 3               ! with these weights
COVER .3 1.51872 64    ! the cover glass will be 0.3 mm thick with this GLM

PLASTIC 1 3 5 7                ! the four elements will be plastic
END                     ! end of goals section

SPECIAL PANT            ! special PANT section starts here
RDR .001                ! these are tiny lenses; reduce derivative increments
TLIMIT 3 .05            ! limits on thicknesses and spaces
SLIMIT 5 .05
END                     ! end of PANT section

SPECIAL AANT   ! start of special AANT section; these go into the merit fn.
ACC 1 1 1               ! center thickness no more than 1.0 mm
ACM .2 .1 .2            ! and no thinner than 0.2 mm
ACA 55 1 1              ! avoid critical angle; 55 degrees from surface normal
AEC .1 .1 .1            ! keep edges over 0.1 mm
ARC 0.2
GTR 0 1 3 P 1
M 1.35 10 A P YA 1      ! target the chief ray at three field points
M .945 10 A P YA .7     ! to control distortion
M .54 10 A P YA .4
END                     ! end of AANT section
GO                      ! DSEARCH runs
TIME                    ! when it is finished, see how long the run took.
```

There are some subtle considerations here. First, these will be very tiny lenses, and the default edge-control target (1 mm) that DSEARCH puts in its optimization MACro is too large; that gets changed with the AEC monitor. Also, the default minimum airspace and thickness monitor of 1 mm is also too thick and is overridden with the ACM of 0.2 mm. The added ACC monitor will not let thickness grow to more than 1.0 mm, overriding the default of 25.4 mm, and the ACA monitor will penalize solutions where rays encounter too steep an angle anywhere.

Those monitors are very weak as entered, which is done on purpose: if you strongly control those items, DSEARCH will tend to favor designs that do not offend them—but we want the program to favor designs with small *image errors*, and don't really care much about the mechanical properties at first. When you get a good design, you can easily modify those monitors later on, increasing the weights to make the design practical.

DSEARCH lets you manage the image distance in two ways: If you just give a distance, such as **BACK 0.4**, the program adds a YMT solve at the end and includes a target in the AANT file to control the resulting value. If you add a weighting factor, such as **BACK 0.4 100**, that weight is applied to the target. The other way is to request an exact value, in this case with **BACK 0.4 SET**. Now the program will simply set the back-focus distance to the entered value, 0.4 in this case, and will *not* add a YMT solve. This is often a good choice for difficult designs, especially when the other options return systems with a virtual image.

Since we are enabling use of aspherics, we have to be careful to give a grid higher than the default **NGRID 4**, and to correct at six field points instead of the default three. Otherwise there will likely be intermediate pupil and field zones that fly away out of control. We ran this job several times, and adjusted the field points to better maintain good correction at intermediate points.

We have also added a new monitor. **ARC** means Automatic Ray Centration, which will keep the chief ray centered between the upper and lower rim rays at full field. While you may think that condition ought to come for free, it can be violated in wide-angle lenses if substantial coma shows up somewhere within the lens. When that occurs, it becomes nearly impossible to correct other aberrations from that point on. This monitor, with a low weight, is convenient insurance that you should keep in mind and use whenever you see that the chief ray is not well centered in the beam.

Also, note that we have added a GTR directive in the SPECIAL AANT section. Without this addition, we got solutions that were great everywhere else but where the TFAN showed large aberrations right at the top edge of the beam. The usual ray-grid targets cover the pupil in a reasonable fashion, but do not go exactly to the edge. The GTR target does go to the edge and helps to avoid this kind of solution.

The bounds on the glass variables also need some attention. Those will be replaced with plastics from the U (Unusual materials) catalog when the design looks good, and we want the model glass to fall in the area where plastics are to be found. This is the purpose of the **PLASTIC** declaration in the input file. Any surface so designated is restricted to the line on the glass map shown in figure 44.1.

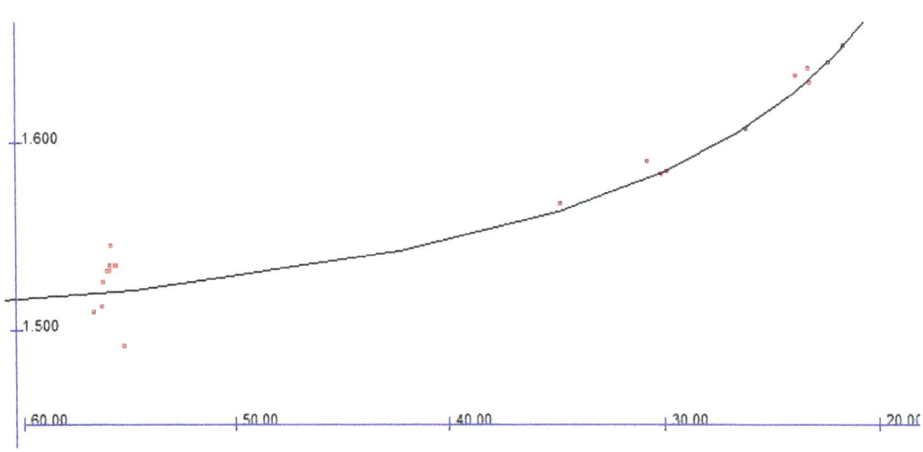

Figure 44.1. Line on glass map valid for plastic materials.

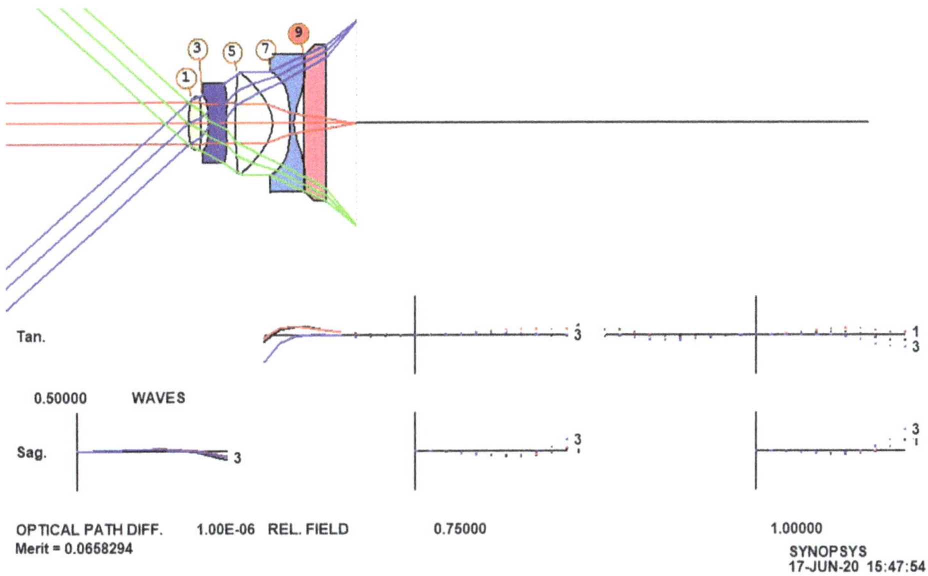

Figure 44.2. Aspheric pinhole lens found by DSEARCH, before optimizing.

The red dots are the plastics currently in the U catalog. The resulting glass models should behave much like real plastics when those are inserted later.

Okay, Run the DSEARCH MACro listed above, and in about six minutes you see the best design the program found, shown in figure 44.2.

This is amazing. The lens is already diffraction limited, straight out of DSEARCH. The OPD errors are less than ¼ wave almost everywhere—but the last lens is too close to the cover plate. The edge-control monitor AEC works well at the edge of the lens, but this problem shows up partway out. (You could also just *move* the cover plate, since the aberrations of a flat window are independent of location, but let's assume the image distance is fixed.) That's why the ray fans are drawn with separate dots instead of continuous lines: to let you know that those rays are feathered.

We have to fix that. Add to the AANT section in the file DSEARCH_OPT, which is in a new editor window, the lines

```
LLL .1 5 .05
A P ZG .8 0 0 0 9
S P ZG .8 0 0 0 8
```

This entry sets a lower limit of 0.1 to the difference between the global Z-coordinates of the chief ray on surfaces 8 and 9 at the 0.8 field.

Now run this file and anneal (**20, 2, 50**). The quality is still excellent, as shown in figure 44.3, and the clearance issue is corrected.

Now let's see if we can improve it some more. At the moment, the aspheres use G-terms three and six only, which vary the fourth and sixth-power terms in the polynomial expansion. Add to the PANT file the variables

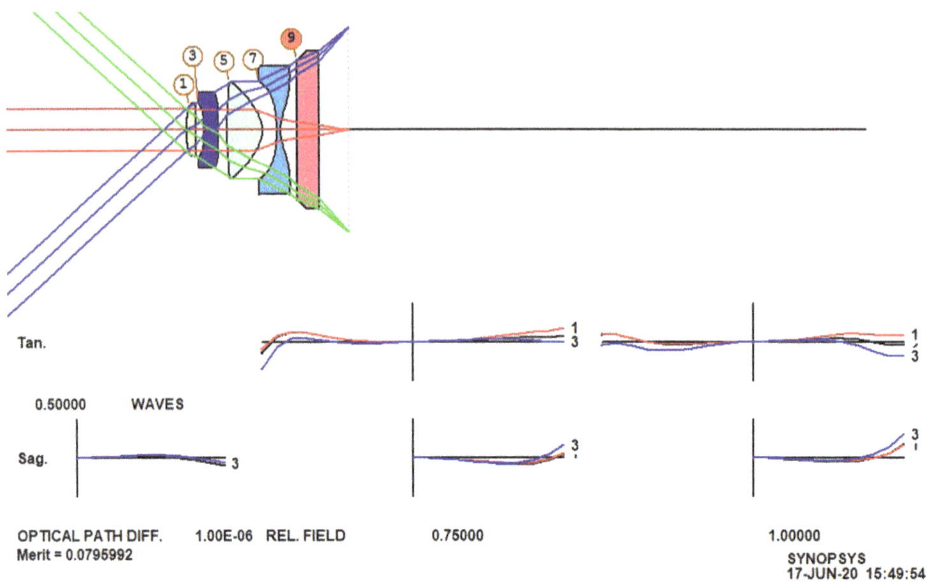

Figure 44.3. Lens reoptimized with altered MF. Clearance issues have been resolved.

```
VY 1 G 10
VY 2 G 10
VY 3 G 10
VY 4 G 10
VY 5 G 10
VY 6 G 10
VY 7 G 10
VY 8 G 10
```

to also vary the eighth-power terms, and reoptimize and anneal. The MF comes down to 0.0171, as shown in figure 44.4.

Now it's time to switch to real plastics—but first change the material on surface 9, the cover plate, to the real glass the customer wants to use: Hoya type BSC7. To do so, open the WorkSheet and type in the edit pane

```
9 GTB H
BSC7
```

click Update and save a checkpoint. The model glass on the cover plate is replaced. Optimize once again and save a checkpoint.

Now open the real-glass menu (**MRG**) and select the U catalog. That catalog does not have ordinary optical glasses, but it does have the plastic materials. When you specify that catalog, the ARGLASS program (which is run from the **MRG** dialog) automatically selects only plastics and only replaces GLMs designated **PLASTIC** in

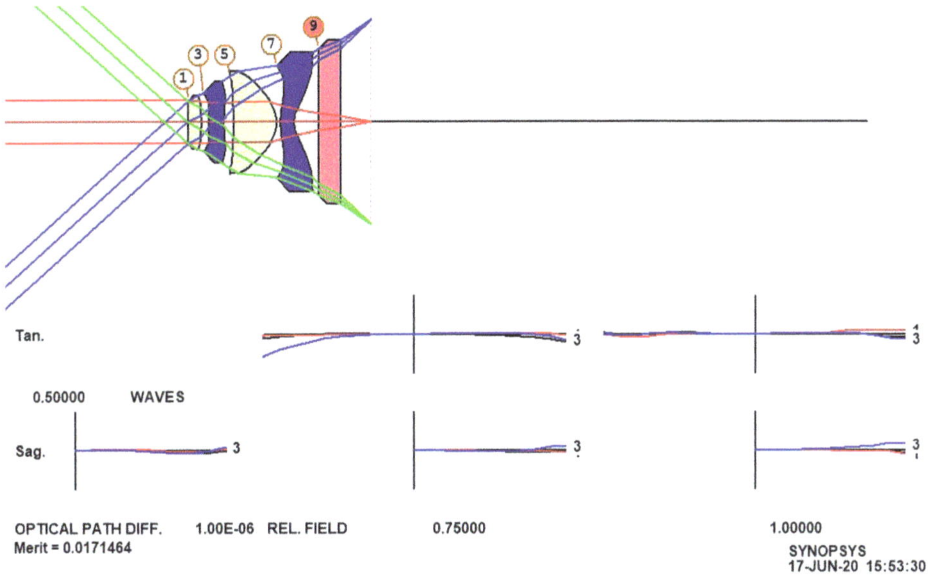

Figure 44.4. Lens optimized and annealed with added eighth-power terms.

the RLE file. It has three modes; it can replace the lenses in numerical order, reverse order, or it can sort them so it replaces the ones furthest from a real material first. The latter option is often better, so check the SORT option. When this run finishes, delete the GLM variables from the optimization file, optimize, and anneal again.

> Sometimes changing to real glass causes ray failures. The program adjusts the curvatures to maintain element powers, but if aspheric terms are present, some rays can still fail. If this happens, run ARGLASS again after the other materials are changed. This usually works. If not, go back to your checkpoint, open the glass map (**`MGT`**), click on the material closest to the surface number that produced the error message, and select Apply + Adjust. That uses a different algorithm to alter the lens surfaces, and the lens may then work properly with that material.

Now there are real materials everywhere as shown in figure 44.5.

The modulation transfer function (MTF) curves for this design (**C44L1**), which is close to perfect, are shown in figure 44.6. (To get these MTF curves, go to the **`MMF`** dialog, select the Multicolor option, and click Execute.)

44.1 Encore

Well, that's a start, and now you understand how to use the program, but what could we have done differently? This design is at the diffraction limit, but the MTF at full field is much lower than on axis. Why is that?

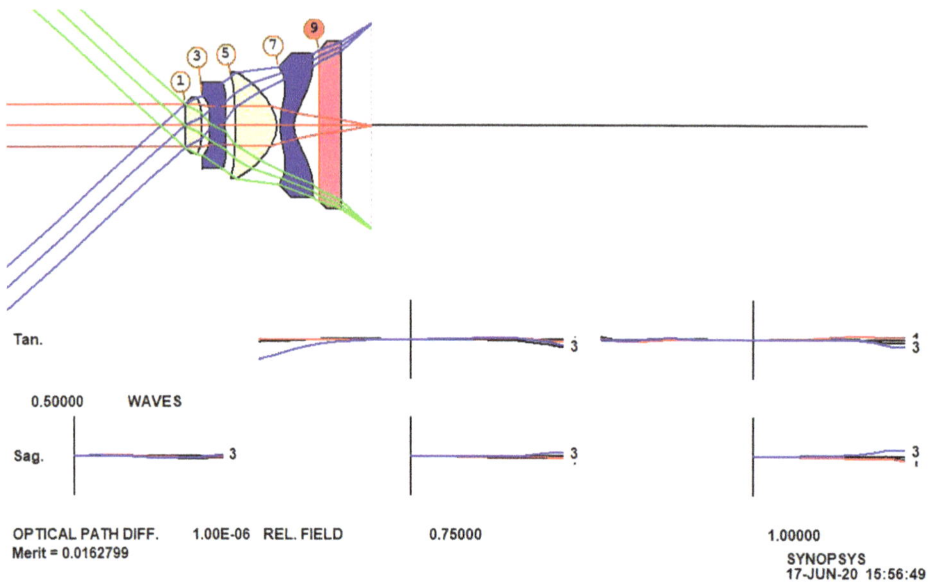

Figure 44.5. Lens optimized with real plastics.

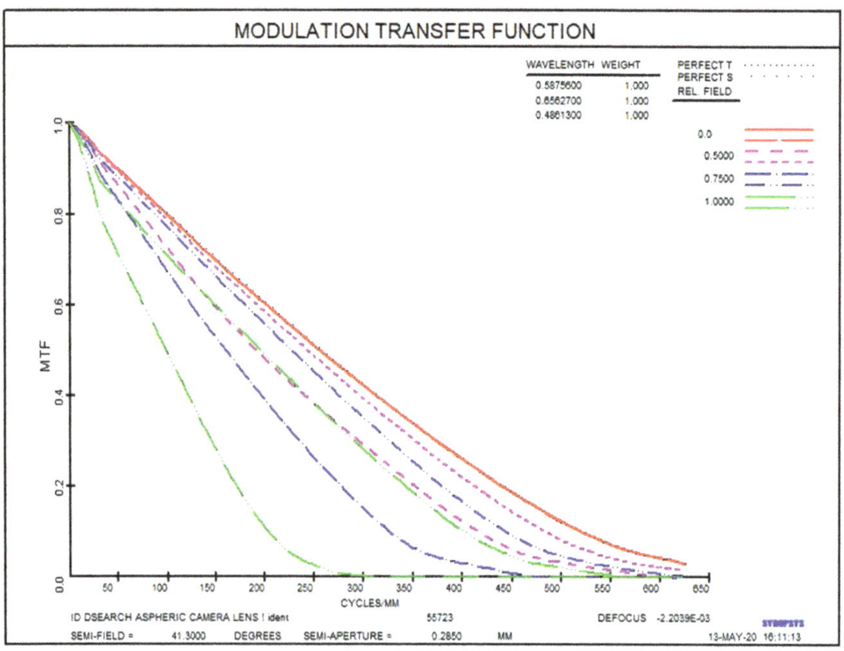

Figure 44.6. MTF curves at four field points for the camera lens.

Well, since the lens has a stop in front and we are correcting distortion, the image necessarily shows $\cos^4$ darkening[1]. In fact, at a field angle of 41.3 degrees, the edge is just under 32% as bright as the center. How does Nature manage to do that? By changing the effective F/number! Type the commands

```
FN 0
FN 1
```

and you see that while the on-axis F/number is indeed about 2.72, at the edge it is 5.99 in the tangential direction and 3.55 in the sagittal. The higher F/number increases the size of the Airy diffraction disk, which lowers the cutoff frequency in the Y direction. That is what the MTF curves tell us.

If that situation is satisfactory, you are done. But let's assume you really want uniform illumination over the field. You can't achieve that unless you let the distortion grow, which may not be a problem if you plan to compensate electronically afterwards. Here's what you do:

1. Rename the MACro. Delete (or comment out) the lines in the SPECIAL AANT section of the DSEARCH input that give targets to the chief-ray YA at three field points. Those were there to control distortion.

```
SKIP
M 1.35 10 A P YA 1
M .945 10 A P YA .7
M .54 10 A P YA .4
EOS
```

2. Add some new requirements in the same section. These will control the relative illumination at five field points, and the distortion will be free to grow to satisfy them.

```
M 1 1 A P ILLUM .2
M 1 1 A P ILLUM .4
M 1 1 A P ILLUM .6
M 1 1 A P ILLUM .8
M 1 1 A P ILLUM 1
```

3. Since the F/number at the edge of the field will now be smaller than before—which is harder to correct, increase the weights on the outer two fields from 3.0 to 4.0 (on the **FWT** line).
4. Change the AEC monitor to **AEC .1 1 .1** to better avoid overlap. Note that DSEARCH always puts in a default AEC monitor, but we want to override that one with this one.

[1] The illumination at the image is reduced by three factors: the obliquity of the entering beam, the obliquity of the chief ray at the image, and the greater distance of the edge of the image than the center from the exit pupil. The first two each contribute a cosine, and the last a cosine squared. The result is four powers of the cosine.

5. Comment out the QUICK directive. Some tasks work better that way, and it's a variation you will want to explore. But since the quick mode will minimize third-order aberrations, including distortion, which we don't want to do now, we defeat that option.

Run this version on DSEARCH, add the 10th-power variables, then optimize and anneal. The construction is different, and the MF comes down to 0.068 as shown in figure 44.7. The lens is excellent, but now the last two elements are too close together.

Add a new requirement to control the difference in the ZG coordinate at the 0.8 field on surfaces 7 and 6, as you did above.

```
LLL .1 5 .05
A P ZG .8 0 0 0 7
S P ZG .8 0 0 0 6
```

Now optimize and anneal once again. The lens is improved, as shown in figure 44.8.

Then again assign the real glass to the cover plate, insert real plastics with the **MRG** dialog, and optimize and anneal, and you get the lens in figure 44.9. (**C44L2**)

This case worked quite well, but sometimes matching real plastics presents a new problem.

If you look at the glass table display in figure 44.1 showing plastics in the U catalog, you see that there is a large gap between a group of plastics to the left and another to the right. If a model plastic happens to fall inside that gap, or if the optimization program moves it there when other models are matched, the program

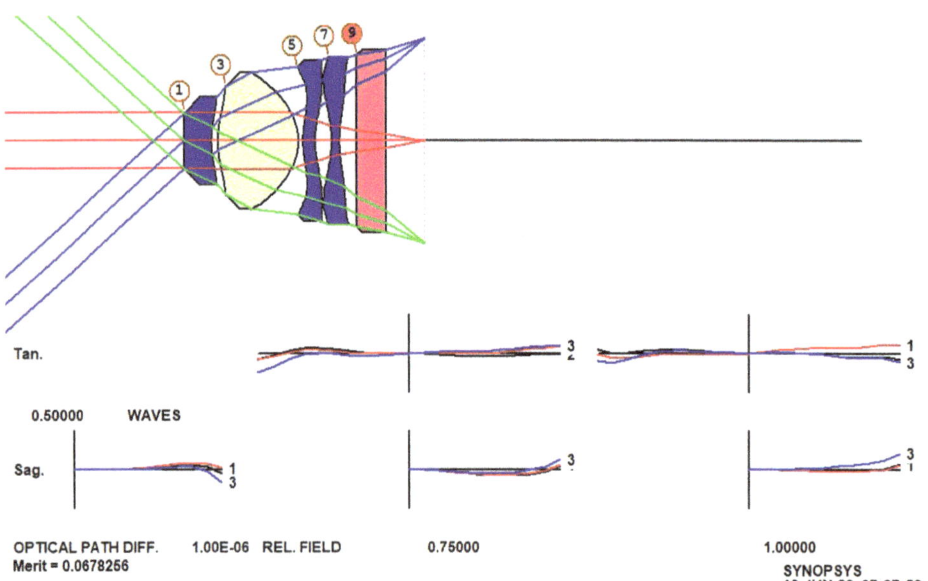

Figure 44.7. Lens returned by DSEARCH with uniform illumination as a goal, then optimized and annealed.

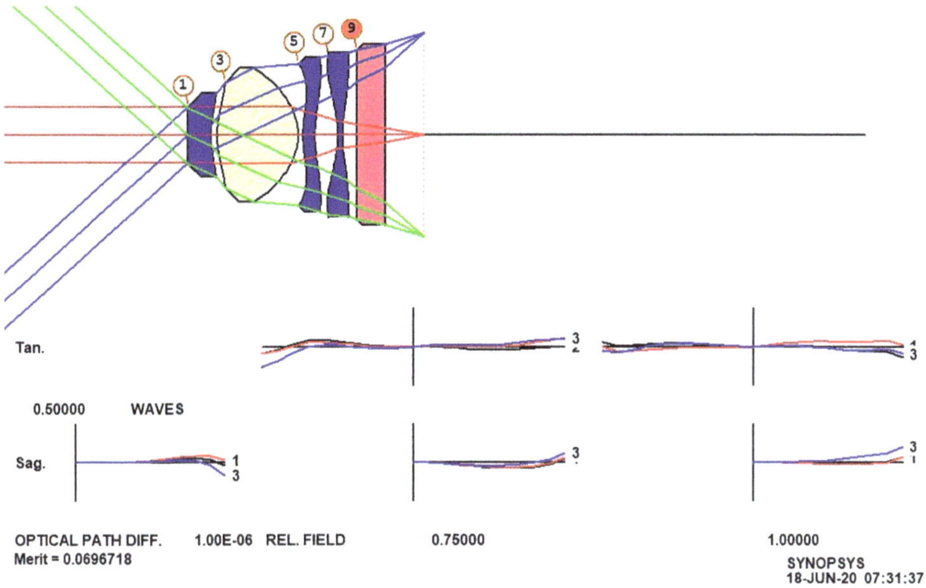

Figure 44.8. Lens reoptimized to control spacing problem.

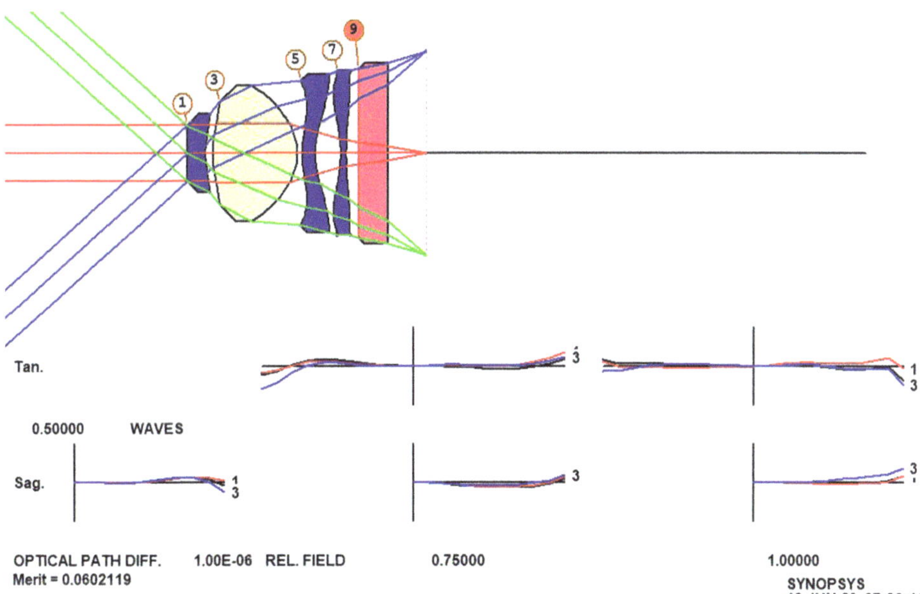

Figure 44.9. Lens optimized for uniform illumination, matched to real plastics.

must make a significant change when it then converts it to a real material. Sometimes the direction it chooses works well and sometimes not. That is why we provide three options in MRG: SEQUENTIAL, SORT, and REVERSE ORDER.

The MTF is superb, as shown in figure 44.10.

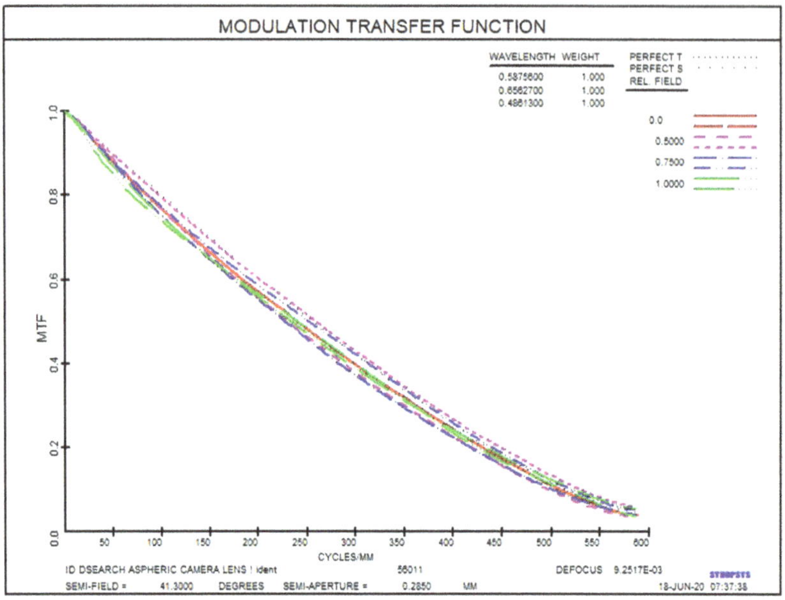

Figure 44.10. MTF curves for the uniform-illumination lens.

And the illumination is quite uniform, shown in figure 44.11, plotted with the command

```
ILLUM 500 P
```

The program has indeed introduced significant distortion, however, as shown in figure 44.12, in a plot produced by the command

```
GDIS 21 G
```

What would the world look like imaged through this lens? The Field Blur menu can show it. First, open a photograph in a graphics program and then crop it square and resample it to 400 × 400 pixels. Copy it to the clipboard. Then open the Spectrum Wizard (MSW), select 10 wavelengths, Get spectrum, and Apply to lens.

Then open **MFB**, click Load from Clipboard, and Process. The photo is distorted as it would be if imaged through the lens, as shown in figure 44.13.

44.2 Coda

We made it look easy, and it is if you follow the steps above. But of course, lens design has pitfalls all over the place, and things do not always work perfectly the first time. Here are some of the problems that you may encounter and how to deal with them:

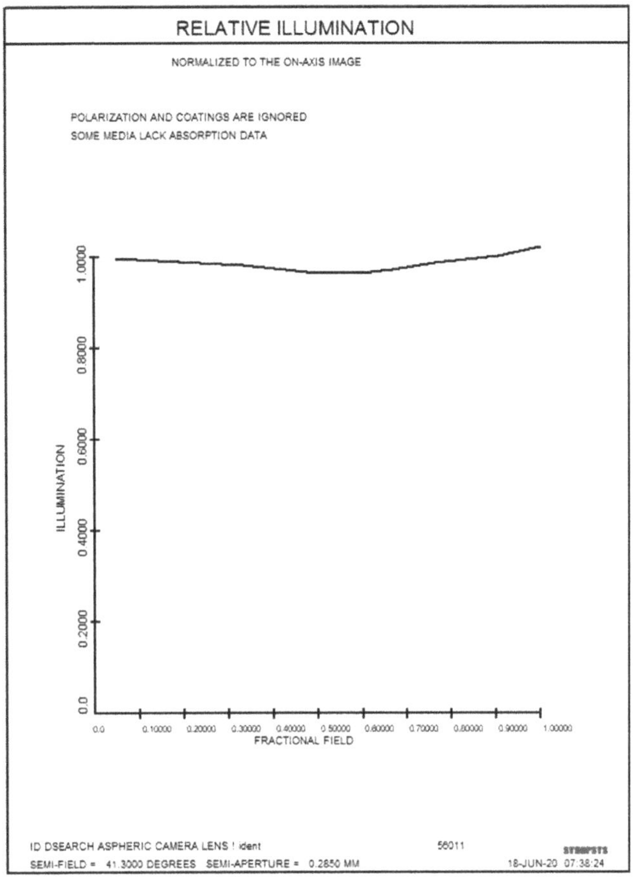

Figure 44.11. Illumination uniformity curve.

1. We specified an aspheric count of 3 in this example, which assigns terms up to R^6 to the surfaces. What happens if you use fewer or more terms than this? As a rule, it is better to start with a smaller number and then add more after you have optimized the result as well as possible, as we did above. Too many terms right at the start can send the design to a region where the terms are fighting each other and become too large. Also, raytracing can prove a problem with many high-order terms, since the beam can exhibit caustics or steep ray angles where you don't want them. Sometimes you can obtain excellent results by starting with only two terms and then adding more later. The **ASPH** directive in the DSEARCH input file tells the program how to use the conic constant and higher-order aspheric terms: (**ASPH Q**) uses a conic constant even during the quick mode, (**ASPH R**) will use all the requested G-coefficients during the real mode (instead of just the conic constant), and (**ASPH Q R**) will do both. Changing any of these will send the program up different branches of the design tree.

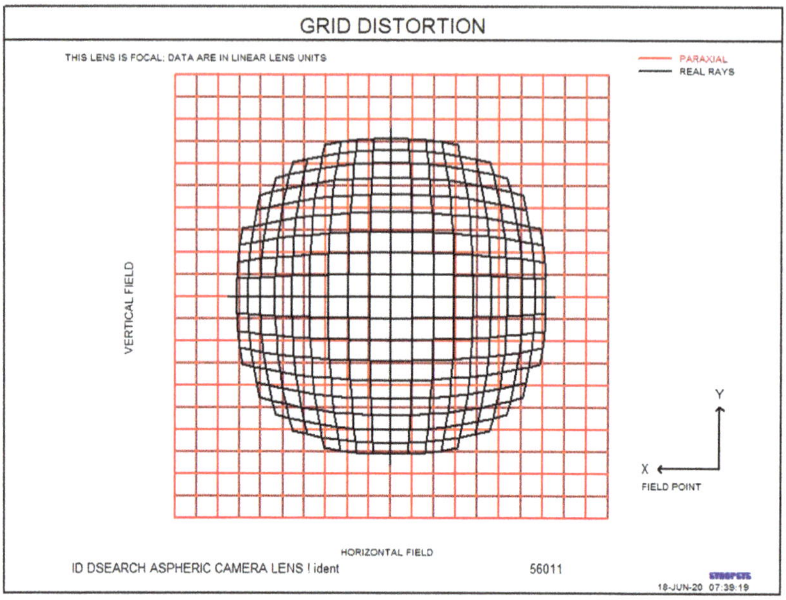

Figure 44.12. Distortion plot of the uniform-illumination lens.

2. Note that the **FNUM** request in the DSEARCH input file specified a weight of 10; this is more important than meets the eye. Chapter 35 explained how, if you left the weighting factor off, the program would control the F/number exactly with a paraxial solve—which can lead to ray failures if the radius that results is too steep. So, for fast lenses such as this one, it is a good idea to add a weight so there will be no UMC solve. Then the program adds a requirement to the merit function to control the F/number, with a starting radius on that surface given by the RSTART value. In the second example, where we did not target the image height, the F/number would probably have grown larger than the target value if we had assigned a lower weight. The program will do absolutely anything to reduce the merit function, and giving up a little on that score would likely bring down the other aberrations significantly, resulting in a great image at a higher F/number. To prevent that, we specified a weight of 10 so that solution would not look so attractive.
3. We chose to give the back focus distance a fixed value in this example. If you instead input a weighting factor on the **BACK** line, the program would assign a YMT solve to the last surface, so the image would always be at the paraxial focus, and then add a target to the AANT file to drive it to the requested value. Both methods work, but when you are targeting the YA of selected rays in order to control the image height, as we are here, it is best to set the value yourself. Otherwise the program may fail to correct a virtual image since the image height has to change when it moves it from inside to outside the lens and it might not want to go there.

Figure 44.13. Photo of a rocket on its transport as distorted by the aspheric lens.

4. Remember that DSEARCH uses the annealing feature (if you request it, which is nearly always a good idea), and that feature makes small random changes to the lens, over and over. This exercise was prepared with switch 98 turned on—so the reader could get similar results by using the same random numbers. But if you run it with that switch turned off, your results will differ and will sometimes be better. It is often a good idea to run DSEARCH more than once, (with *random* random numbers), and look at some of the other configurations it returns each time.
5. These designs met our goals very nicely. But suppose you don't want the cost of a four-element lens. What can you do with three elements? Try it and find out! It will likely not be as good, but then, maybe you don't need that level of resolution with your sensor.
6. Remember that DSEARCH is searching a very bushy design tree and can't examine every branch every time. If you change almost anything in the

DSEARCH input, such as the **RT** parameter, field weights, monitor targets, number of iterations, etc.—the program will search a different set of branches and return different results. The power of this method is that it can search a huge number of branches simultaneously, and most runs return at least one lens that is at or close to your requirements. By all means experiment with the input and keep the better results in the library so you can examine them at your leisure.

This lesson used plastics for all the lens elements except the cover glass. What if you want some elements to be made of glass and the others of plastic? Simple. Just declare which elements are of plastic in the DSEARCH input file, and the program will restrict them to the smaller range where plastics are to be found. Glass elements, on the other hand, will still be free to move over the usual range of the glass map. When the design is satisfactory and you run **ARG**, the program will match only the plastic elements if the 'U' catalog is selected—and will match only glass elements with any other catalog. Simple indeed. An example is given in chapter 45.

Let us also observe that the lenses shown above are *very tiny*, measuring just over two millimeters from first to last element. On paper they work fine, but one must ask just how small an element one can reasonably expect from lens makers. Figure 44.14 shows a design by Irina Livshits of ITMO University in St. Petersburg, Russia. This lens uses glass elements, with no aspherics, and shows that the technology for making microscope objectives, which are also very small, can be applied here as well.

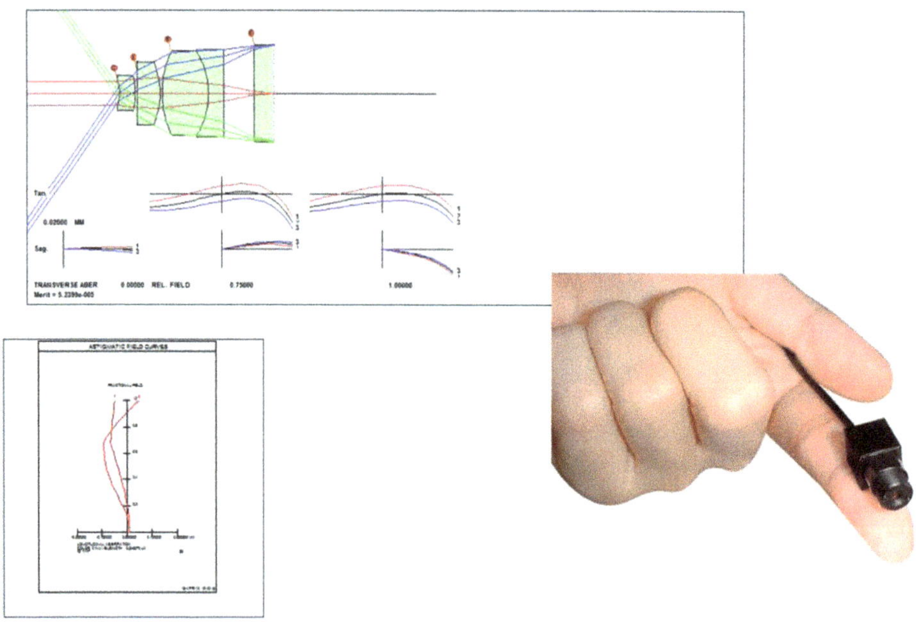

Figure 44.14. Example of a very small spy camera lens, successfully manufactured in Russia. Reproduced with permission from Irina Livshits.

In any case, one always wants to engage the shop and verify that what is on the print is within the capability of their equipment and experience.

Also, in case it is not obvious, the presence of the cover glass matters. Even though all surfaces are flat, a window like that has aberrations of its own, acting in that regard like a negative lens, the effect depending on the thickness, index, and the convergence angle of the beam. So if the application will involve a cover glass—or a beamsplitter prism, for example—it is wise to include a cover glass of that thickness in the DSEARCH input so the correction will take account of the effect.

44.3 Tolerancing the aspheric lenses

It will be instructive to ask for a tolerance budget for the lens we designed above (**C44L2**). To do so, we have to understand how aspheric surfaces are tested. These elements are assigned general aspheric terms G 3, 6, and 10, which modify the shape given by the radius and conic constant—but it makes little sense to calculate a tolerance for the individual coefficients since that is not what the technician measures in the shop anyway.

There are two common techniques for testing aspherics: one can design a null system, or simply measure the surface sag at an array of points and calculate some statistics the tell how far from the nominal shape the actual surface is. The first, the null test, is difficult and expensive to set up. One designs a second system that produces exactly the opposite aberrations of the surface under test. Then one reflects a beam off of that surface and examines the fringes with an interferometer. If done correctly, this test is extremely sensitive and can show figure errors of a small fraction of a wavelength—but the null test must first be designed and then manufactured to its own tolerance budget. In this example, where the aspheric curves are steep, the only practical null test would be the other elements of the design itself—which are no easier to make than the surface you want to test.

So this is a good example of a lens for which one would prefer to use a profilometer. Then one can measure the actual surface sags at many points and compare the results with the ideal shape. Here is our BTOL MACro (**C44M2**). The **PFTEST** directive tells the program to assign this attribute to all aspheric shapes in the lens.

```
BTOL 2
EXACT INDEX 1 3 5 7 9
EXACT VNUM 1 3 5 7 9
ADJ 10 TH 100
PFTEST ALL
TOL WAVE 0.05
GO
```

This MACro produces a budget, part of which Is shown below.

BUDGET TOLERANCE ANALYSIS

EL.	SURF	RADIUS	RADIUS TOLERANCE (RADIUS)	(FRINGES)	-----B----- THICKNESS	THICKNESS TOL
1	1	2.75796	0.00000	0.00000	0.19587	0.00281
1	2	0.84449	0.00000	0.00000	0.07752	0.00162
2	3	0.89769	0.00000	0.00000	0.88341	0.00183
2	4	-0.51009	0.00000	0.00000	0.05000	0.00188
3	5	0.95600	0.00000	0.00000	0.13302	0.00169
3	6	0.71159	0.00000	0.00000	0.26848	0.00109
4	7	-1.00418	0.00000	0.00000	0.05516	0.00227
4	8	3.59226	0.00000	0.00000	0.13091	0.02157
5	9	INFINITE	0.00000	2.05505	0.30000	0.01174
5	10	INFINITE	0.00000	2.28023	0.40000	0.00000
	11	INFINITE	0.00000	0.00000	0.00000	0.00000

In this case, the aspheric surfaces are not assigned a radius tolerance (or an irregularity, conic constant, or rolled-edge tolerance), because those are all figure errors that are included in a later section that applies to the aspherics, shown below.

. . .

SURFACE TESTED WITH PROFILOMETER: TOLERANCE IS STANDARD DEVIATION OF SAG ERROR OVER SURFACE

ELE	SURF	ST. DEV. TOL	G 15 CHANGE
1	1	0.00012	0.00682
1	2	0.00017	0.00505
2	3	0.00023	0.00394
2	4	0.00069	0.00675
3	5	0.00112	0.00981
3	6	0.00112	0.00813
4	7	0.00107	0.00729
4	8	0.00203	0.01359

Here you see the tolerances that BTOL assigned to the aspheric figure of each element. On surface 1, for example, the program found that a value of 0.00682 for term G 15 would be part of a reasonable budget. That term creates an astigmatism error in the shape, which is an error that does not affect focus or magnification but increases the wavefront variance. The standard deviation (**SD**) of the surface error is given as 0.00012 mm. This is what the technician can deduce, once the profilometer measurements are in hand. He calculates the measured SD by subtracting each measured surface sag from that calculated for the ideal shape, then finding the variance of those data (the average of the squares minus the square of the average). The SD is then the square root of the variance.

But what if the figure error is not simple astigmatism? As explained earlier, if the aberrations are small, the effect on the system MTF is a function of the wavefront variance and is not very sensitive to which aberrations may be present. So we use term G 15 as a proxy for any kind of small figure error.

When you run this BTOL analysis, you see many lines scrolling on the Command Window

```
***** WARNING: ZERNIKE FIT ERROR  0.179243     IS LARGE. *****
***** WARNING: ZERNIKE FIT ERROR  0.154225     IS LARGE. *****
***** WARNING: ZERNIKE FIT ERROR  0.238314     IS LARGE. *****
 Reduced deriv. increment for PFTEST RMS   of surf. no.   3 to  2.360486E-01
 Reduced deriv. increment for PFTEST RMS   of surf. no.   3 to  4.720972E-02
 Reduced deriv. increment for PFTEST RMS   of surf. no.   3 to  9.441945E-03
 Reduced deriv. increment for Thickness    of surf. no.   4 to  2.540005E-02
 Reduced deriv. increment for Thickness    of surf. no.   4 to  5.080010E-03
```

Those lines are a running record of how the program adjusts the derivative increments of each parameter in the tolerance budget. If a trial increment causes large wavefront errors, or if the accuracy of a curve fit to the wavefront is not close enough, the program automatically reduces the increment in steps until the effect is small enough to give accurate derivatives. So those lines show that the program is working properly and are not a cause for alarm.

IOP Publishing

Lens Design (Second Edition)
Automatic and quasi-autonomous computational methods and techniques
Donald C Dilworth

Chapter 45

Designing a very wide-angle lens

Creating a wide-angle front end so DSEARCH works over a very wide field

Designing a wide-angle lens with DSEARCH presents a new complication: if you enter a wide-angle object specification of more than 90 degrees in the SYSTEM section of the DSEARCH file, it is likely that none of the candidate configurations will work because no light can get through at the extreme field angle. DSEARCH can correct for some ray failures, but usually cannot optimize such a system. So what do you do?

There is a rather simple trick that works in such cases: Rough out a simple front end that converts the beam into one with a smaller angle, and then go from there, declaring that portion with **USE CURRENT**. Here is an example.

We want a lens with a semi-field angle of 92.4 degrees that works at F/2.0. The added elements will be made of plastic, which can be aspheric. First, we have to create a front end that will trace.

Enter a simple RLE file with two lenses and specify object type OBD, which is used for wide angles, declaring a paraxial stop on 5. Start with a moderate angle, say 50 degrees, and then, using the WorkSheet sliders, give the elements some negative power and bend them to the right. When that looks good, increase the OBD field angle, continuing in this manner until you reach the desired angle of 92.4 degrees. Here is a suitable front end, in figure 45.1.

```
RLE
ID WIDE-ANGLE DESEARCH
WAVL .6562700 .5875600 .4861300
 APS             5
 UNITS MM
```


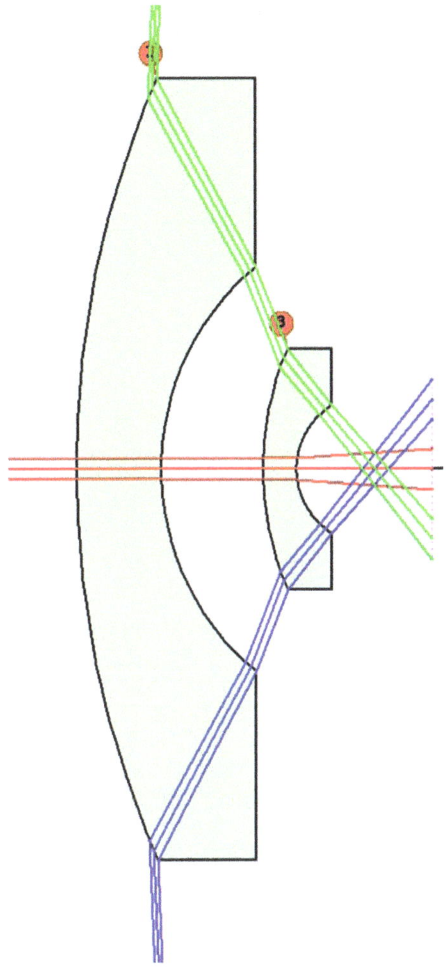

Figure 45.1. The front end roughed out so light can get through at a shallower angle.

```
   0 AIR
   0 CV  1.0000000000000E-09 AIR
   1 CV      0.0356159993000   TH      2.50000000
   1 GLM      1.50000000             55.00000000
   2 CV      0.2018873610000   TH      2.99808431 AIR
   3 CV      0.1145140002814   TH      1.00000000
   3 GLM      1.50000000             55.00000000
   4 CV      0.4600712360000   TH      4.00383115 AIR
   5 CV      0.0000000000000   TH      0.00000000 AIR
 END
```

The 92.4-degree entering beam exits at a reasonable angle. Now create a DSEARCH input MACro: (**C45M1**)

```
RLE
ID WIDE-ANGLE DESEARCH
WAVL .6562700 .5875600 .4861300
 APS                5
 UNITS MM
 OBD  1.00000E+09   92.4 0.2887  -11.0345861 0 0 0.2887

    0 AIR
    0 CV  1.0000000000000E-09 AIR
    1 CV      0.0356159993000   TH       2.50000000
    1 GLM      1.50000000             55.00000000
    2 CV      0.2018873610000   TH       2.99808431 AIR
    3 CV      0.1145140002814   TH       1.00000000
    3 GLM      1.50000000             55.00000000
    4 CV      0.4600712360000   TH       4.00383115 AIR
    5 CV      0.0000000000000   TH       0.00000000 AIR
 END

CORE 32
OFF 1
ON 99
TIME
DSEARCH 2 QUIET
USE CURRENT 5 ALL

GOALS
ELEMENTS 5
FNUM 2 1
BACK 5 SET
STOP MIDDLE
STOP FREE
ASPHERIC 3 5 6 7 8 9 10 11 12 13 14
FOV 0 .2 .4 .6 .8 1
NGRID 6
SNAP 1
RT 0.5
RSTART 100
TSTART 3
ASTART 2

PLASTIC 5 7 9 11 13
! QUICK 30 40
ANNEAL 20 1
NPASS 50
END

SPECIAL AANT
ACC 10 1 1
ACA 55 1 10
ASC 75 1 1
LUL 90 .1 1 A TOTL

END

GO
TIME
```

This file says to use the current system (the two lenses you adjusted above) and start adding elements at surface 5. All surfaces are to be variables, including the current ones. It specifies a back-focus distance of 5 mm, fixed with the **SET** directive. We can free up that thickness later, if it does not have to be exactly that value.

This input requests a maximum element thickness of 10 mm and an upper limit on the total length of 90 mm, to keep things reasonable. Also, it restricts ray intercept angles to no more than 55 degrees. Otherwise, for steep field angles like this, one can get grazing-incidence rays, which are impractical because of coating concerns and can lead to ray failures while optimizing.

Note that we are *not* using the QUICK option in this case. That is a powerful tool for simpler jobs, but this one is not simple. Third-order aberrations have little meaning with so wide a field, and we need the power of the full optimization on each candidate system. That line is commented out above, to emphasize the point.

Okay, your input is prepared, so run this DSEARCH file. In less than four minutes you see the results shown in figure 45.2.

At this stage, the lens has only conic constants assigned to the surfaces, since that is the default unless you enter **ASPH R** in the DSEARCH file, which we did not do here. That makes sense; it is a good idea not to use high-order terms when you are roughing out a lens. Save those for when the lens is as good as you can get it with only spherical or conic surfaces. Even so, this is an excellent start.

But it needs improvement. Run the optimization MACro that DSEARCH created, and the lens is better. The surfaces now have higher-order aspheric terms.

Then change the thickness variable declaration to **VLIST TH ALL**. This will let the back focus vary now, since we are close to a solution.

Optimize and anneal (**20, 2, 50**); the result is shown in figure 45.3. The end of the full-field TFAN looks bad, but the lens drawing shows that the full-field point uses a larger aperture at the stop than does the on axis, with the present paraxial pupil definition. So that issue may go away in the next step.

Now the design has taken shape, and we see that the stop naturally wants to be near the last elements. Assign a real stop to surface 9 with WS and fix up the pupil as well.

```
APS -9
CSTOP
WAP 2
```

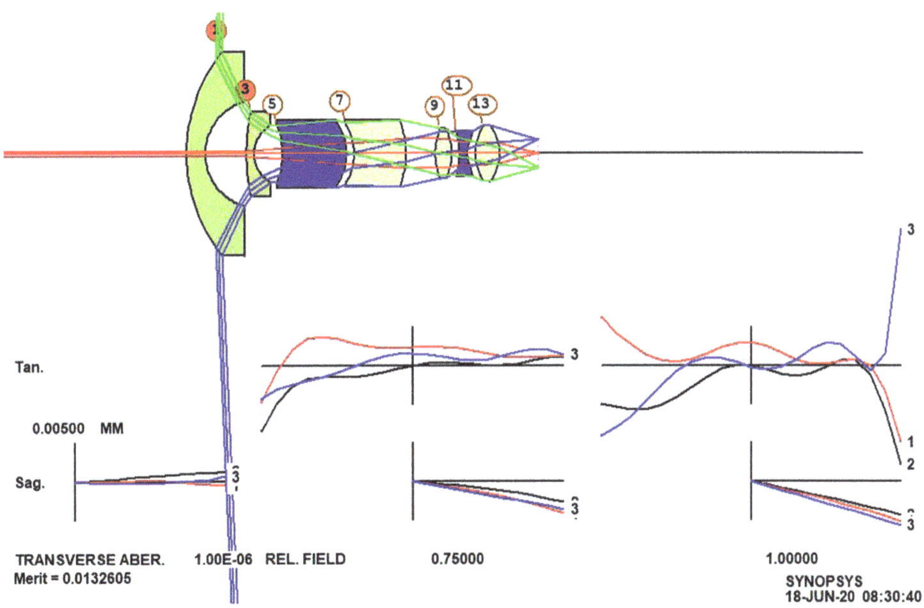

Figure 45.2. First results from DSEARCH for a wide-angle lens.

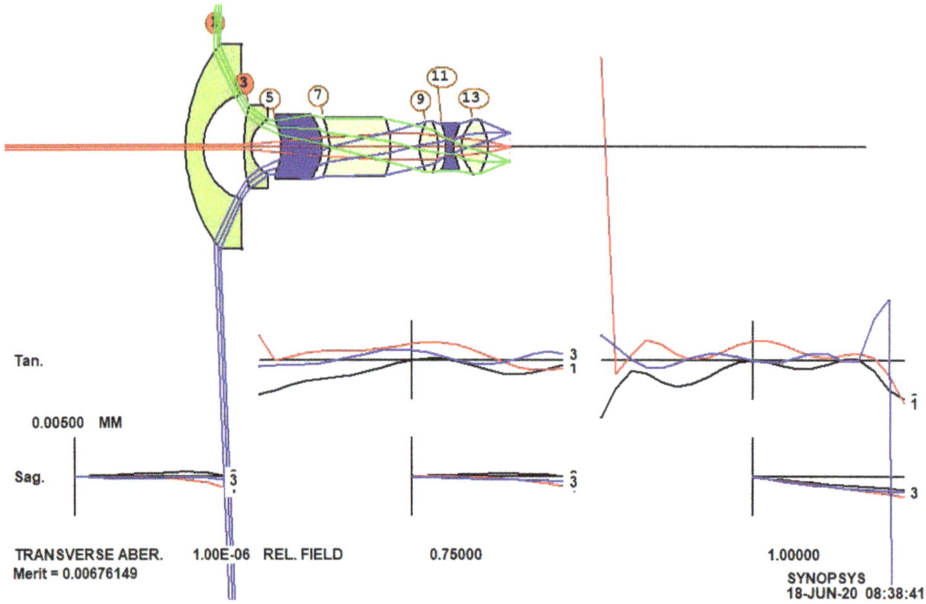

Figure 45.3. Wide-angle lens reoptimized.

Now delete the variable for YP1, reoptimize and anneal. The MF is now at 0.00175, and the fans look great.

Next we will insert real materials. Make a checkpoint, open the **MRG** dialog, select the U catalog (which will match only plastic elements), select QUIET, SORT, and click OK. The lens now has real plastics, shown in figure 45.4. Here we have selected the option in PAD to fill elements matched to real materials with solid color and those assigned a glass model with a cross-hatch pattern. That option makes it easy to keep track of which elements are which.

Now replace the first two elements with glass. Run **MRG** again, this time selecting the Ohara catalog. The program matches only the first two elements, which are glass, not plastic, and the design comes back just as good as before, shown in figure 45.5. (**C45L1**)

How well did we do? Let's look at the diffraction pattern over the field. Go to the **MPF** dialog, select Show visual appearance, magnify 5, and click Execute. The result, in figure 45.6, is perfect over the whole field.

Let me add some words of wisdom. Note that we did not use a curvature or thickness solve in this exercise, since a common problem with very wide-angle lenses is trying to avoid ray failures. While using solves makes perfect sense mathematically, they can cause just this kind of problem with this kind of lens. Also, we did not change to a real pupil until after the lens had its final form. The real-pupil search is robust but not infallible, and with this kind of steep ray angle and power-series aspherics, it is possible to obtain a configuration where there is no solution to the search. And even worse, sometimes there are *two* solutions, and the program might choose the wrong one. All that can be avoided by using the implied pupil until the design is in good shape.

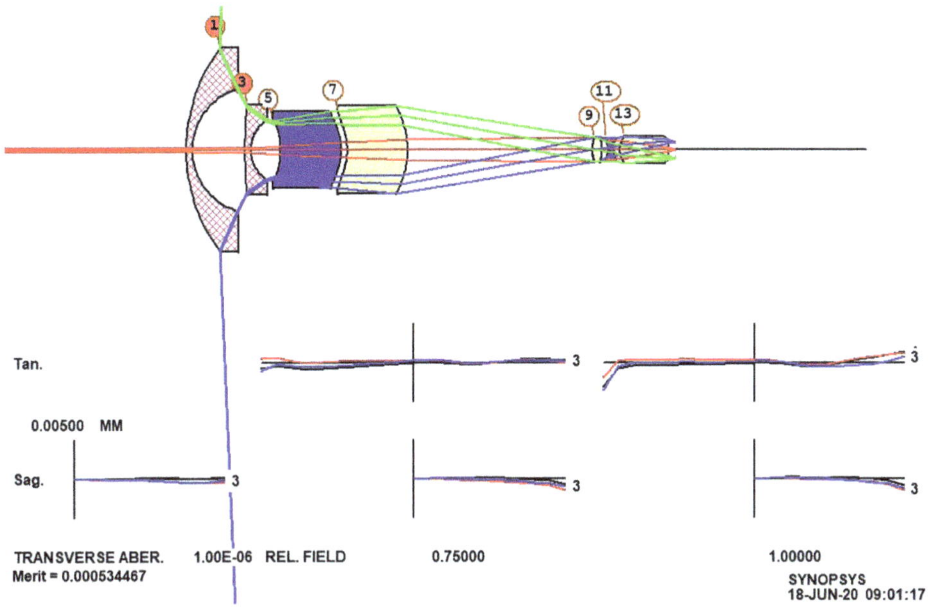

Figure 45.4. Lens with real plastics replacing the glass models on the last five elements.

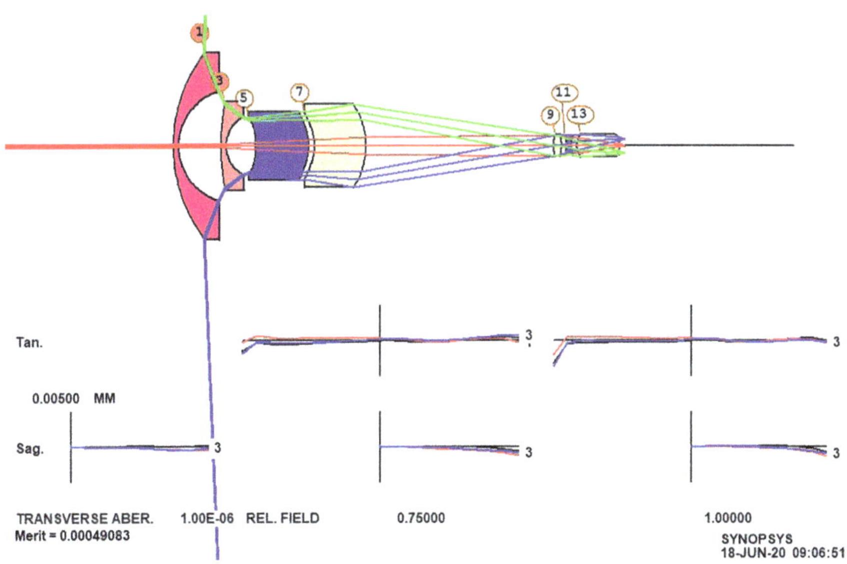

Figure 45.5. Wide-angle lens with all real materials.

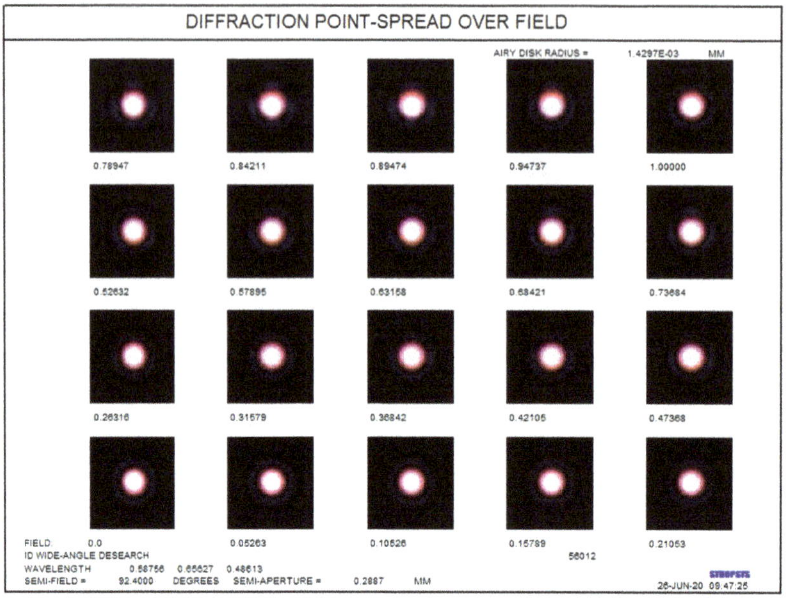

Figure 45.6. Diffraction pattern over field of the final wide-angle lens.

We also used the ASC monitor in this example. That is to avoid steep surfaces, which in this case turned out to be necessary. When we ran it without it, the second surface wanted to come out *hyperhemispheric*, meaning it went past the hemisphere point. While the lens was otherwise excellent, that kind of surface is very difficult to manufacture and to coat and is better avoided. But the monitor was able to avoid it. And if the ASC monitor is not sufficient, look up the RSLOPE operand in the help file. That is intended specifically for hyperhemispherical surfaces.

This lens is obviously very good, but do we really need seven elements? And can we get a shorter lens? How about one with a smaller first element? All these questions normally come up when you design a lens, and all can be quickly answered. Just add the new requirements to the DSEARCH input file—and find out.

Lest it seem too simple, we must repeat that we tried several combinations of RSTART, ANNEAL, and so on, and most of the runs produced a pretty good lens. But the best one resulted from the input quoted above. So if your first results are not pleasing, by all means experiment with the input parameters. That is how you can explore this very bushy bush.

Lastly, do not hesitate to investigate more than just the top lens returned by DSEARCH. We used the top one in this case, but that is not always the best one when you get to the final design. That's why DSEARCH returns more than just one solution.

45.1 Wide-angle lens II

In the first part of this chapter, we observed that an input angle of more than 90 degrees requires a custom-designed front end, the purpose of which is to reduce the chief-ray angle to a more manageable value so that rays can get through. But if the field is somewhat *less* than 90 degrees, the job is much simpler.

To illustrate this, we will design a lens with a maximum field angle of 86 degrees instead of 92.4. We still have to be careful about rays failing, but now there is a different solution. In MACro **C45M2**, we use DSEARCH by itself, with no added front end, and specify a Wide-Angle Correction parameter (**WAC**) of 50 degrees. We ask for seven elements, the same as in the previous lens, when the two front elements were added to the five constructed by DSEARCH.

```
CORE 128
OFF 1
ON 99
TIME
DSEARCH 3 QUIET

SYSTEM
APS               5
UNITS MM
ID MEDIUM-ANGLE DESEARCH
OBD 1.0E9 86. .3
END

GOALS
ELEMENTS 7
FNUM 2 1
BACK 5 SET
STOP MIDDLE
STOP FREE
ASPHERIC 3 5 6 7 8 9 10 11 12 13 14
FOV 0 .2 .4 .6 .8 1
NGRID 6
SNAP 1
RT 0.5
RSTART 50
TSTART 3
ASTART 2

PLASTIC 5 7 9 11 13
! QUICK 30 40
ANNEAL 20 1
NPASS 50
WAC 50 ! wide-angle correction to 50 degrees

END

SPECIAL AANT
ACC 10 1 1
ACA 55 1 10
ASC 75 1 1
LUL 90 .1 1 A TOTL
END

GO
TIME
```

With this parameter activated, as each case is constructed by DSEARCH, it will first apply a paraxial UPC solve on side 2 of the first element, giving a target value, in this case, of 50 degrees. Then the paraxial chief-ray angle will be smaller than the original 86 degrees, and the chances of all the real rays getting through the lens will be much improved. (The program removes the new solve once the proper curvature on side 2 is calculated, so that radius is again a variable.)

When this MACro is run, the program produces the lens in figure 45.7. (**C47L2**)

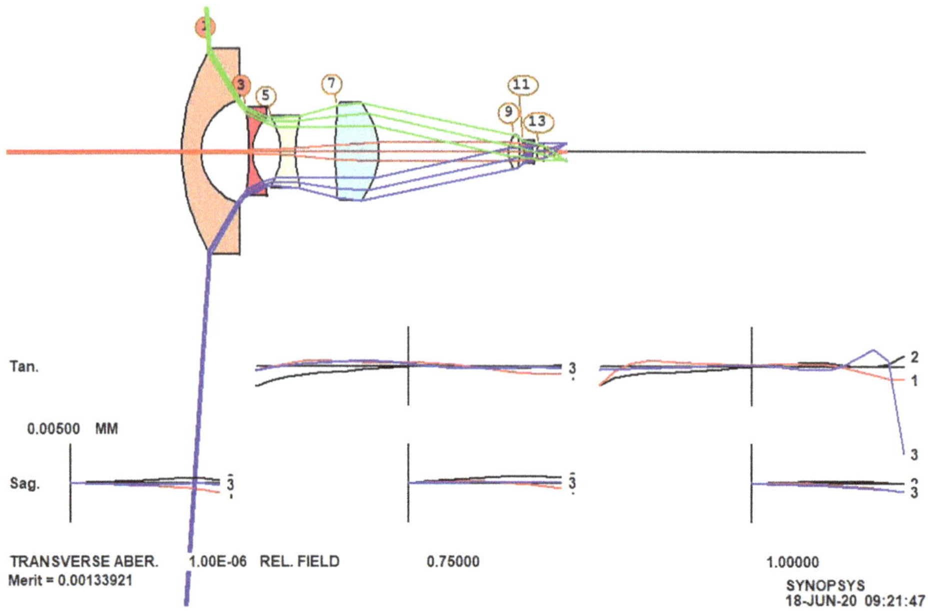

Figure 45.7. Lens returned by DSEARCH when the maximum field angle is 86 degrees and WAC has been activated.

In this case, the search statistics report that of 128 cases to be analyzed, a total of 59 were skipped due to ray failures. If this MACro is run *without* the WAC parameter, the program skips 125 of the 128 cases. It is always a good idea not to skip cases, since a promising configuration might then be missed. The WAC parameter is another one you can experiment with, as you search for the best lens design with DSEARCH.

The lens still needs additional work, as was done previously, but the point is that with this slightly reduced field angle the job is simpler and does not require the preliminary calculation of the front-end elements as we did above.

IOP Publishing

Lens Design (Second Edition)

Automatic and quasi-autonomous computational methods and techniques

Donald C Dilworth

Chapter 46

A complex interferometer

Setting up an interferometer

Interferometers have two channels, and the beams are combined at a beamsplitter. One often wants to see the difference in the shape of the two wavefronts, as when testing aspheric mirrors. To work properly, the shapes should be very similar.

In this example, fringes between the two channels give spectral information as the position of one of the mirrors is moved back and forth. In this configuration, the instrument is called a *Fourier-Transform* spectrometer. Here one is not concerned with the shape of the wavefront, as long as the two channels match, but with its absolute phase.

We will set up one channel at first, entering those data that are easy to figure out, and then let the program calculate the rest for us. Here is the input for the first step (**C46M1**):

```
RLE
ID INTERFEROMETER EXAMPLE
WAVL 4.6 4.25 3.9
OBB 0 1 30 0 0 0 30
1 TH 100                ! DUMMY SURFACE FOR REFERENCE
2 AT -45 0 100          ! SCAN MIRROR
2 REFL
2 TH 0
3 AT -45 0 100          ! FOLD AXIS
3 TH -200               ! TO BEAMSPLITTER
4 TH -3 GTB U           ! THROUGH 3 MM THICK GERMANIUM
GE
5 REFL                  ! REFLECT AT BEAMSPLITTER
5 PTH -4 PIN 4          ! COMING BACK AGAIN
4 AT 30 0 100           ! TILT OF BEAMSPLITTER
7 AT 30 0 100           ! TILT AXIS
7 TH 70                 ! TO REFERENCE MIRROR
8 REFL                  ! HERE
8 PTH -7                ! BACK TO BEAMSPLITTER
```

doi:10.1088/978-0-7503-3695-6ch46

```
9 AT -30 0 100          ! ENTER IT AGAIN
9 PTH 4 PIN 4           ! SAME SIZE AS BEFORE
10 TH -.1               ! SMALL AIRSPACE
11 PTH 4 PIN 4          ! COMPENSATOR DUPLICATES GEOMETRY
12
13 AT 30 0 100          ! DUMMY TO FOLLOW BEAM
13 TH -200              ! DISTANCE TO FOLD MIRROR 1
14 AT -30 0 100         ! RIGHT HERE
14 GID
FOLD 1                  ! IDENTIFY IT
14 REFL                 ! REFLECT THERE
15 AT -30 0 100         ! DUMMY TO FOLLOW BEAM
15 TH 250               ! DISTANCE TO PRIMARY MIRROR
16 RD -180 CC -1 TH -90   ! PARABOLOID HERE
16 REFL                 ! REFLECT THERE
16 GID
PRIMARY M               ! IDENTIFY IT
17 AT 45 0 100          ! SMALL FOLD MIRROR
17 REFL                 ! REFLECT HERE
17 GID
SECONDARY M             ! IDENTIFY IT
18 AT 45 0 100          ! DUMMY TO FOLLOW BEAM
18 TH 90                ! DISTANCE TO TERTIARY
19 RD -180 TH -350      ! DISTANCE TO FINAL IMAGE
19 REFL                 ! REFLECT AT TERTIARY
APS 19                  ! THE STOP IS HERE
19 GID
TERTIARY M
20
END
```

Run the MACro above, and you get the PAD picture in figure 46.1.

To get this display, click on the PAD Top button, select Custom rayset, HBAR 0.0 and 11 rays. Also select the Solo top display option and turn on switch 38, which shows numbers for all surfaces, including dummies.

So far, you have the basic elements in place but do not yet know the details of the tertiary mirror at surface 19. You want a sharp image on surface 20, and you will insert additional fold mirrors to separate three wavelength regions onto different detectors later when you get to that step. But now you need to know the radius and conic constant on 19. Type the following into a new editor:

```
PANT
VY 19 ASPH
END
AANT
GSR 0 1 4 P
END
SYNO 10
```

After running this file, the system looks just the way it should, as shown in figure 46.2.

The command ASY now shows the shape of surface 19:

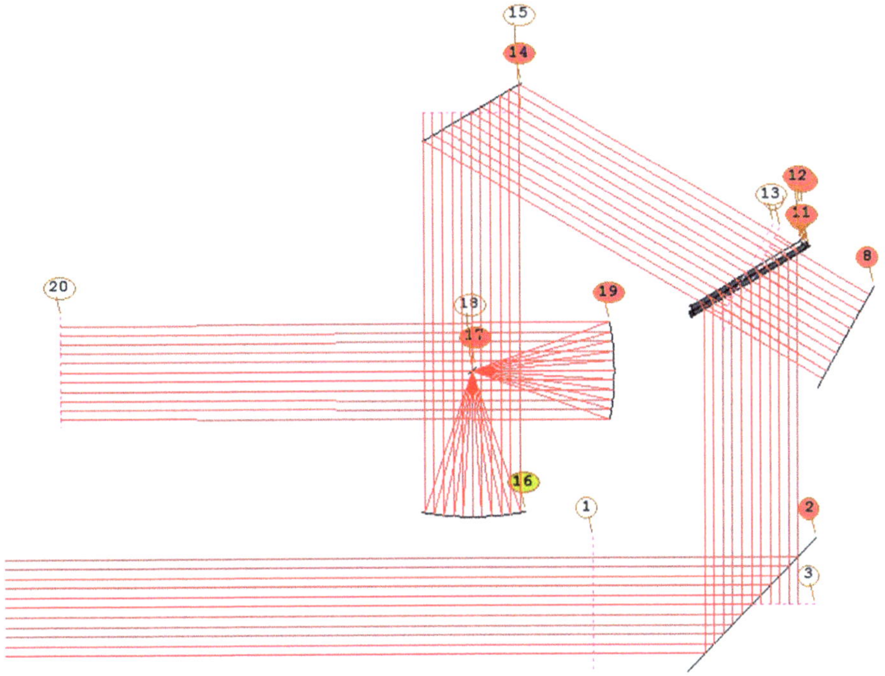

Figure 46.1. Interferometer, original setup.

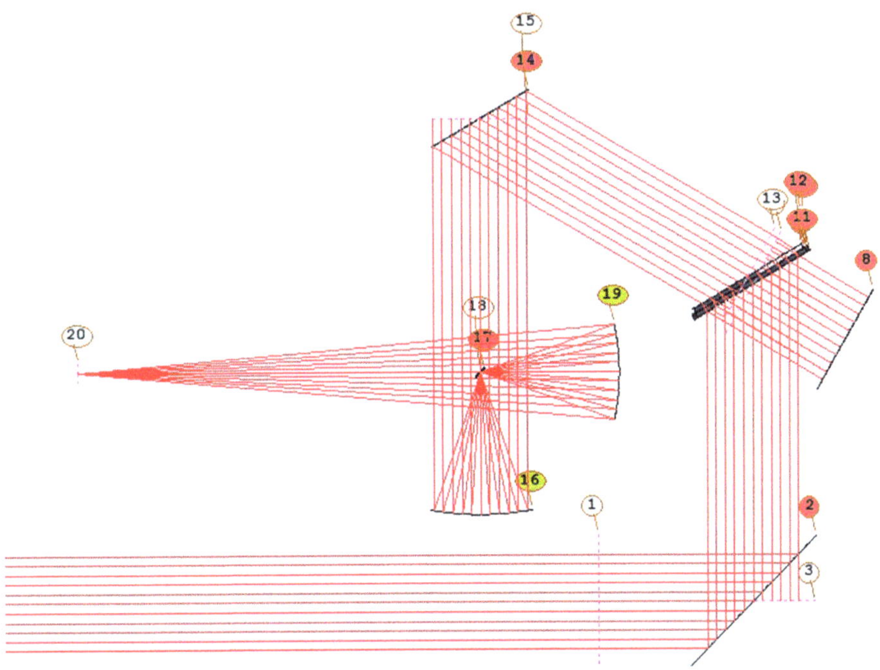

Figure 46.2. Interferometer after optimizing the shape of surface 19.

```
SYNOPSYS AI>ASY

SPECIAL SURFACE DATA
-------------------------------------------------------------------------------
 SURFACE NO.  16 -- CONIC SURFACE
 CONIC CONSTANT (CC)      -1.000000
 SEMI-MAJOR AXIS (b)  INFINITE       SEMI-MINOR AXIS (a)  INFINITE

 SURFACE NO.  19 -- CONIC SURFACE
 CONIC CONSTANT (CC)      -0.349174
 SEMI-MAJOR AXIS (b)    -219.999999  SEMI-MINOR AXIS (a)     177.482393

TILT AND DECENTER DATA
LEFT-HANDED COORDINATES
--------------------------------------------------------------------------------
SURF TYPE          X            Y            Z       ALPHA       BETA      GAMMA
--------------------------------------------------------------------------------
  2 REL        0.00000      0.00000      0.00000  -45.0000     0.0000     0.0000
  3 REL        0.00000      0.00000      0.00000  -45.0000     0.0000     0.0000
  4 REL        0.00000      0.00000      0.00000   30.0000     0.0000     0.0000
  7 REL        0.00000      0.00000      0.00000   30.0000     0.0000     0.0000
  9 REL        0.00000      0.00000      0.00000  -30.0000     0.0000     0.0000
 13 REL        0.00000      0.00000      0.00000   30.0000     0.0000     0.0000
 14 REL        0.00000      0.00000      0.00000  -30.0000     0.0000     0.0000
 15 REL        0.00000      0.00000      0.00000  -30.0000     0.0000     0.0000
 17 REL        0.00000      0.00000      0.00000   45.0000     0.0000     0.0000
 18 REL        0.00000      0.00000      0.00000   45.0000     0.0000     0.0000

KEY TO SURFACE TYPES
--------------------
GLB  GLOBAL COORDINATES                 LOC  LOCAL COORDINATES
REL  RELATIVE COORDINATES               REM  REMOTE TILTS IN RELATIVE COORD.
SYNOPSYS AI>
```

Okay, one channel looks good; now let's set up the second. We can start with the above setup and just modify it as required. First, bump this setup into ACON 2, with the ACON copy button, and then modify the geometry at the beamsplitter. Make a CHG file (**C46M2**):

```
CHG
13 SIN              ! NEED TWO ADDITIONAL SURFACES
13 SIN
4 NAS               ! REMOVE TILTS THERE NOW
7 NAS
9 NAS
13 NAS
4 TH -3 GTB U
GE
4 AT 30 0 100        ! TILT OF BEAMSPLITTER
5 TH -.1 TRANS
6 PTH 4 PIN 4
7 TH 0
8 AT -30 0 100
8 TRANS              ! NOT REFLECTIVE ANYMORE
8 TH -70
```

```
9 REFL
9 PTH -8 AIR
10 AT 30 0 100
10 PTH -4 PIN 4
11 PTH -5
12 REFL
12 PTH 5
13 PTH 4 PIN 4
14 TH 0
15 AT 30 0 100
15 TH -200
END
```

This file first deletes most of the declarations assigned to the beamsplitter in channel 1, since reflections and tilts now occur on different surfaces. Then it replaces them with the data for the other channel. The new system is shown in figure 46.3.

(Here we have turned off switch 38 to make a cleaner picture. Now it will not show surface numbers on dummy surfaces.)

Okay, both channels are defined, and they are both current, in ACONS 1 and 2. Now make a perspective drawing showing both at the same time. Create a MACro:

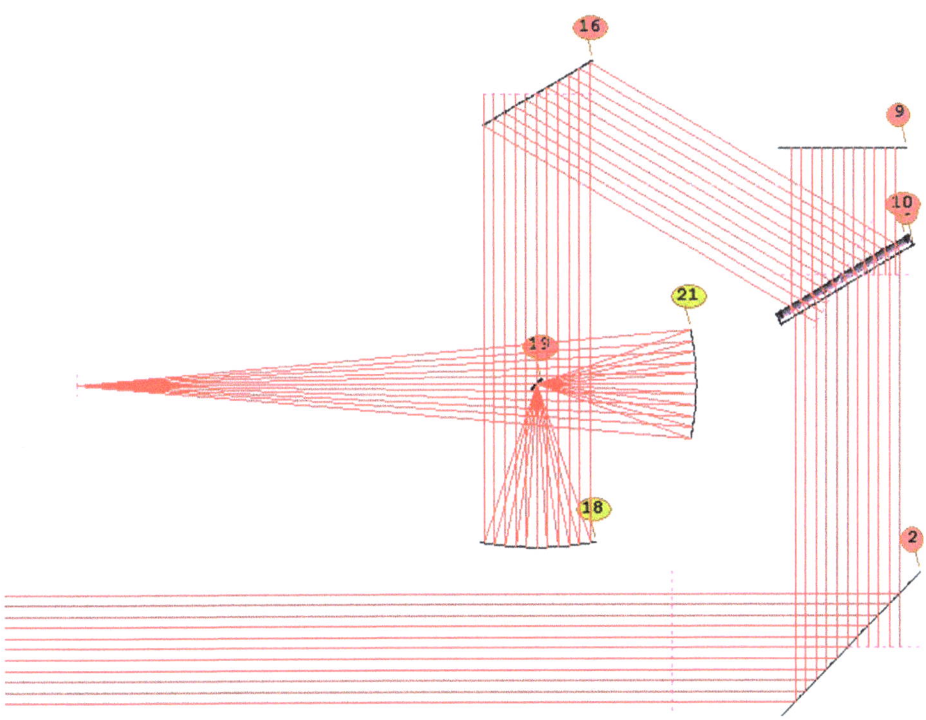

Figure 46.3. Second channel of the interferometer.

```
ACON 1
HPLOT 1
PER 0 0 .015 1 123
PUP 2 1 10
PLOT
RED
TRACE P 0 0 10
END

ACON 2
APLOT 1
PER 0 0 .015 1 123
PUP 2 1 10
PLOT
BLUE
TRACE P 0 0 10
END
```

This gives us the picture in figure 46.4.

Not bad. Let's improve it some more. Open the Edge Wizard (**MEW**) and select Create All. Do this for both ACONs. This creates edge data on all elements. Now

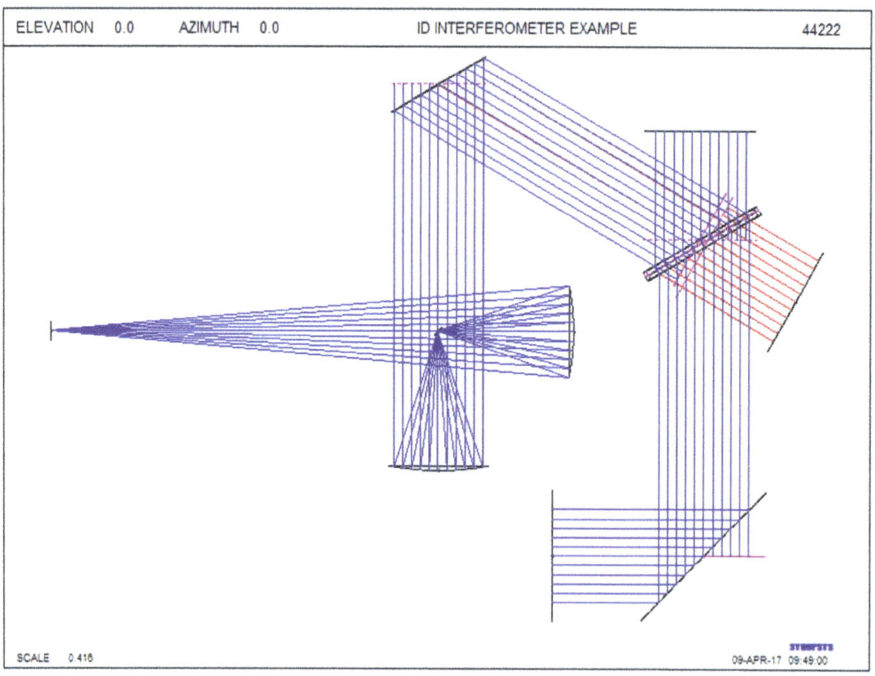

Figure 46.4. Both channels of the interferometer, shown by PERSPECTIVE.

run the MACro above, after turning on switch 20, adding TRA requests for HBAR = 1 and −1, and changing the PER requests to RSOLID. The picture is in figure 46.5. We are off to a very good start (**C46L1**). (Switch 20 controls drawing of dummy surfaces.)

This is a brief lesson on how to set up systems like this. SYNOPSYS can show the system and the image quality very nicely—but it will not model the interference between the two channels. The program can model the difference in *shape* between a wavefront and a reference beam (see **HELP IFR**), but the information in a Fourier-transform interferometer arises from the relative *phase* of the wavefronts, which is not calculated by SYNOPSYS. Of course, this is no matter since the theory is already very mature, and all we care about is wavefront quality in these systems.

The next step will be to add additional fold mirrors and detector optics in the space before the final image. If beamsplitting cubes are to be used to separate different wavelengths, the systems should be designed with blocks of glass of the equivalent thickness before the image. But now that you know how these things are done, we leave that as an exercise for the student.

If you are especially perceptive, you will have noticed that there is a small decenter in the beam as it goes through the beamsplitter—which we have ignored.

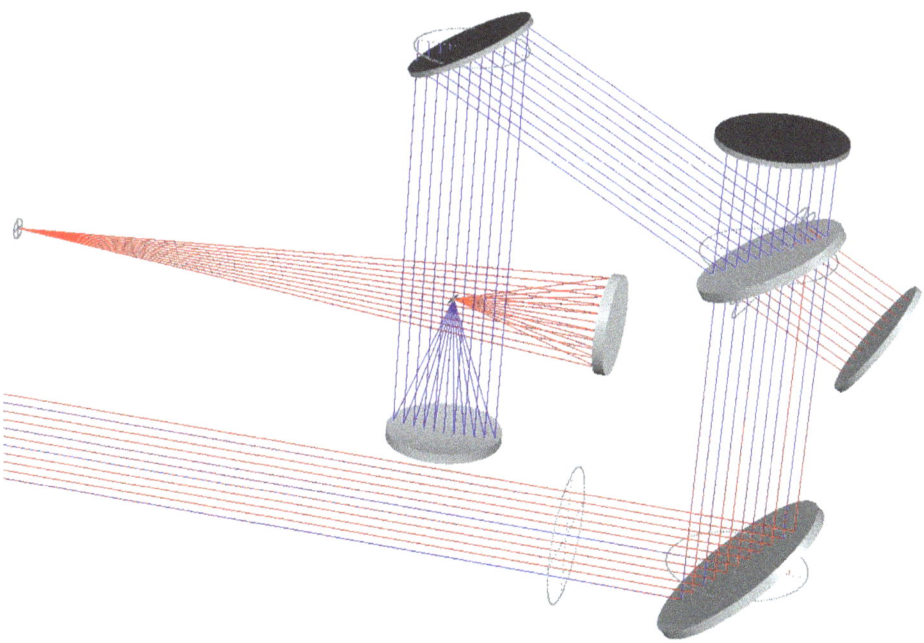

Figure 46.5. Both channels of the interferometer, with mirror thicknesses added, shown by RSOLID.

But that's not hard to model either; just adjust a decenter on the primary mirror to compensate, if you really want to get that precise.

Lastly, we want to observe that, in SYNOPSYS, you can design up to six configurations at the same time, and they are completely separate systems unless you declare otherwise. This is in contrast to the practice in some other codes, where there is actually only a single configuration, and when you ask for a different configuration you get the same system unless you spell out the differences. The result can be much the same, but the philosophy is reversed.

IOP Publishing

Lens Design (Second Edition)
Automatic and quasi-autonomous computational methods and techniques
Donald C Dilworth

Chapter 47

A four-element astronomical telescope

Global search for a telescope design with no secondary color

The goal in this lesson is to design a very good astronomical telescope for advanced amateur astronomers. The input file for DSEARCH is as follows (**C47M1**):

```
CORE 16
ON 1
ON 99
TIME
DSEARCH 1 QUIET
SYSTEM
ID DSEARCH TELESCOPE
OBB 0 0.7 75
WAVL 0.6563 0.5876 0.4861

UNITS MM
END
GOALS
ELEMENTS 4
FNUM 8
TOTL 0 0
STOP FIRST
STOP FIX
RSTART 200
TSTART 10
ASTART 10
RT 0
OPD
```

doi:10.1088/978-0-7503-3695-6ch47

```
FOV 0.0 0.75 1.0
FWT 5.0 3.0 1.0
NPASS 40
ANNEAL 50 10 Q
COLORS 3
SNAPSHOT 10
QUICK 50 50
END
SPECIAL PANT

END
SPECIAL AANT
ADT 7 .1 10
LUL 300 .1 1 A TOTL
END
GO
TIME
```

The reader should need no coaching by this time, but some points may not be obvious. We want a telescope with extremely good image quality at F/8. This DSEARCH input will target only OPD errors in the real merit function (plus the usual monitors and first-order goals). Run this MACro, and DSEARCH returns the lens in figure 47.1 in about 9 s.

This is essentially perfect already, but it has model glasses, and we need to replace them with real glass. Run the optimization MACro prepared by DSEARCH, make

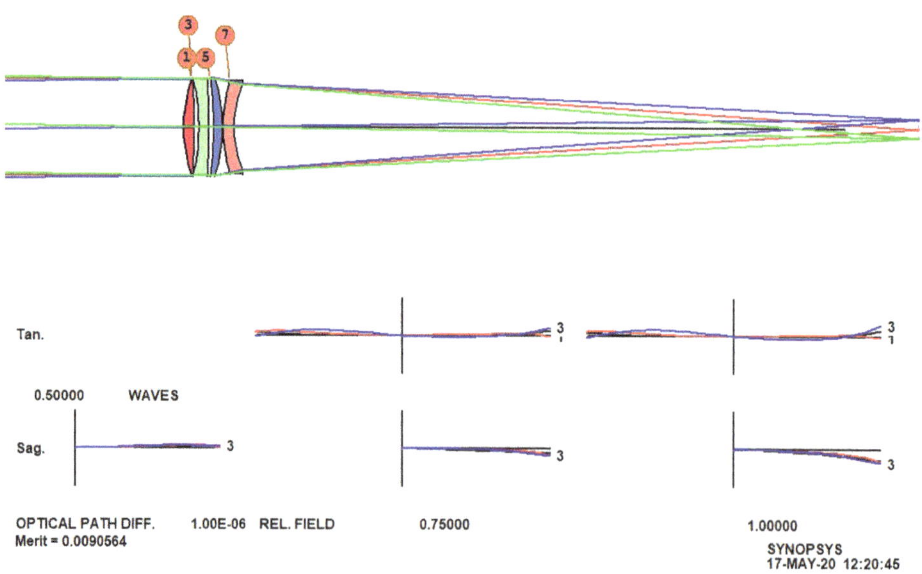

Figure 47.1. Lens returned by DSEARCH for a telescope example.

a checkpoint, and then use **MRG** to replace them with glass from the Guangming catalog. The lens is just as good, but we wonder if we can do even better.

It's time to use another tool. Restore the checkpoint and save the optimization MACro with the name GSOPT.MAC.

Now make a new MACro to run GSEARCH, containing the following:

```
GSEARCH 3 QUIET
SURF
1 3 5 7
END
NEAREST 5 P 10
G
END
GO
```

Now the program will search 81 combinations of glass (3^4) and only investigate glass types not more than 10 times the price of BK7. The result (**C47L1**) is superb, after more optimization and annealing, shown in figure 47.2.

If you have been reading carefully, you will notice that here we have violated some of the rules we mentioned in previous chapters. We optimized this lens with OPD errors but not TAP. Yes, that sometimes falls into a hole where the output is collimated instead of focusing, but not very often. One has to keep one's eyes open, and don't be afraid to experiment.

Also, we put an **ADT** monitor in the DSEARCH file, which is not always a good idea, but we got excellent results anyway. Running SYNOPSYS is a lot like driving a Maserati: if you know how to drive it, it works better. To prepare this lesson we

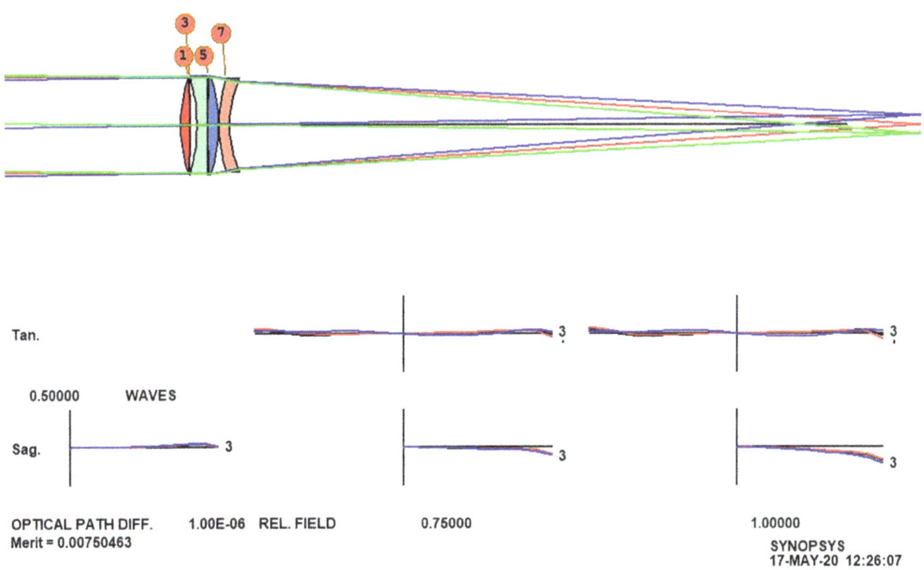

Figure 47.2. Telescope design with real glass.

tried a variety of input variations, many of which produced a lens with about one wave of secondary color. The trick here seems to be the positive flint element—but not all of the DSEARCH runs found that solution. We have a very bushy bush here, and it is a good idea to run the program several times, as we did, and select the best lens of the lot. (We also did this lesson again with switch 98 turned off, which yields different results each time, and got different lenses that also had no secondary color.) There are lots of solutions out there, and if you master these search tools you will be an expert explorer in no time.

We also ran this exercise with an F/number of 7 instead of F/8 and got very similar results. How low do you think you can go? Try 6 and see how close you can come. Then use AEI to try to improve that design. That's what lens design is all about: see what works, find a way to improve it, and be very pleased when you succeed.

IOP Publishing

Lens Design (Second Edition)
Automatic and quasi-autonomous computational methods and techniques
Donald C Dilworth

Chapter 48

A sophisticated merit function

The earlier chapters tell how one can (and should) incorporate all the goals that have to be satisfied by the lens into the merit function, mechanical as well as optical. This lesson gives an example of such a design, one that required extensive control of both characteristics. It will be instructive for the reader to study how this was achieved. The system (**C48L1**) is shown in figure 48.1.

This lens works in the thermal infrared, from 8 to 12 μm, and must be corrected to ¼ wave or better. The merit function that optimized it was rather complicated, owing mostly to the need to keep the obscuration to a minimum. You saw in section 8.1 how introducing an obscuration into the entrance pupil affects the modulation transfer function and learned that one does not want it to be any larger than absolutely necessary. The lens in this lesson is a *catadioptric* design, meaning it has both lenses and mirrors, and the effect of the obscuration is just as important here.

Such systems can be arranged in one of two ways: axially symmetric (which is easier to design and build but is obscured) or with off-axis components, which are more difficult and expensive but can avoid an obscuration. A symmetric design is obscured since one has to collect the light from the first mirror and send it back through a hole in that mirror. In this example, we want to keep that obscuration to not more than 45%. Here is our MACro (**C48M1**):

```
LOG
STO 2
AWT: 0
BB: 0.45

PANT
VLIST RAD  2  4 5 6 7 8 9 10
VY 2 CC 100 -100
VY 4 CC 100 -100
VY 1 TH
VY 2 TH 399 -300
VLIST TH 6 8 10
END
```

doi:10.1088/978-0-7503-3695-6ch48

```
AANT
OBS .1 BB
AEC 1 .2
!M 95 .1 A TOTL
M 10.16 1 A GIHT

NAME BULGE
M 0 .1 A P YA 1 0 1 0 6
S P YA 1 0 1 0 7

NAME BULGE2
M 0 .01 A P YA 1 0 1 0 6
S P YA 1 0 1 0 9

NAME OBSC
M BB 50 A P YA 1 0 1 0 4
DIV YMP1

NAME SETBACK
M 19 .1 A ZG 14
S ZG 2

NAME FOCUSAIR
M 2 1 A TH 10
S P ZA 1 0 1 0 10

NAME CLEARCONE
LL 0 1 1 .01
A P YA 0 0 BB 0 3
S P YA 1 0 1 0 5

M 1.5 1 A BACK


GSR AWT 10 6 P
GSR AWT 10 6 1
GSR AWT 10 6 3
GNR AWT 6 6 P .7
GNR AWT 6 6 P 1
GNR AWT 6 6 1 1
GNR AWT 6 6 3 1
M 0 50 A 1 YA 1
S 3 YA 1
END

SNAP
SYNO 20
```

First, we have to explain how the system is set up.

1. Surface 1 is a dummy used for controlling geometry.
2. Surface 2 is the primary mirror, a conic.
3. Surface 3 is another dummy, whose position is variable and will be used to locate surface 5.
4. Surface 4 is declared coincident with surface 1, whose thickness is variable. This is the secondary mirror, also a conic.

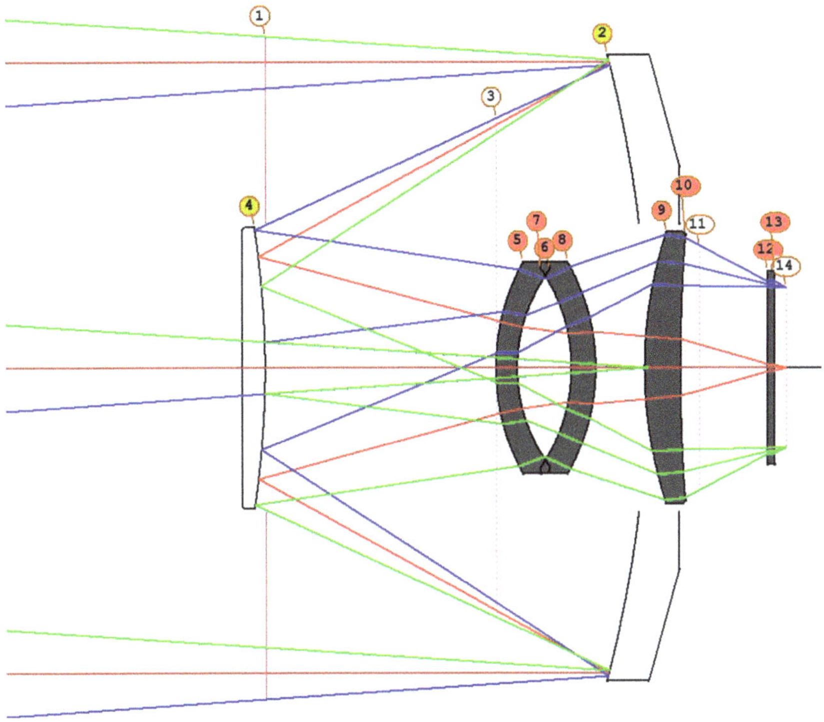

Figure 48.1. Catadioptric system requiring a sophisticated merit function.

5. Surface 5 is declared coincident with surface 3. As 3 moves, 5 will move with it.
6. Surfaces 5 through 10 are a set of Germanium corrector elements.
7. Surface 11 represents the location of a shutter.
8. Surfaces 12 and 13 are a Germanium window in front of the detector at 14.

Look at the items in the above MACro.

1. The symbol **AWT** is assigned a value of 0. This is the aperture weighting and can be altered if you want to do tradeoff studies.
2. Symbol **BB** is assigned the value 0.45. This is the allowed obscuration, and can also be varied for tradeoff studies.
3. The AANT file declares **OBS .1 BB**, which causes the ray grids generated by the **GSR** and **GNR** requests to delete any rays falling in a circle of radius **BB** (0.45 in this case) and decentered by 0.1 in the Y direction at full field. That portion will be obscured anyway, and it makes no sense to correct rays that will not get through.
4. The **AEC** monitor keeps edges to no less than 1 mm.
5. The target on **GIHT** controls the focal length.
6. The next aberration is assigned the name **BULGE**. This name will show up on aberration listings, making it easy to identify the values later.

7. The **BULGE** aberration is defined as the difference between the Y coordinate of the upper rim ray (URR) at full field on surface 6, and the same ray on surface 7. This keeps the first two lenses about the same size. We don't want either one to be much larger than the other.
8. The aberration named **BULGE2** is assigned a low weight (and is not met exactly) and is there to ensure that the lens at 9 is not very much larger than the lens at 6. The weight is adjusted to give the best balance. We don't want the hole in the primary to be larger than the secondary mirror, and this is one of several ways we might control that.
9. Aberration **OBSC** takes the URR (upper rim ray) on 4 and divides by the paraxial **YMP1**, assigning a target value of **BB** (which is the number 0.45) and a high weight of 50. Now, if the extreme ray on 4 wants a larger aperture than **BB**, the MF gets a penalty. This will keep the obscuration due to the secondary mirror under control.
10. Aberration **SETBACK** gives a target of 19 mm to the difference between the global Z-coordinate of surface 14 and that of surface 2. This controls the distance from the primary mirror to the image plane.
11. Aberration **FOCUSAIR** gives a target of 2 mm to the difference between thickness 10 and the Z-coordinate of the URR on that surface. If that lens were strongly curved toward the image, it would encroach on the airspace between 10 and 11, and we have to leave room in that space for the shutter. This will control that.
12. Aberration **CLEARCONE** gives a lower limit to the difference between the Y-coordinate of the on-axis ray at a fractional entrance coordinate of **BB** on surface 3 and the URR (Upper Rim Ray) on surface 5. The first of these rays comes in at the edge of the obscuration, and we don't want it obscured after the first reflection by the lens at 5, whose aperture is given by the second ray.
13. We want to maintain a clearance of 1.5 mm between the cover plate and the sensor. Thickness 13 is governed by a YMT solve and can change.
14. The rest of the MF consists of rays to be traced and corrected. Lateral color is corrected by taking the difference between the chief rays in colors 1 and 3, with a target of 0.

With a merit function like this, one can optimize the system and see how it works with a given obscuration. Then you can reduce the obscuration, in stages, by redefining symbol **BB**, and optimize each case. When the results start to degrade, you know how far you could push that parameter.

You see how it is sometimes necessary to control individual rays on individual surfaces, along with global positions of selected surfaces. If your project has complex requirements such as these, don't be afraid to create a complex merit function to design it. That's what those options are there for.

Readers with sharp eyes will notice that surfaces 6 and 7 overlap as shown in figure 48.1. Those edges result from the default calculation, and of course one has to adjust them before the lenses are fabricated. This is easily done with the procedures outlined in Chapter 40. We leave this as an exercise for the student. But here's a hint: type

```
FEATHER 6
```

and the program reports the Y-coordinate of the feathering point as well as the surface sags at 6 and 7. Assign that coordinate to the points E on 6 and A on 7. Then the sag tolerance on those surfaces determine the airspace tolerance too.

IOP Publishing

Lens Design (Second Edition)
Automatic and quasi-autonomous computational methods and techniques
Donald C Dilworth

Chapter 49

When automatic methods do not apply

The previous chapters of this book presented numerous examples of how autonomous methods can find excellent lens configurations more quickly and easily than was possible using classic techniques. The design process has been shortened from weeks to minutes, and you might think there is no need for the kind of expertise that was important a generation ago. But you would be wrong. These new methods cover a wide range of design problems—but there are situations where they are not appropriate. Then it is up to you. Here is an example. Call it a Final Exam.

Some years ago, I was asked to devise a problem to be given to an advanced lens design class, one that would test their knowledge of optics. I came up with a task that cannot be solved by a computer program. You have to think. You have to be clever. You have to know your optics. I urge the reader to first try to solve it yourself, before reading the solution. If you can do it, congratulations. If not, go back to the first chapters and study them again. Here is an updated version of that problem, shown in figure 49.1.

The rules are as follows:

1. The images must fall on the slits of a spectrometer. They must be narrow in one direction but long in the other… and
2. The orientation of the two images must be at 90 degrees, as shown in figure 49.2.

And there are other requirements:

1. Images must be narrow. 90% of the energy must pass through a slit of 13 μm width.
2. The image length must be between 0.1 and 0.2 mm.
3. Chromatic aberration must be corrected. Analysis will be done at the C, d, and F Fraunhofer lines.
4. The stars are at the top and bottom of a field of 20 degrees total angle. Pupil diameter is 20 mm.
5. All surfaces must be spherical (no cylinder lenses).

doi:10.1088/978-0-7503-3695-6ch49

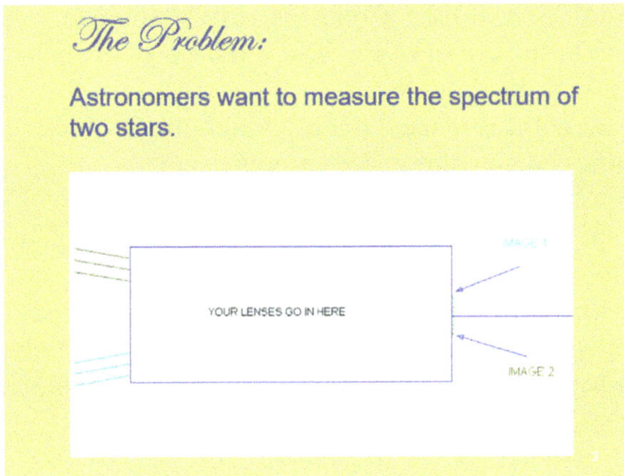

Figure 49.1. Outline of the final exam problem.

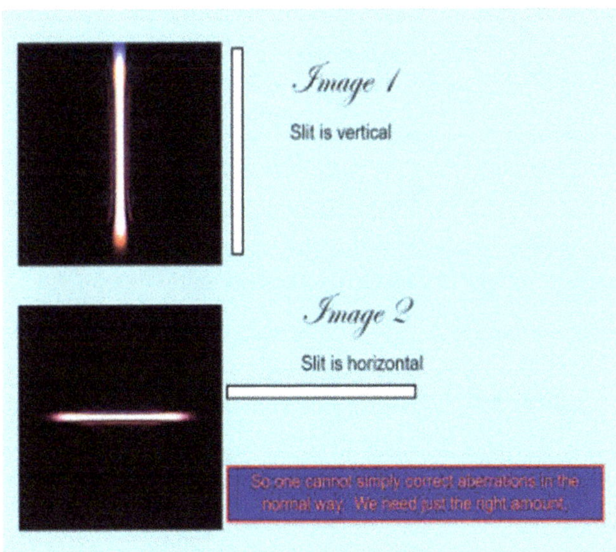

Figure 49.2. The orientation of the two slits.

6. The simplest lens system that meets the requirements is the winner. If a tie, then the highest energy wins.
7. Two versions will be submitted, at F/10 and F/5.

So that's the problem. Can you solve it? How many elements do you need? Give it a try before you read the solution below.

The solution

Understanding the source of aberrations and how they are affected by design variables is the key. What we need is a controlled amount of astigmatism. What does that aberration look like? Figure 49.3. shows an example.

What properties of astigmatism should one know about?

- Is always zero at the center of the field if the lenses are spherical and centered.
- Is symmetric with field angle.
- Well, that's no good. We need the top and bottom of the field to be different.
- Therefore, the system cannot be centered.

In figure 49.4 you see an example where tangential rays are defocused, while sagittal rays are almost in focus. The difference in *angle* is the astigmatism. In figure 49.5 you see that if you take only a short, decentered section of the spherical aberration curve, what you wind up with looks a lot like astigmatism. That's a clue.

- If you *decenter* the pupil, then spherical aberration looks like astigmatism.
- So a decentered element somewhere might be of use.
- What about astigmatism of a *centered* lens? Figure 49.6 shows that, away from the optical axis, the sagittal and tangential focus surfaces separate.
- But then the top and bottom of the field get the *same* astigmatism.
- That's no good; they have to be different.

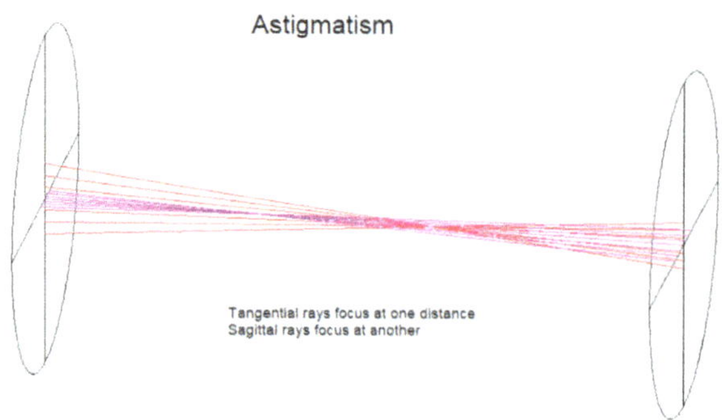

Figure 49.3. Illustration of the properties of astigmatism.

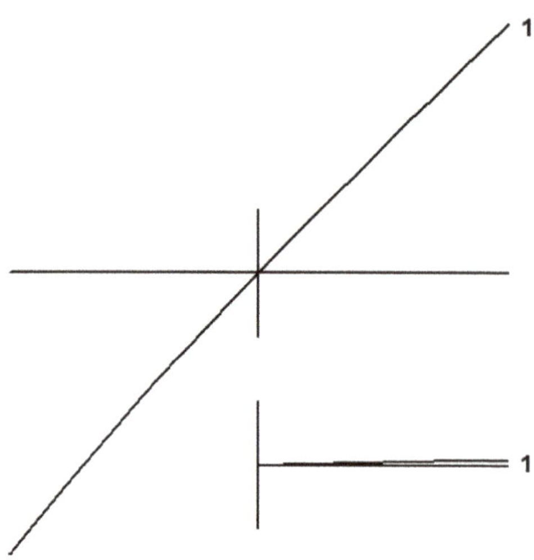

Figure 49.4. TFAN and SFAN plots showing severe astigmatism.

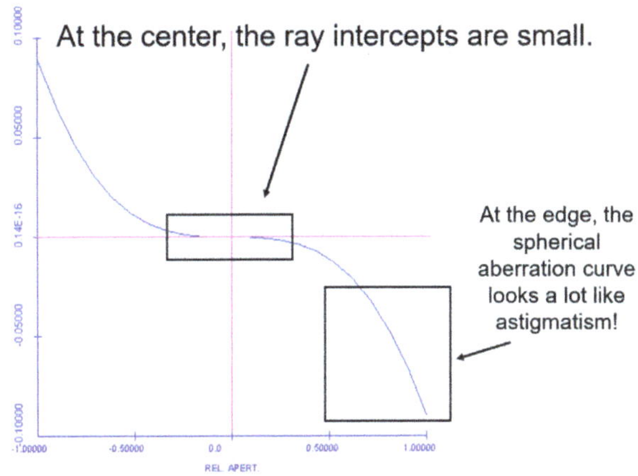

Figure 49.5. A decentered plot of spherical aberration looks a lot like astigmatism.

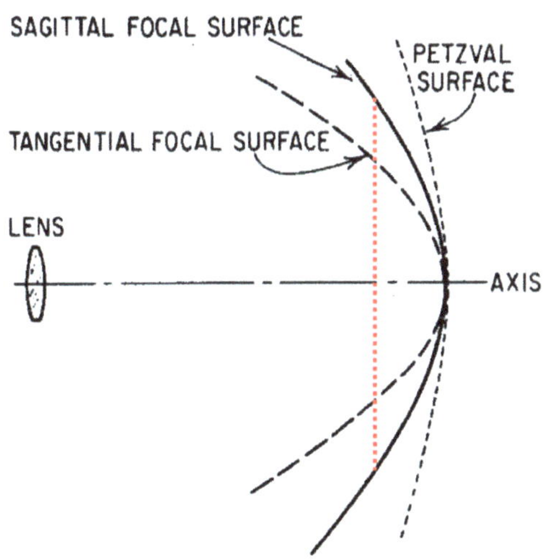

Figure 49.6. Astigmatism of a centered lens shows up off axis.

But if you *tilt* the focal surface, this lets one see the tangential image at one side of the field and the sagittal image at the other. So tilting something might work (figure 49.7).

And here is another possibility: Shifting the stop *decenters* the off-axis beam in the pupil, as shown in figure 49.8. Look at the path of the full-field chief ray. It is decentered in the box to the left. If the stop were put there, you'd get a different chief ray. That's another clue.

So we have identified three possible ways to make the sagittal and tangential astigmatism show up at the focal plane as we want:

- Decenter an element
- Tilt something
- Shift the stop surface.

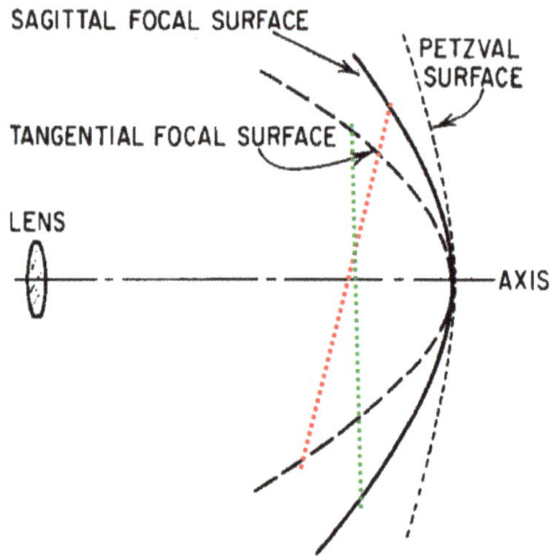

Figure 49.7. Tilting the focal surface.

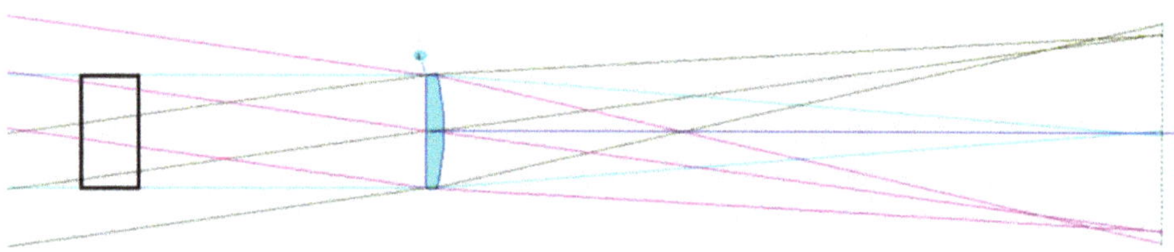

Figure 49.8. Shifting the stop causes the beam to be decentered in the pupil.

So far, so good. Let us guess that we can do the job with a four-element lens. We run DSEARCH with the following input file (**C49M1**),

```
CORE 16
OFF 1 99
DSEARCH 3 QUIET
 SYSTEM
 ID FINAL EXAM PROBLEM
 OBB 0 10 10
 WAVL 0.6563 0.5876 0.4861

 UNITS MM
 END
```

```
GOALS
ELEMENTS 4
FNUM 10
BACK 0 0
TOTL 100 .1
STOP MIDDLE
STOP FREE
RT 0.5
FOV 0.0 0.75 1.0 0.0 0.0
FWT 5.0 3.0 1.0 1.0 1.0
NPASS 44
ANNEAL 200 20 Q
COLORS 3
SNAPSHOT 10
QUICK 44 44
END
SPECIAL PANT

END
SPECIAL AANT

END
GO
```

and it comes back with 10 designs, well corrected but rotationally symmetric at the moment. The top one shown in figure 49.9.

Now we have to modify this lens so it meets our specs. We will go in steps. Here are the variables:

```
PANT
VY 0 YP1
VLIST RAD ALL
VLIST TH ALL
SKIP
VY 3 AT 2
VY 5 YDC 2
VY 9 AT 1
EOS
VLIST GLM ALL
END
```

This will vary the stop position, vary radii, thicknesses and airspaces, and the glass models, and we will activate the tilt and decenter variables later.

The trick now is to create a merit function that leads to the solution we want. Here is an outline (**C49M2**):

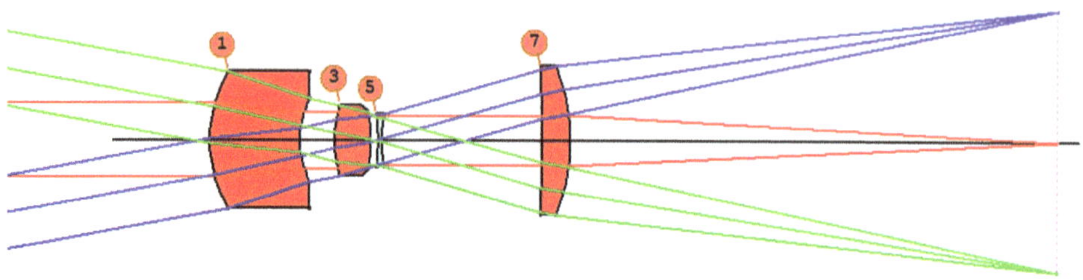

Figure 49.9. Top lens returned by DSEARCH.

```
AANT
AEC 3 1 1
ACM 3 1 1
LLL 1 1 1 A ETH 8
M 0 1 A 1 YA 1
S 3 YA 1
SKIP
LLL .11 1 .1
A P YA 1 0 1
S P YA 1 0 -1

LLL .11 1 .1
A P XA -1 -1
S P XA -1 1

LUL .19 1 .1
A P YA 1 0 1
S P YA 1 0 -1

LUL .19 1 .1
A P XA -1 -1
S P XA -1 1

(*** magic ***)
EOS
GXR 0 2 4 P 1
GXR 0 2 4 1 1
GXR 0 2 4 3 1
GYR 0 1 4 P -1
GYR 0 1 4 1 -1
GYR 0 1 4 3 -1
END
SNAP
SYNO 30
```

Ignoring for the moment the (*** magic ***) portion, we try this MF, optimize and anneal, and find…it doesn't work. We need to get 90% of the energy through our 13 micrometer slits, but this design only gets 89% through. Why? What can we do?

Well, it's easy to show that, at F/10, a perfect image at these wavelengths *would* get only 89% of the energy through. So, on the face of it, the problem is impossible!

Except, it isn't. Here is where your knowledge of optics is essential. The software has corrected the aberrations to the degree we specified, the image is diffraction limited in the cross-slit direction, and we still do not meet the specs. What can we do?

We have to be clever. Here is the thought process:

- The specification says F/10.
- But then the diffraction pattern is already too large to fit inside the slit.
- (Clever trick) the spec says nothing about distortion!
- What does *distortion* have to do with the size of the diffraction pattern?

Can you answer that last question? Figure 49.10 offers a clue.

That should be enough of a hint. From chapter 2 you learned about the Lagrange Invariant. You learned that, if you change the value of Y_B, you also change the value of U_A. Got it yet?

Well, as figure 49.11 shows, a steeper angle creates a smaller Airy diffraction disk. And a smaller disk can get more energy through a small slit.

So the trick is to get just the right amount of distortion.

- Then the cone angle of the rays will change at the edge of the field.
- Then the radius of the Airy diffraction disk will change.
- So we can make it smaller.
- So we can get more energy inside the slit.

...and that is the (*** magic ***) in the AANT file: ask for some barrel distortion. So we add to the AANT file the lines

```
M .90 1 A P YA 1
S P YA -1
DIV CONST 2
DIV GIHT
```

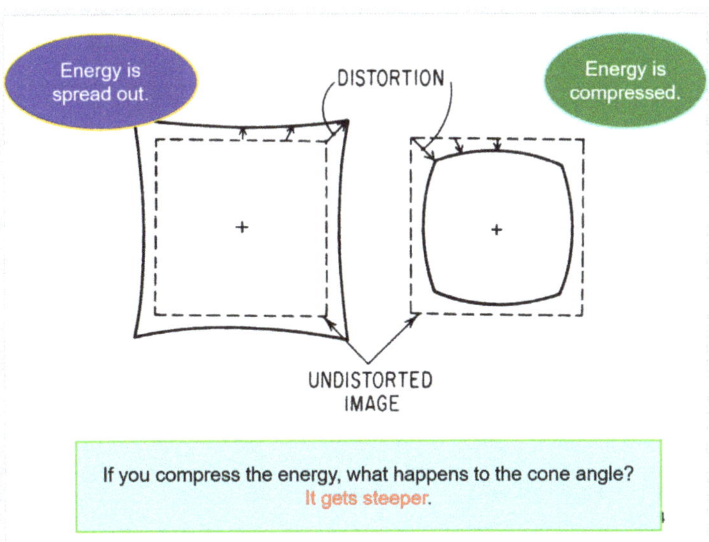

Figure 49.10. Distortion alters the energy density at the image.

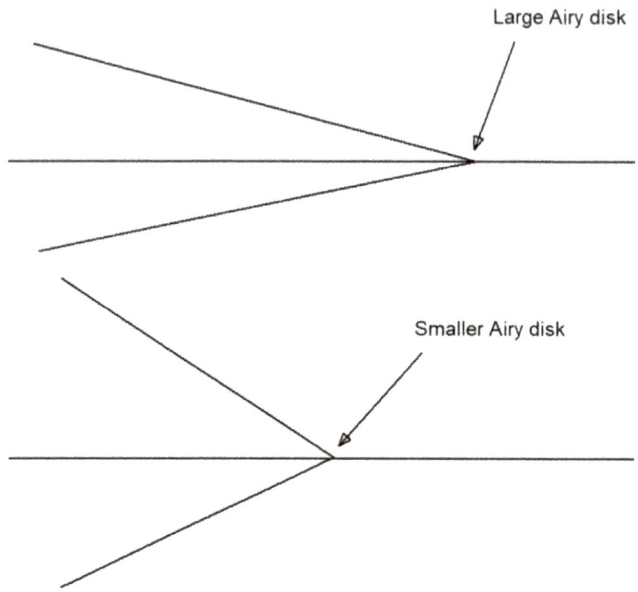

Figure 49.11. Relation between cone angle and Airy diffraction disk.

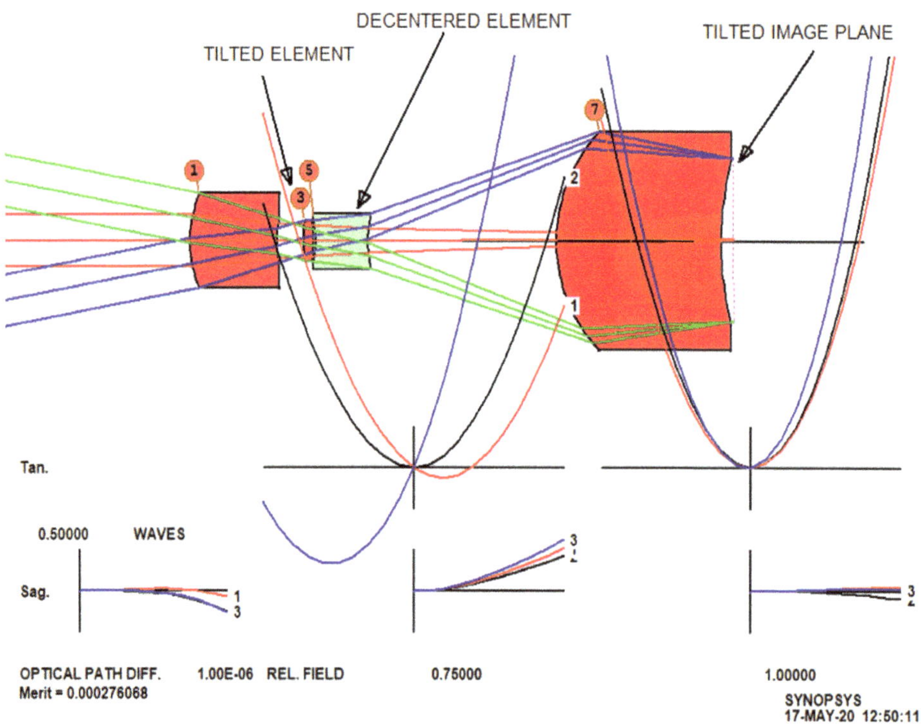

Figure 49.12. Final solution to the Final Exam problem.

This says the actual image scale should be 90% as large as the paraxial scale. Free up the tilt and decenter variables and the slit-length aberrations and optimize again. Then anneal (**55, 2, 50**).

The solution to the F/10 case is shown in figure 49.12. (**C49L1**) This lens puts 90.7% through the Y slit and 92.7% through the X.

The distortion of this lens is about −0.3.37 mm, as is apparent in figure 49.13.

With this trick, all the goals were met. Figure 49.14 shows the images at top and bottom of the field, as displayed by the Image Tools Menu, MIT.

The other half of the problem, meeting the same goals at F/5 is actually easier, since the Airy disk is already much smaller. Figure 49.15 shows the wavefront fringes for both designs, a classic example of astigmatism.

Part of this problem is to calculate the fraction of the energy that gets through the slit. Here's a simple way to get it:

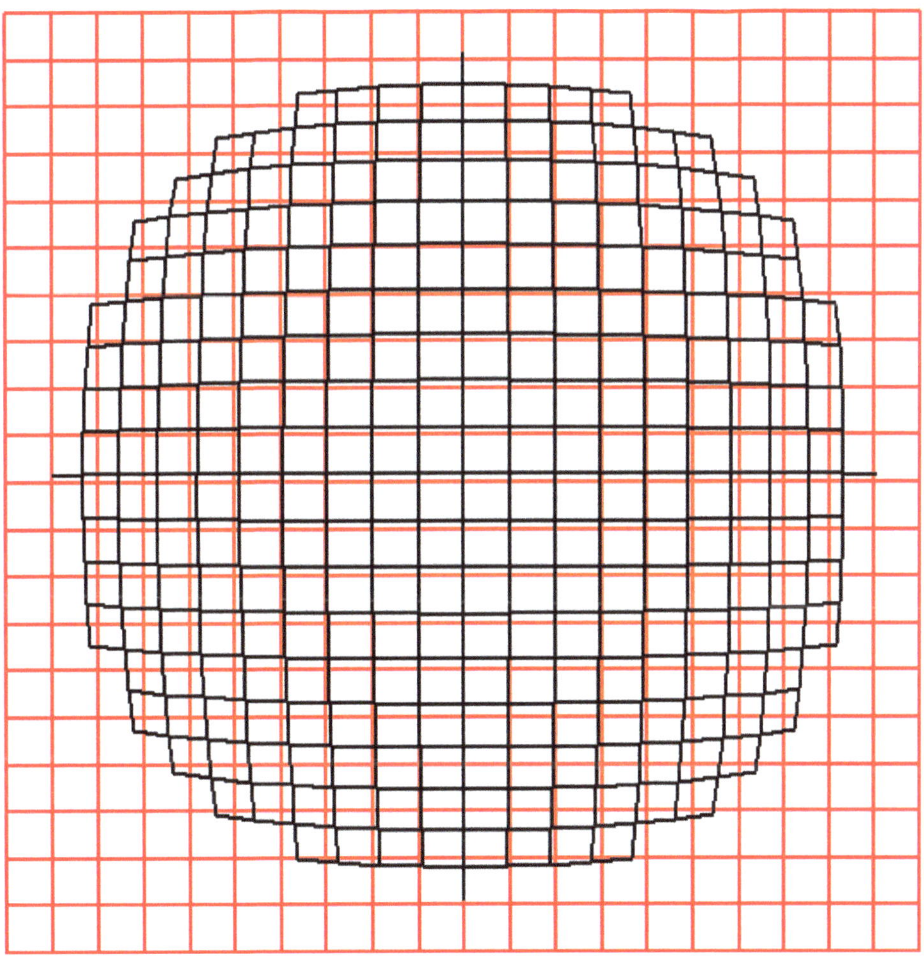

Figure 49.13. Distortion of the final lens.

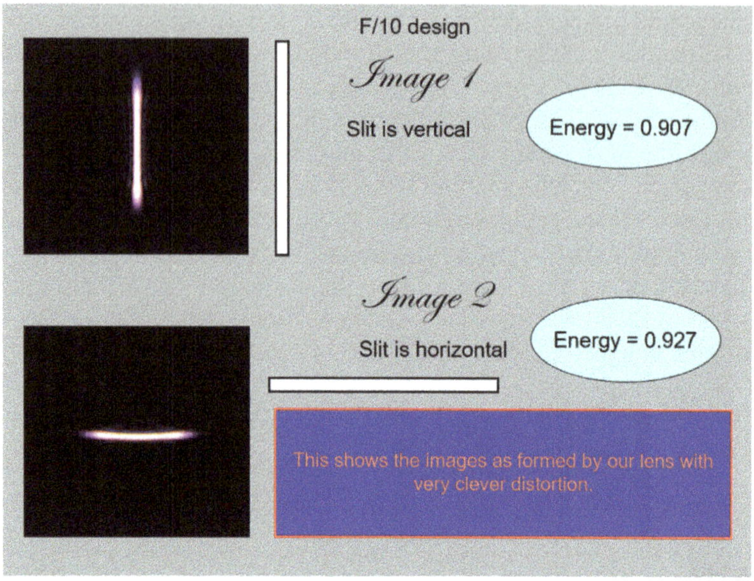

Figure 49.14. Image at top and bottom of the field of the final design.

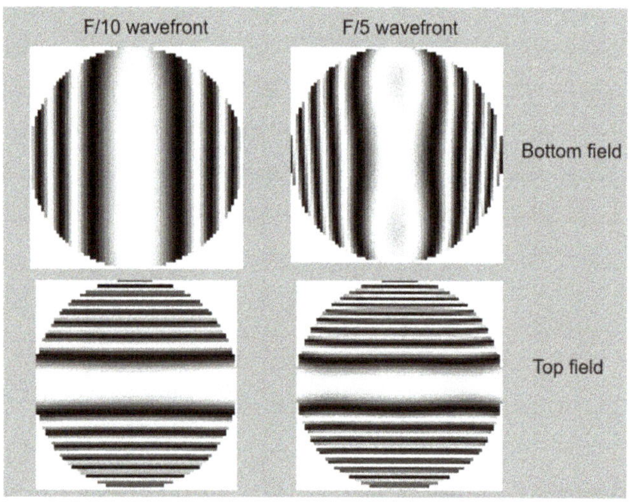

Figure 49.15. Wavefront fringes of both solutions.

```
WMODEL M 1 9999
FOR SLIT
SIZE .3 .013
VARY X POS FROM -.05 TO .05
PLOT
```

This produces a table and a plot of the energy as the slit is moved past the image. A similar calculation, substituting Y for X and calculating the model at a field of –1.0 gives the slit trace at the other side of the field.

So that's how one approaches a problem that does not lend itself to the powerful search tools we have used in earlier chapters. To summarize,

- Always look for items that are not important, and set them free.
- Understand why your lens is not working.
- See if you can trade off something that is not important in order to gain something that is important.
- Check with the customer. He may have forgotten to tell you about some other requirements!

And, by the way, none of the students in the class found this solution. If you figured it out, great. I hope this book helped you to learn what you needed to know.

IOP Publishing

Lens Design (Second Edition)

Automatic and quasi-autonomous computational methods and techniques

Donald C Dilworth

Chapter 50

Testplate matching

In chapter 34, we designed a very good wide-band objective lens. The next step, before asking for a tolerance budget and making element drawings, is to match the design to the testplates of a selected vendor. Chapter 4 explained why this step is important. Here is how to proceed:

Fetch the lens (**C34L2**) and open the optimization MACro once more. (**C50M1**) The testplate-matching program will reuse the most recent parameter definition and merit function, so those have to be current. Add a requirement to the AANT file:

```
M 8 1 A FNUM
```

Since we are going to delete the curvature solve on surface 6, the F/number will no longer be fixed at that value. This new operand will constrain it.

Now delete the curvature solve on surface 6. You want to match all surfaces, but cannot match one where the radius keeps changing. Then open the dialog MMT and enter the data shown in figure 50.1. For this lens we will use the testplates of JML, Inc. Click the OK button.

The program runs **TPMATCH**, which matches every surface, lists the radii it found, then shows the results. These testplates are measured in MM, and this lens is in units of inches, so the radii are first scaled by 0.039 37.

```
RESULTS OF TEST-PLATE FIT

 SURF. NO. FINAL RADIUS ACTION

 1 0.242840E+02 SUCCESSFUL MATCH FOUND
 2 0.756337E+01 SUCCESSFUL MATCH FOUND
 3 0.834198E+01 SUCCESSFUL MATCH FOUND
 4 0.959876E+01 SUCCESSFUL MATCH FOUND
```

doi:10.1088/978-0-7503-3695-6ch50

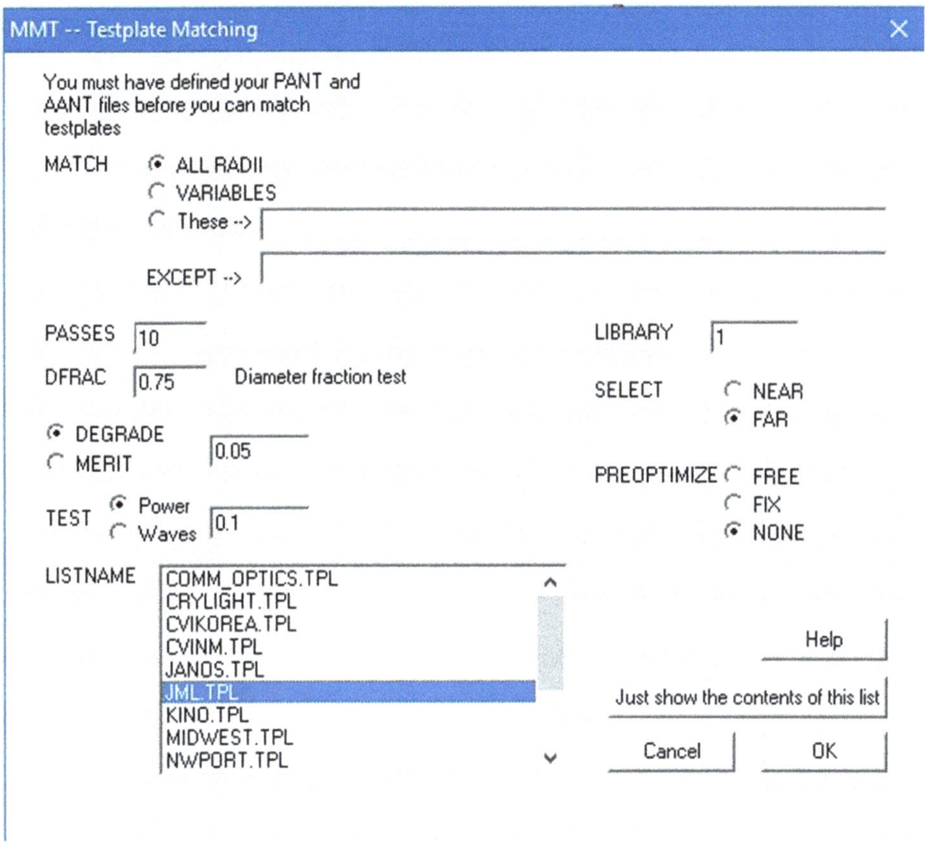

Figure 50.1. The MMT dialog for matching testplates.

```
5 0.818951E+01 SUCCESSFUL MATCH FOUND
6 -0.398767E+02 SUCCESSFUL MATCH FOUND
7 BYPASSED
```

Now, when you run BTOL, be sure to declare those surfaces matched, with the **TPR** designation, so BTOL will assign a tighter radius tolerance to them and everything else will be looser.

IOP Publishing

Lens Design (Second Edition)

Automatic and quasi-autonomous computational methods and techniques

Donald C Dilworth

Chapter 51

Automatic thin-film design

Thin films are widely used in optics, both for antireflection coatings on lenses and high-reflection coatings for mirrors. They also show up in dichroic mirrors, which separate spectral bands, and for other uses. They are deposited onto an optical surface in a vacuum chamber where the material is heated and evaporated, or sputtered onto, the substrate.

Here we will create a custom thin-film design of 30 layers to reflect red light and transmit blue when the filter is used at an angle of 45 degrees. The input file is as follows: (**C51M1**)

```
FILM
DESIGN
BUILD 30 1 1.62
ID TEST CUTOFF FILTER
AANT
GRW 0 45 25 .4 .6
GRW 1 45 25 .62 .8
END
FIX
SYNO 10
ANALY
LAM .4 .8 100 45
PLOT
RETURN
```

This file requests an average reflectance at 45 degrees of zero from 0.4 to 0.6 μm, and 1.0 from 0.62 to 0.8.

After running this file, the program plots the characteristics of the film it designed, shown in figure 51.1. The **FIX** directive causes the program to match the design to a database of commonly used materials, which it also displays with the output. Note

doi:10.1088/978-0-7503-3695-6ch51

that the performance of the film stack varies somewhat with the polarization state of the light.

The design and manufacture of thin films is a mature discipline and much expertise has been developed by many vendors. Of particular note is the fact that the effective index of refraction of the materials when deposited as a thin film is not exactly the same as in the bulk material, so these results, while adequate for computer modeling, should be adjusted by the vendor according to his own proprietary material properties database, and the design adjusted, before manufacture. In the event you have measured data for the index as applied, you can create a database of custom materials and then apply those to the film stack. Section 16.5 in the User's Manual describes this feature, and an example is given in chapter 56.

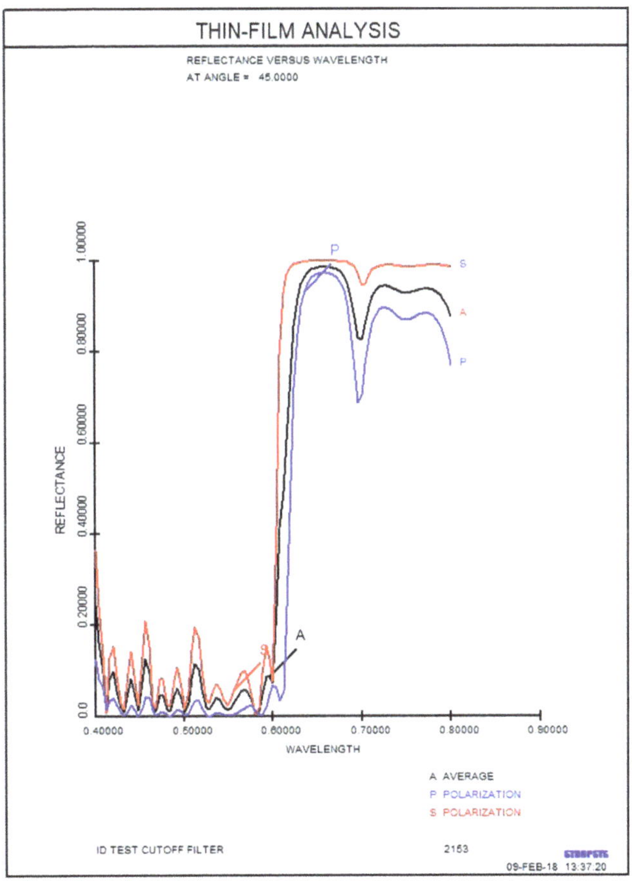

Figure 51.1. Example analysis of custom thin-film stack designed automatically.

ID TEST CUTOFF FILTER 2153

STACK DATA

CONTROL WAVELENGTH = 0.5876 MICRONS

CONTROL ANGLE = 0.000 DEG.

SURF NO.	OPTICAL THICKNESS (WAVES)	PHYSICAL THICKNESS (MICRONS)	INDEX	IMAG. INDEX	
INCIDENT MEDIUM			1.0000		
2	0.4096	0.120339	1.9729		HFO2
3	0.2863	0.121923	1.3655		MGF2
4	0.3838	0.104692	2.1535		TAO5
5	0.3005	0.127965	1.3655		MGF2
6	0.3720	0.101486	2.1535		TAO5
7	0.3028	0.128944	1.3655		MGF2
8	0.3525	0.096174	2.1535		TAO5
9	0.3220	0.128695	1.4585		SIO2
10	0.3282	0.089524	2.1535		TAO5
11	0.3990	0.169883	1.3655		MGF2
12	0.8142	0.222101	2.1535		TAO5
13	0.2027	0.068438	1.7479		CEO2
14	0.4088	0.120109	1.9729		HFO2
15	0.1895	0.063979	1.7479		CEO2
16	0.3956	0.116242	1.9729		HFO2
17	0.1973	0.066629	1.7479		CEO2
18	0.3956	0.107916	2.1535		TAO5
19	0.1999	0.067490	1.7479		CEO2
20	0.3976	0.108471	2.1535		TAO5
21	0.1994	0.067346	1.7479		CEO2
22	0.4097	0.114626	2.0980		ZRO2
23	0.2042	0.073164	1.6210		CEF3
24	0.4209	0.123648	1.9729		HFO2
25	0.1743	0.062443	1.6210		CEF3
26	0.3726	0.125824	1.7479		CEO2
27	0.1865	0.066831	1.6210		CEF3
28	0.4156	0.122097	1.9729		HFO2
29	0.2285	0.077178	1.7479		CEO2
30	0.4248	0.118851	2.0980		ZRO2
31	0.0000	0.000000	1.6200		
SUBSTRATE			1.6200	0.0000	

IOP Publishing

Lens Design (Second Edition)

Automatic and quasi-autonomous computational methods and techniques

Donald C Dilworth

Chapter 52

Automatic clocking of wedge errors

Another automatic feature comes in handy when the elements of a lens are manufactured and measured, and each is found to have a small wedge error. Good shop practice can minimize such errors, but they never go to zero. And more accurate 'dewedging', as it is called, is more expensive. So you want to see if you can compensate at assembly for such errors. The Monte-Carlo evaluation program **MC** can model cases where lenses are mounted with the wedges alternating, up and down, between elements, and that usually helps a good deal. The **UCLOCK** program can do even better.

Here is an example. Get out the lens stored as 1.RLE, run UCLOCK, and specify a small wedge error on each of the four elements. (**C52M1**)

```
FET 1
UCLOCK
WEIGHT 1 1 1
2 1
4 2
6 3
8 4
GO

UCLOCK LIST
UCLOCK PLOT
```

Here, element 1 has a wedge of 1 arcminute assigned to surface 2, element 2 has 2 min of wedge on surface 4, and so on. The shop must accurately measure these wedges and mark the edge where it is thickest so they know how the wedge is to be oriented when the lens is assembled.

Run this job, and the program finds the best clocking angles, shown in figure 52.1. Now the program has assigned an alpha tilt to the second side of each element, modeling the wedge, and a gamma tilt of two surfaces to the first side, modeling the clocking angle of that element in the cell. Element 1 is unclocked, giving a reference direction to the others. The program lists the results:

```
--- UCLOCK LIST

OPTIMUM CLOCKING OF WEDGED ELEMENTS IS AS FOLLOWS:
BORESIGHT WEIGHT =    1.0000
DISPERSION WEIGHT =    1.0000
AXIAL COMA WEIGHT =    1.0000

  No.  SURF  WEDGE, MIN     RADIANS       DEGREES
   1     2     1.00000      0.00029       0.01667
   2     4     2.00000      0.00058       0.03333
   3     6     3.00000      0.00087       0.05000
   4     8     4.00000      0.00116       0.06667

RESIDUAL BORESIGHT ERROR, IN LENS UNITS:    0.04879882
RESIDUAL DISPERSION ERROR, IN LENS UNITS:    0.00111729
RESIDUAL AXIAL COMA, IN WAVES:    0.02229963

UNCLOCKED BORESIGHT ERROR, IN LENS UNITS:    0.10480944
UNCLOCKED DISPERSION ERROR, IN LENS UNITS:    0.00174071
UNCLOCKED AXIAL COMA, IN WAVES:    1.13678441


 TILT AND DECENTER DATA
 LEFT-HANDED COORDINATES
```

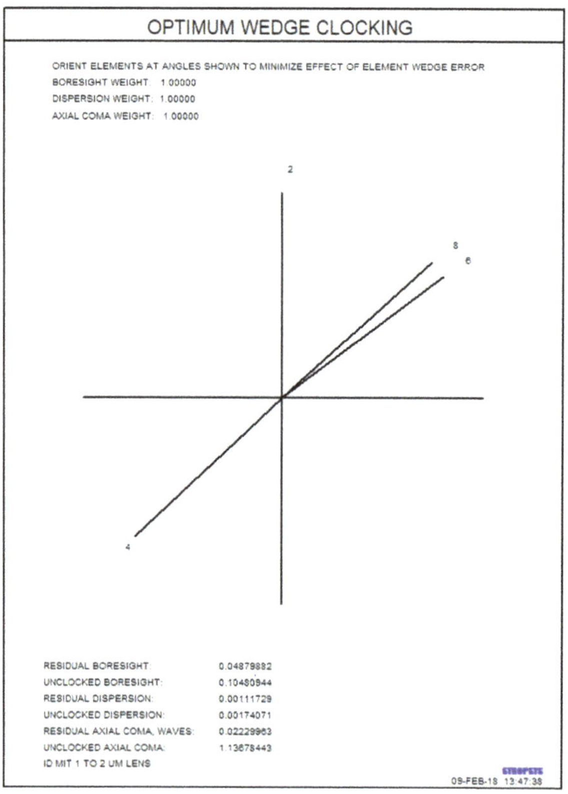

Figure 52.1. Output from UCLOCK, showing optimum clocking angle of each element of the lens.

SURF	TYPE	X	Y	Z	ALPHA	BETA	GAMMA
2	REL	0.00000	0.00000	0.00000	0.0167	0.0000	0.0000
3	REL	0.00000	0.00000	0.00000	0.0000	0.0000	-132.4438
4	REL	0.00000	0.00000	0.00000	0.0333	0.0000	0.0000
5	REL	0.00000	0.00000	0.00000	0.0000	0.0000	53.4173
6	REL	0.00000	0.00000	0.00000	0.0500	0.0000	0.0000
7	REL	0.00000	0.00000	0.00000	0.0000	0.0000	48.4099
8	REL	0.00000	0.00000	0.00000	0.0667	0.0000	0.0000

```
KEY TO SURFACE TYPES
______________________
GLB  GLOBAL COORDINATES                LOC  LOCAL COORDINATES
REL  RELATIVE COORDINATES              REM  REMOTE TILTS IN RELATIVE COORD.

SURF MESSAGES
______________________________________________________________
  3  UNDO TILTS/DECENTERS OF SURFACE NO.     2
  5  UNDO TILTS/DECENTERS OF SURFACE NO.     4
  5  UNDO TILTS/DECENTERS OF SURFACE NO.     3
  7  UNDO TILTS/DECENTERS OF SURFACE NO.     6
  7  UNDO TILTS/DECENTERS OF SURFACE NO.     5
  9  UNDO TILTS/DECENTERS OF SURFACE NO.     8
  9  UNDO TILTS/DECENTERS OF SURFACE NO.     7
--- UCLOCK PLOT
```

Now you know that element 2 should be oriented so that the thickest edge is −132.4438 degrees rotated with respect to element 1, and so on.

The program also finds that, if you do *not* clock the elements in this way, you will get axial coma of 1.14 waves—which is pretty bad—but if you clock the elements as directed above, this goes down to 0.022 waves. The new geometry is shown in the perspective drawing in figure 52.2. Note how the surface curves, which are lined up in the local Y direction at each surface, are all rotated with respect to each other. This is an example of how the power of automatic lens design methods can improve yields.

One final note: Dewedging is done by mounting the lens on a precision spindle, running a dial indicator on both sides, and adjusting the centration until both run

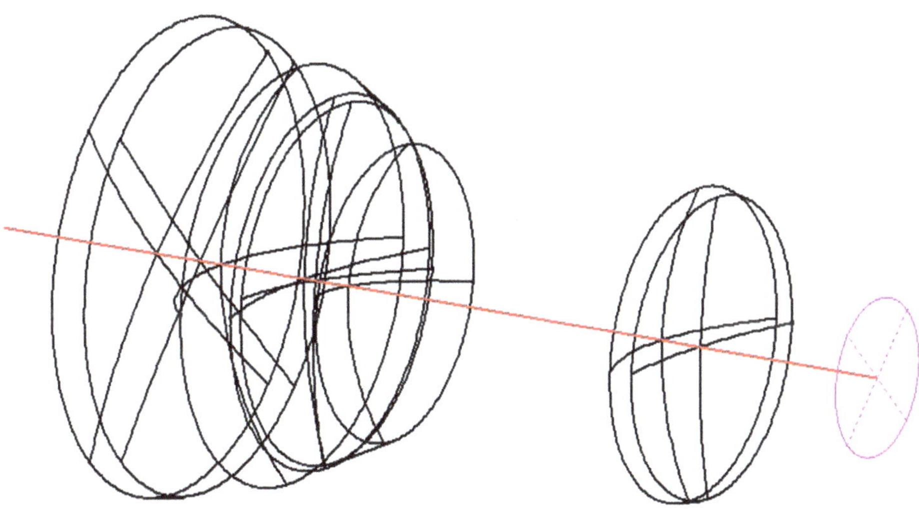

Figure 52.2. Rotation of elements shown in perspective after calculating with UCLOCK.

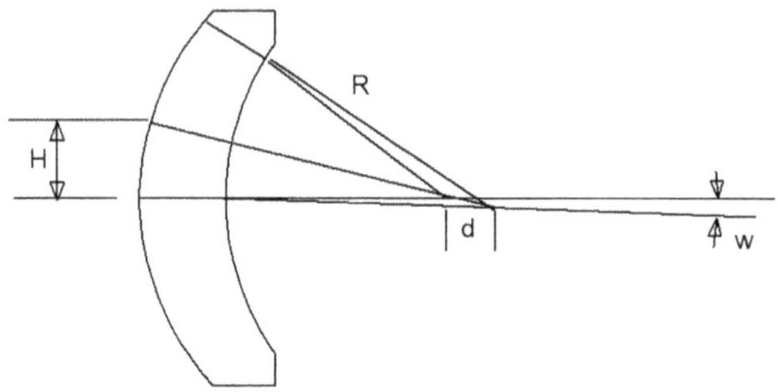

Figure 52.3. Removing the wedge of a meniscus lens becomes more difficult as the centers become closer together.

true. Then the edge is ground down until the desired element diameter is reached. At this point the wedge error should be very small, but of course still nonzero.

But this does not work for some meniscus elements, where the centers of curvature of the two sides are located close together on the optical axis. Such elements have to be ground and polished with careful attention to the wedge error right from the start, which is difficult and expensive. Keep this in mind, and try to avoid such elements. The monitor **AMS** can control the separation.

In figure 52.3, a lens has two radii with almost the same center location, the centers separated by a small distance d, and a wedge error of W. In order to remove the wedge, the lens has to be edged by an amount H, given by

$$H = W{*}R^2/d$$

If d goes to zero, H becomes infinite. Since one usually has only a small amount of extra glass that can be removed in edging, the quantity d has to be greater than some amount. That amount is the target for **AMS**.

IOP Publishing

Lens Design (Second Edition)

Automatic and quasi-autonomous computational methods and techniques

Donald C Dilworth

Chapter 53

XSYS an expert-systems approach to lens design

The previous chapters have shown many examples of how to use the global search routines DSEARCH, ZSEARCH, and GSEARCH. This section describes yet another approach, utilizing expert knowledge to construct candidate lens forms.

Normally, in the field of AI, an 'expert systems' program is one that employs a tree-structured logic wherein decisions are derived from the responses of a number of experts in a particular field to a lengthy debriefing. Remarkable performance has been achieved in some areas, rivaling that of a human expert.

The SYNOPSYS program takes a somewhat different approach: the program is presented with a number of finished lens designs, the products of expert designers, which represent the state of the art. Using these lenses as models, the program analyzes the optical properties in great detail, determining for each lens the first and higher-order properties, and the aberrations that are present in the beam before and after each element. In so doing, the program 'learns' how a particular optical problem was solved. The more examples it is given, the more it learns.

Then, when presented with a new problem or a lens that is not well corrected, it can determine if the current problem resembles one for which it knows a solution.

This feature is unique among all the programs in SYNOPSYS in that its performance is not built in. When the program is first installed, and without an example file, **XSYS** is ignorant. Most engineering programs give a certain output in response to a certain input, but **XSYS** has to be educated before it will work. And if the example file contains only microscope objectives, for example, it will probably not be of use designing infrared telescopes: the examples should represent more or less the kind of systems for which the program will be used. Here we will use XSYS to design a forward-looking infrared radiometer lens, commonly called a FLIR.

doi:10.1088/978-0-7503-3695-6ch53

XSYS can be used for three broad purposes:

1. Any of the example lenses may be retrieved intact for use as a starting point. (Example: **USE 4**.)
2. A general set of requirements may be input, and **XSYS** will find the lens (or portion of a lens) that most closely matches them (according to various weights input by the user) when the scale and aperture are adjusted as well as possible. This lens can then be used as a starting point. (**START**)
3. If you provide a starting lens, **XSYS** can analyze the beam everywhere in the lens and determine if a similar set of aberrations or first-order properties exists in any of the examples—and if so, how the aberrations were corrected or how the first-order properties were achieved. It will then suggest alternate starting points based on this information. This feature is perhaps the most powerful ability of **XSYS**. (**ALTER**)

The program is started with the command **XSYS**, which displays the following choices:

```
Welcome to the SYNOPSYS expert systems program.
Select one of the following input commands:
```

```
M                                  (this menu again)
START                              (to start a new lens from scratch)
ALTER                              (find alternate starting point)
REENTER                            to re-use the last match list)
LIST                               (to list example lenses on file)
USE                                (to use one of the example lenses)
INITIALIZE                         (to generate a file of examples)
ADD                                (to add a lens to the example file)
END
```

To illustrate the use of this feature, we must first create the database. The required files are found in the installation directory SYNOPSYS/USER. To go there, assuming you are in the DBOOK-ii directory at the moment, type

```
CHD
```

click Browse, select **User** and click OK.

Now load a MACro that will create a database of thermal infrared designs, which are provided in that directory.

```
LM XSYSINIT
```

A file opens where you can see the names of the 33 files it will study.

```
; XSYSINIT.MAC
; THIS FILE CREATES THE DATABASE THAT WILL BE
; USED BY XSYS.  LIST HERE ALL LENS FILES THAT
; YOU THINK MAY CONTAIN USEFUL INFORMATION

XSYS
INIT
X1
X2
X3
X4
X5
X6
X7
X8
X9
X10
X11
X12
X13
X14
X15
X16
X17
X18
X19
X20
X21
X22
X23
X24
X25
X26
X27
X28
X29
X30
X31
X32
X33

END
```

Run this MACro, and then display the ID lines of all of the database lenses with the commands

```
XSYS
LIST
```

```
---- LISTING OF LENSES IN EXAMPLE FILE ----

LENS NUMBER   1 ID 3 X WITH SILICON                              28304
LENS NUMBER   2 ID F/3.111 TRIPLET                                 124
LENS NUMBER   3 ID 8X GERMANIUM AFOCAL                             124
LENS NUMBER   4 ID WIDEBAND IR 4X AFOCAL                           124
LENS NUMBER   5 ID 15X IR AFOCAL                                   124
LENS NUMBER   6 ID DETECTOR LENS                                   124
LENS NUMBER   7 ID IR 1X AFOCAL                                    124
LENS NUMBER   8 ID FOLDED IR 4X AFOCAL                             124
LENS NUMBER   9 ID COMPACT IR 7X AFOCAL                            124
LENS NUMBER  10 ID EYEPIECE FOR NON-GALIEAN 1 OF P0265             124
LENS NUMBER  11 ID FOUR ELEMENT INFRARED OBJECTIVES 1 O            124
LENS NUMBER  12 ID AFOCAL ZOOM REFR TELESC 3 OF P1488            32816
LENS NUMBER  13 ID SELETABLE FOV INFRARED LENS 1 OF P38            124
LENS NUMBER  14 ID ACHROMATIC DOUBLET FOR IR 3 OF P4330            124
LENS NUMBER  15 ID AFOCAL ZOOM TELESCOPE   3 OF P7084              124
LENS NUMBER  16 ID INFRARED OPTICAL SYSTEM 1 OF P2596              124
LENS NUMBER  17 ID INFRARED OPTICAL SYSTEM 5 OF P2596              124
LENS NUMBER  18 ID ZOOM SYSTEM FOR INFRARED 1 OF P5315             311
LENS NUMBER  19 ID AFOCAL TELESCOPE WITH 2-M 8 OF P6069            124
LENS NUMBER  20 ID AFOCAL REFRACTOR TELESC 1 OF P7520              124
LENS NUMBER  21 ID AFOCAL TELESCOPE WITH 2 M 3 OF P6069            124
LENS NUMBER  22 ID INFRA-RED OPTICAL SYS   2 OF P9217              124
LENS NUMBER  23 ID ATHERMALIZED IR ZOOM LENS 7 OF 10 P9            124
LENS NUMBER  24 ID TRIPLET                                           7
LENS NUMBER  25 ID FIVE-ELEMENT LENS                               124
LENS NUMBER  26 ID SIX-ELEMENT LENS                               8908
LENS NUMBER  27 ID INFRARED OPTICAL SYSTEM 5 OF P2596              124
LENS NUMBER  28 ID SEVEN-ELEMENT LENS                              124
LENS NUMBER  29 ID 7-ELEMENT LENS                                  124
LENS NUMBER  30 ID FOUR-ELEMENT LENS                               124
LENS NUMBER  31 ID FOUR-ELEMENT LENS                               124
LENS NUMBER  32 ID 8-ELEMENT LENS                                  124
LENS NUMBER  33 ID 8-ELEMENT TELEPHOTO                             124

XSYS>
```

Let us assume you already have a lens that you would like to improve. We will illustrate by starting with the example LENS NUMBER 9, and then alter the design. Create a new MACro (**XSYS.MAC**):

```
; XSYS.MAC
; HERE IS A SIMPLE EXAMPLE OF HOW THE EXPERT-SYSTEMS PROGRAM
; XSYS CAN HELP YOU FIND STARTING POINTS.  WE OPEN A PRELIMINARY
; LENS (USE 9) FROM THE XSYS DATABASE, AND THEN ASK IT
; TO FIND COMBINATIONS WHERE THE AFOCAL MAGNIFICATION IS 0.2
; WITH A WEIGHT OF 10

XSYS
USE 9
XSYS
ALTER
AFMAG .2 10
TOTL 300 .01
CWT 0
PASSES 5
PLOT AUTO
snap
GO
```

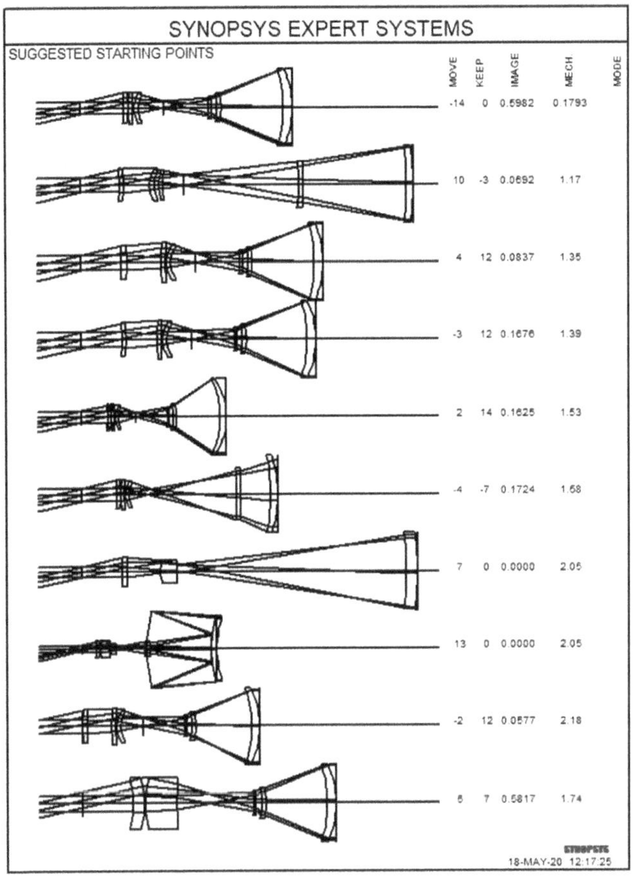

Figure 53.1. Suggested lens forms returned by XSYS.

The program examines all combinations according to its own logic and displays a plot of the 10 best configurations it found, shown in figure 53.1. Note that number 8 is reflective, while the others are all refractive. This is something that XSYS can do (assuming there is at least one reflective example in the database) and the other search programs cannot (unless you give a reflective front end to DSEARCH; see chapter 45 for an example of a wide-angle front end).

Now the program asks you for input. If you enter a number, say 2, and press the Enter key, the program loads the second example and displays it in PAD, as shown in figure 53.2. According to the results on the XSYS drawing, this lens was created by keeping three surfaces, turned around, and moving 10 surfaces from one of the examples.

Now you have a choice of actions:

```
This is match number 2
 Enter "Y" to optimize, <CR> to skip, <ESC> TO END "L"
for list, or number.
```

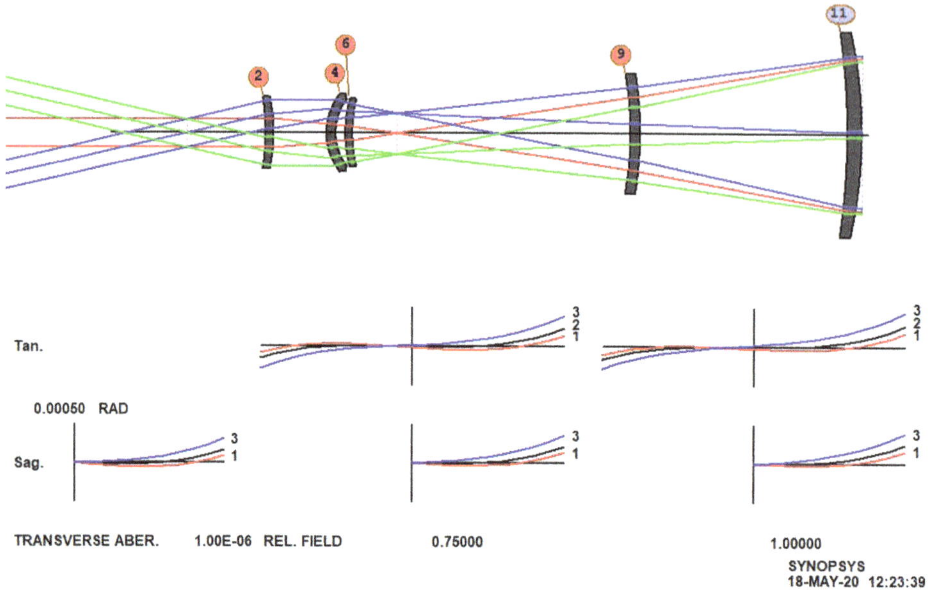

Figure 53.2. The second example returned by XSYS.

Assuming you like this starting lens, enter Y and press Enter. The lens is optimized according to the original requirements you put in the ALTER file section. Again you have a choice:

```
Enter "K"eep this lens, <CR> for next, "L"ist, number,
STORE
 XSYS>
```

Let us keep this lens and look at it more closely. Type K. The lens is not bad, as shown in figure 53.3, but the program has removed all paraxial pickups and solves, and we have to assign those where we want them.

Type the following at the SYNOPSYS AI> prompt:

```
YPP1 = 0
CHG
7 YMT 0
END
```

This lens will be used with a scanning prism located at surface 1, and the stop has to remain there, so the paraxial value of YPP1 should be zero; and we want surface 8 to stay at the intermediate image. Save a checkpoint.

Now open the MACro XSYSFILE.MAC, which the program has created. This will be our starting optimization file, but we have to edit it somewhat. In the PANT file, delete the variable for YP1 so the stop stays fixed. Also, add a variable for the conic constant on surface 12, which was aspheric in the example lens.

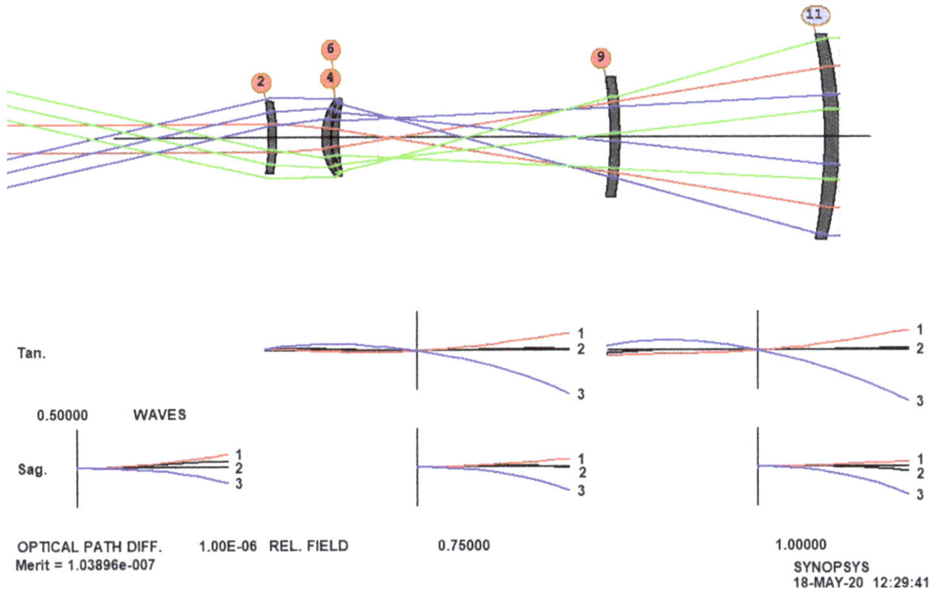

Figure 53.3. Lens optimized by XSYS. It needs to be customized for our purposes.

```
PANT P
 !VY      0 YP1
 VY      2 RAD
 VY      3 RAD
 VY      4 RAD
 VY      5 RAD
 VY      6 RAD
 VY      7 RAD
 VY      9 RAD
 VY     10 RAD
 VY     11 RAD
 VY     12 ASP
 VY      3 TH
 VY      5 TH
 VY      7 TH
 VY     10 TH
VY 12 CC
END
```

We have to alter the AANT file to better correct those things that are important.

```
AANT P
AEC
GNO 0 1 4 M 0
GNO 0 1 4 M 1
M 0 1 A P YA 1 0 0 0 12
 M    0.151948E-01    658.120      A FNUM
 M    300.000         0.333333E-04 A TOTL
```

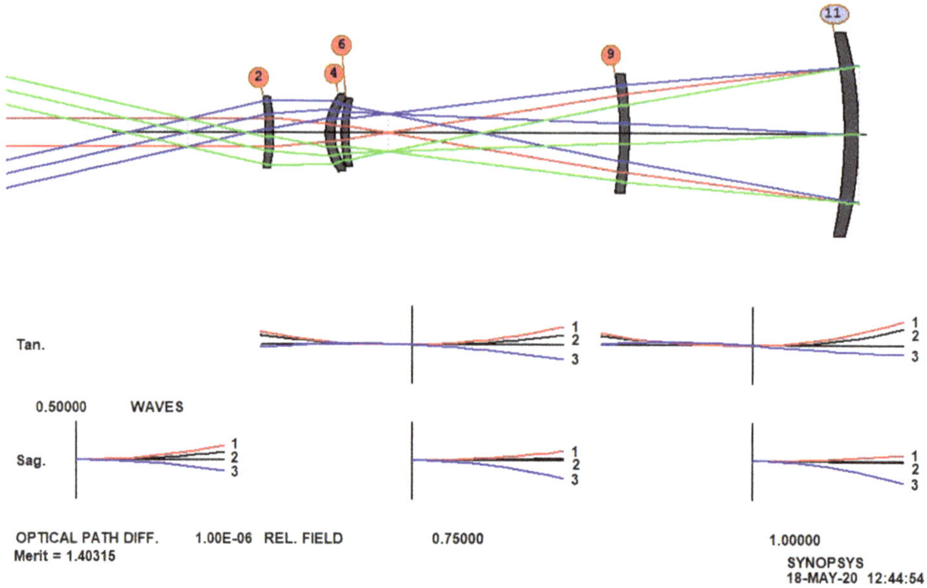

Figure 53.4. Infrared lens optimized with custom goals.

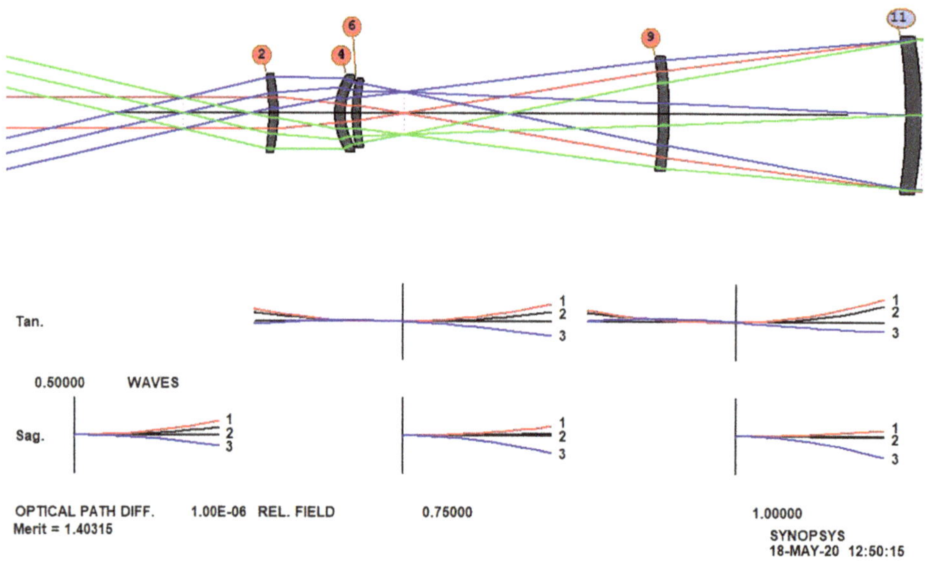

Figure 53.5. Final infrared lens design.

These additions will minimize the OPD errors on axis and at full field and try to bring the full-field chief ray to the center of surface 12. These are Germanium lenses, and we want to minimize the size of those expensive elements.

Run this MACro, and the lens starts to look pretty good, as shown in figure 53.4.

The lens came from the example file with EFILE edges already assigned, but since the magnification has changed, the edges should be modified as well.

Click the Edge Wizard button on the PAD toolbar . Click Erase all, and then Create all. A new set of edges has been created. You can modify these if you wish, but the lens is essentially ready for tolerancing and preparation of drawings, as you have learned to do in previous chapters. Here is the final design, shown in figure 53.5.

This is a brief introduction into what XSYS is and what it can do. If you happen to have a collection of lenses with which to build the database, you may find it a fast and easy way to create new lens forms.

IOP Publishing

Lens Design (Second Edition)
Automatic and quasi-autonomous computational methods and techniques
Donald C Dilworth

Chapter 54

DUV system with quarter-wave plate

This chapter discusses an interesting system that utilizes features not yet mentioned.

Figure 54.1 shows a lens corrected for a single wavelength in the deep ultraviolet, at 0.266 12 μm. **(C54L1)** At this wavelength, most optical materials have poor transmission, so all elements are made of fused silica, one of the few materials that work well in this region.

This is an inspection system used to examine wafers used for computer chips for defects. Resolution has to be nearly perfect and the design has to be free of ghost images. The details etched on the wafers are smaller than a micron, which calls for the short wavelength.

To satisfy the ghost image requirement is actually the more difficult in this instance, so heroic measures were taken. In the picture, the wafer is at the left end, and the imaging optics is off to the right. Laser light comes up from the bottom (not shown) into the cube at surface 27, which has a polarizing reflection at a 45-degree angle. It comes in polarized in the Y-direction and is strongly reflected in the cube. Then it proceeds to the left until it reaches the element at surface 7, shown enlarged in figure 54.2.

That element consists of two pieces, forming a quarter-wave plate (QWP), made of crystal quartz, which is strongly birefringent, meaning the index of refraction is different in two polarizations. The idea is that the wavefront should be retarded by ¼ wave in one polarization with respect to the other.

But the thickness of a quartz crystal that yields just the desired quarter-wave retardance turns out to be extremely thin, much too thin for manufacturing, so the element has to be made of two pieces, with the crystal axes rotated by 90 degrees from each other, with practical thicknesses that differ between the two by a very small amount. So the net retardance is just what is desired even though the element itself is not excessively thin.

Type the command **LE**, which opens the lens file in the editor and note the three surfaces starting at number 7:

doi:10.1088/978-0-7503-3695-6ch54

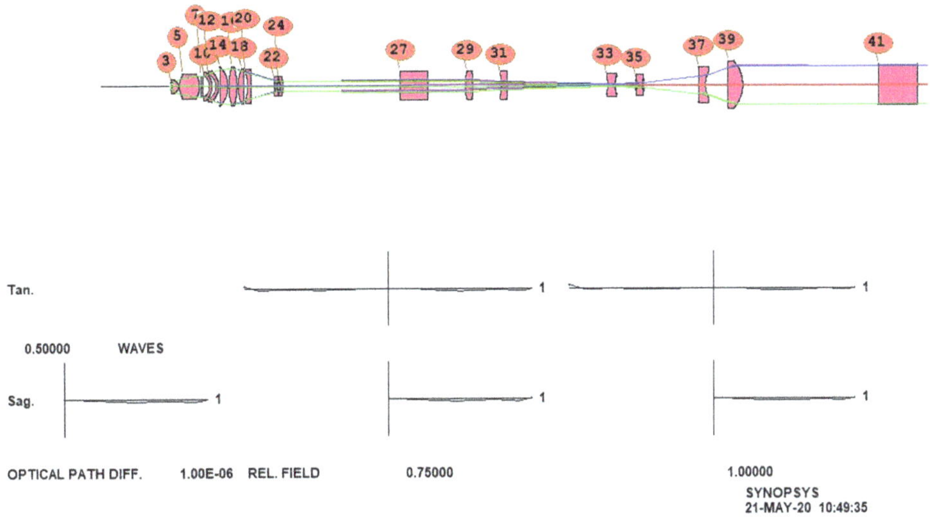

Figure 54.1. A deep UV design.

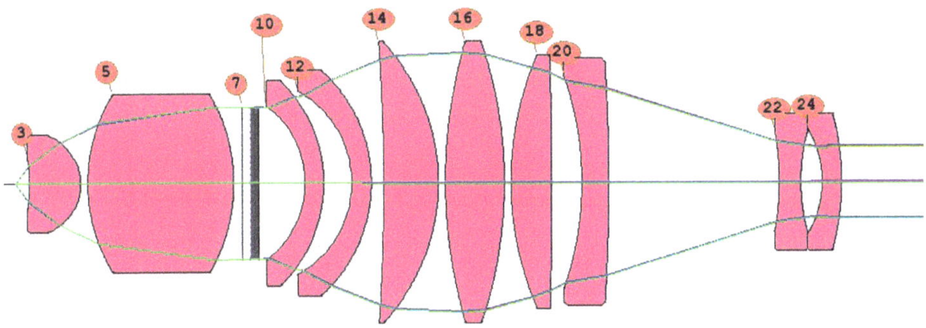

Figure 54.2. Objective portion enlarged. Surface 7 is the location of a quarter-wave plate.

```
    7 CV       0.0000000000000   TH       1.24974585
    7 N1 1.59131727
    7 WPLATE       2.50550000     0.25000000      0.0
 CRQUARTZ
    8 CV       0.0000000000000   TH       1.25575415
    8 N1 1.59131727
    9 CV       0.0000000000000   TH       7.04884779 AIR
```

These data specify that the waveplate should be 2.5055 mm thick and retard the beam by 0.25 waves. Surface 8 is at the first side of the second half. The program adjusts both thicknesses so they total to the specified value but differ by the amount required for the retardation.

Why do we want the retardance of the beam to differ in X and Y polarizations by exactly ¼ wave? The reason is very interesting. That retardation converts the Y-polarized light going to the left into circular polarization, and after the beam reflects from the wafer, it is still circular, but oriented in the opposite direction. Now, when it passes once more through the QWP, it becomes linear polarized again, but now in the X direction. Thus when it again encounters the beamsplitter cube, it passes through with very low reflectance and proceeds to the right.

Why so complicated? It turns out that all ghost image light reflected from surfaces between the QWP and 27 remains Y-polarized and is reflected *downward* by the cube and never reaches the image analysis optics. So this simple trick has eliminated most ghost images. Ghosts between the wafer and the QWP are not eliminated in this way, however, so their contribution has to be minimized in the manner discussed in chapter 36. And for the same reason, the QWP is located as close to the wafer as practical so there will be fewer lenses in between.

How close? Since the polarizing properties of a QWP work best in collimated light, which is approximately true in this design, it cannot be placed closer to the wafer, where the beam is strongly convergent. This is a significant constraint. The object is declared with type OBF, which is used for fast entering beams, and here the entering marginal rim ray hits surface 1 at a height of 2.7316 mm, which implies an angle tangent of 1.333 33, corresponding to a half-angle of 53.129 degrees. That is rather steep.

In the editor, you see the object declaration:

```
POLARIZATION CIRCULAR RIGHT
UNITS MM
OBF   2.0487    -3.62100E-02   2.7316  -0.0362   0.0000        0.0000   2.7316
```

When the object is declared OBF and you request a fractional aperture for a ray, you get a fractional *angle* instead of fractional height. That affects the ray distribution in ray grids and models what one would get if the lens is corrected for the Abbe sine condition, which you read about in chapter 3. Since one cannot convert the beam from collimated to F/0.623 with fewer than two elements, this is as far as the design could be pushed.

This design is set up starting from the wafer, so it models the light after it reflects from the wafer and goes to the right. We can examine the polarization state as the beam propagates from that point on.

For this, we will use the **PSURF** command. But first we must select the New Windows button on the top toolbar. PSURF shows up to 40 surfaces on a single plot, but this lens has 43, so it will make two pictures. Now the program will create a new window for each plot instead of reusing the previous one. Type that command and look at the first of the two new windows created by the **PSURF** command, shown in figure 54.3. Light starts out circularly polarized and becomes linear in the X-direction at surface 9, as desired. In this example, the surfaces have not yet been assigned antireflective coatings, so the ray intensity gets smaller as it goes through the system due to reflection losses. This is evident in the changing length of the polarization vector on the plot from surface to surface.

You also observe in the editor that the lens has been declared in external coordinates:

```
EXTERNAL POSITION            0.00000000        0.00000000        2.04870000
EXTERNAL ANG        0.00000000        0.00000000        0.00000000
```

This says that surface 1 is at a Z-coordinate of 2.0487 mm with respect to some external measuring point. Now we can examine the location of ray intercepts in that system, instead of the system of each surface itself.

Type the command

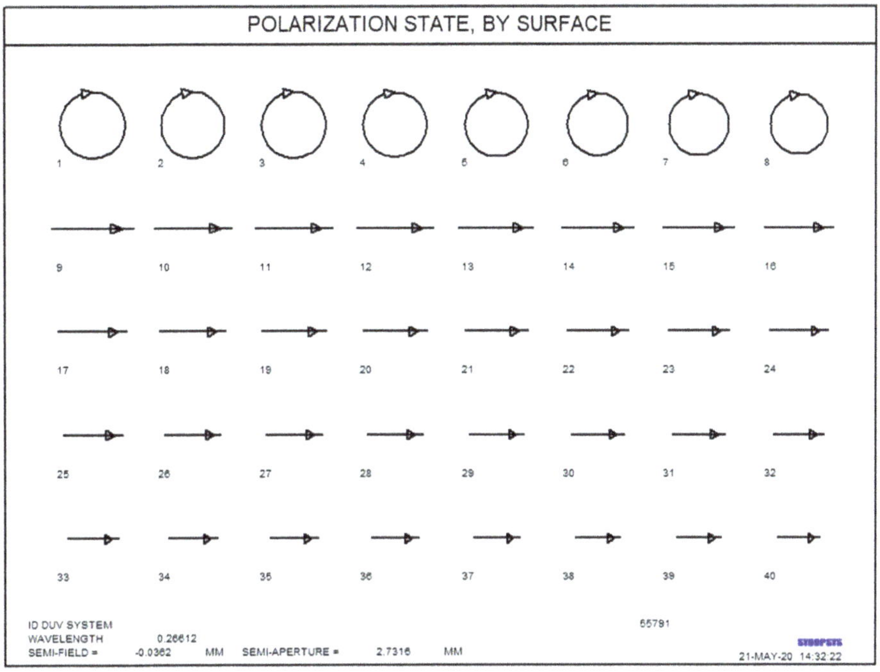

Figure 54.3. Polarization state of the light as it passes through the system. Polarization changes from circular to linear at surface 9, and the size of the diagrams shows that the intensity of the ray is reduced as it goes through the lens.

```
ERAY P 0 0 0 SURF
```

and you see the path of the chief ray. This comes in handy when you want to see the path of a ray in the coordinate system of a telescope pier, for example.

```
SYNOPSYS AI>ERAY P 0 0 0 SURF

 ID DUV SYSTEM                               55791           21-MAY-20    14:22:13

 GLOBAL RAYTRACE ANALYSIS

 RAY DATA IN EXTERNAL COORDINATE SYSTEM

 FRACT. OBJECT HEIGHT           HBAR       0.000000    GBAR        0.000000
 FRACT. ENTRANCE PUPIL COORD. YEN          0.000000    XEN         0.000000
 COLOR NUMBER                    1

                              RAY VECTORS              (X DIR TAN)     (Y DIR TAN)
 SURF            X              Y             Z             ZZ              HH
 ------------------------------------------------------------------------------
    1       0.000000       0.000000      2.048700      0.000000       0.000000
    2       0.000000       0.000000      2.048700      0.000000       0.000000
    3       0.000000       0.000000      2.047561      0.000000       0.000000
    4       0.000000       0.000000      9.768947      0.000000       0.000000
    5       0.000000       0.000000     10.768947      0.000000       0.000000
    6       0.000000       0.000000     32.528528      0.000000       0.000000
    7       0.000000       0.000000     33.771598      0.000000       0.000000
```

8	0.000000	0.000000	35.021344	0.000000	0.000000
9	0.000000	0.000000	36.277098	0.000000	0.000000
10	0.000000	0.000000	43.325946	0.000000	0.000000
11	0.000000	0.000000	46.072626	0.000000	0.000000
12	0.000000	0.000000	50.203414	0.000000	0.000000
13	0.000000	0.000000	53.109393	0.000000	0.000000
14	0.000000	0.000000	54.950645	0.000000	0.000000
15	0.000000	0.000000	63.126439	0.000000	0.000000
16	0.000000	0.000000	64.126439	0.000000	0.000000
17	0.000000	0.000000	72.962786	0.000000	0.000000
18	0.000000	0.000000	73.962786	0.000000	0.000000
19	0.000000	0.000000	79.928891	0.000000	0.000000
20	0.000000	0.000000	84.324157	0.000000	0.000000
21	0.000000	0.000000	88.382043	0.000000	0.000000
22	0.000000	0.000000	113.971387	0.000000	0.000000
23	0.000000	0.000000	117.189760	0.000000	0.000000
24	0.000000	0.000000	120.425477	0.000000	0.000000
25	0.000000	0.000000	123.425477	0.000000	0.000000
26	0.000000	0.000000	566.297086	0.000000	0.000000
27	0.000000	0.000000	250.000000	0.000000	0.000000
28	0.000000	0.000000	280.000000	0.000000	0.000000
29	0.000000	0.000000	320.000000	0.000000	0.000000
30	0.000000	0.000000	328.000000	0.000000	0.000000
31	0.000000	0.000000	359.097337	0.000000	0.000000
32	0.000000	0.000000	365.097337	0.000000	0.000000
33	0.000000	0.000000	472.933411	0.000000	0.000000
34	0.000000	0.000000	481.163590	0.000000	0.000000
35	0.000000	0.000000	504.974983	0.000000	0.000000
36	0.000000	0.000000	512.187993	0.000000	0.000000
37	0.000000	0.000000	572.680553	0.000000	0.000000
38	0.000000	0.000000	579.510663	0.000000	0.000000
39	0.000000	0.000000	603.355740	0.000000	0.000000
40	0.000000	0.000000	620.049040	0.000000	0.000000
41	0.000000	0.000000	769.000000	0.000000	0.000000
42	0.000000	0.000000	811.000000	0.000000	0.000000
43	0.000000	0.000000	910.000000		

SYNOPSYS AI>

Chapter 55

Lens coatings, polarization

Fetch the six-element lens you designed in chapter 38, a lens that replaced an earlier Petzval design, shown in figure 55.1 (**C38L2**).

At the moment, this lens is entirely uncoated, and polarization mode has not been turned on. Let us examine the transmission. Type the AI question

```
TRANS?
```

```
SYNOPSYS AI>TRANS?

POLARIZATION MODE IS TURNED OFF; COATINGS IGNORED
(APODIZATION 1.00000 POLARIZATION 1.00000 ABSORPTION 0.97726)

 The transmission of the ray from
 Relative field (HBAR, GBAR) =        0.00000000        0.00000000
 Relative pupil coordinate (XEN, YEN) =        0.00000000        0.00000000
 in color number  2
 on surface number  13
 is equal to        0.97726440
 SYNOPSYS AI>
```

Transmission is very high—but the program alerts you that coatings (and reflection losses) are ignored. How important are they?

We have to put the lens in polarization mode to find out, since the effect of coatings depends on the polarization of the light.

```
CHG
POL LIN Y
END
TRANS?
```

```
SYNOPSYS AI>TRANS?

(APODIZATION 1.00000 POLARIZATION 0.43575 ABSORPTION 0.97726)

 The transmission of the ray from
 Relative field (HBAR, GBAR) =        0.00000000        0.00000000
 Relative pupil coordinate (XEN, YEN) =        0.00000000        0.00000000
 in color number  2
 on surface number  13
 is equal to        0.42584288
 SYNOPSYS AI>
```


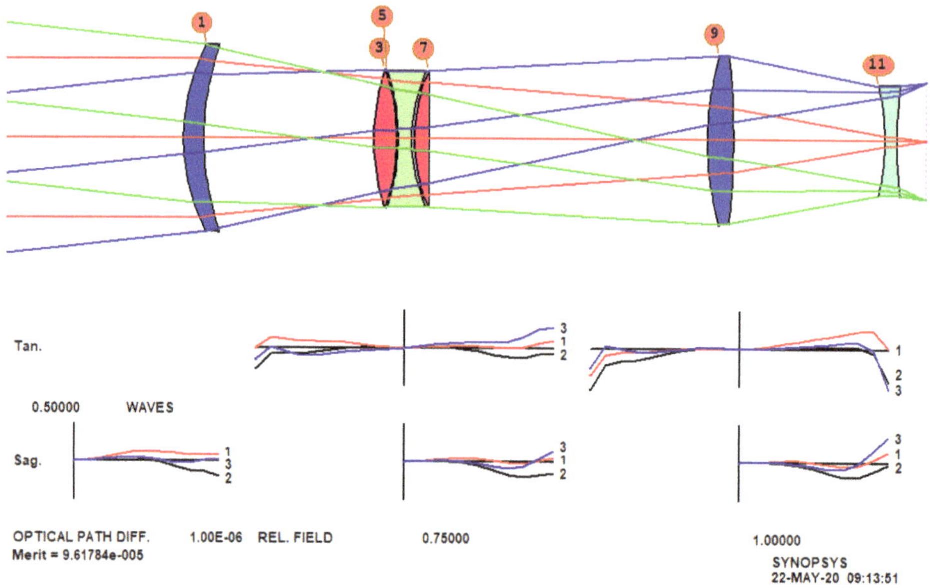

Figure 55.1. Six-element lens.

Indeed. When reflection losses are considered, transmission is reduced to only 42.6%. Let us apply an antireflection coating to the lens surfaces.

```
CHG
COAT QMD ALL
END
TRANS?
PCOAT
```

```
SYNOPSYS AI>TRANS?

(APODIZATION 1.00000 POLARIZATION 0.95749 ABSORPTION 0.97726)

 The transmission of the ray from
 Relative field (HBAR, GBAR) =       0.00000000      0.00000000
 Relative pupil coordinate (XEN, YEN) =       0.00000000      0.00000000
 in color number  2
 on surface number  13
 is equal to          0.93571967
 SYNOPSYS AI>PCOAT

 SURF. NO.   COATING
 -------------------
    1 COATING QMD
    2 COATING QMD
    3 COATING QMD
    4 COATING QMD
    5 COATING QMD
    6 COATING QMD
    7 COATING QMD
    8 COATING QMD
    9 COATING QMD
   10 COATING QMD
   11 COATING QMD
   12 COATING QMD
 SYNOPSYS AI>
```

Transmission is much improved; now we get 93.57% of the light through at the axis ray. **PCOAT** verifies that all surfaces are now coated.

What is this coating, 'QMD'? Ask the program to print the coating on surface 1.

```
SYNOPSYS AI>PCOAT 1

 COATING STACK, SURFACE No.                1
 STD. COAT.                                        2
         STACK DATA
         CONTROL WAVELENGTH =     0.5876 MICRONS
         CONTROL ANGLE =      0.000 DEG.
                          OPTICAL    PHYSICAL
          SURF NO.       THICKNESS  THICKNESS    INDEX     IMAG. INDEX
                          (WAVES)    (MICRONS)
        INCIDENT MEDIUM                          1.0000
                2         0.2500     0.107580    1.3655                    MGF2
                3         0.0000     0.000000    1.6714
          SUBSTRATE                              1.6714      0.0000

SYNOPSYS AI>
```

Here you see that coating QMD is a quarter-wave layer of MgF2, or magnesium fluoride. Ideally, if a quarter-wave layer has in index of refraction that is the square roof of that of the glass, the reflection loss would be zero, but there are no materials with so low an index, and this is the best we can do. This is a standard coating, routinely applied to lenses for use in the visible spectrum.

In the days before the development of antireflection coatings, which are applied in a vacuum chamber by evaporating or sputtering the desired material, a designer would likely redesign this lens so that elements at surfaces 3, 5, and 7 could be cemented together, in order to reduce the reflection losses from four surfaces.

We can learn more about such losses by tracing a polarized ray and examining what happens at each surface.

```
SYNOPSYS AI>PRAY P 0 1 0 SURF


    SURF. NO.    MAGN.             X               Y               Z            PHASE
 -------------------------------------------------------------------------------------
 Incid. S   1 0.100000E+01  0.000000E+00  0.100000E+01  0.000000E+00  0.000000E+00
 Incid. P   1 0.000000E+00  0.000000E+00  0.000000E+00  0.000000E+00  0.000000E+00
 Refr.  S   1 0.997210E+00  0.000000E+00  0.997210E+00  0.000000E+00 -0.837452E+02
 Refr.  P   1 0.000000E+00  0.000000E+00  0.000000E+00  0.000000E+00 -0.838313E+02
 Incid. S   2 0.997210E+00  0.000000E+00  0.997210E+00  0.000000E+00 -0.837452E+02
 Incid. P   2 0.000000E+00  0.000000E+00  0.000000E+00  0.000000E+00  0.000000E+00
 Refr.  S   2 0.995497E+00  0.000000E+00  0.995497E+00  0.000000E+00 -0.170790E+03
 Refr.  P   2 0.000000E+00  0.000000E+00  0.000000E+00  0.000000E+00 -0.870540E+02
 Incid. S   3 0.995497E+00  0.000000E+00  0.995497E+00  0.000000E+00 -0.170790E+03
 Incid. P   3 0.000000E+00  0.000000E+00  0.000000E+00  0.000000E+00  0.000000E+00
 Refr.  S   3 0.992247E+00  0.000000E+00  0.992247E+00  0.000000E+00 -0.258086E+03
 Refr.  P   3 0.000000E+00  0.000000E+00  0.000000E+00  0.000000E+00 -0.873006E+02
 Incid. S   4 0.992247E+00  0.000000E+00 -0.992247E+00  0.000000E+00 -0.780860E+02
 Incid. P   4 0.000000E+00  0.000000E+00  0.000000E+00  0.000000E+00  0.000000E+00
 Refr.  S   4 0.979970E+00  0.000000E+00 -0.979970E+00  0.000000E+00 -0.153885E+03
 Refr.  P   4 0.000000E+00  0.000000E+00  0.000000E+00  0.000000E+00 -0.763096E+02
.
.
.
```

Here you see a portion of the output from the polarization raytrace, PRAY. The incident ray is polarized in the Y-direction, so the incident S polarization has magnitude 1.0 and the P polarization has magnitude 0.0. Lower down in the listing you see the details at the end of the lens.

```
Incid. S 11 0.957981E+00  0.000000E+00 -0.957981E+00  0.000000E+00  0.632493E+02
Incid. P 11 0.000000E+00  0.000000E+00  0.000000E+00  0.000000E+00  0.000000E+00
Refr.  S 11 0.957817E+00  0.000000E+00 -0.957817E+00  0.000000E+00 -0.238204E+02
Refr.  P 11 0.000000E+00  0.000000E+00  0.000000E+00  0.000000E+00 -0.870813E+02
Incid. S 12 0.957817E+00  0.000000E+00 -0.957817E+00  0.000000E+00 -0.238204E+02
Incid. P 12 0.000000E+00  0.000000E+00  0.000000E+00  0.000000E+00  0.000000E+00
Refr.  S 12 0.957700E+00  0.000000E+00 -0.957700E+00  0.000000E+00 -0.111379E+03
Refr.  P 12 0.000000E+00  0.000000E+00  0.000000E+00  0.000000E+00 -0.875631E+02
V1: (X,Y,Z),Phase  0.000000E+00 -0.957700E+00  0.000000E+00 -0.111379E+03
V2: (X,Y,Z),Phase  0.000000E+00  0.000000E+00  0.000000E+00 -0.875631E+02
INTENSITY: Apod., Polar., Prod.  0.100000E+01  0.917189E+00  0.917189E+00


           THIS ANALYSIS CONSIDERS REFLECTION LOSSES ONLY, NOT ABSORPTION
          SYNOPSYS AI>
```

For the on-axis upper rim ray, the magnitude coming out is 0.9577, and the intensity of the ray is the square of this amplitude, or 0.9172. This is somewhat lower than the transmission of the axis ray, owing to the steeper angle of incidence at each surface, which increases reflection losses there.

The above rays were all traced in the primary color. How does the transmission vary with wavelength? The **XCOLOR** analysis can tell us. A transmission analysis is performed at each of 21 wavelengths and plotted as shown in figure 55.2.

```
SYNOPSYS AI>XCOLOR

(APODIZATION 1.00000 POLARIZATION 0.79043 ABSORPTION 0.89779)
(APODIZATION 1.00000 POLARIZATION 0.82237 ABSORPTION 0.91384)
(APODIZATION 1.00000 POLARIZATION 0.85076 ABSORPTION 0.92768)
(APODIZATION 1.00000 POLARIZATION 0.87551 ABSORPTION 0.94079)
(APODIZATION 1.00000 POLARIZATION 0.89664 ABSORPTION 0.95128)
(APODIZATION 1.00000 POLARIZATION 0.91425 ABSORPTION 0.96043)
(APODIZATION 1.00000 POLARIZATION 0.92852 ABSORPTION 0.96780)
(APODIZATION 1.00000 POLARIZATION 0.93967 ABSORPTION 0.97348)
(APODIZATION 1.00000 POLARIZATION 0.94793 ABSORPTION 0.97754)
(APODIZATION 1.00000 POLARIZATION 0.95358 ABSORPTION 0.97869)
(APODIZATION 1.00000 POLARIZATION 0.95688 ABSORPTION 0.97869)
(APODIZATION 1.00000 POLARIZATION 0.95810 ABSORPTION 0.97804)
(APODIZATION 1.00000 POLARIZATION 0.95749 ABSORPTION 0.97726)
(APODIZATION 1.00000 POLARIZATION 0.95528 ABSORPTION 0.97577)
(APODIZATION 1.00000 POLARIZATION 0.95171 ABSORPTION 0.97425)
(APODIZATION 1.00000 POLARIZATION 0.94697 ABSORPTION 0.97299)
(APODIZATION 1.00000 POLARIZATION 0.94124 ABSORPTION 0.97225)
(APODIZATION 1.00000 POLARIZATION 0.93470 ABSORPTION 0.97238)
(APODIZATION 1.00000 POLARIZATION 0.92749 ABSORPTION 0.97283)
(APODIZATION 1.00000 POLARIZATION 0.91975 ABSORPTION 0.97348)
(APODIZATION 1.00000 POLARIZATION 0.91158 ABSORPTION 0.97421)
 SYNOPSYS AI>
```

Transmission is reduced in blue and red light, so the image will appear with a slight yellow cast, which is not a problem with these aerial cameras.

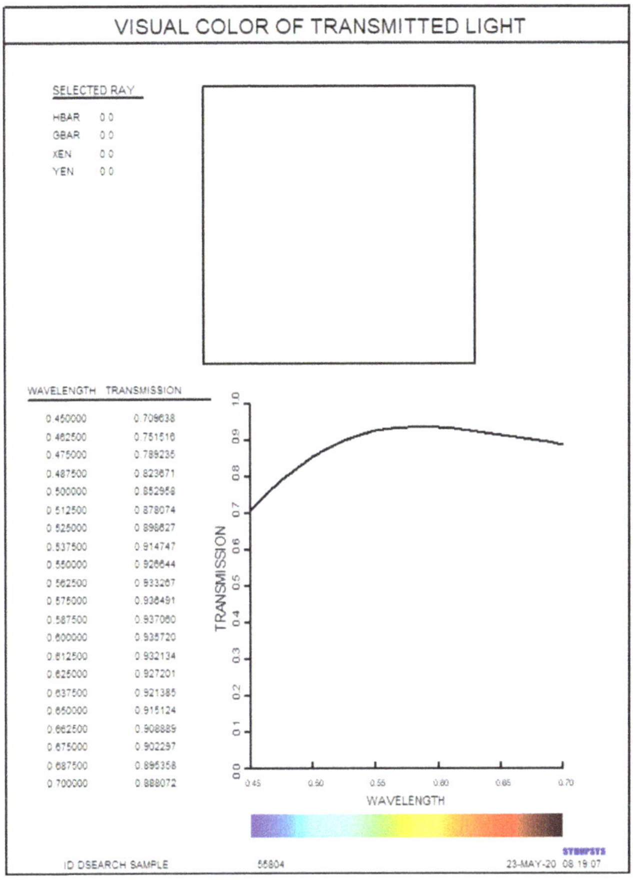

Figure 55.2. Transmission of lens with QMD coatings, over the visual spectrum.

What happens if we assign a high-efficiency coating instead of the quarter-wave MgF2 we did above? The stock coating `HEA` consists of three layers and is corrected for a somewhat wider spectrum. With that coating assigned to all surfaces, the `XCOLOR` plot is different, as shown in figure 55.3. This is somewhat better in the red and blue, but lower in the green.

Using the tools described in chapter 51, we designed a custom coating optimized for the actual range of this lens, and when it was applied to the lens surfaces, transmission was higher. This is explained in chapter 56.

A variety of features are available for analyzing the effects of absorption, coatings, vignetting, and apodization, all found in the dialog **MTR** (Menu, Transmission), shown in figure 55.4.

Why do we have to activate polarization mode to analyze lens transmission? Due to the compound angle of rays hitting a lens surface where both X and Y are nonzero, polarization is rotated with respect to the axis ray, and a thorough analyses has to take this into account. Here is a good illustration. Create a singlet lens in between two polarizers, shown in figure 55.5. (**C55L1**)

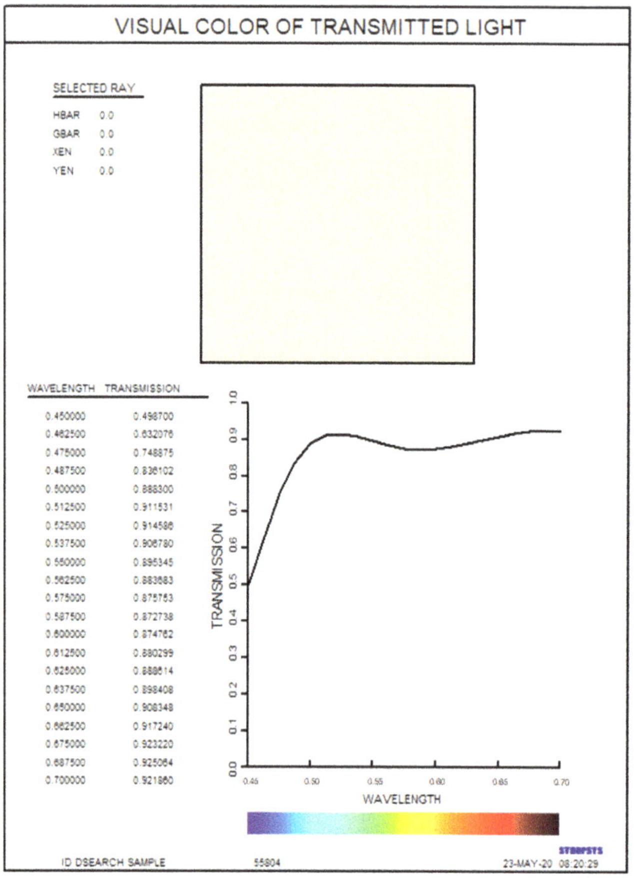

Figure 55.3. Transmission with standard HEA coatings.

```
RLE
ID CROSSED POLARIZERS
WA1 .6328
OBA 100 1 15
POL LIN Y
UNI MM
1 TH 1 GTB S
N-BK7
1 POLARIZER
2 TH 5
3 RD 50 TH 7 PIN 1
4 PCV -3 TH 5
5 TH 1 PIN 1
5 POLARIZER
6 TH 1
5 GT 90 0 2
8
END
```

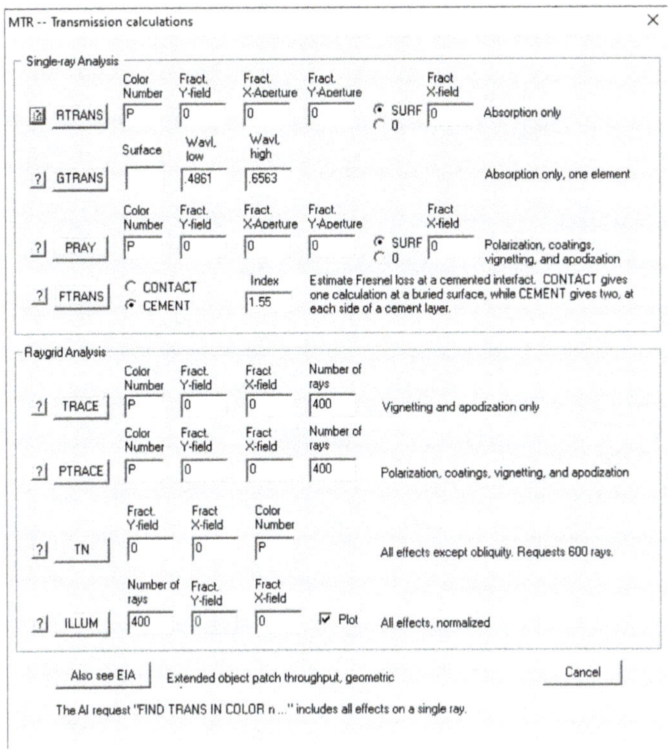

Figure 55.4. Options for analyzing lens transmission.

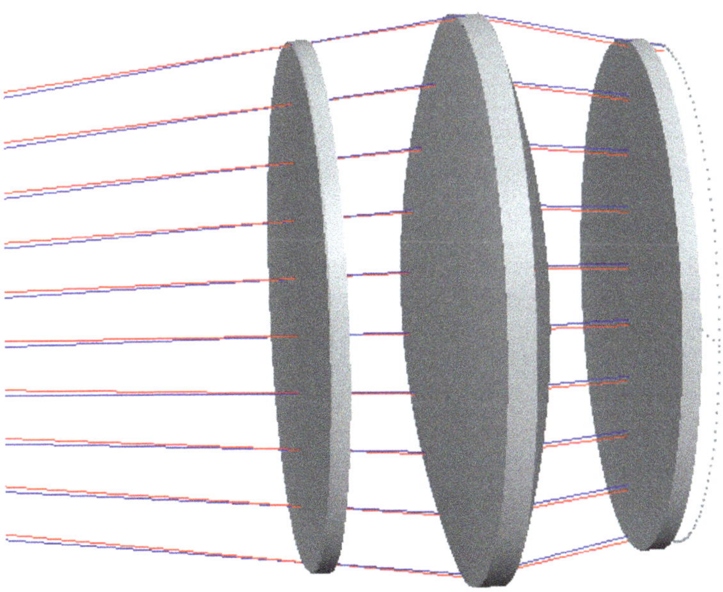

Figure 55.5. Simple lens between crossed polarizers.

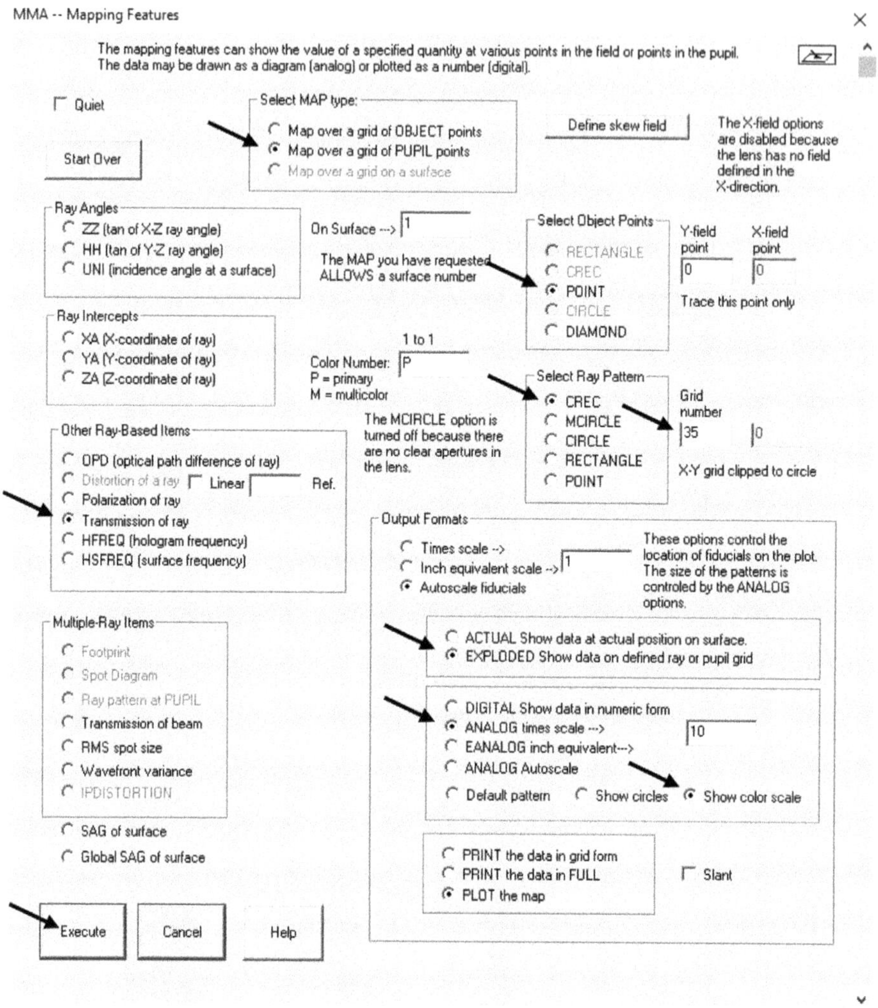

Figure 55.6. MMA dialog, with selections to map transmission over the pupil.

Surfaces 1 and 5 have been declared polarizers, and the element at 5 is rotated by 90 degrees in gamma, which is a rotation about the local Z-axis. One might expect that all light would be blocked since the two polarizers are now crossed.

Open the mapping dialog MMA, and select the options shown in figure 55.6. This will plot a map of ray transmission over the aperture and show the results in a color scale; the result is in figure 55.7. Rays in the meridional plane ($x = 0$) and skew plane ($y = 0$) are indeed totally blocked by the polarizers–but rays with a

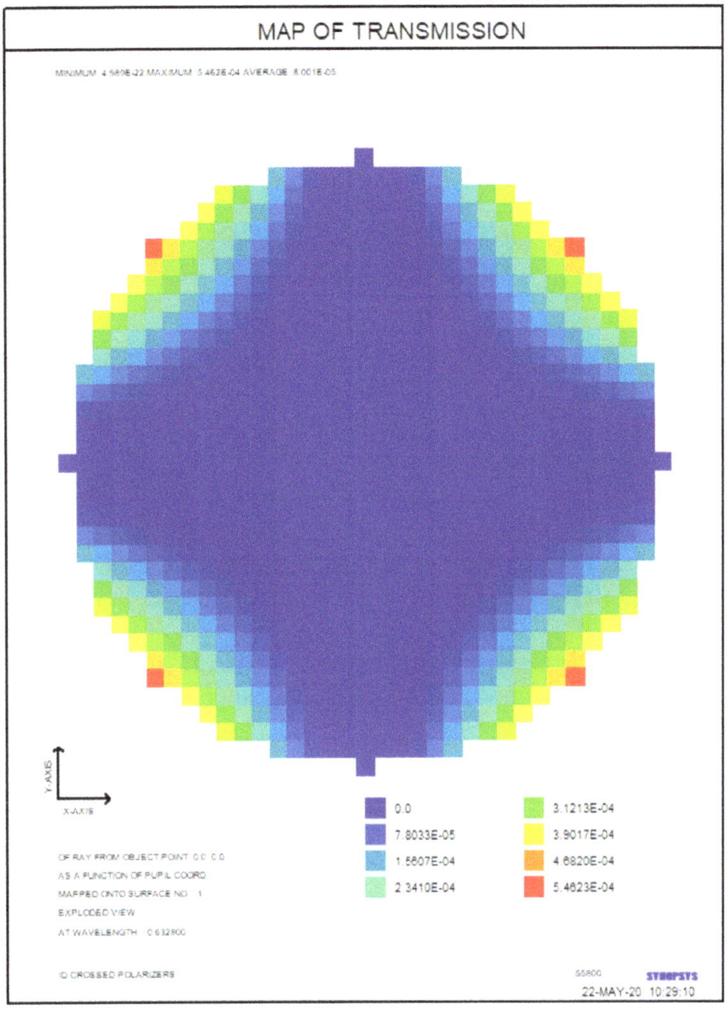

Figure 55.7. Transmission over pupil for lens between crossed polarizers. Light passes through the quadrants but not in the meridional or skew planes.

compound angle of incidence, in the quadrants of the pupil, are not. This is a very pretty sight, and if you ever get a chance to see it on an optical bench be sure to take a look.

IOP Publishing

Lens Design (Second Edition)
Automatic and quasi-autonomous computational methods and techniques
Donald C Dilworth

Chapter 56

A custom coating with custom materials

Chapter 55 showed how to assign standard coatings to a lens. Types QMD and HEA were both pretty good, but we wonder if we can do better with a custom coating.

Following the instructions in chapter 16 of the User's Manual, we create a MACro that asks for high efficiency over the range of 0.45–0.7 μm. Enter the following lines in an editor. (**C56M1**)

```
FILM
DESIGN
ANG 0
BUILD 4 1 1.671
ID TEST
FIX
AANT
RAVE .45 0 0 1
RAVE .5 0 0 1
RAVE .55 0 0 1
RAVE .6 0 0 1
RAVE .65 0 0 1
RAVE .7 0 0 1
END
SYNO 30
SPE
ANA
TRANS
LAM .45 .7 100 0
PLOT
```


When you run this file, the program branches to the thin-film section, which has its own input formats that use many of the same commands you have seen in the lens design section. Here, we ask for a coating of four layers, with a substrate index of 1.671, and ask for transmission of 1.0 over the range. Then we ask for an analysis of that coating. The program creates a coating stack, which is listed with the SPEC command (since we are in the FILM section, that command applies to the current film stack). The FIX directive makes the program fit the coating stack to a library of standard materials, which are listed along with the stack data.

```
FILM>SPEC

 ID TEST                                          55805
          STACK DATA
          CONTROL WAVELENGTH =     0.5876 MICRONS
          CONTROL ANGLE =      0.000 DEG.
                              OPTICAL    PHYSICAL
            SURF NO.         THICKNESS  THICKNESS    INDEX     IMAG. INDEX
                              (WAVES)    (MICRONS)
          INCIDENT MEDIUM                            1.0000
                 2            0.2307     0.099291    1.3655                    MGF2
                 3            0.4720     0.132196    2.0980                    ZRO2
                 4            0.6613     0.222305    1.7479                    CEO2
                 5            0.0000     0.000000    1.6710
            SUBSTRATE                                1.6710      0.0000

 FILM>
```

How did it come out? The plot in figure 56.1 shows transmission above 0.998 over most of the visual range. This looks like a very promising coating.

But the performance shown is based on the index of refraction as taken from the database in SYNOPSYS–which applies to the bulk material. When deposited as a thin film, the index is usually different, and here we want to take that difference into account.

To do so, we have to create a library of custom materials. MACro **C56M2** shows the input that creates a new library, LIB.CFL, with interpolation coefficients for the three materials.

```
FILM
NLIB
1 "C-MGF2" 0
0.179416E+01  -0.718242E-02   0.434546E-01  -0.908349E-02   0.109341E-02  -0.474168E-04
2 "C-ZRO2" 0
0.4332306E+1 -0.8256538E-3 -0.3023147E-1 0.2664098E-1 -0.3074757E-2 0.1172724E-3
3 "C-CE02" 0
0.2923834E+1 -0.3986741E-2 0.1077115E+0 -0.7148323E-1 0.2204815E-1 -0.1638747E-2
END
```

In this example, we have entered the power-series coefficients for the bulk materials, but in practice you would replace those data with the coefficients obtained from the coating vendor. The coefficients can apply either to a power-series expansion (as calculated by the GFIT feature of SYNOPSYS), or to the Sellmeier formula, widely used for index interpolation by the optical glass industry.

Now we create a new coating stack, based on the data listed above, but specifying the three materials we just created. (**C56M3**) This stack is saved as coating CUSTOM01.

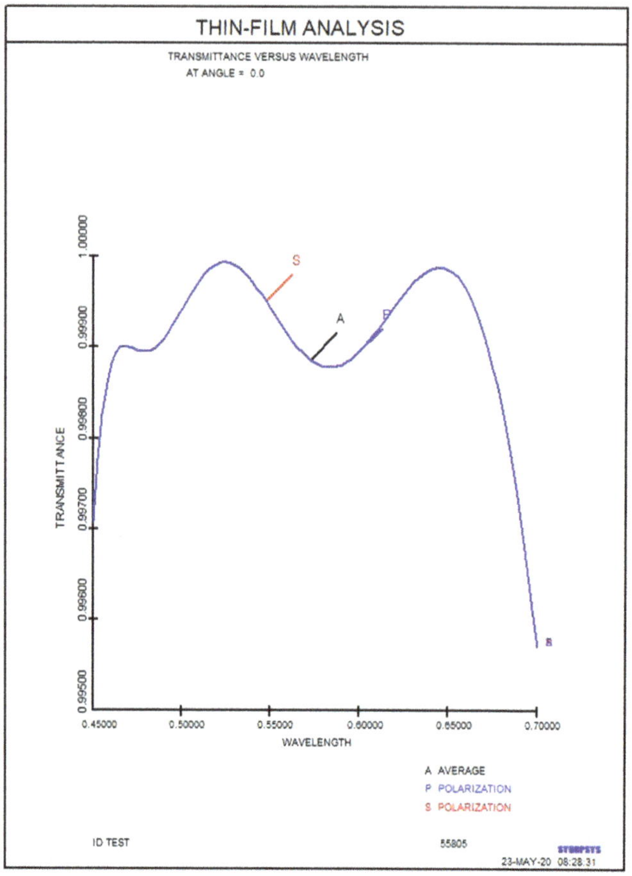

Figure 56.1. Transmission curve for a custom coating stack.

```
FILM
RFILM
ID CUSTOM HEA COATING
WAVL .5876
ANGLE 0
0 1
.2307 C01
.472 C02
.6613 C03
0 1.671
END
SAVE CUSTOM01
```

Now we are ready to apply that coating stack to the lens elements.

```
RETURN
CHG
COAT 01 ALL
END
```

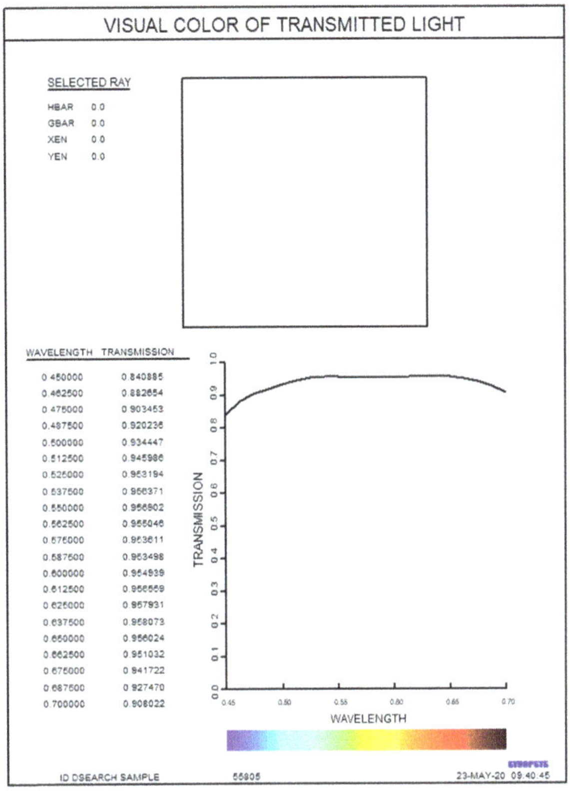

Figure 56.2. Transmission over wavelength with custom coating.

The RETURN command returns from the FILM section to SYNOPSYS, and then coating 01, which was saved as CUSTOM01, is applied. Verify this with the PCOAT command:

```
SYNOPSYS AI>PCOAT

 SURF. NO.   COATING
 ------------------
   1 CUSTOM COATING  1
   2 CUSTOM COATING  1
   3 CUSTOM COATING  1
   4 CUSTOM COATING  1
   5 CUSTOM COATING  1
   6 CUSTOM COATING  1
   7 CUSTOM COATING  1
   8 CUSTOM COATING  1
   9 CUSTOM COATING  1
  10 CUSTOM COATING  1
  11 CUSTOM COATING  1
  12 CUSTOM COATING  1
 SYNOPSYS AI>
```

Now XCOLOR in figure 56.2 shows better transmission over the spectral range. Compare this with figures 55.2 and 55.3. Yes, this is a better coating.

IOP Publishing

Lens Design (Second Edition)

Automatic and quasi-autonomous computational methods and techniques

Donald C Dilworth

Chapter 57

Focusing x-rays

You have already seen how to design and analyze optical designs used in the ultraviolet, visible, and thermal infrared portions of the electromagnetic spectrum. Now we will design a system of two mirrors that will focus a beam of x-rays at a wavelength of 0.000 63 μm.

There are no optical glasses that work in this region, and mirrors reflect very little unless the incidence angle is very small. We call these *grazing-incidence* mirrors, and they are very different from anything discussed earlier. Figure 57.1 shows an example **(C57L1)**.

Light (if we can call radiation that is invisible to the eye 'light') enters from the left, reflects from the two mirrors, and exits to the right. A look at the RLE file shows how this system is set up.

```
RLE
ID X-RAY 10X
 WA1 .0006200
 WT1 1.00000
 APS              1
 CFOV
 GLOBAL
 UNITS MM
 OBA   51.269     0.44350        0.0298    0.3028 0.44350        0.3028    0.0298
 MARGIN       1.126490
 BEVEL        0.225299
   0 AIR
   1 CV       0.0000000000000    TH      97.57000000 AIR
   2 CV       0.0000000000000    TH       0.00000000 AIR
   3 RAO       4.43500177       17.74000000       0.00000000        0.00000000
   3 CV -3.0981772400899E-05    TH       0.00000000 AIR
   3 TORIC      -9.82796000
   3 DECEN       0.00000000       0.00000000       0.00000000    101
   3 AT      89.00000038      0.00000000    101
   3 EFILE EX1       8.870000       8.870000       8.870000       0.000000
   3 EFILE EX2       8.870000       8.870000       0.000000
   3 EFILE MIRROR     2.661000
   3 REFLECTOR
   4 CV       0.0000000000000    TH     -17.74000000 AIR
   4 DECEN       0.00000000       0.00000000       0.00000000    101
   4 AT      89.00000018      0.00000000    101
   5 CV       0.0000000000000    TH       0.00000000 AIR
   6 RAO       4.43500177       17.74000000       0.00000000        0.00000000
   6 CV  3.0981772400899E-05    TH       0.00000000 AIR
```


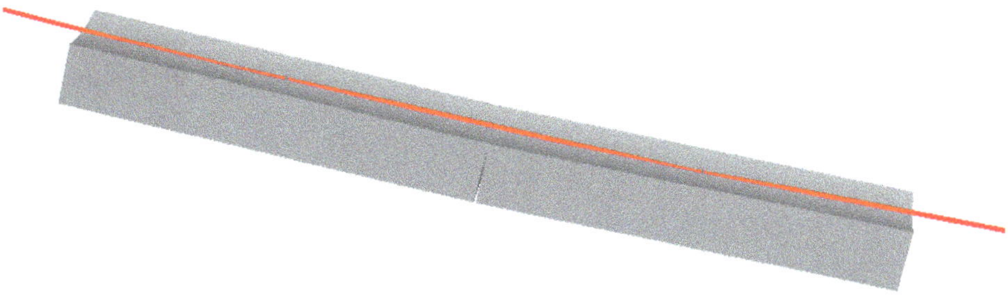

Figure 57.1. Two concave toroidal mirrors used to focus x-rays.

```
  6 TORIC       9.82796000
  6 DECEN       0.00000000        0.00000000        0.00000000    101
  6 AT      89.00000028       0.00000000    101
  6 EFILE EX1     8.870000      8.870000      8.870000      0.000000
  6 EFILE EX2     8.870000      8.870000      0.000000
  6 EFILE MIRROR    -2.661000
  6 REFLECTOR
  7 CV        0.0000000000000    TH      88.70000000 AIR
  7 DECEN       0.00000000        0.00000000        0.00000000    101
  7 AT      89.00000022       0.00000000    101
  8 CV        0.0000000000000    TH    1501.69100000 AIR
  9 CV        0.0000000000000    TH       0.00000000 AIR
END
```

Surfaces 3 and 6 are declared TORIC, with a Y-plane curvature given by the CV data and *x*-plane radius of curvature by the TORIC data. These data can be listed as shown in the sequence below.

```
SYNOPSYS AI>3 RAD?

 The specification of surface  3  is toric.
 The base radius of curvature of  3  is  -32277.04300000
 SYNOPSYS AI>ASY

 SPECIAL SURFACE DATA

  SURFACE NO.   3 -- TORIC SURFACE
  RX     -9.827960


  SURFACE NO.   6 -- TORIC SURFACE
  RX      9.827960


 TILT AND DECENTER DATA
 LEFT-HANDED COORDINATES

 SURF TYPE       X            Y            Z        ALPHA      BETA      GAMMA    GROUP

   3 REL      0.00000      0.00000      0.00000  89.0000    0.0000    0.0000     101
   4 REL      0.00000      0.00000      0.00000  89.0000    0.0000    0.0000     101
   6 REL      0.00000      0.00000      0.00000  89.0000    0.0000    0.0000     101
   7 REL      0.00000      0.00000      0.00000  89.0000    0.0000    0.0000     101

 KEY TO SURFACE TYPES

 GLB  GLOBAL COORDINATES                      LOC  LOCAL COORDINATES
 REL  RELATIVE COORDINATES                    REM  REMOTE TILTS IN RELATIVE COORD.
 SYNOPSYS AI>
```

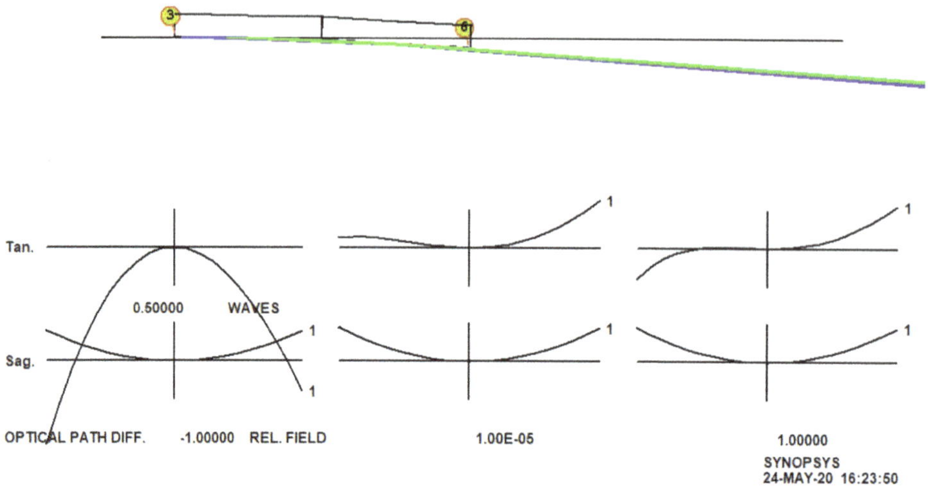

Figure 57.2. Grazing-incidence x-ray focusing cylindrical mirrors.

Note the alpha tilts applied to four surfaces, each with a group size of 101. That size exceeds the total number of surfaces, so the tilt will not activate a reverse tilt later in the system, which is a powerful feature of the relative tilt option but is not called for here.

Systems of this nature work over a small field of view, in this case only 0.4435 mm in radius at a distance of 51.269 mm. This system is promising, although the OPD fans at the –1 field point are rather large. as shown in figure 57.2.

A goal of this design is to keep the incidence angle no greater than about one degree. If the angles get steeper, less light is reflected. The output from the RAY command shows the angle, which is 89.0 degrees at surfaces 3 and 6, the toroidal mirrors.

```
SYNOPSYS AI>RAY P 0 0 0 SURF

 INDIVIDUAL RAYTRACE ANALYSIS

 FRACT. OBJECT HEIGHT                HBAR      0.000000   GBAR       0.000000
 FRACT. ENTRANCE PUPIL COORD.         YEN      0.000000    XEN       0.000000
 COLOR NUMBER                           1

                          RAY VECTORS                  (X DIR TAN)   (Y DIR TAN)     (INC. ANG.)
 SURF          X              Y              Z              ZZ            HH              UNI
 ------------------------------------------------------------------------------------------------
 OBJ       0.000000       0.000000       0.000000       0.000000       0.000000
   1       0.000000       0.000000       0.000000       0.000000       0.000000       0.000000
   2       0.000000       0.000000       0.000000       0.000000       0.000000       0.000000
   3       0.000000 -3.784326E-08 -2.218469E-20       0.000000      57.289986      89.000000
   4       0.000000 -6.604557E-10       0.000000       0.000000  3.488314E-09       0.000000
   5       0.000000 -6.254314E-08       0.000000       0.000000  3.488314E-09       0.000000
   6       0.000000 -3.625268E-06  2.035900E-16       0.000000      57.289969      89.000000
   7       0.000000 -6.326964E-08       0.000000       0.000000 -2.216482E-09       0.000000
   8       0.000000 -2.598716E-07       0.000000       0.000000 -2.216482E-09       0.000000
 IMG      0.000000-3.588342E-06     0.000000
 SYNOPSYS AI>
```

Let us try to improve this system. MACro **C57M1** should do the job.

```
PANT
VY 3 RAD
VY 3 G 1
VY 6 RAD
VY 6 G 1
VY 8 TH
END

AANT
M 4.435 10 A P YA 1
LUL 1700 1 1 A ZG 9
GNO 0 1 4 P 0
GNO 0 1 4 P 1
GNO 0 1 4 P 1
GNO 0 1 4 P 0 1
GNO 0 1 4 P 1 1
GNO 0 1 4 P 1 1
END

SNAP
SYNO 20 0 FIX 20
```

Here we are varying the radius of curvature in the Y-direction with the RAD variables, and in the X-direction with the G 1 variables. We want a magnification of 10x, as requested by the target on the full-field chief ray, and the total length of the system should be no more than 1700 mm, as requested by the upper limit on the ZG (global Z-coordinate) of surface 9. Note that we cannot just give a target to the paraxial TOTL plus BACK, as we often do with lenses, due to the steep folding angles.

After running this MACro, the system is somewhat better, as shown in figure 57.3. **(C57L2)**

A plot showing the diffraction pattern over the field is shown in figure 57.4. This was obtained from the dialog MPF, where we selected the visual appearance and FULL for the number of fields. The command sequence

```
PSVISUAL 2 5
OFP 6000 1 0
FIELDS FULL
ICOL M
PLOT
```

produces the same picture.

Let us examine the central image more closely. Open the MDI dialog, select the Solid option and click the PSPRD button. This displays the image as a solid model that you can spin around and look at from any angle, shown in figure 57.5. The

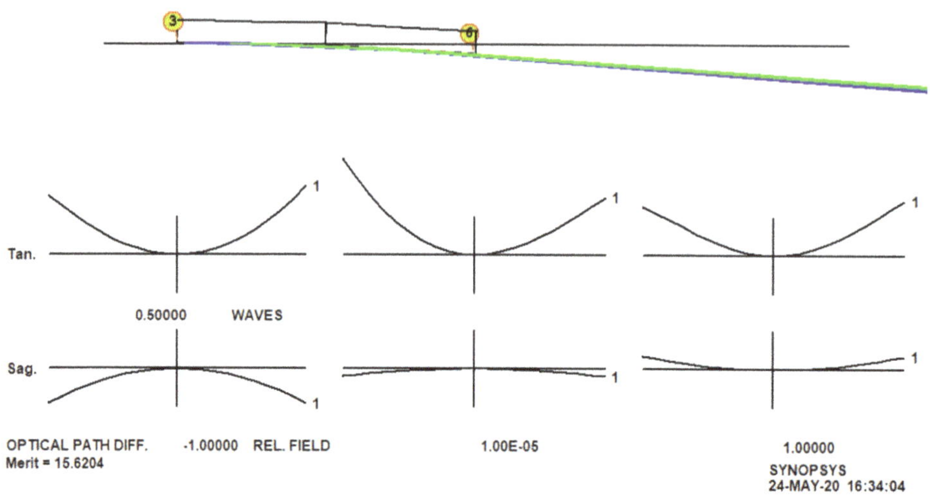

Figure 57.3. X-ray system after reoptimization. OPD errors are now more uniform.

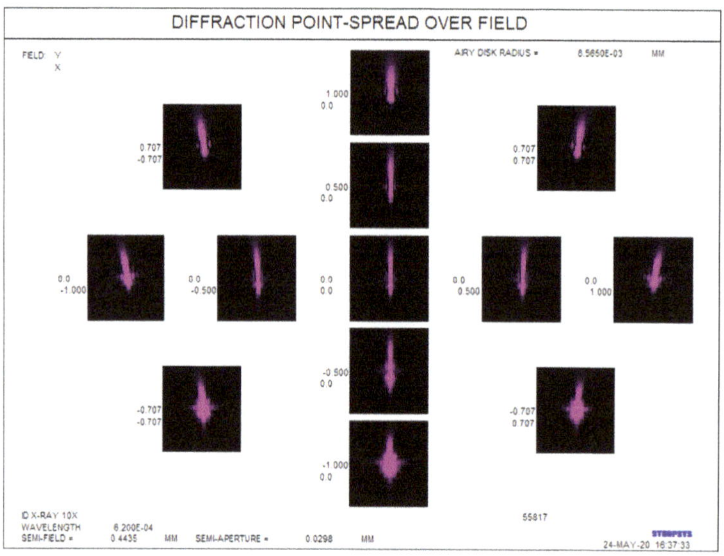

Figure 57.4. Diffraction pattern over field of the x-ray system.

image is displayed in violet, since the actual radiation is invisible and therefore has no color. An infrared image would be shown in red for the same reason.

Let us suppose the user wishes to model the extended image from a region at the object that is imaged into a square of 0.05 mm on a side at the focal plane. They want to know how much of the energy in that image spills out of a detector of that size. The MIT dialog can provide the answer. Open MIT and select the options shown in figure 57.6.

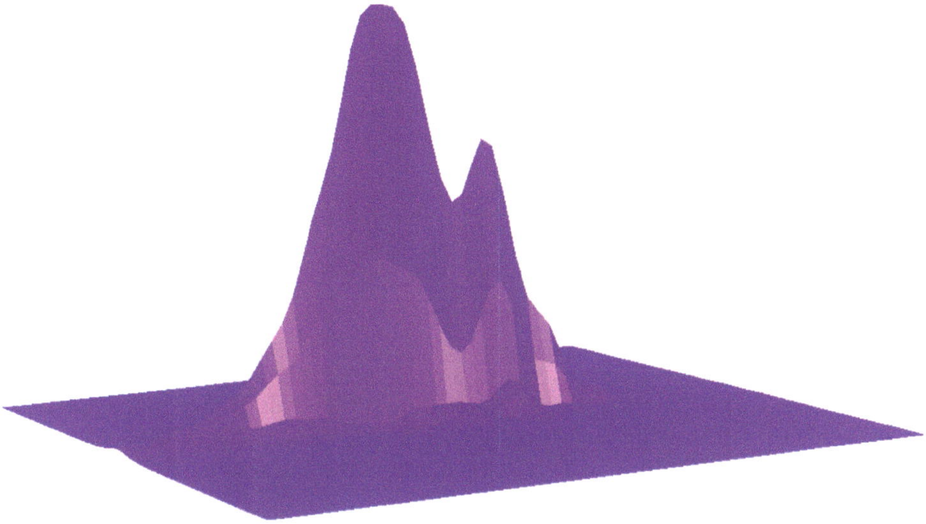

Figure 57.5. Central image from x-ray system, shown as a solid model.

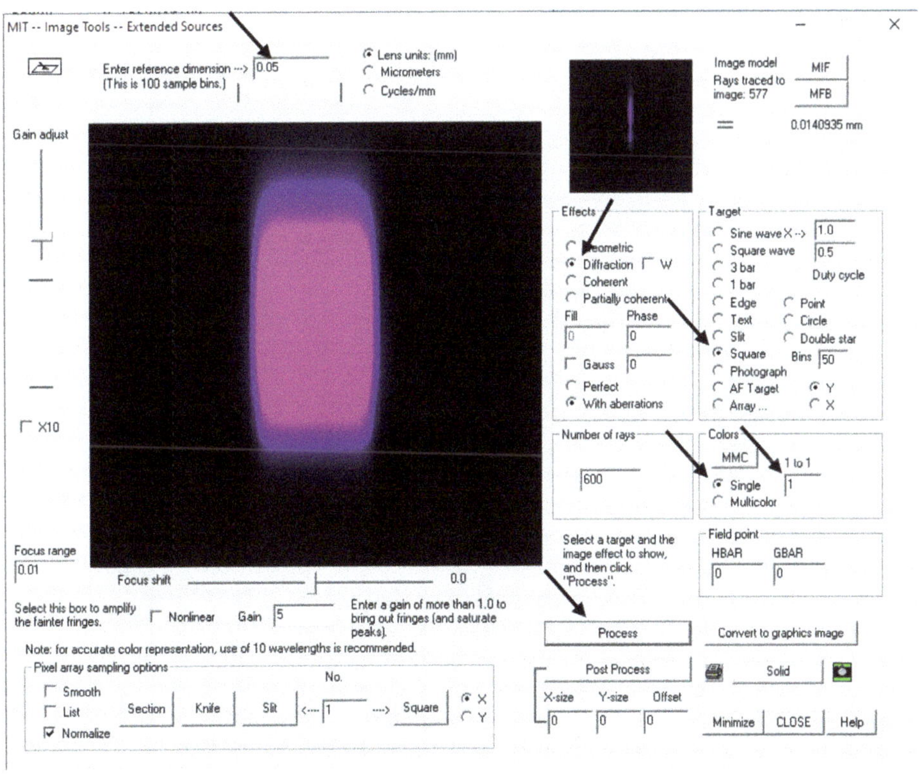

Figure 57.6. Image tools display of image of a square patch in the image plane.

The program samples a perfect square and then convolves it with the diffraction pattern shown in the small box at the upper right.

To determine the fraction of energy seen by the square detector, deselect the Normalize box, enter a dimension of 100 bins, select the y-direction, and click the Square button.

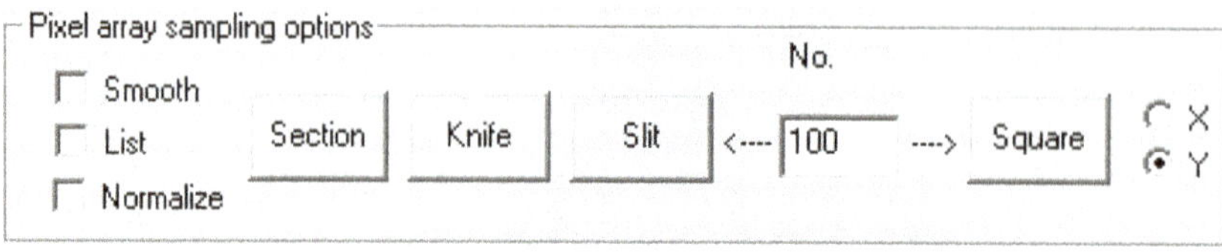

The reference dimension is 0.05 mm, so that is the area of the image plane that was convolved with the diffraction pattern. The size of the area of that image to be sampled is given as 100 bins, which in this case is the same size as the square. The

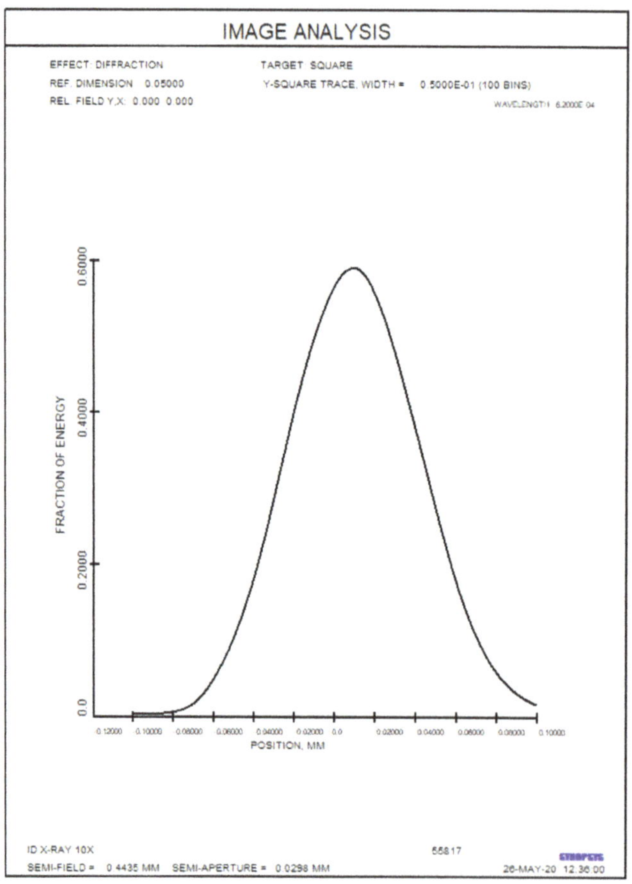

Figure 57.7. Plot of energy on a square as it is scanned over the image of a patch.

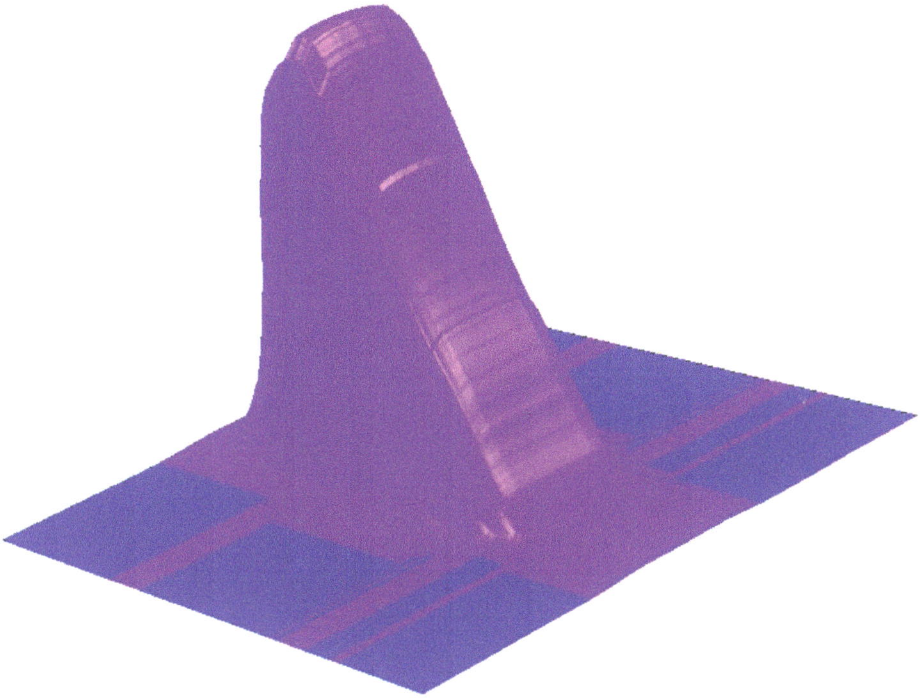

Figure 57.8. Solid model of the convolved diffraction image with the image patch.

program produces a plot of the energy on the square as the latter is shifted in y, shown in figure 57.7. The peak is about 0.6, meaning that 40% of the energy is lost due to aberrations and diffraction. If you scan in the x-direction, the figure is 0.568, since the patch is not symmetric in y.

This MIT analysis assumes that the image structure does not vary significantly over the sampled patch, a condition known as *isoplararity*. Figure 57.4 shows little change over parts of the field near the center, so this assumption is likely valid.

The convolved image patch can also be viewed as a solid model, if one clicks the Solid button in the MIT dialog, as shown in figure 57.8.

In earlier chapters, we selected the Coherent effect for the MIT analysis–which is valid for a point source but not advisable for an extended one. Since the object patch is not assumed coherent, to get an accurate model you would have to specify the partially coherent effect with a fill factor of 2.0. The result is much the same as the diffraction model, but takes much longer to calculate.

The above analysis takes diffraction into account, but if a lens has geometric aberrations larger than the diffraction pattern, one can get a purely geometric analysis in a different way. The image patch of 0.05 mm on a side corresponds to an object patch whose normalized size in object space is 0.011 69. The Extended Image Analysis feature can then determine the geometric throughput, as follows:

```
--- CHG
--- 9 RAO .05 .05
--- END
        GIHT          FOCL          FNUM          BACK          TOTL          DELF
     -4.27665    8344.60407    -876.68701    1463.75733    168.53000    1722.04677
Lens number     10 ID X-RAY 10X
---
--- EIA P 2000
--- IOBJ REC .01169 .01169
--- CENTER 0 0
--- GRID 11
--- GO
NUMBER OF RAYS ATTEMPTED        239096
NUMBER OF RAYS THROUGH          134840
GEOMETRIC EFFICIENCY       0.5640
```

Here we specify a rectangular aperture at the image and then ask EIA to analyze an object of the required dimensions, centered at the axis, and create a grid of 11 × 11 spot diagrams at the image, each with 2000 rays. Of this ensemble, 56.4% of the energy gets through the aperture, a figure not much different from the diffraction analysis from MIT. Why are they similar? For this system, it turns out the geometric spot diagram is about the same size as the diffraction pattern (which is not usually the case).

IOP Publishing

Lens Design (Second Edition)

Automatic and quasi-autonomous computational methods and techniques

Donald C Dilworth

Chapter 58

A singlet achromat

You have seen how to correct aberrations by combining two or more optical elements, and previous chapters have shown how lenses can be improved by the addition of a diffractive optical element (DOE). Now we will design a single element with a DOE that corrects chromatic aberration. The lens supplies the power, while the DOE does the correction.

MACro **C58M1** starts out with a simple starting point, opens the PAD display and initializes the format, then optimizes the lens, varying some aspheric terms on the DOE. Notice how PAD can accept user input: if you type the **PAD** command yourself and follow the prompts—or enter the response to those prompts in the input file, as we do here—the display is set up as requested; in this case, the blank lines correspond to hitting the Enter key when prompted with no other data.

```
RLE
ID SINGLET DOE
WAVL CDF
UNI MM
OBB 0 1 1
1 RAD 10 TH .3 GTB U
ZEON480R
2 USS 16
CWAV .6328
HIN PICKUP
2 TH 15
3
END
PAD
D
```

doi:10.1088/978-0-7503-3695-6ch58

```
T
1
123

O
B
M


PANT
VY 1 RAD
VY 2 G 26
VY 2 G 27
VY 2 G 28
END
AANT
GSO 0 10 5 2
GSO 0 5 5 1
GSO 0 5 5 3
END
SNAP
SYNO 20
ASY

DMASK 2 PROFILE 2
```

Run this file, and the lens is corrected to about 0.1 waves over the visible spectrum, as shown in figure 58.1. **(C58L1)** Here we have varied three of the terms in the power-series expansion of the phase of the DOE, along with the first radius of curvature.

Figure 58.2 shows the zone profile of the resulting DOE. This should not present a problem to the lens maker.

58.1 Single-element achromat with no DOE

Now we will show yet another concept that can correct chromatic aberration in a singlet element. Figure 58.3 shows how this works (**C58L2**).

The principle is rather simple: the two aspheric surfaces have almost equal and opposite power, and each zone of the input beam hits surface 2 at an aperture higher than at surface 1, and the power almost cancels out. But short wavelength light is refracted more steeply at surface 1 and hits surface 2 even higher up, where the negative power is stronger than for light at shorter wavelengths. So the higher positive power in blue is cancelled out by the higher negative power at that location.

This kind of element does not have reasonable first-order properties, owing to the short central radii. Also, since the central ray does not behave as do zonal rays, we have implemented a special program option,

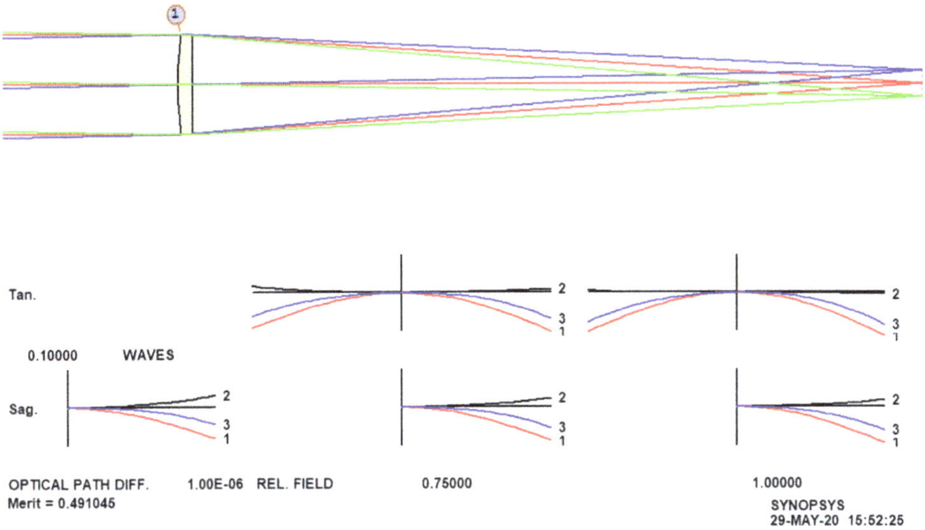

Figure 58.1. Singles lens corrected for chromatic aberration via a DOE on surface 2.

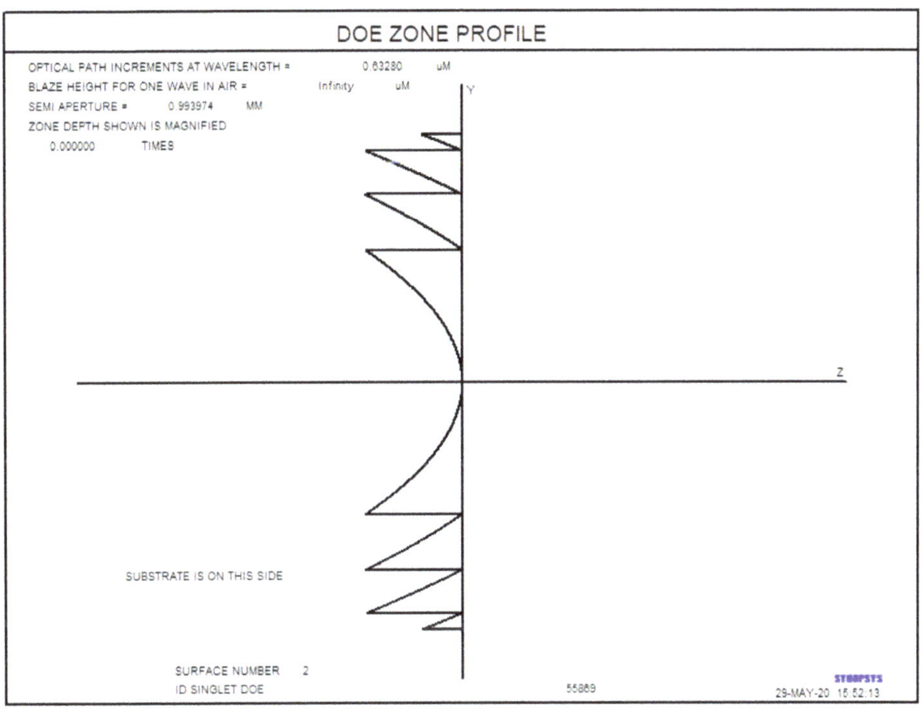

Figure 58.2. DOE zone profiles of the singlet achromat.

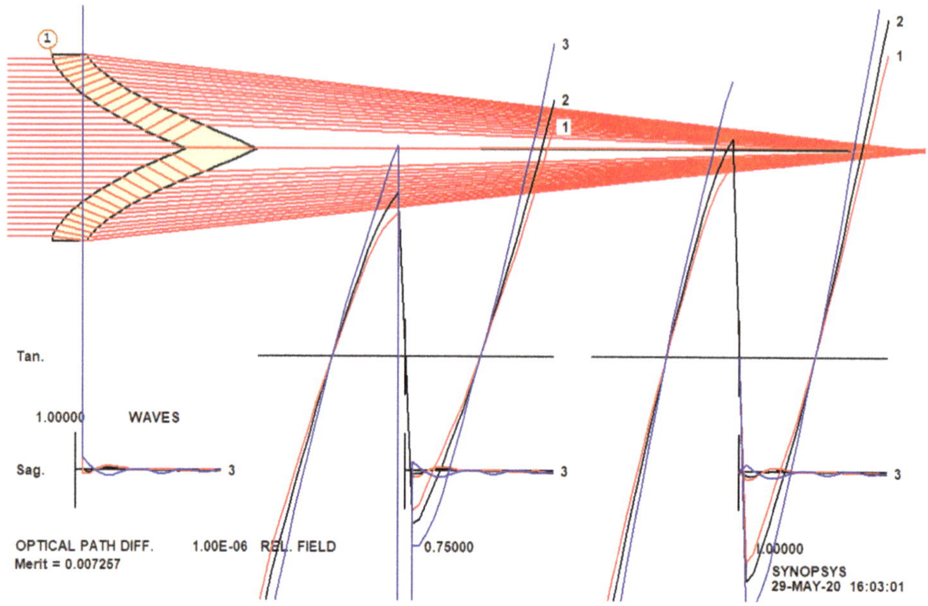

Figure 58.3. Method of correcting color with two steep aspheric surfaces.

```
ICR .5
```

in the lens file, which declares that the chief ray is not taken at the center of the pupil, as is normally the case. Instead, the program will trace the ray at the point (0.5, 0.0) in the pupil whenever it needs a chief ray for reference—as it does for OPD calculations.

To design this lens, we first roughed out a system with two steep hyperbolic surfaces, and then ran the MACro **C58M2**, shown below.

```
ON 68
PANT
RDR 1.0E-5
VY 1 CV 100 -100
VY 1 CC 1000 -1000
VY 1 G 3
VY 1 G 6
VY 1 G 10
VY 1 G 16
VY 1 G 18
VY 1 G 19
VY 1 G 20
VY 1 G 21

VY 2 CV 100 -100
VY 2 CC 1000 -1000
VY 2 G 3
VY 2 G 6
```

```
VY 2 G 10
VY 2 G 16
VY 2 G 18
VY 2 G 19
VY 2 G 20
VY 2 G 21

VY 1 TH
END

AANT
GTR 0 1 10 M
GTO 0 .1 21 M
END

SNAP
SYNO 50
```

To make this work, we override the default limits on radius of curvature, since the axicons want a very short radius at the vertex.

A drawback to this design is the poor field correction, shown in figure 58.4. Even at a field angle of 0.1 degrees, the image is not well corrected.

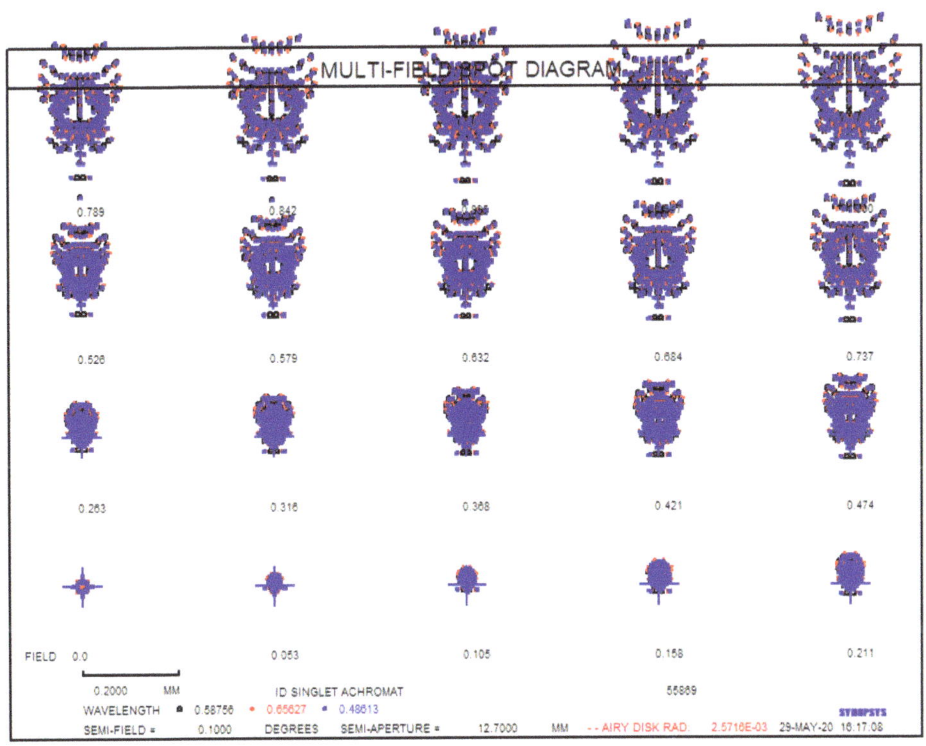

Figure 58.4. Spot diagrams over the field of the lens in figure 58.3.

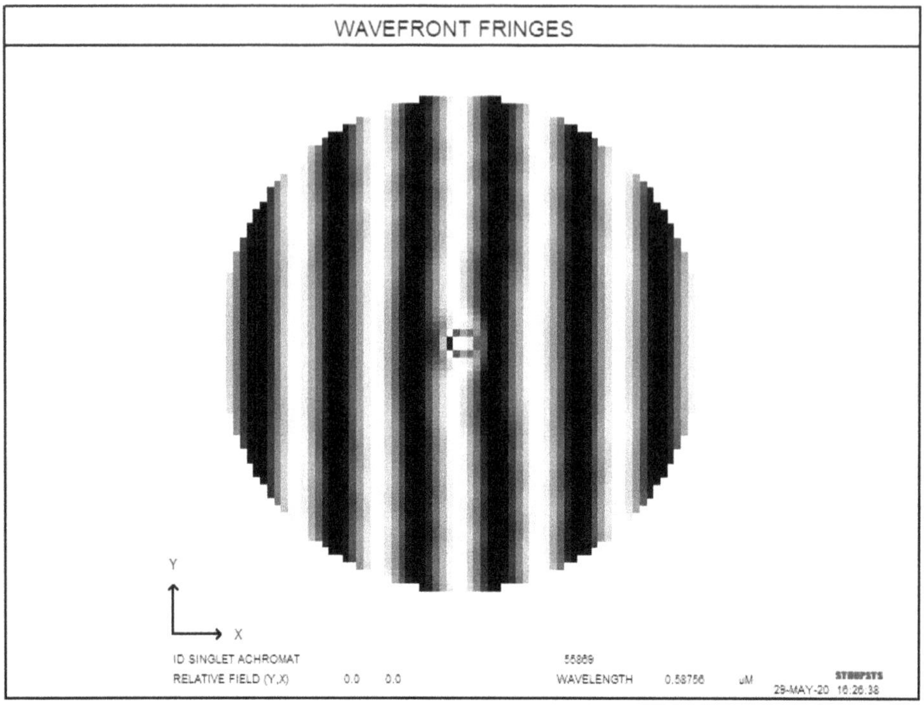

Figure 58.5. Wavefront fringes of the second variety of singlet achromat.

Still, over a very small field, the lens is excellent. Figure 58.5 shows the OPD errors on axis, with three fringes of wedge added. Except for a small area right at the vertex, the wavefront is almost perfect. This illustration can be created from the MDI dialog, or with the command form:

```
FRINGES
PUPIL P 0 0 5 0 3 0
```

IOP Publishing

Lens Design (Second Edition)

Automatic and quasi-autonomous computational methods and techniques

Donald C Dilworth

Chapter 59

Pupil aberrations and the optical image

You have seen how the obscuration of a Newtonian telescope lowers the modulation transfer function (MTF) at mid frequencies. Now consider the lens from the previous chapter (C58L2): there is no obscuration as such, but the beam has a prominent hole in the center, shown in figure 59.1. What effect does that have on the image?

The effect is striking, as shown by the diffraction point-spread function in figure 59.2. The first diffraction ring is more intense relative to what one would get with a perfect image. One can also examine this with the Image Tools Menu (MIT), with a reference dimension of 0.005, diffraction effect, Point source, Process, and then Smooth and Normalize Section in X. This cross-section of the diffraction pattern is shown in figure 59.3. This is not what one expects from a perfect wavefront. (The coherent option in MIT is not appropriate for this system, since it assumes by default that the exit pupil ray grid is geometrically similar to that of the entrance pupil.)

What is the effect on the MTF of this ray distribution, one might ask? Here the answer is not simple, and understanding the result requires some background knowledge.

Ask for an MTF plot at the on-axis point, in two ways: the first, shown in figure 59.4 is calculated with the Hopkins convolution method, referred to in previous chapters, with the command `MTF P 0 0 0 0 P`, while figure 59.5 shows the results from the Fourier-transform MTF with `DMTF P 0 6000 0 0 P`.

The two are very different. Which one is correct?

This is where it gets complicated. The concept of the MTF embodies the assumption that the lens is a *linear filter*, which means that if a lens forms an image of a sine-wave target, the image is also a sine wave, with a frequency given by the magnification and a contrast given by the current aberrations and diffraction. But what happens if the image is not a pure sine wave? That is the case if the lens exhibits strong pupil aberration, as does this lens, as we shall now show.

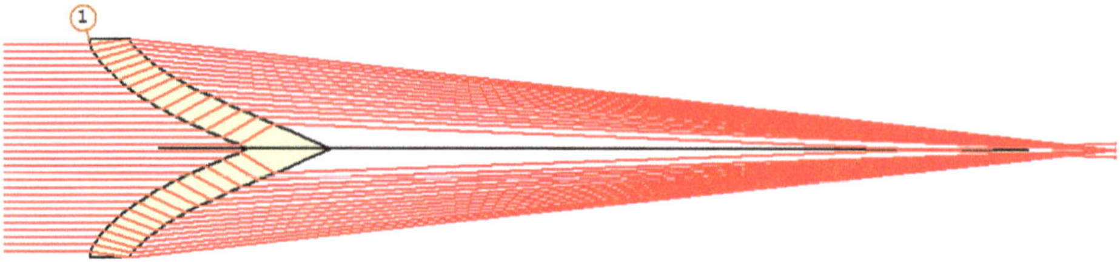

Figure 59.1. This system has small OPD errors but large pupil distortion.

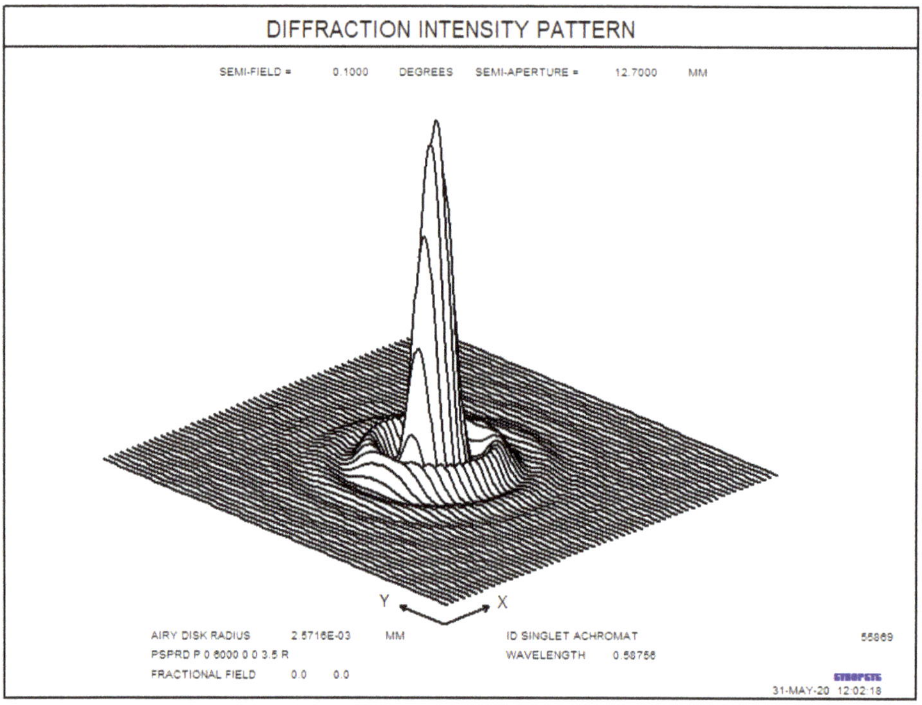

Figure 59.2. PSPRD image of the system with large pupil distortion.

59.1 Convolution MTF

According to the Hopkins algorithm, one calculates the MTF by comparing the OPD errors in pairs of points sheared with respect to each other by an amount that depends on the frequency to be analyzed. Consider the idealized case in figure 59.6.

Imagine a cone of light that comes into the lens but only passes through point B in the entrance pupil. If there were no target, that light would be imaged at point E and go on fill the image plane. (The pupils are images of the aperture stop, so they are also images of each other.)

Now add the sinewave target, which acts like a diffraction grating and produces copies of point B located at A and C; those points correspond to the +1 and −1

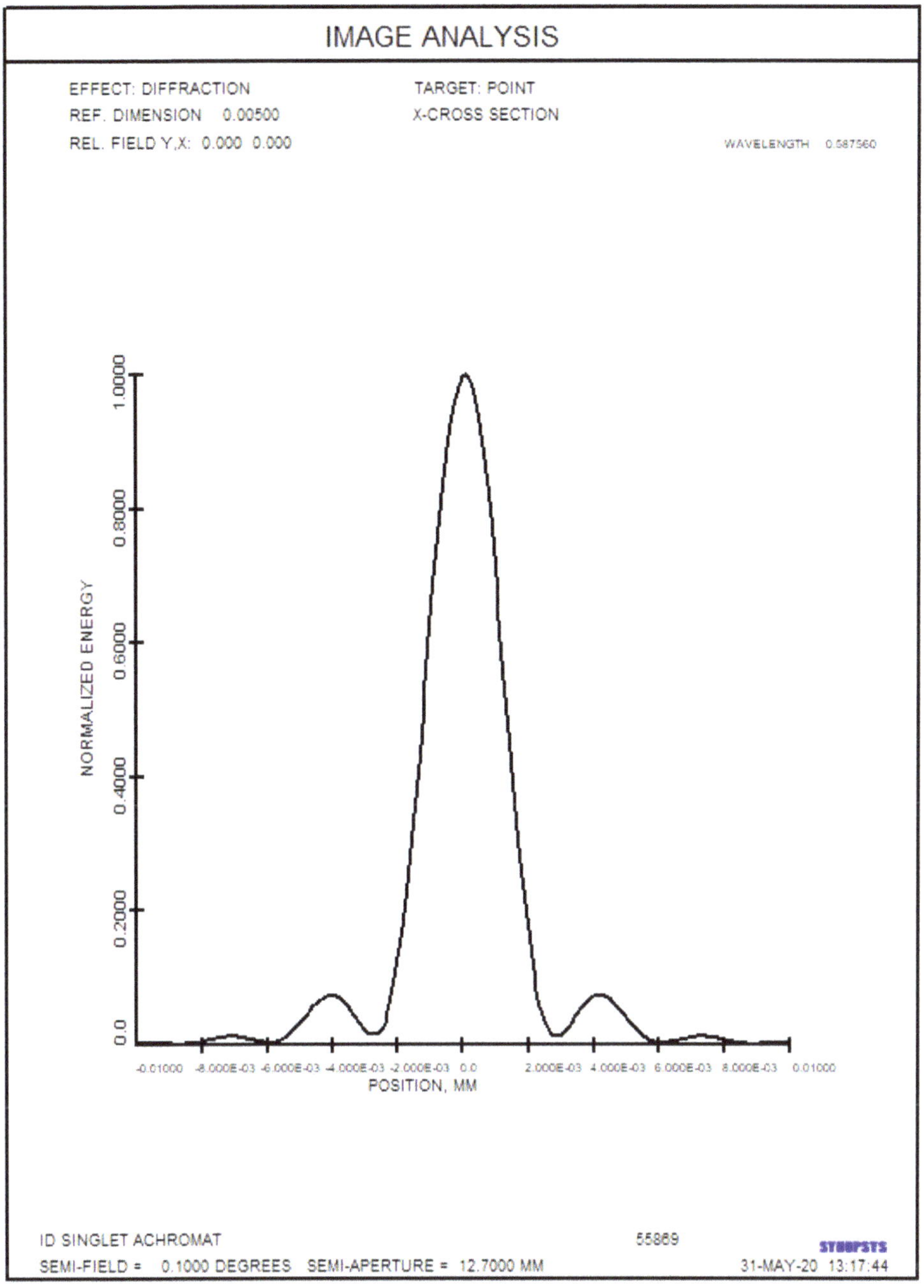

Figure 59.3. Image Tools analysis of the diffraction pattern.

diffraction orders. If the lens has no aberrations, the images of those points, at D and F, are in phase and equidistant, and together with light from point E, the zero order, they combine to produce a sine wave at the image plane. That, in essence, is how images are formed. But what happens if the spacing from D to E is different from

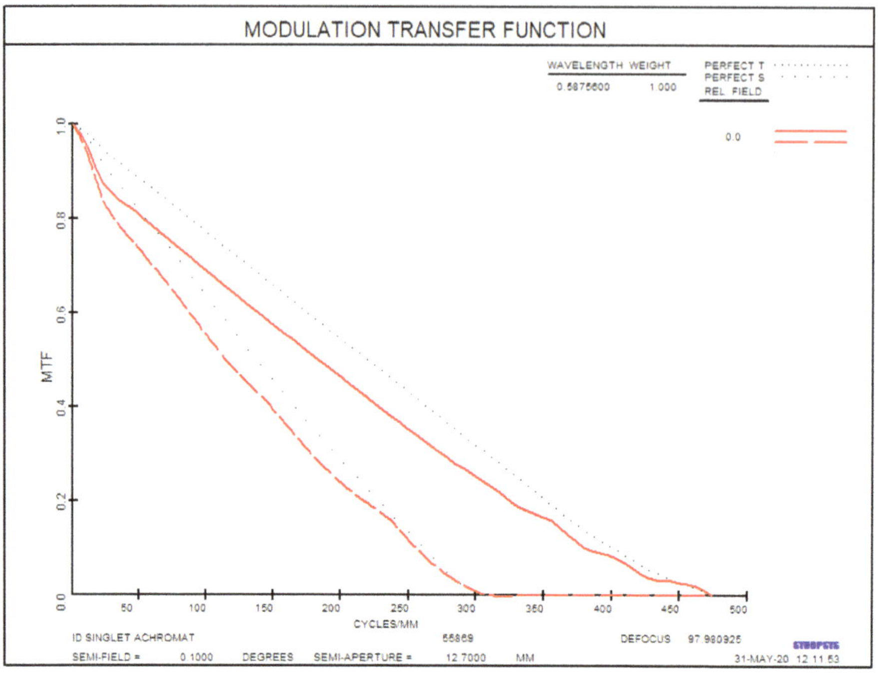

Figure 59.4. MTF plot produced by the convolution algorithm.

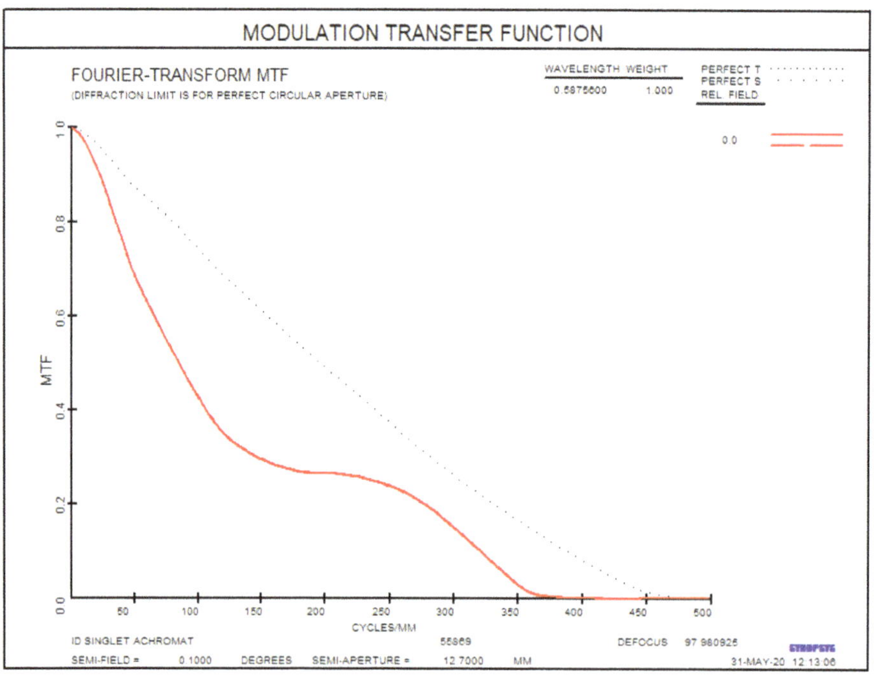

Figure 59.5. MTF plot produced by the Fourier-transform method.

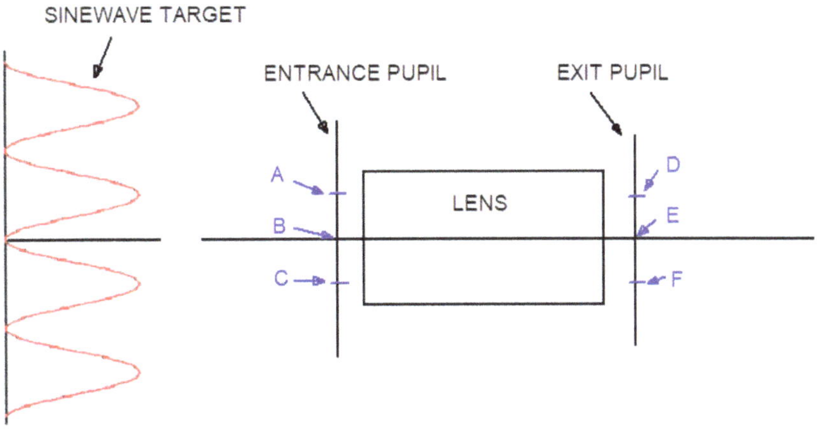

Figure 59.6. Schematic illustration of lens pupils with sinewave object.

that from E to F? In the case of strong pupil aberrations, those spacings can be quite different. Figure 22.6 gives a good picture of how nonuniform the spacing can be.

In that case, the spatial frequency at the image corresponding to the spacing from D to E will be different from that calculated from the spacing from E to F. Therefore, the two points create two sinewave images of two different frequencies, and the combination cannot be a pure sine wave anymore. So the concept of the MTF is violated. In effect, that concept implies a condition that does not exist, and one can say that the 'MTF' is *undefined* in this case. So our question, which one is correct, has a clear answer: *neither one*! In practice, there are an infinite number of points in the entrance pupil that are filled with light, and each set of diffracted points is imaged in the exit pupil with a spacing that creates its own frequency at the image. The actual image is the scalar superposition of the sine waves from all such pupil points[1], but only with zero pupil aberration is that image itself a pure sine wave. Higher spatial frequencies diffract point B at a greater distance than do lower frequencies, and some of those points are diffracted entirely out of the pupil for many locations of the zero-order point, which is why the MTF curve falls off at high frequencies.

On the other hand, the Fourier-transform MTF shown in figure 59.5 bypasses the assumption that the image is a sine wave and instead calculates the frequency content of the image of a point source. The source of course contains all spatial frequencies, but only those that get through the lens contribute to the final image. That image always has a shape and a Fourier transform, so the MTF still has meaning, just not the same meaning as that of the convolution MTF. (The *geometric* MTF, which ignores diffraction and is applicable only if the spot diagram is large compared to the diffraction disk, also gives the frequency content of the image of a

[1] Strictly speaking, diffraction from the target occurs in *amplitude* space, not intensity space, and the amplitude function of a sinewave target is not a pure sine wave. This complication has little effect and is widely ignored.

point source, which in that case is a spot diagram.) If pupil aberrations are all zero, the convolution and Fourier transform MTFs produce essentially the same curve, the main difference arising from the fact that the diffraction pattern, which extends to infinity, is sampled over only a finite area.

Some years ago there was a meeting of one of the optical societies where the discrepancy between the two MTF methods was the subject of lively debate. I pointed out how these considerations isolate the two issues, and one cannot always point to a single correct answer.

59.2 Coherent imaging

Coherent imaging is an issue in many modern lens-design problems and requires special tools for accurate modeling. The basics can be understood by referring again to figure 59.6.

In the fully coherent case, one in fact deals only with radiation coming into point B in the figure, and the situation is then more complicated.

In that case, when the image is analyzed by a coherent algorithm such as found in the Image Tools Menu (MIT), the program performs a 2-D fast Fourier transform

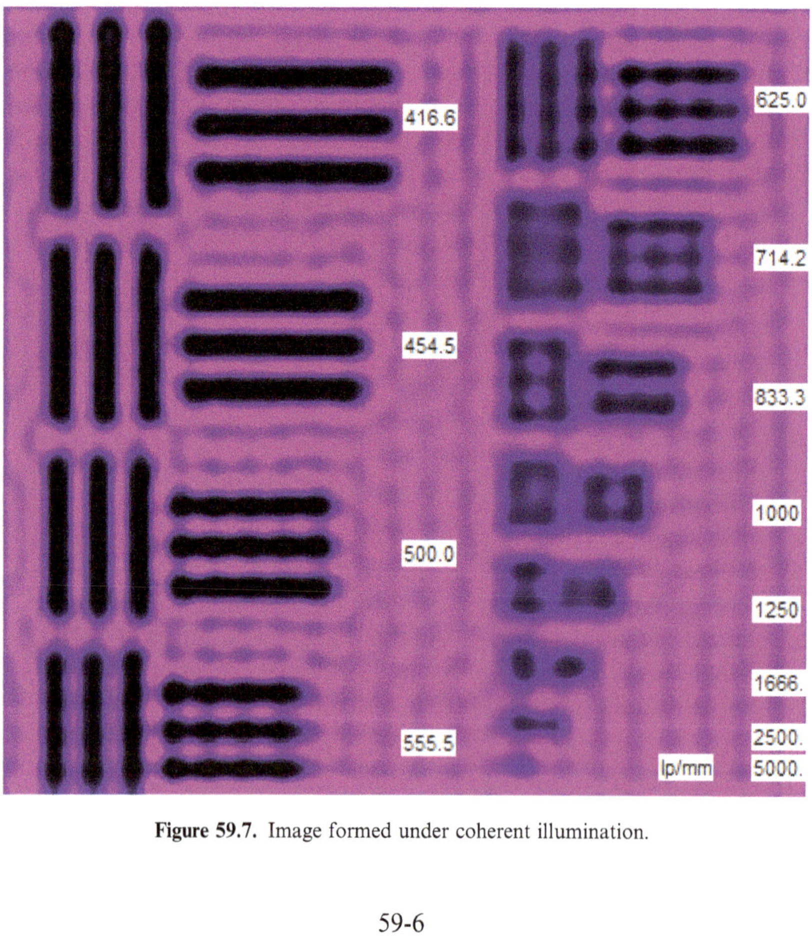

Figure 59.7. Image formed under coherent illumination.

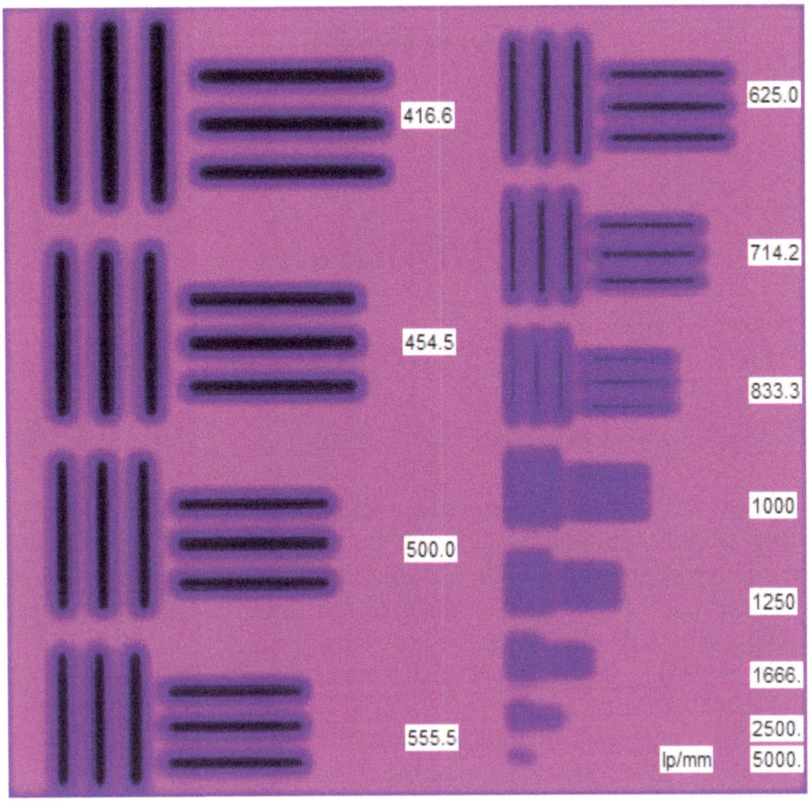

Figure 59.8. Image formed with incoherent illumination.

over the *entrance pupil*, since it assumes equal spacing of the sampled points, which happens only there, and that analysis also ignores pupil aberrations and therefore is not appropriate for systems such as the above lens.

And that is not all. Under coherent illumination, neighboring areas of the image actually affect each other, and the image structure can be complicated indeed. figure 59.7 shows the image of the Air Force resolution target as formed with coherent light by a microlithography lens. figure 59.8 shows the same target as imaged with incoherent light. The structure in the two examples is quite different. In practice, such lenses are often designed to work with *partially* coherent light, where the pupil is partly filled, typically to about 40% of the diameter, and image analysis must take this fraction into account.

IOP Publishing

Lens Design (Second Edition)
Automatic and quasi-autonomous computational methods and techniques
Donald C Dilworth

Appendix A

A brief history of computer-aided lens design

Newcomers to the field of computer-aided lens design are likely familiar with but two or three commercial programs, which are the survivors of a long development effort by many researchers at many institutions. The author's career has spanned almost 60 years and has utilized a wide variety of computers, operating systems, and programming languages, including the following:

- 1961: At MIT, the IBM 650, a vacuum-tube processor, batch mode with punched cards.
- 1962: the Honeywell 800, a solid-state CPU, also batch mode. Programming was in the 'MAC' language, developed at MIT.
- 1963: the Honeywell 1800, a faster CPU.
- 1967: an IBM 1130, an early minicomputer, punch-card input, and 8K words of 16 bits. First use of Fortran.
- 1971: the CDC 3300 and then 6600, batch mode via telephone connection. Programmed in Fortran.
- 1977: the Altos PC came out, with an 8080 cpu chip. Programmed in assembly language.
- 1983: the VAX 11/730, with 8K of memory, programmed in Fortran.
- 1987: the PC became available, and, with a version of Unix installed, could do interactive lens design.
- 1992: SYNOPSYS program was ported to DOS, programmed in Fortran.
- 1999: the first native Windows version, fully interactive, programmed in both C++ and Fortran.

Of course, we were not the only ones developing lens design software. Other programs were under active development by other authors as listed below; some were written for proprietary use in industry.

- Slams (C G Wynne)
- Ordeals (Tropel)
- Flair (Radkowski)
- COP (Grey)

doi:10.1088/978-0-7503-3695-6ch60

- Lead (Kodak)
- Father (B & L)
- Spade (Sperry)
- Optik V (Texas Inst.)
- Alsie (Osaka, Suzuki)
- SIGMA (Kidger)
- Bathos (Blandford)
- ACCOS (Spencer)
- CERCO (French)
- Cool Genii (Genesee)
- CODE n (Harris)
- Oslo (Sinclair)
- ZEMAX (Moore)
- SYNOPSYS (Dilworth)

Those programs employed a variety of optimization methods, of which the following are noteworthy:

- Correction (Itek)
- Orthonormalization (Grey, Unvala)
- Damped least-squares (Levinberg)
- Steepest descent
- Simplex (Bathos)
- Random search (Texas Instruments)
- Adaptive (Glatzel)
- Metric schemes
- Solution scaling
- PSD (Dilworth)

IOP Publishing

Lens Design (Second Edition)
Automatic and quasi-autonomous computational methods and techniques
Donald C Dilworth

Appendix B

Optimization methods

B.1 Mathematical methods of lens optimization

The program at Itek, mentioned in the previous section, seems to be unique; it used a correction algorithm instead of the minimization methods employed by almost everyone else. In that method, the number of targets in the merit function could not exceed the number of variables. So you selected a few carefully chosen rays, gave target values slightly smaller than the current ones, and submitted your batch run. If that converged, you reduced the targets and tried again. With much user intervention and many iterations, one could come up with a good design. That program was used on the recently declassified Corona project, which designed aerial reconnaissance cameras during the Cold War. The first illustration in chapter 38 shows a lens from that project.

The technique of orthonomalization was interesting to some researchers early on. This is a method that works via linear algebraic manipulations of the Jacobian matrix, with the goal of mapping the current set of variables into a new set, where the derivatives would all be the same magnitude and the effects of each variable independent of the others. While interesting from a mathematical standpoint, we observe that the process does not introduce any new information into the problem. At best, it might avoid numerical difficulties due to matrix conditioning—but more recent methods have avoided that problem in other ways.

The PSD methods, developed by the author, began in the 1980s with the goal of improving the then-standard Damped-Least-Squared method, which converges very slowly for many problems. This difficulty arises because that method calculates only the first derivatives of the items in the merit function with respect to the design variables. The text below explains the mathematics of the Least-Squares method and why it performs poorly, and how newer methods work much better.

B.2 The DLS method and descendants

The merit function φ is the sum of the squares of the errors to be corrected, defined by the vector *f*. One creates the matrix *L* from the derivatives, and then the required changes in the variables are easily calculated after the matrix is inverted.

But this solution is usually far off the mark, owing to the nonlinear nature of the design landscape. To improve its performance, the concept of 'damping' was introduced, the effect of which is to reduce the length of the solution vector. If the solution stays in the region of approximate linearity, it should be an improvement, and with many iterations one hopes to find a good result.

But the improvement resulting from the use of a damping factor was only modest, and the Damped-Least-Squares method still converged very slowly. In an effort to speed things up, many schemes were devised for utilizing the damping, *D*, in various ways. These include the following:

- Additive (many)
- Multiplicative (Meiron)
- Search for best (Dilworth)
- Different classes of variables
- Homogeneous 2nd derivatives (Buchele, Feder)
- Pseudo Second Derivatives PSD (Dilworth)

The following mathematics summarizes this development. The merit function φ is the sum of the squares of the image defects, f_i; the gradient G_j is one-half the derivative with respect to the design variable x_j, and the set of derivatives of G_j with respect to variable X_k gives us the Jacobian matrix, L_{jk}.

To find the optimum, set the gradient to zero and solve for the variable change δ_j:

$$\varphi = \sum f_i^2$$

$$G_j = \frac{1}{2}\frac{\partial\varphi}{\partial x_j}$$

$$L_{jk} = \frac{\partial G_j}{\partial x_k}$$

$$0 = G_j + L_{jk}\Delta x_k$$

$$\delta_j = -\, L_{jk}^{-1} G_k$$

Adding a damping term, *D*, as shown below, reduces the magnitude of δ_j, and one hopes the solution will then remain in the region of approximate linearity. This is the classical DLS method. Note that all variables get the same damping, which is applied to the diagonal of the matrix L_{jk}.

$$\varphi = \sum f_i^2 + \sum D^2\delta_j^2$$

$$G_j = \sum f_i \frac{\partial f_i}{\partial x_j} + \sum D^2\delta_j$$

$$L_{jk} = \sum \frac{\partial f_i}{\partial x_j}\frac{\partial f_i}{\partial x_k} + D^2\Big|_{j=k}$$

B.3 The PSD methods

Some of the methods mentioned above were markedly better than the raw DLS algorithm, but most failed to deal with the underlying source of the problem, which can be seen if we expand the Jacobian matrix to two derivatives, as follows.

$$L_{jk} = \sum \frac{\partial f_i}{\partial x_j}\frac{\partial f_i}{\partial x_k} + \sum f_i \frac{\partial^2 f_i}{\partial x_j \partial x_k}$$

$$\frac{\partial^2 f_i}{\partial x_j^2} \approx \frac{\left.\frac{\partial f_i}{\partial x_j}\right|_{\Delta x_j} - \frac{\partial f_i}{\partial x_j}}{\Delta x_j + \varepsilon}$$

Here it is apparent that the second derivative, were its value known, should be added to the matrix L in precisely the place where the damping D appears on the diagonal in the older DLS method. In other words, the purpose of introducing D is to replace the unknown values of the second derivatives. That insight lead to the first form of the PSD method, called PSD I[1].

The idea is quite simple; keep track of the changes in the first derivatives from one iteration to the next, divide by the variable change, and the result is just the second derivative (ignoring higher and mixed orders). Experience showed that the method was remarkably better than DLS, but only if a stabilizing factor ε was added where shown. That insight lead to the PSD II method, which is based on the statistically expected influence of the higher-order derivatives[2] and worked better than did PSD I.

$$\left.\frac{\partial f_i}{\partial x_j}\right|_{\Delta x_j} - \frac{\partial f_i}{\partial x_j} \approx \frac{\partial^2 f_i}{\partial x_j^2}\Delta x_j + \frac{\partial^2 f_i}{\partial x_j^2}\left.\sqrt{\sum_k \Delta x_k^2}\right|_{k \neq j},$$

$$\frac{\partial^2 f_i}{\partial x_j^2} \approx \frac{\left.\frac{\partial f_i}{\partial x_j}\right|_{\Delta x_j} - \frac{\partial f_i}{\partial x_j}}{|\Delta x_j| + \left.\sqrt{\sum_k \Delta x_k^2}\right|_{k \neq j}},$$

$$L_{jk} = \sum \frac{\partial f_i}{\partial x_j}\frac{\partial f_i}{\partial x_k} + \sum f_i \left.\frac{\partial^2 f_i}{\partial x_j \partial x_k}\right|_{k=j}.$$

A further refinement leads to the PSD III method. On average, we assumed that the mixed second partial is roughly the same as the homogeneous second partial—but the latter involves both j and k. Which one should we use? Define $\sec_j$ as shown below, and then combine the influence of both j and k by the ratio of the second-order terms from the previous iteration. Now we have a better approximation to the second partials:

---

[1] Dilworth D C 1978 Pseudo-second-derivative matrix and its application to automatic lens design *Appl. Optics* **17** 3372.

[2] Dilworth D 1983 Improved convergence with the pseudo-second-derivative (PSD) optimization method *Proc. SPIE* **399** 159.

$$\frac{\partial^2 f_i}{\partial x_i \partial x_j} \approx \frac{\partial^2 f_i}{\partial x_j^2}$$

$$\sec_j = \sum \left(f_j \frac{\partial^2 f_j}{\partial x_j \partial x_k} \right) \bigg|_{k=j}$$

$$\frac{\partial^2 f_i}{\partial x_j \partial x_k} \approx \sqrt{\frac{\partial^2 f_i}{\partial x_j^2} \frac{\partial^2 f_i}{\partial x_j^2} \frac{\sec_k}{\sec_j}}$$

$$\frac{\partial^2 f_i}{\partial x_j^2} \approx \frac{\left.\frac{\partial f_i}{\partial x_j}\right|_{\Delta x_j} - \frac{\partial f_i}{\partial x_j}}{|\Delta x_j| + \sqrt{\sum_k \Delta x_k^2 \frac{\sec_k}{\sec_j}}\Bigg|_{k \neq j}}$$

This PSD method adds a term to the matrix where the previous D was added, but now the value of that term differs, in practice, from one variable and the next, by as much as 14 orders of magnitude—a far cry from the constant D used in the standard DLS method.

Since the derivation involves some crucial approximations and assumptions, one must test it before declaring success. To that end, we did a simple design job (a triplet) and plotted the logarithm of the merit function φ as a function of iteration number, repeating the exercise with several different algorithms. The result is shown in figure B.1.

The convergence rate of the PSD III method, shown in curve A, is orders of magnitude faster than the classic DLS method, in curve I. and the assumptions therefore appear to be valid.

B.4 Global search algorithms

Many researchers have attempted to devise a method by which the computer could find the 'global optimum', a task both extremely complicated and fortunately unnecessary. As when searching for the end of a rainbow, you never reach your goal. One always suspects there may be a better solution yet to be found. The most popular search method involves defining a multi-dimensional grid of designs, where each radius, thickness, airspace, index, and Abbe number takes on a discrete set of values, resulting in a search space of perhaps 200,000 or more cases in all, which, when optimized with the ordinary DLS method, can take many days to evaluate. Although in principle this method can find the best of that large number of possibilities, it is too slow to be practical.

The algorithms in `DSEARCH` and `ZSEARCH` work differently; the default method uses a binary search tree: for a lens of N elements, one can generate a binary number of N bits and then create candidate lenses in which each element is assigned either a positive or negative power according to the value of the corresponding bit. Thus, for a lens of seven elements, there are only 2^7 cases, or 128 in all—a far cry from the

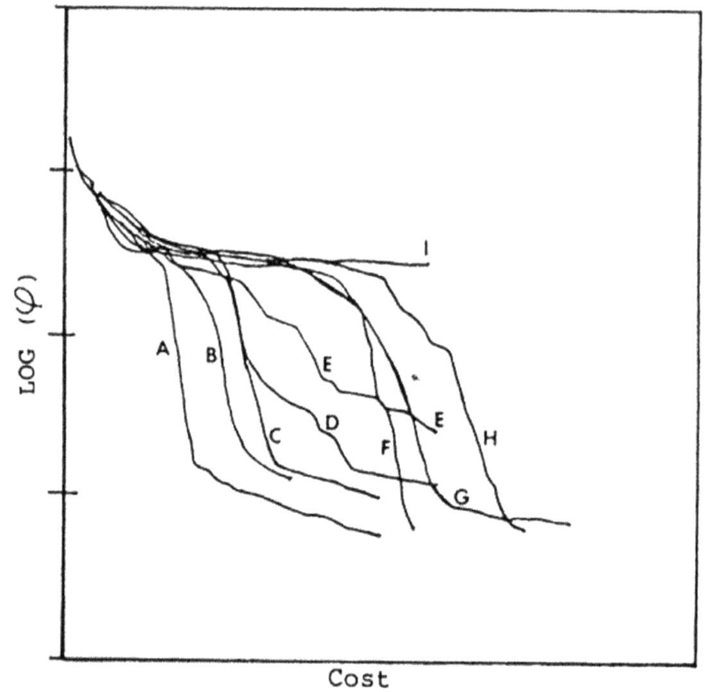

Figure B.1. Comparison of convergence rate for several optimization algorithms. Curve I is classic DLS, curve C is PSD I, and curve A is for PSD III.

huge search space mentioned above. But one has to ask: does this simple method work, and what should be the initial powers of those elements?

Referring again to our original mountain-range metaphor, each case starts at the top of a tall mountain (with plane-parallel plates), selects a direction given by the particular value of the binary number representing that case, sets the curvatures according to that direction, and then jumps downhill and begins optimizing. But how far down should the algorithm jump? That jump controls the initial powers. An interesting result came out when I examined a range of initial radius values and ran DSEARCH for each one, as shown in figure B.2. I discovered that, if the initial radius is too long, the lens often fails to trace, due to the curvature solve on the last element which becomes so steep that rays experience TIR errors. And if the initial radius is too short, many other elements have the same problem. The program automatically corrects most of those ray failures, but in the process it might move the design toward a better or a poorer solution, which accounts for the chaotic behavior at both ends of the curve. Fortunately, there seems to be a wide range where the initial value is not critical. So the initial powers can be assigned more or less at will, and perhaps more than one value can be tested.

Another question is, how many iterations should one request for each candidate solution? Fewer passes will run faster, but one does not want to miss a good solution that might be found with more. The goal is to devise a method that works reliably and very quickly. Figure B.3 shows the convergence history for two different lenses.

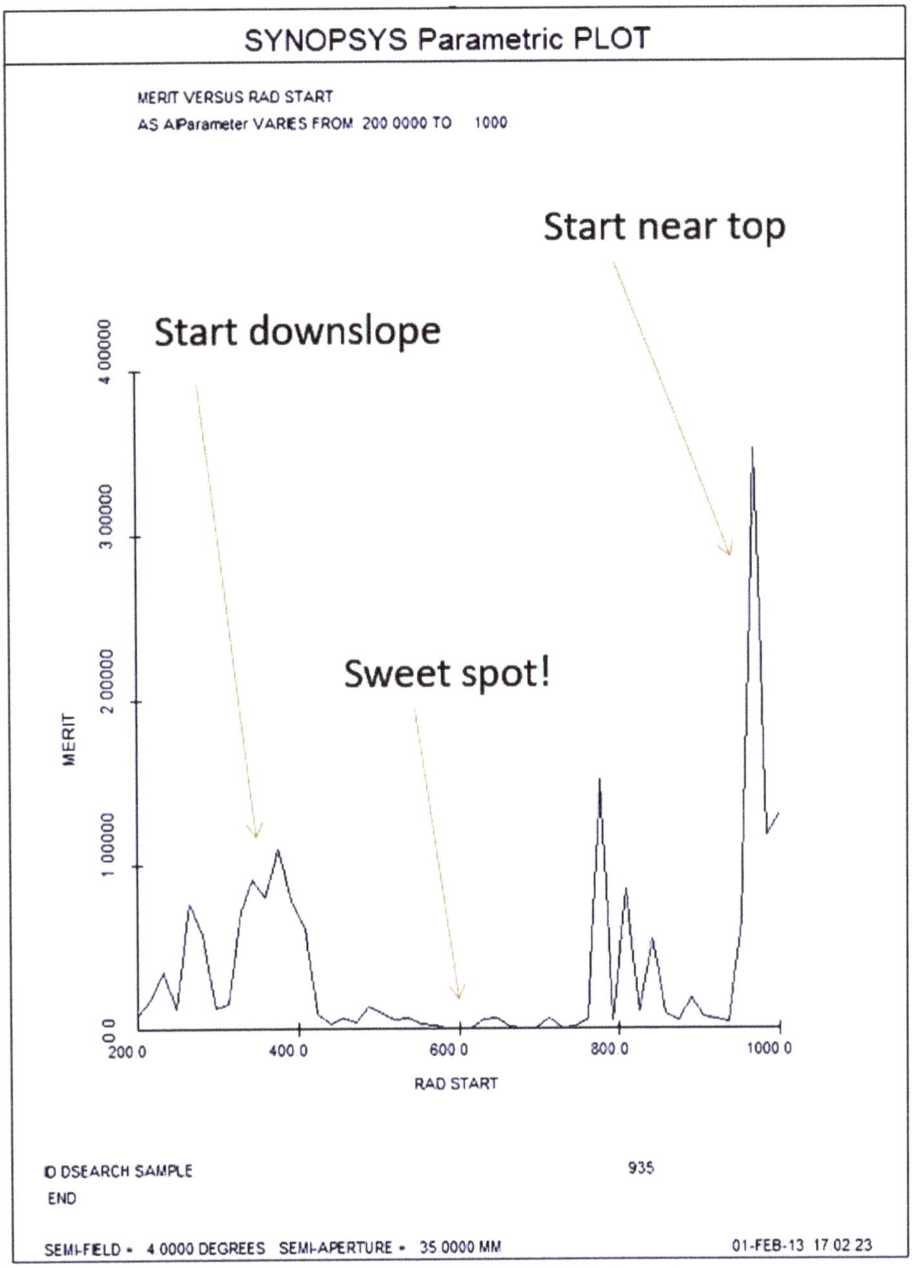

Figure B.2. Best merit function as a function of initial element radii.

On the left, the lens reached an excellent solution in only 30 or so iterations. The case on the right was more erratic; the MF came down in several steps, and if one stopped optimizing after 30 iterations that lens would not score as highly as the previous one and the search algorithm might reject it—even though after 80 iterations it is clearly

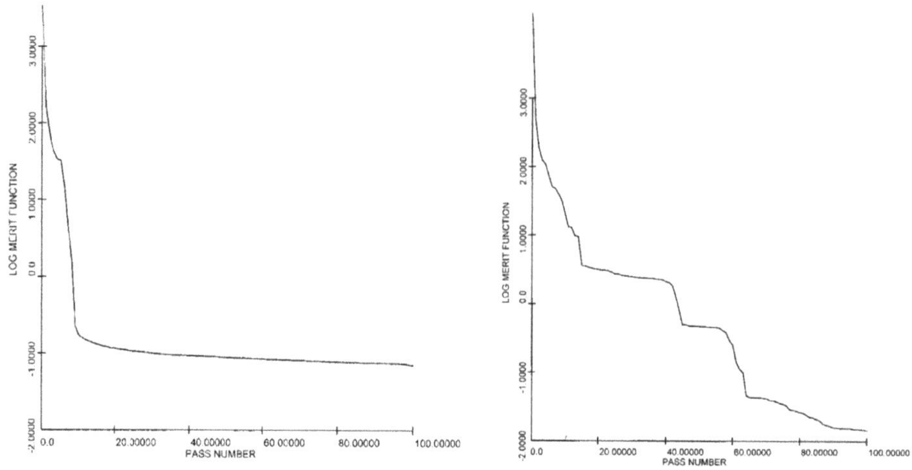

Figure B.3. Convergence rates for a well-behaved and an erratic lens; the latter is the better design.

superior. These results can influence the parameters one submits to the search routines.

A major goal of the search methods is to achieve maximum speed, and the optional `QUICK` mode enhances it significantly. That step optimizes each candidate lens with a special MF containing only the first-order goals and all of the third and fifth-order aberrations plus three real rays (plus any `SPECIAL AANT` requirements the user may have submitted). This can be evaluated many times faster than can a MF with a grid of real rays, and it quickly weeds out the candidates that perform poorly on that score. Then there are fewer remaining cases to be optimized thoroughly with real rays.

B.5 Why are DSEARCH and ZSEARCH so powerful?

It is the speed of convergence of the PSD III method that makes the search programs `DSEARCH` and `ZSEARCH` practical. Each case can be optimized in a matter of seconds or less, and within a few minutes one can explore hundreds or thousands of different branches of the lens design tree. That algorithm seems to be key to the success of this new paradigm, and together with the binary search protocol, it is a very efficient way to explore the landscape.

It is remarkable that, if one actually *calculates* the second derivatives—which was not practical decades ago with more primitive computational tools—and optimizes the lens with those instead of the 'pseudo' second derivatives approximated by the PSD algorithms, the results are *not as good as those from PSD*! This tells us that the logic of the PSD calculation applies to *all* the higher-order derivatives, not just the second, and the matrix in fact more closely approaches what you would get with more than just two terms, as ideally it should. Nature is not always so kind to computer programmers, and it is wonderful when a program works better than one expects. Lens design is a very rewarding field.

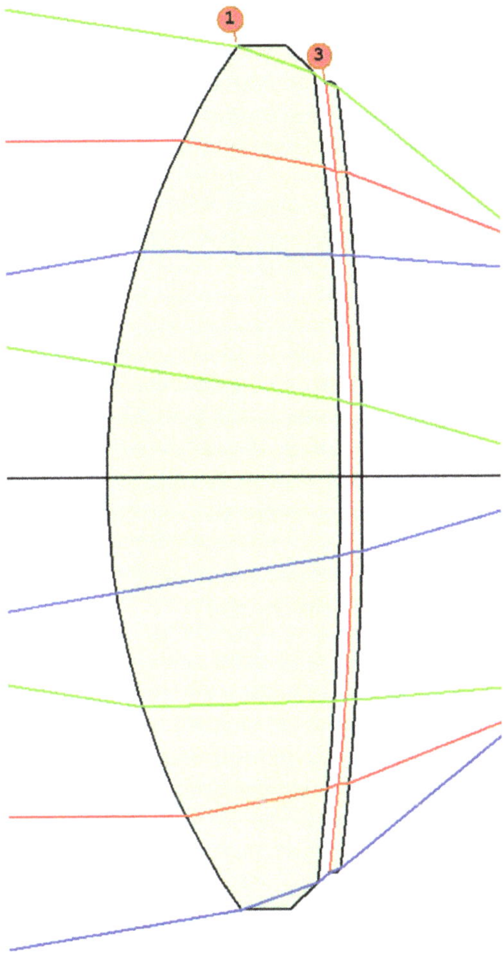

Figure B.4. A thin shell added adjacent to a lens element, part of the AEI algorithm.

B.6 Adding and deleting lens elements automatically

If a lens does not perform adequately, a time-honored strategy is to add an element somewhere. Then the powers of the other elements can be reduced as well as their aberration contributions. But where should one add the new element? This seems like a very complex problem, requiring deep insight into lens design theory, but it can in fact be solved by a rather simple algorithm that started with an idea of Florian Bociort[3], called the *saddle-point theory*.

The idea is that, if one adds a thin shell adjacent to an existing lens element, as shown in Figure B.4, the ray paths do not change, so the MF is also unchanged—but

[3] Bociort F, Serebriakov A and van Turnhout M 2004 Saddle points in the merit function landscape of systems of thin lenses in contact *Proc. SPIE* **5523** 174–84

now there are six new degrees of freedom, and it is likely that optimizing with these additional variables will lead to an improved design. No deep theory is required.

That is the principle behind **AEI**, which you have used in several of the previous chapters. The program can test where adding an element does the most good, and the result is usually a better lens. Here is another instance where pure number crunching can yield results as good or better than can be obtained by a human expert.

AEI is a special case of the more general Saddle-Point Build (**SPB**), which can build up an entire lens one element at a time using the same method. That feature is similar in some respects to DSEARCH, but where it is applicable, the latter is often better since it can draw on a larger set of possibilities.

And the reverse also works: try to reduce each lens element in turn to a thin shell with zero power, and if the MF is not seriously degraded, that shell can simply be removed with little loss of quality. This is exactly what **AED** does, often yielding a lens that is almost as good as before but requiring fewer elements. These tools, together with the search routines, help you explore the complex lens design landscape quickly and easily.

B.7 What about traditional methods?

Open any of the classic texts on lens design and you will find pages of math, equations that can lead to a lens configuration with certain desirable properties, perhaps suitable for input to an optimization program, along with many examples of classic design forms that include specific data for those lenses. The goal has long been to provide starting points that have a good chance of yielding a good design when optimized. All this made sense in the days when having that starting point was the key to success, but today it is less useful since the new search methods can produce many excellent configurations in minutes, given just the design goals. This book has illustrated the importance and power of these new tools and justifies our assertion that the new paradigm has radically changed what lens designers do and how they do it. We believe that careful study of the examples presented here will prepare the student to fully utilize these new tools and to be more effective at their task than any of the old masters ever were.

Appendix C

The mathematics of lens tolerances

The subject of lens tolerances is as important as that of lens optimization, although the mathematics is somewhat simpler. A classic method of generating a tolerance budget was first to get a table of *inverse sensitivities*, which gives the amount by which every manufactured parameter can be in error, assuming everything else were perfect, while just meeting the imaging requirements. Then, if there are N such parameters, the practice was to divide each of the sensitivities by the square root of N, and that became the tolerance budget. This procedure was widely used and worked very well.

In fact, it worked *too well*. It can be shown that the budget so prepared is appropriate for the case when every parameter is always to be found exactly *at one end or the other* of its tolerance range. But most parameters are likely to be found at a random place *within* the budget, not exactly at the end, and the image degradation is therefore usually less than the budget allowed for, and that is why it worked so well. But there was a price: the lenses were more expensive than they had to be.

The whole purpose of generating a budget is, of course, because everything made by human hands is imperfect. Every dimension of a finished lens is slightly different from the numbers on the drawing. How much different can they be? The budget provides an answer to that question.

The tolerance budget prepared by BTOL first calculates a set of *standard deviations* (**SD**) of each parameter so that, if the lens were made to that budget, it would meet the design goals. Then, when it prints the actual budget, it accounts for the difference between the standard deviation and the actual tolerance range of that parameter. Let me explain.

For a one-dimensional parameter such as element thickness, it can be shown that the SD equals the tolerance limit divided by the square root of 3. So the printed budget gives the desired SD multiplied by that factor. The budget is therefore somewhat looser than that given by the root-of-N rule, the image will be as desired within the stated confidence level, and the lens will be less expensive. For two-dimensional parameters, such as lens decenter, the adjustment factor is the square root of 2 instead of 3.

doi:10.1088/978-0-7503-3695-6ch62

Figure C.1. shows how the SD of the toleranced parameters influences the final image quality. It turns out that the only quantity of interest is the SD array. What is the rule for calculating those values?

There are an infinite number of budgets, all mathematically correct, that predict the same image quality, and the task is to find one that minimizes overall lens cost. Any parameter can be assigned a tighter tolerance, and some of the others loosened—but how do you know which ones? Figure C.2. shows the logic of that calculation. The program proceeds in steps, initially with all tolerances as loose as economically

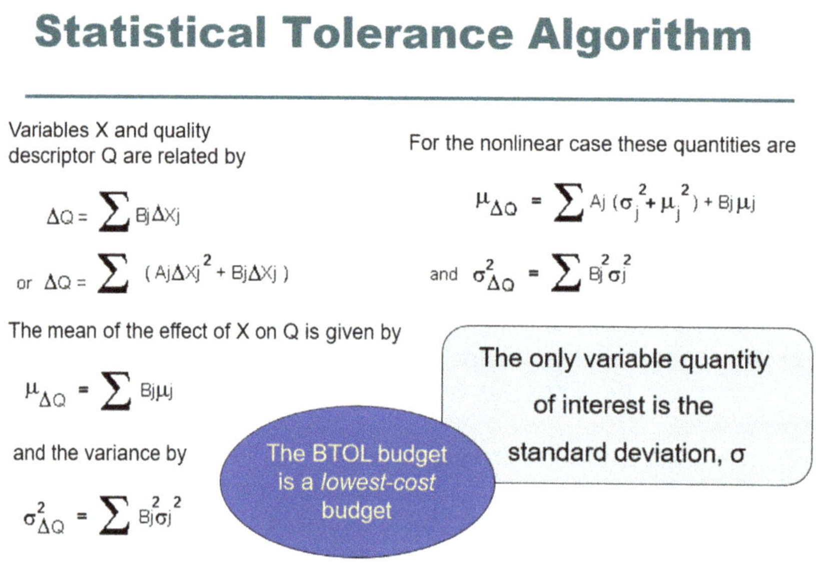

Figure C.1. Relation between the standard deviation of variables and degradation of image quality.

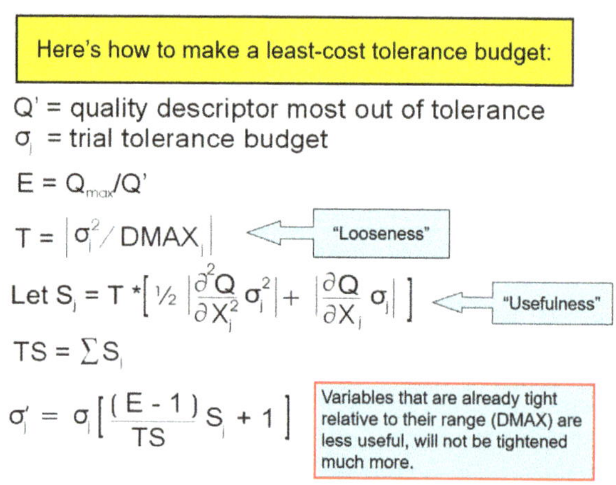

Figure C.2. Rules for deriving a least-cost tolerance budget.

practical (given by the value of **RANGE** in the BTOL input, which defines the 'range' of each variable), evaluates the image quality, finds the image point most out of spec, identifies those parameters that most strongly affect that image point, and reduces the tolerance on those parameters. Then it iterates. If a given parameter starts to get excessively tight compared to its range, the program tries to leave it alone and tighten up others instead. The result is a least-cost budget. As the figure shows, one calculates the 'looseness' and 'usefulness' of each parameter and then modifies the SD according to a simple formula.

The budget produced by BTOL tells you that the lens will perform at the requested statistical confidence level, assuming every parameter is somewhere within its tolerance budget, preferably at a random location, and provided all modeled adjustments are performed at assembly. This calculation is not as mystical as it may seem at first; the program makes the allowed adjustments every time it calculates the derivatives of the quality with respect to the parameters. So the adjustments are built into the derivatives and the result comes about automatically.

Appendix D

Things every lens designer should understand

Here is a list of concepts and practices every lens designer should be aware of and follow. Those who want to explore the topics in more depth are encouraged to read any of the recent books on the topic, which develop the mathematical theory behind these concepts. A practicing designer does not need to dive that deep, in my opinion.

1. Lenses are designed by observing the paths of "light rays", which do not really exist but are useful and quite accurate as long as one does not examine details with dimensions comparable to the wavelength of light.
2. The performance of a complex lens depends on the aberration contributions of individual elements. Positive and negative elements contribute aberrations with differing signs, so a combination of powers is usually required.
3. Aberrations arise because of three situations:
    a. A steep angle of incidence of a ray relative to a surface normal contributes high-order aberrations, because Snell's law then departs more strongly from the paraxial version. Such angles are usually to be avoided. Those aberrations are more difficult to correct and require a more complicated lens. Sometimes they cannot be avoided, and one must then strive to balance many orders of aberrations. That usually requires many lens elements.
    b. The bending of individual elements influences whether the Abbe sine condition is satisfied and is a useful variable for correcting many aberrations.
    c. The dispersion of the glass creates, and can be used to correct, chromatic aberration.
4. If your lens shows aberrations that vary over the aperture but are relatively constant over the field, correct them with elements near a pupil or stop.
5. If aberrations vary strongly with field angle but not with aperture position, correct them with elements near an image.
6. Both of the above situations can be detected and handled by the Automatic Element Insertion feature (AEI).

7. If your lens has good performance, it is sometimes possible to remove an element with little loss of quality. The Automatic Element Deletion feature (AED) can test this possibility.
8. One can sometimes improve a lens by running AEI and AED repeatedly, thereby altering the lens construction in stages.
9. If an element is strongly curved, you can sometimes find a different solution region by flipping the bending. The Bend-Flip Optimization (BFO) feature can do this automatically.
10. If the elements of your lens are each contributing large amounts of aberration, even though the final image looks great, you will likely see tight tolerances because even a small misalignment of such a design will throw the balance off. The techniques of tolerance desensitization are then called for. You want the elements as weak as possible, as long as the imaging goals are met. The command **`THIRD CPLOT`** will show the third-order contributions of every surface, and you can see where they are largest. Desensitization targets in the merit function can often relax tolerances, as discussed in chapters 10 and 13.
11. Distortion is corrected by elements not very close to the stop, per item 5 above.
12. Try to use DSEARCH or ZSEARCH, if possible. Those features will discover lens configurations quickly and well. (Unless you already have a good configuration and the problem is simple.)
13. If the geometry of the problem makes it possible, aim for a lens with some degree of symmetry. That will make many field aberrations easier to correct.
14. If secondary color is an issue, try to favor glass types as described in chapters 12 and 34. The search routines will sometimes find those combinations automatically, and you can steer the process as instructed in those chapters.
15. Forget much of what the early texts advise. One of them says to select the glass types first when starting a design. Today, one lets the program find the best region of the glass map with the GLM variables or the search routines, and then fits the design to real glasses in the *last* step. Another text says that when designing a four-element lens one should design two doublets, correct them individually, and then combine them. This is nonsense. DSEARCH can design the lens by itself, and it will be much better.
16. If performance at or near the diffraction limit is required, be sure your merit function contains OPD targets. Sometimes one gets the best results with a combination of OPD and TAP targets. Sometimes GO2 targets will yield a better image than GNO. That option targets the square of the OPD, which tends to ignore small errors and concentrate on larger ones. One cannot predict which will work the best, and you simply have to try them.
17. To peak the performance at a given MTF frequency, you should first get the design as close as you can with OPD aberrations, and then use the GSHEAR ray grid option. When the MTF is very close to your target,

you can switch to MTF aberrations to see if things improve. GSHEAR targets the difference in the OPD at separated points in the pupil, and that difference is what governs the MTF at that shear frequency.

18. Do not try to correct 3rd-order aberrations to zero; you need them to balance the higher orders. You can sometimes reduce spacing and alignment sensitivity by reducing the contributions of a given element or group, but proceed with caution because the aberration balance may be thrown off.
19. If aspheric surfaces are allowed in your design, creep up on them gradually. Use the Automatic Aspheric Assignment (**AAA**) to determine which surface should be aspheric. That feature will add a conic constant where it will do the most good, and then you can run the Automatic G-variable Test (**AGT**) to determine which aspheric coefficients will be most useful on that surface. It makes no sense to use a high-order aspheric term to correct defocus, for example, so only add aspherics when the lens is already as good as possible without them. The exception is when designing molded plastic elements, which are usually aspheric early on—but even then, start off with just a few terms and add more as you need them.
20. Your merit function should contain a complete description of the problem, and this includes mechanical requirements as well as optical. Some designers only want image quality in their MF—but it makes no sense to get a great image if the lens will not fit in the required box. If the program knows about all your goals, it will favor designs that tend to meet them. That is exactly what you want.
21. One cannot always predict the results of a given optimization run, and often a new problem develops that was not anticipated beforehand. The process of lens design consists mostly of modifying the MF as required when one discovers a shortcoming. So change the MF and keep going. If you run into a blind alley and nothing seems to work, it's time to try a different search result.
22. Make frequent use of checkpoints, and save intermediate versions of your lens whenever you make a substantial improvement. You may well want to go back to that version later if unanticipated problems show up in later versions.
23. Your job is not complete until you make a table of tolerances. I know an expert designer who sends his design to a customer—expecting the *customer* to calculate the tolerance budget. This is unprofessional. If tolerances can be relaxed by a suitable change to the design, it is your job to figure that out. The customer will not know how. BTOL is the tool to use in most cases.
24. It is a good idea to become familiar with shop practice. Watching an optician make a precision surface has a way of humbling the designer—who may then be more sensitive to the challenges inherent in the designs he sends to the shop. Learn what they find easy and what hard, and try to make their life less stressful. Very thin edges, for example present a challenge to the lens

maker, even though the optimization program has no problem with them. Watch for meniscus lenses where the centers of curvature of the two sides are very close to each other. Such lenses are very difficult to manufacture since the methods used to eliminate wedge errors do not work very well, as explained in chapter 50. The **AMS** monitor can help in this situation. If a surface is almost flat, make it exactly flat. If a lens has two radii that are almost equal, make them exactly equal. Then there is no chance of the lens being inserted into the cell backwards. If two adjacent surfaces have almost the same radius of curvature, with a small airspace, try to cement the two. That will eliminate some tolerances.

25. If your DSEARCH or ZSEARCH run does not return a design you like, it's time to vary some of the input parameters. Even a small change can have a big effect. Items to consider include the following:
    - STOP FIX or Free
    - RSTART value, maybe more than one value
    - TSTART value
    - ASTART value
    - RT value
    - Include OPD, TOPD, OPSHEAR, or TOSHEAR
    - Number of FOVs
    - Weights on the fields
    - Number of rays in the grid
    - Number of passes
    - Enable or disable QUICK mode
    - Number of annealing passes
    - Try changing switches 95 and 67. They generally take a different path.

    It might seem daunting to think of exploring all the potential combinations of these parameters—but you should not have to. In our experience, most combinations return excellent starting points, and the purpose of trying still others is to give you more options. We generally get excellent results in from one to four attempts.

Appendix E

Useful formulas

One degree = 0.017 453 29 radians

One arcminute = 0.000 290 888 radians = 0.016 666 67 degrees

The human eye can resolve about one arcminute

One arcsecond = 4.848 14E-6 radians = 2.777 777E-4 degrees

One milliradian = 0.052 9578 degrees = 3.437 75 arcminutes = 206.264 8 arcseconds

One micrometer = 0.000 039 37 inches

For a lens in the visible spectrum, the diameter of the Airy diffraction disk = the F/number, approximately, in micrometers

The radius of the first dark ring in the Airy Disk = 1.22 λ F/number, in air, with an unobscured aperture.

The numerical aperture (NA) of a lens = $n \sin(\theta)$, where θ is the convergence angle of the marginal ray and n is the index in image space.

F/number = 0.5/NA; $\sin \theta = -0.5$/(F/number) (in air)

Longitudinal aberration = 2 × transverse aberration × F/number

Focus shift from plane parallel plate of index n = thickness × $(1-1/n)$

Cutoff frequency F_{co} = 1743/(F/number) in lines/mm at $\lambda = 0.574$ um

Density of a filter D = log(1/transmission)

Diameter of geometric spot giving MTF cutoff of F_{co} = 0.039 37 inches/ F_{co}(c/mm)

Sag of sphere of radius R, at height s: $z = R - \sqrt{R^2 - s^2}$

Radius of sphere with sag z, at height s: $R = \frac{s^2}{2z} - \frac{z}{2}$

For lens of magnification m and focal length f, focal point S1 = $(m+1)f/m$; S2 = $(m+1)f$

Reflection loss, uncoated surface of index n, $Refl = \left(\frac{n-1}{n+1}\right)^2$

Magnifying power, lens of focal length f, M = $(10 + f) / f$

To defocus a lens of focal length f by one diopter, the image shift in inches $\Delta S = f^2/39.37$

Strehl ratio = exp($-4\,\pi^2$ variance), approximately

Thin lens; $s1$ and $s2$ measured from lens $1/s1 + 1/s2 = 1/f$

Thin lens; $s1$ and $s2$ measured from focal points $s2\ s2 = f^2$

Angular resolution of a telescope in visual light = 4.66/diameter of objective in inches, in arc seconds.

Longitudinal magnification = square of lateral magnification.

Appendix F

Bibliography

These are all classic texts on lens design, and most contain extensive mathematical developments that support some insight. That insight was important in the days before routines like DSEARCH and ZSEARCH became available but is less so today. Some of the information is still useful, however, especially in Warren Smith's books. Others are listed for those who want a deeper picture of the complex landscape that is optics and lens design. As a fringe benefit, you will gain an appreciation for the amount of labor that was once required of the lens designer—and be gratified at the power of the new automatic features, which relieve you of most of that labor.

References

Benford J R, Cook G H, Hass G, Hopkins R E, Kingslake R, Lueck I B, Rosin S, Scott R M and Shannon R R 1965 *Applied Optics and Optical Engineering* ed R Kingslake Vol. III (New York: Academic)

Born M and Wolf E 1980 *Principles of Optics* 6th edn (Oxford: Pergamon)

Cox A 1964 *A System of Optical Design* (Waltham, MA: The Focal Press)

Dilworth D C *SYNOPSYS Supplement to Joseph M. Gary's Introduction to Lens Design* (Richmond, VA: Willmann-Bell)

Flügge J 1955 *Die Wissenshaftliche uund Angewandte Photographie* ed K Michel (Berlin: Springer)

Geary J M 2011 *Introduction to Lens Design* (Richmond, VA: Willmann-Bell)

Kingslake R 1978 *Lens Design Fundamentals* (New York: Academic)

Kingslake R 1983 *Optical System Design* (New York: Academic)

Kingslake R 1989 *A History of the Photographic Lens* (New York: Academic)

Kingslake R and Johnson R B 2010 *Lens Design Fundamentals* (Bellingham, WA: SPIE Press)

Laiken M 1991 *Lens Design* (New York: Marcel Dekker)

O'Shea D C 1985 *Elements of Modern Optical Design* (New York: Wiley)

Rutten H G J and van Venrooij M A M 2002 Telescope Optics ed R Berry (Richmond, VA: Willmann-Bell)

Smith G H 2007 *Practical Computer-Aided Lens Design* (Richmond, VA: Willmann-Bell)

Smith W J 1966 *Modern Optical Engineering* (New York: McGraw-Hill)
Wolf W and Zissis G *The Infrared Handbook* (Arlington, VA: Office of Naval Research, Department of the Navy)
Yoder P R 2002 *Mounting Optics in Optical Instruments* (Bellingham, WA: SPIE Press)

www.ingramcontent.com/pod-product-compliance
Ingram Content Group UK Ltd.
Pitfield, Milton Keynes, MK11 3LW, UK
UKHW051911100926
8556IPUK00005B/842

9 780750 336932